Cut-and-Cover Metro Structures

Cut-and-Cover Metro Structures

Geo-Structural Design:
An Integrated Approach

Krishan Kaul

BE, MSc, Eur Ing, C Eng, FICE, FIStructE, FGS, FIE (Ind.)
Formerly Technical Director, Hyder Consulting, UK

CRC Press
Taylor & Francis Group
Boca Raton London New York

CRC Press is an imprint of the
Taylor & Francis Group, an **informa** business

A SPON PRESS BOOK

CRC Press
Taylor & Francis Group
6000 Broken Sound Parkway NW, Suite 300
Boca Raton, FL 33487-2742

First issued in paperback 2019

© 2010 Krishan Kaul
CRC Press is an imprint of the Taylor & Francis Group, an Informa business

No claim to original U.S. Government works

Typeset in Sabon by Glyph International

ISBN-13: 978-0-415-46907-4 (hbk)
ISBN-13: 978-0-367-86444-6 (pbk)

British Library Cataloguing in Publication Data
A catalogue record for this book is available from the British Library
Library of Congress Cataloging in Publication Data
Kaul, Krishan.
Cut-and-cover metro structures : geo-structural design, an integrated
approach / Krishan Kaul.
    p. cm.
Includes bibliographical references and index.
1. Subways. 2. Underground construction. 3. Excavation. I. Title.
TF845.K38 2010
625.1–dc22
2009032088

Visit the Taylor & Francis Web site at
http://www.taylorandfrancis.com

and the CRC Press Web site at
http://www.crcpress.com

Dedication

It is beyond the ken of man to 'invent', for he cannot tap what already exists not in the universe. The so-called inventions are merely attempts by man to comprehend but a fraction of the Majesty, Grandeur and Excellence of the Supreme. And, through Divine Compassion, when man is so inspired, the ego is made to flee by the enormity of its ignorance and the head bows down in grateful homage.

This humble effort is dedicated at the Lotus Feet of Mother Sai, the motivator, the inspiration and the very source of strength, with all my love, reverence and a heartfelt prayer that those practitioners and students of engineering whose curiosity takes them through the pages of this book may find it useful and worthy of their attention. I sincerely hope that the structural and geotechnical engineers engaged in the design of cut-and-cover structures in general and metro structures in particular enjoy the book as much as I have done writing it.

I offer my loving and respectful salutations to the memory of my late father, Shri Dina Nath Kaul, who was an inspiring example to me and many others in self-discipline, self-sacrifice and self-less service; to my late mother, Shrimati Padmavati Kaul, whose continuous flow of love, prayers and blessings throughout her life have given me strength and much more besides; and to my late uncle and aunt, Shri Radha Krishan and Shrimati Gunwati Kaul, to whom I owe a great deal for their blessings, prayers and love. It was my late uncle, a renowned civil engineer of his time in Kashmir, India, who inspired me to take up civil engineering.

PKK

Contents

Acknowledgements

I am privileged to have been associated, over a period of three decades, with Hyder Consulting (originally known as Freeman Fox and Partners, and later, Acer Consultants), a firm of multidisciplinary consulting engineers, in a variety of very interesting and challenging projects in the cut-and-cover field. Over the years, this association has taken me to various parts of the world, including Hong Kong, Iraq, Portugal, Egypt, Vietnam, Taiwan and Singapore. The valuable interactions with my colleagues, associates and other experts in the field during these specialist assignments have deeply enriched me and are gratefully acknowledged.

I am deeply indebted to all those members of the 1979–80 teaching and research staff in the Geotechnical Engineering faculty, at the Department of Civil Engineering, University of Surrey, UK, especially the late Professor Noel Simons, Dr. Bruce Menzies, Dr. Chris Clayton, Mike Huxley and others, at whose hands I received instruction during my study for the master's degree. It was during this period that I became convinced about the special importance of the application of the geotechnical principles in the structural design, in particular, of cut-and-cover structures. As a civil/structural engineer, it made me realize that, in order to achieve a sound, sensible and complete structural design of cut-and-cover metro structures, it was imperative to take on board, not in isolation but integrally, the consideration of the geotechnical aspects as well. My search for a book with such an approach, however, drew a complete blank. And so, the seeds of writing a book with an integrated geo-structural approach to the design of cut-and-cover structures were sown.

In the course of writing this book, my researches took me through an extensive study of the available volume of literature directly or indirectly connected with the construction and design (both structural and geotechnical) of cut-and-cover metro and similar structures. In recognition of this fact, I duly record my gratitude to all the authors of the books and the papers appearing in the various journals and conferences that have been referred to in the preparation of this book.

My special thanks are due to Professor G. G. Meyerhof, Professor J. S. Tsai, Dr. Y. M. A. Hashash and Dr. D. I. Bush for taking time out to provide clarifications on certain technical matters concerning their work. I am also grateful to Dr. Bindra Thusu, MSc, PhD, FGS, Senior Research Fellow, University College London, for his valuable suggestions.

I owe a deep debt of gratitude to Mr **Terrence Hulme**, BA, BSc, ARSM, Eur Ing, CEng, FICE, FIStructE, FASCE, FCIArb, Consulting Engineer, Victoria, Australia, for going through the manuscript and writing the Foreword.

My heartfelt thanks go to Rajeev Kumar Singh, B.Tech. (Mech.), Senior Engineer, and Mohammad Qasim Alvi, Technician, both of Shalcot Mechanique (P) Ltd., for displaying a great deal of patience while helping to prepare the diagrams for this book. My thanks are also due to Messrs Ashish Raina and Autar Jalla for their help in facilitating the preparation of the diagrams. I am most grateful to Robert Thomas, Manager, Library and Information Services, Institution of Structural Engineers, London, and to the staff at the library of the Institution of Civil Engineers, London, including Mike Chrimes, Head of Knowledge Transfer, and Debbie Francis, the Librarian, for their ever-valuable assistance, offered always with courtesy and kindness.

During the production of this book, I received sincere and invaluable support and advice from the staff of Glyph International, Bangalore, and Sithivinayagam and Kavitha Ashok definitely merit special mention here, for the long hours they spent with me to polish this book into this final form.

To my wife Dr. Shibni Kaul, MSc, PhD, I shall remain eternally grateful for displaying extreme patience and understanding, and for looking after me and keeping me well-nourished so lovingly during all hours that I would be glued to my laptop. My undertaking became all the more light and enjoyable for the constant flow of love, encouragement and timely technical assistance from my nephew Sudhir Kaul, B.Tech., Director, Shalcot Mechanique (P) Ltd., my son Dr. Prashant Kaul, MB BChir (Cantab.) and my daughter-in-law, Dr. Mala Shaykher Kaul, MD. To crown it all, the beaming smile on the face of that little bundle of joy, our grandson, Master Āditya Kaul, made all the effort worthwhile.

Finally, my heartfelt thanks go to all my colleagues, friends and well-wishers whose good wishes and faith in me have been a constant source of encouragement throughout my life and, in particular, during the writing of this book.

Krishan Kaul
London

Foreword

Krishan Kaul, formerly Technical Director, Hyder Consulting, London, has, during his illustrious career, been involved in the design of many major transportation projects. Having worked in a firm that can justly lay claim to the design of such iconic works as the Sydney Harbour Bridge and the Hong Kong Mass Transit Railway amongst its earlier projects, it is understandable that the author should choose, from the field of transportation, the design of underground railway structures as the subject of his book.

There has, over the second half of the twentieth century, been a revival of interest in railways, particularly metropolitan railways, and, as it becomes increasingly more apparent that surface transportation cannot solve the problems of the urban sprawls that we call cities, the need for the design and construction of underground railways will intensify.

Responding to these increasing demands, the underground engineering profession has seen almost half a century of innovation and invention. Improvement in earth-retaining structures began in the late fifties when slurry diaphragm walls were first used in England on the Hyde Park Corner underpass development. Design has been revolutionized by the availability of the computer and, more particularly, the personal computer. At the start of construction of the Victoria Line in London in the early sixties, only hand-operated calculating machines were available. Yet, 20 years later, at the start of construction of the Singapore Metro, a pocket calculator had the capacity to calculate the entire route alignment. I count myself fortunate that, with a career starting in 1951 and enduring into the twenty-first century, I have been able to witness many wide-ranging advances and even participate in a number of them.

One of the major difficulties accompanying the rapid advancement in the design and construction of underground railway structures lies in transmitting the newly acquired knowledge to the upcoming generations of engineers. The publication in England of *The Art of Tunnelling* (Szechy, 1966) was an eye opener to many in the West of the developments that had taken place in Eastern Europe since 1945. The problem is that those at the forefront of these developments are generally too busy to break off and make time to record and publish their work. Nevertheless, it is those who are aware of the reality and are intimately involved in the dirty and dangerous nature of underground work that can contribute most to the profession.

It is therefore pleasing to welcome such a comprehensive book on the design of underground railway structures by an engineer who has been intimately and very practically connected with the process for many years and is now willing, while still at the vanguard of the engineering profession, to make the experience of those years available to others. Confining his interest to the specific topic of the design of cut-and-cover metro structures has allowed the author the freedom to cover the subject in the fullest detail, from the initial planning through to the necessary interaction with the ground, which is an essential and unique feature of the design of underground structures and their construction. Full awareness of the need for this interaction, as also highlighted by Professor Burland (2006), is amply displayed by the author in making the integrated geo-structural design the focus of attention in this book.

The coverage in the book of the varied topics that come up during the design of cut-and-cover structures is comprehensive, logically structured and well presented, with a clear emphasis on the practical implications involved. The critical commentary on the top-down *vis-à-vis* the bottom-up methods of construction presents one of the most explicit comparisons I have ever seen. The inter-relationship between the design and construction, a unique feature of the design of underground structures, makes the process of design both complex and challenging. The implications of the reality that large underground structures cannot be simply wished into place have eluded many aspiring designers and resulted in much anguish during the construction phase. The author seeks to make the reader aware of all the intricacies of design, complexities of construction together with their impact on design, and the difficulties of combining the varying, and often conflicting, aspirations of the many disciplines involved in planning an underground railway station and the need for tactful compromise to reach a workable solution, and has achieved that goal competently and in a practical way.

The interaction inherent between the temporary earth retaining works and the surrounding ground and structures, and therefore necessarily between structural and geotechnical engineers, calls for synergy between the two disciplines. This stamps the design of underground structures as a markedly different and unique branch of structural design. It also sets this book apart from those that deal more generally with structural design and fail to highlight the need to enlist the assistance of the ground to ensure an economical solution.

Throughout the book, the need for cost-effectiveness as the cornerstone in design is duly recognized discouraging, at the same time, the use of an overly simplified model that allows for the worst case everywhere. While such an approach would ensure that the structure did not fail, it could, however, lead to it never being built, as few promoters would be willing to come forward to finance the escalated cost that such types of design would entail. On the other hand, the warning to ensure the adequacy of ground support systems in view of the potentially catastrophic effects in the event of their failure, and particularly in the light of certain recent events, is timely and should not go unheeded.

This book is an expert and comprehensive guide to the practical design of cut-and-cover structures in general and metro structures in particular. It leads the reader clearly and logically from the initial planning and the choice of the structural form through the design process, from concept, assessment of loads to be supported, via specific problems such as seismic, heave, flotation, etc., to the structural analysis. Furthermore, the influence of the method and sequence of construction on the design process of these structures is palpable all the way through, and the need for it to be taken into account is rightly emphasized. The case history illustrations and the quality of the worked examples further materially enhance the usefulness of the book in guiding its users to a realistic and economical solution to the design problems. I welcome this book and commend it to the engineering profession as an informed guide to the practical design of these major underground structures and believe it will prove to be an invaluable resource for the designers.

T. W. Hulme
BA, BSc, ARSM, Eur Ing, CEng, FICE, FIStructE, FASCE, FCIArb
Consulting Engineer
Victoria, Australia

Preface

This book is primarily aimed at practising structural and geotechnical engineers engaged in the design and construction of cut-and-cover structures in general and metro structures in particular. It should also prove useful to postgraduate students and engineers who have interest in, or wish to pursue, advanced training leading to specialization in the cut-and-cover field.

The author, in his experience spanning nearly four decades in the structural and geotechnical engineering fields generally and well over three of those in the design of cut-and-cover structures worldwide particularly, has not come across a book that deals comprehensively with all the major aspects of structural and geotechnical design related to cut-and-cover structures under one cover.

Drawing upon his many years of experience in the structural and the geotechnical engineering fields, it is for the first time as far as the author is aware, that a book has been presented that combines the salient features of both the disciplines, taking on board, as appropriate, the mutuality of their effect on the design and construction of cut-and-cover structures. This approach to the design of cut-and-cover structures represents one of the unique features of this book.

In the design of cut-and-cover structures particularly, soil–structure interaction figures prominently. In dealing with such structures, therefore, sound appreciation of geotechnical principles is as important as a clear understanding of structural concepts. Therefore, to evolve the best design solution possible under a given set of conditions and constraints, ideally, the designer must have a thorough understanding of both the structural and the geotechnical principles. He must not only have a sound appreciation of their relevance to, and impact on, the construction, but also a clear grasp of the influence of the method and sequence of construction on the design.

Furthermore, the designer must also combine within himself the ability to deal with both the structural and the geotechnical aspects with consummate ease. However, the reason why this is generally not so is that each discipline represents a separate specialist field in its own right, and engineers combining within themselves specializations in both the disciplines are not easy to come by. Therefore, the success of carrying out the design and construction of cut-and-cover structures has to rely, largely on how best the respective inputs from the two disciplines can be unified and used in a complementary manner.

With discrete inputs from the two disciplines, it is possible to arrive at an acceptable solution; however, this requires blending harmoniously the structural and geotechnical requirements. It is also important to ensure that, in so doing, the integrity of the fundamental principles on which each input is based is not compromised. With close liaison between the specialists, appreciation of each other's viewpoint and readiness to reach an acceptable compromise, it is no doubt possible to achieve 'a' solution.

However, experience has shown that, by and large, the attitudes across the disciplines tend to be speciality-centric, which often lead to, what could be termed, the 'discipline polarity'. This can be seen to manifest itself, occasionally, in terms of certain reluctance on the part of the specialists to give way and venture outside their perceived domains of familiarity and comfort. Such an approach reflects a preferential, discipline-oriented interest. This can, at times, subscribe to a lack of appreciation of the principles, and breed a certain indifference to the requirements, of each other's disciplines.

Besides, the inability to fathom how much margin of comfort the 'other side' is keeping up its sleeve and the apprehension of being 'caught out' should one allow one's own perceived cushion of safety to be eroded in any way, do not help the situation either. When invited to soften their stance, each specialist is often too ready to reel off a stream of reasons, which may not always make much sense to, nor be entirely understood by, the other as to why he is unable to do so – all in the interest of the structure, of course!

Well, be that as it may, one thing is clear – without an in-depth insight into the technical compulsions of each other's disciplines, one lacks the ability to appreciate, first hand and in their entirety, the extent and severity of the problems likely to be faced. Consequently, the solutions proposed by each discipline specialist, however sound they might appear from their own respective standpoints, tend to be conditional, incomplete and therefore not always entirely satisfactory.

It is also difficult to meaningfully judge how far the reasons advanced by the 'other side' in favour or refutation of a proposal are prompted by genuine understanding or legitimate concern, to be able to question these accordingly and to suggest improvements or alternatives. As a result, the design and construction proposals get tossed to and fro between the two disciplines until some sort of compromise is reached. It is not unusual for such consultations, at times, to be protracted, and even then the end result may, at best, turn out to be no better than just a tolerable compromise.

Given the persistence of such a polarity of attitudes, the only way to cut loose from its limiting shackles is for the engineer to develop the ability to deal with the requirements of both the specialities by himself and with equal ease. This, however, is easier said than done.

There can be no guarantee that the combined efforts of the two individual specialists, howsoever well coordinated and harmonized, can lead to anything better than a compromise solution. On the other hand, a specialist who combines within himself both the specialisms and has the ability to appreciate and address the various implications comprehensively is, on balance, better placed to come up with a complete, unified and an altogether ideal solution. Such a specialist is also likely to have the ability to reach that solution relatively easily and quickly. It was this belief and the desire to apply an integrated approach to the design of cut-and-cover structures that inspired the author, even as a structural engineer, to specialize in the geotechnical engineering field also.

For a number of years, the need for a design manual with an integrated approach that combined within it the application of both the geotechnical and the structural principles has been felt. In recognition of this need, the author has attempted, through the pages of this book, to make available to and share with other engineers in the industry the benefit of his dual expertise. It is believed that, for those structural and geotechnical engineers whose curiosity, or need, takes them through the pages of this book, its contents will afford them a deeper insight into, and a far better appreciation of, each other's disciplines insofar as they have a bearing on the design and construction of cut-and-cover structures in general and metro structures in particular. It will also add depth and dimension to the insights of their own respective disciplines, infuse a new realism in their thinking and, hopefully, lay to rest some of the ghosts keeping them apart.

Krishan Kaul
London

Glossary of symbols

A	Huder's factor to allow for arching in trenches; free-field ground displacement response amplitude of an idealized shear wave
A, B	Skempton's porewater pressure parameters; Poisson's ratio factors; constants of integration
$\overline{A}$	Porewater pressure parameter $(= B \times A)$
A_C	Cross-sectional area of typical column; cross-sectional area of tunnel structure
A_{EQ}	Equivalent average area of cross-section of structure per unit length
A_f	Skempton's porewater pressure parameter at failure
A_s, A_{s1}, A_{st}	Areas of tensile reinforcement
A_W	Cross-sectional area of wall
B	Width of cut; overall width of structure
B'	Clear width of structure
C_P, C_S, C_R	Apparent velocities of P, S and Rayleigh-wave propagations
C_c	Compression index
C_s	Apparent velocity of S-wave propagation in soil
C_V	Seismic earth pressure coefficient
COF	Carry-over factor
CK_oU	K_o Consolidation undrained
$C\text{-}U;U;U\text{-}U$	Consolidated undrained; undrained; unconfined undrained
D	Seat of heave; overall depth of structure; depth to lowest point of failure surfaces; outside diameter of caisson
D_W	Wet depth; slurry–water differential head
E	Kinetic energy; Young's modulus of elasticity; undrained modulus for soil
E_c	Initial tangent modulus of elasticity for concrete
E'	Modulus appropriate to drained samples of soil
E_I	Modulus after isotropic consolidation of soil sample
E_L	Modulus of sample from laboratory curve
E_t	Elastic modulus of tunnel material
E_{K_o}	Modulus from simulated field curve of anisotropically consolidated sample
E_o	Initial tangent modulus of soil
$E'_{(e)}$	Expansion modulus for soil appropriate to drained conditions

$E_{u(e)}$	Expansion modulus for soil appropriate to undrained conditions
$E_{uo(e)}$	Expansion modulus at surface of soil for undrained conditions
EI	Bending stiffness
$E_c I_c$	Rigidity of cracked section
$E_c I_o$	Rigidity of uncracked section
EI_{equiv}	Equivalent rigidity
ESP	Effective Stress Path
F	Force; factor of safety; flexibility ratio in seismic design
F_1, F_2	Functions of geometry and depth of influence for stress diffusion
F_A, F_{ass}, F_{cal}	'Available', 'assumed' and 'calculated' factors of safety
F_{ave}, F_{crit}	Average and critical factors of safety against piping
F_r	Flexibility ratio
F_C	Factor of safety against quicksand; centrifugal force per track
F_R	Factor of safety required; reduction factor
F_S	Component of factor of safety available from frictional resistance; sinking force available
F_T	Factor of safety against heave
FEM	Finite Element Method
G	Shear modulus or modulus of rigidity
G'	Drained shear modulus of soil
G_m	Shear modulus for medium
G_s	Specific gravity of soil particles
G_S	Shear modulus of soil
H	Depth of trench; horizontal force; escalator rise; height of box structure
HA, HB	Highway loading
H_C, H_P	Net heights of columns and platform walls between floors
H_e	Exposed height of caisson
H_{crit}	Critical depth of trench
H_f	Slurry head
H_o	Thickness of stratum
HDC	Hand-Dug Caisson
H_S	Submerged depth of caisson
H_T	Height of caisson over which friction is considered
H_W	Water head
HWT	High Water Table
I	Moment of inertia about neutral axis; scale factor; influence factor
I_B, I_R, I_W	Moments of inertia of base slab, roof slab and walls of box structure
I_t	Moment of inertia of tunnel cross-section
I_c	Moment of inertia of cracked concrete section
I_o	Moment of inertia of uncracked concrete section
I_P, I_σ	Stress influence factors based on geometry
K	Earth pressure coefficient; Meyerhof coefficient; absolute rotational stiffness; intercept of failure line on maximum shear stress axis; factor related to geometry of structure; rate of change of modulus of soil with depth; Jacobson coefficient; $M/bd^2 f_{cu}$

K_{ba}, K_{cd}	Absolute rotational stiffness
K_4, K_5, K_6	Factors for rotation of reinforced concrete derailment barrier in plastic phase incorporating contribution of reinforcement and concrete
K'	Drained bulk modulus
K, K'	Factors related to geometry of structure
K_A, K_B, K_D	Factors related to geometry of structure
K_P, K_V	Factors related to geometry of structure
K_R, K_R'	Factors related to the geometry of the structure
K_a	Coefficient of active earth pressure
K_a, K_t	Longitudinal and transverse spring coefficients of medium
K_{ae}	Modified coefficient of active earth pressure due to earthquake activity
K_C	Pressure coefficient for wet concrete
K_o	Coefficient of earth pressure at rest
$\left.\begin{array}{l} K_o^H, K_o^L \\ K_o^N, K_o^{OC} \end{array}\right\}$	Coefficients of earth pressure at rest for heavily, lightly, normally and overconsolidated soils
K_m^N	Earth pressure coefficient for normally consolidated soil at strain m
K_p	Coefficient of passive earth pressure
K_X	Racking stiffness of structure
K_X, K_Y	Lateral stiffness of structure
KEL	Knife Edge Load (component of highway loading)
L	Overall length of structure; span; wavelength of an ideal sinusoidal shear wave
L'	Clear length of structure
L_P	Length of platform
LI	Liquidity Index
LL	Liquid Limit
LWT	Low Water Table
M	Mass; bending moment
$M_{AB}^S, M_{BC}^S \ldots$	Moments in spans AB, BC, etc.
M_B	Characteristic moment at root of cantilever wall due to compaction loads
M_{cr}	Flexural cracking moment after initiation of first crack
M_d	Design moment after cracking of concrete but before yielding of steel
M_{K_a}	Moment due to active earth pressure
$M_{K_{ave}}$	Moment due to average of active and at-rest earth pressures
M_{K_o}	Moment due to at-rest earth pressure
M_L	Local or Richter earthquake magnitude
M_S	Surface wave earthquake magnitude
M_e	Expansion secant strain modulus of soil
M_{er}	Equivalent expansion secant strain modulus of soil
M_{ez}	Equivalent strain modulus of soil
M_{max}	Maximum bending moment
M_o	Seismic moment

M_u	Ultimate bending moment
M_x	Bending moment at any cross-section x
M_w	Moment magnitude for an earthquake
N	Normal component of resultant force on failure plane; bearing capacity factor for deep strip footing; number of segments
N_b	Non-dimensional stability number
N_c	Coefficient based on Skempton's bearing capacity factors
N_{cb}	Non-dimensional bearing capacity factor
N_α	$= \tan^2 \left(45° + \dfrac{\alpha}{2} \right)$;
N_ϕ	Flow Factor
NDP	Normalized Deformation Parameter
NSP	Normalized Soil Parameter; Normalized Strength Parameter
NSF	Negative Skin Friction
OCR	Overconsolidation Ratio
P	Axial load; point load; design impact load; pole
P_A	Holding down force
P_{ae}	Modified active earth pressure force incorporating earthquake activity
P_{eq}	Equivalent upward load on structure due to heave
P_f	Force exerted by fluid pressure
P_H	Horizontal load
PI, I_P	Plasticity Index
PL	Plastic Limit
$P - \Delta$	Load-deformation
Q_P	Point load surcharge
Q_{max}	Maximum frictional force
R	Racking coefficient; reaction; radius of track
RDG	Residual Decomposed Granite
R_F	Frictional resistance mobilized
S	Shear force; shape factor in flow nets $= n_f / n_d$
S_B, S_G	Buoyant and gross sinking weights per m height of caisson
$SHANSEP$	Stress History and Normalized Soil Engineering Properties
SSR	Shear Strength Reduction technique
S_n	Stability number
S_R	Settlement ratio
S_t	Sensitivity
S, T	Geometry of Pavlovsky fragment II
T	Scale factor; mobilized shear resistance; translational stiffness; support thickness; predominant period of shear wave in soil deposit
T_B, T_R, T_W	Thicknesses of base slab, roof slabs and perimeter wall
$TCOF$	Translational Carry-Over Factor
T_F, T_P, T_S	Thicknesses of floor slabs, platform slabs and platform walls
T_S	Tensile force in steel
TSP	Total Stress Path
U	Porewater pressure component of force; uplift force due to buoyancy

U_B	Force of buoyancy or net artesian pressure
$U\text{-}D$	Undrained
V	Velocity of flow; vertical force; gross volume
V'	Volume of the portion of structure below water level
V_P, V_S, V_R	Peak particle velocities associated with P, S and Rayleigh-waves in seismic design
W	Self-weight of structure; single point load; width of box structure
W_A	Weight of backfill above water table
W_B	Effective weight of backfill below water table
W_C	Total live load on concourse slab
W_G	Gravity load
W_K	Gross weight of kentledge
W_P	Total live load on platform slabs
W_R	Required minimum weight of structure in flotation calculations
W_S	Effective weight of total soil backfill
W_T	Total live load on track slabs
Y	Distance from neutral axis of the section at which shear is required
a	Lever arm
a_S, a_S, a_R	Peak particle accelerations associated with P, S and Rayleigh-waves
a_1, a_2	Stiffness coefficients of box structure
a, c	Functions of soil characteristics
$\bar{a}$	Area of excavation per pile; lever arm
a_d	Ground motion at tunnel depth
b_T, d_T	Width and depth of a typical trench
c, c'	Undrained and effective stress shear strength parameters
c_w	Wall adhesion
d	Height of rectangular box structure
e, e_o	Voids ratio and initial voids ratio
f	Ultimate soil–structure interface friction
f_{cu}	Characteristic cube strength of concrete
f_{tu}	Uniaxial tensile strength of concrete
f_{tub}	Tensile strength of concrete in bending
f_y	Yield stress in steel
g	Acceleration due to gravity; non-dimensional shear factor
h	Thickness of soil deposit
h, h_e, h_p, h_v	Total, elevation, pressure and velocity (or kinetic) heads
h_m	Head loss across fragment
h_r	Reduction in hydrostatic head
i, i_c	Hydraulic and critical hydraulic gradient
i_{ave}	Average exit gradient
i_{fc}, i_{fT}	Critical gradient and exit gradient for filter material
i_T	Terzaghi gradient
$k; k_x, k_y, k_z$	Coefficient of permeability and its components in 3 coordinate axes
k_{eq}	Coefficient of permeability for an equivalent isotropic layer

$k_{h(eq)}, k_{v(eq)}$	Coefficients of permeability for equivalent single anisotropic layer
k_H, k_V	Coefficients of horizontal and vertical seismic inertial forces
m	Inverse gradient
$m^f_{AB}, m^f_{BA}\ldots$	Fixed-end moments at AB, BA, etc.
m_B	Long-period body wave earthquake magnitude
m_v	Coefficient of volume change from odometer test
m, n	Load geometry
n	Factor for axial load; number of layers
n_d, n_f	Number of head (or equipotential) drops and flow channels
p, p'	Mean total and effective stresses
p_a, p_p	Active and passive earth pressures
p_e, p_s	Total earth and slurry pressures
p_c	Gross contact pressure
$\bar{p}_f$	Effective stabilizing pressure
$p_{O/B}$	Weight of backfill or overburden pressure
q	Rate of flow or seepage; stress relief upon unloading
q, q'	Deviator stress
q, q_s	Intensity of loadings at surface and surcharge load
q_{ex}	Vertical stress relief
q_L, q_S	Intensity of line and strip load surcharges
r	Half height of rectangular box structure
s_u, s_{us}	Undrained shear strength and as measured on sample
$s_{u(b)}$	Undrained shear strength of soil below base of cut
s_{u_o}	Shear strength of clay at surface
s_{u_I}, s_{u_F}	Average shear strengths of intact clay and at fissures
s_u^{PSA}	Undrained shear strength (Plain Strain Active mode)
s_u^{PSP}	Undrained shear strength (Plain Strain Passive mode)
s_z	Skin friction mobilized around pile
t	Spring stiffness
$(t)m^f_{AB}$	Translational fixed-end moment at A in span AB
$u; u_o, u_{ss}$	Porewater pressure (*pwp*); *pwp* with initial hydrostatic distribution and at steady-state-seepage
$y; y_{max}$	Deflection or displacement; maximum mid-span displacement
z	Elastic section modulus
Δ	Inward displacement; amount of translation; maximum free-field lateral displacement of soil
Δe	Change in voids ratio
Δ_F	Shortfall in factor of safety $= F_R - F_A$
Δp_E	Seismically induced dynamic lateral earth pressure
Δ_{f-f}	Lateral free-field shear displacement
Δ_h	Head drop
Δ_{max}	Maximum deformation
Δ_{str}	Maximum racking displacement of structure
$\Delta B, \Delta S$	Extension of roof slab width, extension of roof slab all round perimeter
ΔH	Thickness of a typical layer; slurry head

$\Delta L_a, \Delta L_p$	Active movement of soil away from and passive movement towards wall
ΔP_W	Increase in penetration of perimeter wall
$\Delta T, \Delta T_B, \Delta T_R$	Increase in thickness of perimeter wall, base slab, roof slab
$\Delta u, \Delta u_{(max)}$	Excess and maximum excess porewater pressures induced
$\Delta \sigma_a'$	Relief in effective stress
$\Delta \sigma, \Delta \sigma'$	Change in total and effective stresses
$\Delta \sigma_{oct}, \Delta \tau_{oct}$	Mean changes in octahedral normal and octahedral shear stresses
$\Delta \sigma_z$	Total vertical stress relief
$\Delta \sigma_1, \Delta \sigma_3$	Changes in total major and minor principal stresses
Σ	Sum of
Φ_m	Dimensionless form factor for Pavlovsky fragment
∇	Laplacian operator 'del'
α	Adhesion factor; dilation factor; $\alpha = \tan^{-1}\left(\dfrac{\tan\phi'}{F}\right)$
α, β	Aspects of strip load geometry related to depth, z; correction factors for differential movements
β	Angle; slurry depth as a fraction of failure depth; ratio of shear and bending deflections
γ	Bulk unit weight of saturated soil
γ_{str}	Angular distortion of box structure
γ'	Effective unit weight of week sub-layer
γ_C	Unit weight of wet concrete
γ_d	Dry unit weight of compacted sand backfill behind wall
$\gamma_{d(min)}$	Dry unit weight of sand in its loosest state
γ_s	Maximum free-field strain or angular distortio in soil element
γ_s, γ_{sub}	Saturated and submerged unit weights of soil
γ_e'	Equivalent effective unit weight of soil
γ_f, γ_f'	Unit weight and equivalent effective unit weight of slurry fluid
$\bar{\gamma}_f, \gamma_f^c$	Effective unit weight of slurry fluid; critical unit weight of slurry fluid
γ_t	Unit weight of soil
γ_w	Unit weight of water
δ	Angle of wall friction
δ_e, δ_p	Elastic and plastic deflections
δ_e, δ_c	Initial undrained or elastic, and 'consolidation' or progressive heave
$\delta_{e(ave)}$	Average elastic heave
δ_{ep}	Reduced elastic heave with the use of piles
δh	Thickness of stratum
δ_{max}^h	Maximum heave
δ_{max}^s	Maximum surface settlement
δ_{max}^w	Maximum lateral wall deflection
δ_{TF}	Total final deformation
$\delta x, \delta y, \delta z$	Variation in stress due to excavation
$\varepsilon; \varepsilon_1, \varepsilon_2, \varepsilon_3$	Strain; strain in the three coordinate axes
$\varepsilon_1', \varepsilon_3'$	Axial and lateral strain under drained conditions

$\varepsilon^a_{\max}$	Maximum axial strain caused by $45°$ incident shear wave
$\varepsilon^b_{\max}$	Maximum bending strain caused by $0°$ incident shear wave
ε^{ab}	Total free-field axial strain
ε_{cc}	Maximum compressive strain in concrete
$\varepsilon_{co}, \varepsilon_{cu}$	Compressive strain and limiting strain in concrete
ε_s	Tensile strain in steel
ε_V	Volumetric strain
$\varepsilon_o, \varepsilon_{oc}$	Expansion and recompression strains under complete unloading
ε_r	Expansion strain due to partial stress relief
ε_{rc}	Compression strain under partial stress increase
θ	Inclination of failure plane with the horizontal
θ_p, θ_u	Rotation in plastic phase and at ultimate limit state
ϕ	Undrained shear strength parameter; angle of incidence with respect to tunnel axis
ϕ, ϕ'	Angles of shearing resistance with respect to total and effective stress
$\phi'_o, \phi_m, \phi'_{mob}$	Mobilized values of the angles of shearing resistance $= \tan^{-1}\left(\dfrac{\tan\phi'}{F}\right)$
$\tan\phi'$	Effective stress shear strength parameter
μ	Correction factor for induced porewater pressure; Poisson's ratio; rock rigidity
μ_0, μ_1	Dimensionless parameters for embedment and deforming layer
v, v'	Poisson's ratios appropriate to undrained and drained conditions
v_c	Poisson's ratio factor
v_m	Poisson's ratio of soil medium
ρ	Undrained elastic heave; density
ρ_e	Stress relief expansion factor
ρ_s	Density of soil
$\rho_s, \rho_{s'}$	Percentage of tensile and compression reinforcement
σ, σ'	Total and effective stresses
$\sigma_0, \sigma_1; \sigma'_0, \sigma'_1$	Initial and final total and effective stresses
$\sigma_a, \sigma_b, \sigma_c$	Cell pressures under quick undrained conditions
σ_{co}	Maximum stress in concrete
σ'_b	Vertical effective stress at formation
σ'_c	Effective lateral pressure due to compaction
σ_H, σ_V	Total horizontal and vertical stresses
σ_{h_o}	Initial total horizontal stress
σ'_H, σ'_V	Effective horizontal and vertical stresses
σ_n, σ'_n	Total and effective normal stresses
σ_o	Stress relief due to complete unloading as determined in laboratory on unconfined undisturbed samples; initial vertical total stress
σ'_o	Initial vertical effective stress
σ_r	Stress relief
σ'_r	Partial stress relief or increase
σ_s	Steel stress

σ_v^*	Vertical stress below current formation level
σ_v', $\sigma_{v\max}'$	Effective current and maximum past overburden pressures
σ_{v_o}'	Initial vertical effective stress
σ_{V_c}'	Vertical effective consolidation stress (laboratory)
σ_{V_b}'	Horizontal effective stress
σ_{V_m}'	Maximum effective past pressure
σ_{V_o}'	*In situ* vertical effective overburden pressure
σ_z	Vertical stress at depth z
τ, τ_s	Shear stress in soil
τ_f	Shear strength of soil; gel strength of slurry fluid
τ_{str}	Applied shear stress across structure
τ_{TD}, τ_{VD}	Triaxial compression strength and vane shear strength at depth D
τ_{xy}	Complementary shear stress on gel element
ϕ, ψ	Stability functions
ψ	Function of stiffness coefficients
ψ'	Angle of dilatancy

A note on units

The SI (metric) system of units has been adopted throughout this book. In accordance with this system, the basic units of measurement, the multiples and the derived units used are as follows:

Basic units

Length	m (metre)
Force	N (Newton)
Pressure	Atmosphere (atm), Bar
Displacement	mm (millimetre)
Mass	kg (kilogram)
Time	s (Second)

Multiples

1 Atmosphere = 1 Bar $\approx 103 kN/m^2$
kilo $= 10^3$, e.g. kilo Newton; $1\ kN = 10^3 N$
Mega $= 10^6$, e.g. Mega Newton; $1MN = 10^3 kN = 10^6 N$
$1\ tonne = 10^3 kg$

Derived units

Area	mm^2, m^2
Volume	m^3
Section modulus	m^3
Moment of inertia	m^4
Pressure, stress	kN/m^2
Strength	N/mm^2
Unit weight	kN/m^3
Elastic modulus	kN/m^2
Bending moment	kNm, Nmm
Spring stiffness	kN/m
Permeability	m/s
Energy	mkN
Velocity	m/s
Acceleration	m/s^2
Leakage	m^3/s
Coefficient of volume compressibility	m^2/kN

1 Introduction

The expression 'cut-and-cover' represents an abbreviated version of what is essentially a three-stage process – 'cut, construction and cover' – of installing an underground structure. In the first stage, a 'cut' (i.e. excavation) of the appropriate size is made to make way for the 'construction' of the structure in the second stage. This is followed, in the third stage, by the replacement of the backfill to 'cover' the structure in a manner that reinstates the features on the surface. An underground structure constructed in this manner, utilizing these three stages, although not necessarily always following this order is commonly referred to as a 'cut-and-cover' structure.

The structures generally constructed by the cut-and-cover method mostly comprise multi-level basements, deep underground car and water parks, underpasses, culverts, pumping stations and, of course, metro stations and running tunnels. Of these, the metro station structures are by far the largest and most complex to design and construct. Therefore, to give the subject of the design related to cut-and-cover construction generally the coherence, extent and quality of coverage it deserves, these structures have been chosen as the core reference structures throughout the text. Treatment of the multifaceted aspects of design of such complex structures, it is expected, will more than cover the principles of design for other, relatively simple cut-and-cover structures also. It is with this belief that the book has been structured to deal with the various aspects of the design principally related to cut-and-cover metro station structures. These structures are also referred to variously, in different parts of the world, as the 'underground' or the 'tube', the 'metro', the 'subway', etc.

Cut-and-cover structures are, by definition, confined underground. The principal considerations for the design of such structures need to include, among others, the effect of the following:

- Lateral earth and hydrostatic pressures on perimeter walls acting as retaining elements;
- Lateral pressures on structural floor elements acting as horizontal permanent struts;
- Vertical loads (downward and upward as appropriate) on the horizontal floor elements and the vertical support elements such as columns and walls;
- Seismic loads; and
- Buoyancy on the stability of the structure as a whole.

Magnitude of all these aspects, individually and collectively, together with satisfying the requirements of stability, serviceability, durability and minimizing maintenance, demands a certain minimum bulk of the structural elements and the structure as a whole. This then points to the use of concrete as the most preferred material. Accordingly, in the book, reinforced concrete has been assumed as the principal material of construction for cut-and-cover structures.

Metro station structures are generally located in heavily built-up urban environments and often in difficult ground conditions; as such, certain aspects of the method and sequence of their construction often have a dominant influence on their design. For the design of such structures, therefore, appreciation of such aspects and making appropriate allowances for these are essential. Accordingly, aspects of construction of cut-and-cover structures, insofar as they appertain to or have a bearing on their design, have also been addressed, albeit briefly, in this book.

The process of engineering design with regard to cut-and-cover metro structures is broad based and far wider in scope than might be implied by the mere mathematics of structural analysis, derivation of stress-resultants, i.e. assessment of bending moments, shear and axial forces and the calculation of the size, amount and distribution of reinforcement required in the various elements of the structure. Besides satisfying the structural requirements, both in terms of design and construction, metro station structures must also be designed to take account of and make accommodation for the plant layout with all its associated ducting and cabling, requirements for signalling and telecommunications, as well as the distribution of the motive power, air conditioning, exhaust systems, etc.

The design of an aboveground structure is generally based on well-defined parameters and established principles. The loadings too can be accurately calculated and applied at known discrete locations and in defined directions. The boundary conditions and the response of the structure to the applied loads and the various load combinations can be established with a reasonable degree of certitude. In comparison, however, the design of an underground structure has generally to be based on parameters that are unlikely to be precisely quantified. For example, in a given soil environment, the behaviour of the soil may vary from being nonlinear, linear-elastic to plastic. The *in situ* stress history, anisotropy, heterogeneity, drainage characteristics and the changing porewater pressure boundaries during the various stages of construction are also some of the factors that are difficult to quantify accurately. This makes the analysis approximate and less tractable in comparison with that of the aboveground structures.

Exaggeration of or underestimation in the applied loads and overestimation or softening of the shear strength of the soil with time or strain, can also complicate computations further and make modelling of the soil–structure interaction more difficult. Besides, the soil samples are often relatively small and potentially unrepresentative of the soil mass in which the works are to be executed. In the wake of all such uncertainties, the design is largely based on certain idealizations and a degree of empiricism, which calls for reliance and drawing upon previous experience of the design and construction of similar structures in similar environments.

Construction of a typical cut-and-cover metro structure will, inevitably, disturb the *in situ* stress environment and the porewater pressure regime in the existing soil mass. To a large extent, this disturbance will be dependent, among other things, upon the type of the soil medium, whether free-draining, fine-grained or intermediate, *vis-à-vis* the sequence of construction of the temporary and the permanent works. With the relatively free-draining soils, such as sands and gravels, the effects of the various stages of construction are likely to be immediate. In the case of fine-grained soils such as clays, on the other hand, due to the viscous retardation to free flow of water, the effects can be expected to be time-dependent. With intermediate soils such as silts, however, the response is likely to be mixed, i.e. both short term (undrained) as well as long term (drained). (The terms 'drained' and 'undrained' are discussed in Chapter 11.)

Starting with the ambient *in situ* stresses, the soil pressures will experience a drop in their values resulting from ground relaxation during excavation, followed by redistribution during the various stages of construction of the temporary and permanent works. Thereafter, a gradual climb back towards their eventual at-rest values can be expected. The design must demonstrate that the structure can satisfactorily withstand this cycle of soil pressure change in combination with all the other coexistent dead and live loads and any other incidental loads likely to act, both during the stages of construction and its operational lifetime.

In the light of what has been discussed so far, it is clear that, particularly in the case of cut-and-cover structures, design and construction are inter-related and cannot be isolated from each other. Consequently, for achieving a satisfactory structural design solution, sound knowledge of the aspects of construction is also essential. Ideally, therefore, the engineer must combine within himself the necessary skills to deal with both the geotechnical as well as structural aspects of design. He must not only have a sound knowledge of the various methods and techniques of construction and a clear understanding of the principles of soil and structural behaviour and soil–structure interaction, but also a thorough appreciation of their implications on design. Drawing upon many years of experience in the structural as well as geotechnical engineering fields, the salient features of both the aspects of design have been brought together within this book.

In order to facilitate easy comprehension, each chapter is laid out and structured to develop the topic from its fundamentals. Furthermore, the overall sequencing of the chapters in the book is also designed to maintain a logically evolving progression and continuity between the various topics related to construction and design. Accordingly, the first three chapters can be seen to be representing the preamble in a manner that sets the scene for what follows. The next six chapters deal, sequentially, with the various aspects of construction of cut-and-cover metro structures, highlighting their impact on design and in a manner that makes it possible to identify the specific requirements of design necessary during the various stages of construction. The last 18 chapters then follow, dealing with other aspects of design sequentially.

Starting with the synoptic introduction of the contents of the various chapters, the basic principles of planning associated specifically with the design and construction of a typical metro station structure are outlined and the guiding considerations leading to the threshold of the structural form discussed. All these aspects are sequentially covered in the first three chapters.

In view of the ready availability of a large volume of literature on the various construction processes, and in order to keep the scope essentially specific to design, giving in to the temptation of embarking on a detailed treatise on the construction of metro structures has been deliberately resisted. However, for completeness and to enable the book to stand alone, an overview of the construction process is presented in Chapter 4. It is covered briefly yet in sufficient detail to bring into focus the salient features of construction in such a manner as to enable their likely effect on the design to be appreciated. A conscious effort has also been made to present this overview in a naturally evolving manner and apace with the logical sequence of construction.

At every stage of the construction overview, the necessary design requirements are identified, which help put into perspective the subsequent treatment on design.

The stages of excavation, the way the ground is supported during these stages and the sequence of construction have a dominant influence on the design of cut-and-cover structures. The suitability and the inherent limitations of the various ground and wall support systems available for different ground and groundwater conditions and the implications of their use on design are discussed in Chapter 5.

Control of groundwater during the various stages of excavation and construction constitutes an important aspect for the successful construction of cut-and-cover metro structures. This is discussed in Chapter 6. Control of groundwater through exclusion and removal and the factors influencing it are discussed. A brief commentary on ground movements usually accompanying dewatering operations is presented, and the principal measures that can be employed to limit these are also discussed.

The size and severity of the problems likely to be encountered during the design and construction of a typical cut-and-cover metro structure depend, to a large extent, upon the likely problems of interaction arising out of the proximity of the proposed metro works with the existing or planned future structures or works in the vicinity. As an aid to design, a clear and an in-depth appreciation of the various factors likely to give rise to interaction problems and the resulting constraints is vital. In recognition of this need, various aspects of interaction are laid out in detail, discussed and, where appropriate, put into perspective with the help of relevant case histories gathered from various metros and similar other projects around the world. In view of the major importance of construction interaction on the design of cut-and-cover structures, three chapters (Chapters 7, 8 and 9), have been devoted exclusively to this topic. This brings to conclusion the aspects dealing primarily with construction and construction-related design. The remaining 18 chapters are devoted mainly to other design issues.

The main stages making up the design development process are outlined in Chapter 10. In this chapter, the various steps involved in the development of a conceptual design, arguably the most important stage in the design development process, are discussed. A brief qualitative appreciation of some of the principal constraints is also presented.

Identification and assessment of the various loads applicable to cut-and-cover structures constitutes an important element of the design process. Since the bulk of the principal loadings derives from the ground environment, an appreciation of the engineering properties of soils, their behaviour and response during the various stages of excavation and construction and the assessment of critical earth pressures assume special importance. Ground and groundwater pressures invariably represent the dominant loadings acting on an underground structure. For a reasonably accurate assessment of these, an understanding of the engineering properties of soils and the principles of how the critical loads can develop at various stages of construction is important. All these aspects have been discussed in sufficient detail in Chapters 11, 12 and 13.

Procedures for the assessment of surcharge loads on the basis of the elastic theory are outlined in Chapter 14. Chapter 15 deals with the assessment of seismic loads demonstrating the use of both the modified classical enhanced earth pressure (Mononobe-Okabe) approach as well as the seismic soil–structure interactive (*SSI*) approach using the Seismic Deformation Method (*SDM*).

Construction of the perimeter-wall/cofferdam invariably represents the very first stage in the main construction process. It is important to ensure that, during their excavation, the integrity of the faces of the wall panels is not undermined. Using limit equilibrium theory and incorporating recent advances in design techniques, the stability of trenches in different types of soils is investigated. Assumptions employed are critically examined and, where appropriate, measures to increase the factor of safety are also discussed. Chapters 16 and 17 are devoted to these aspects.

During the construction of a cut-and-cover structure, movement of the surrounding ground and, by implication, also the settlement and stability of the existing structures in the vicinity can be of particular concern to the engineer. Permeability of the ground mass and the rate of flow of subsurface water, which are closely linked with ground movements, therefore, constitute very important considerations both during construction and design. In view of this, Chapter 18 is devoted to the various aspects associated with the seepage of water during construction and its implications on design.

Construction of a typical cut-and-cover metro structure generally entails the removal of large volumes of soil and water. This causes pressure relief, which can lead to the inward movement of the sides of the cut and the upward movement or heave of the formation. In the case of soft soils, particularly, such movements can, potentially, be a source of serious problems for the designer as well as the constructor as has been amply demonstrated by some catastrophic tunnel failures in certain parts of the world. In view of the paucity of ready reference material on this topic, what is presented in Chapters 19, 20 and 21 is the result of an extensive literature search on heave carried out by the author as part of a separate dissertation on the subject. Chapter 19 deals with such aspects as the mechanics of heave and the factors influencing it, whereas Chapter 20 outlines the various steps and methods leading to the prediction of heave. Discussion on some of the available measures to contain heave, particularly in soft ground, and the validity of some of the assumptions used in heave calculations are presented in Chapter 21. A method of design incorporating considerations that can subscribe to safety during excavation particularly in soft ground is also presented.

In the case of underground structures, examining the stability under the influence of buoyancy during the stages of construction and the operational lifetime of the structure can, under certain circumstances, present an important design consideration. Accordingly, Chapter 22 is devoted to various aspects of flotation. Typical flotation calculations are presented for different cases of structural geometry and groundwater conditions. Make-up of the required factor of safety is critically examined. Various antiflotation measures are identified and discussed and their appropriateness and limitations under various situations assessed.

In the light of the principal operations of construction and the associated requirements of design, Chapter 23 deals with the selection of the necessary design parameters. Cut-and-cover structures are, by definition, in intimate contact with the surrounding soil mass, involving significant soil–structure interaction. For design, therefore, knowledge of both the structural and the geotechnical parameters is necessary. However, the design parameters for the structures are mostly invariant, reasonably accurately known and readily available, whereas those for the soil medium are not. Owing to their potential variability, the process of the selection of the appropriate design parameters in this book has been confined to that for the soil medium alone.

Chapter 24 deals with the selection of load factors appropriate to the nature of the various load elements and the identification of the most onerous but realistic load combinations. It is recognized that some of the conventional load factors have, as a result of unchallenged usage over time, come to be regarded, in some cases perhaps for doubtful reasons, as the accepted norm. However, in the selection of the appropriate load factors in this book, a conscious effort has been made, particularly in relation to the design of cut-and-cover metro structures, to resist being drawn into blind conformance with such usage, which could possibly be open to question. Every loading element is examined on its merits and, in the light of sound engineering principles and using engineering judgment, values for load factors deemed to be sensible are proposed. Wherever appropriate, the author has not shied away from making out a case in favour of a departure from any usage that appears inappropriate.

Modelling of a typical cut-and-cover metro structure with its changing boundary conditions and loadings at various stages of excavation and construction, implications of structural idealizations and the response of the differential support system, the effects of the method and sequence of construction and of the locked-in stress resultants, the effects of load transfer *vis-à-vis* the chosen sequence of construction, etc., are all covered in Chapters 25, 26 and 27, the last three chapters.

For better understanding and appreciation, the engineering principles and procedures adopted throughout the book have been amplified, where appropriate, through worked out numerical examples. Wherever they appear, any assumptions made in the analytical solutions have been discussed and, where appropriate, critically examined for their appropriateness, inherent limitations and the suitability of application. Most of the solved examples used to illustrate the principles have been drawn and adapted from real-life problems faced by the author over the years. The illustrations have been so chosen as to bring into focus the impact of the chosen method and sequence of construction on design that is characteristic of cut-and-cover structures and sets these apart from other structures. Every opportunity has been taken to draw conclusions, wherever appropriate, to help highlight the influence on design.

While the book deals with the design aspects related principally to underground metro structures, the principles outlined are equally valid for and applicable to all other types of cut-and-cover structures also.

To maintain the principal themes of the topics in focus without unduly cluttering the contents of the chapters, additional information complementing the basic text and some useful aids to design have been included separately at the end under the various appendices. These include, among others, indicative design loadings and parameters, design of derailment barriers, stability of slopes, etc.

A comprehensive glossary of symbols listing the various notations used and a note on the units adopted throughout the text have also been presented at the beginning of the book for easy reference.

2   General Planning

2.1 Introduction

Planning of a metro system involves a multidisciplinary process. It requires inputs from the various disciplines such as those related to transportation and traffic, route location and track alignment, surveys, permanent way; design of the space and the structure, electrical and mechanical systems and environmental control systems; geotechnical investigation and assessment, practicalities of construction, accommodation of services, etc.

Metro station structures are called upon to meet a broad range of planning requirements in a manner that ensures, among other things, a safe and efficient flow of passengers with the minimum of conflicting movements, the spatial integration of engineering elements and operational facilities and the fulfilment of the specific environmental needs. In the overall design of such structures, coming to terms with the requirements of the various disciplines introducing their own special requirements and constraints, and their interrelationships must be recognized as an essential feature of the design development process. All constraints and parameters and such other factors as are likely to influence the design and construction of the structures need to be identified and assembled at an early stage.

Train size constitutes a key parameter that determines the overall planning of a metro system and is critical for design optimization. The cross-section of the train must be known before the cross-section of the running tunnels and, in turn, the station structures can be finalized. The performance characteristics of the trains also constitute essential input in deciding the line profile, power supply and other characteristics necessary for the system. It is important for the basic car and train dimensions to be defined at the earliest possible stage so that the planning and the design of the structures on the system can proceed on an agreed basis.

In this chapter, only those aspects of the planning inputs are considered that are likely to have a direct bearing on the form, design and construction of a metro station structure itself.

2.2 Constituent elements

A typical cut-and-cover metro station structure comprises a number of basic elements that include:

- Access points from and to the street level;
- Concourse, also variously referred to as Ticket Hall, Booking Hall or Mezzanine;

- Platform(s) for boarding and alighting of passengers, Plant Rooms, Vent Shafts;
- Passages, Stairs, Escalators and Lifts for movement of passengers within the station structure;
- The Tracks for the movement of trains; and
- Nonpublic areas for station staff facilities.

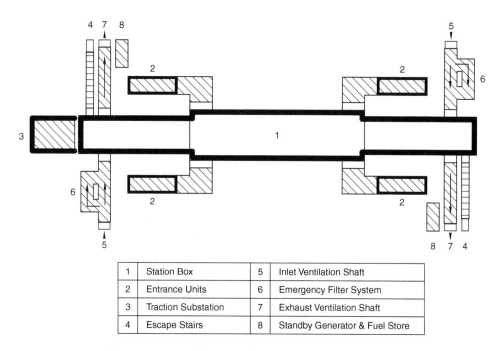

1	Station Box	5	Inlet Ventilation Shaft
2	Entrance Units	6	Emergency Filter System
3	Traction Substation	7	Exhaust Ventilation Shaft
4	Escape Stairs	8	Standby Generator & Fuel Store

Figure 2.1 Typical Layout of Station and Ancillary Units

A typical layout of ancillary units in relation to a cut-and-cover metro station structure is shown in Figure 2.1.

Access points or entrances

Designated areas for gaining access into the station are located on the surface and strategically disposed around the station in such a manner as to maximize catchment and ensure convenience for the metro users. Access points represent both the passenger entry and egress points into and from the station. Although they also constitute the exit points from the station, they are, by commonly recognized usage, identified as Entrances.

Entrances leading into the station need to be protected against the incidence of local flooding from burst pipes, etc. Raising the entrance threshold by 150*mm* or so above the existing or the planned local ground level may be sufficient to cover this risk. However, in tropical cities where monsoons and cloudbursts can put the streets under a metre or more of water as a result of which a more severe risk of flooding can exist, the entrance threshold may need to be raised above the potential flood level; this generally necessitates the inclusion of a few steps rising up from the road level. Alternatively, additional facilities for flood boards or full-scale flood gates may be required. This may require building-in appropriate allowances in the structure in the form of slots to receive the flood boards or gates. However, these are severe measures which can put entrances out of commission temporarily. In areas susceptible to sand storms, the entrances may also need to be protected against these and the ingress of dust and dirt.

Concourse

Concourse is usually located between the entrances and the platform levels. Its purpose is to allow orderly collection of passengers from the entrances, provide facilities for issuing tickets to them and channel them towards the control point, the 'Ticket Barrier'.

Depending upon the fare-system adopted, the ticket barrier may comprise either manned gates where collectors will check tickets or automatic gates where passengers insert coins, tokens or encoded tickets for automatic scrutiny as they pass through the barrier.

The entrance side of the ticket barrier is referred to as the 'free' area and the track side as the 'paid' area. Both these areas are generally contained within the concourse, the 'free' side being larger in area than the 'paid' side. Travel information, waiting space, left-luggage lockers and other public facilities such as shops, telephones, toilets, etc. are sited in the free area of the concourse. Paid area, on the other hand, provides clearly defined access routes to and from the platforms via escalators, lifts and stairways.

Platforms and tracks

Within the general constraints of station positioning, the arrangement of platforms mainly derives from the engineering considerations of alignment and track geometry, operational and, where necessary, interchange requirements, and the volume and the intensity of the anticipated passenger flow.

The basic arrangements of tracks and platforms are generally as follows:

- Twin central tracks with side platforms – normally deriving from a line constructed in cut-and-cover;
- Central or island platform with tracks either side – deriving from separate bored running tunnels entering the station;
- Island platforms with central concourse and tracks either side – adopted in larger stations, again with bored running tunnels.

Island platform, consistent with its design passenger capacity, is appropriate for use with bored running tunnels. Such a layout offers the advantage over that of the side platform in terms of functioning with a tidal passenger flow and also platform supervision. The concourse is located above the platform and tracks, with vertical circulation provided by means of escalators and stairs. At platform level, escalators and stairs, depending upon their numbers, are so located as to offer an even distribution of passengers boarding the trains along the platforms. Large openings within the concourse floor allow a visual communication between the concourse and the platform, which avoids possible confusion and helps in the directional orientation of the passengers. Besides, they also assist in the overall ventilation of the station.

Side platforms, on the other hand, are ideal for use where the twin central tracks enter the station as close together as the gauges permit within cut-and-cover construction. In this case too, the concourse slab is located above the platforms. Side platforms give directional segregation of passengers. However, in comparison with that of the island platform layout, their access requirements, including escalators and stairways, may be greater, particularly where the passenger movement is tidal. Besides, they also require some duplication of operational facilities and equipment. All this may necessitate, potentially, additional platform width, which may make the overall station box structure somewhat wider.

Plant rooms are generally located towards the ends of the station structure beyond the platforms. Two ventilation shafts rising vertically above the track level dealing with the intake of fresh air, circulation of conditioned air and collection and removal of exhaust air are generally provided per station. These are also located, ideally, one at each end of the station structure.

Escalators

Escalators (as also staircases) not only introduce discontinuities but their accommodation also subscribes to sizeable cut-outs in the floor slabs that they connect. A machine chamber, for example, is required in the upper floor underneath the top of the escalator to house the drive machinery. The minimum plan dimensions of the chamber can be approximately $3.5m \times 3m$. Appropriate access through the floor into this area is required for the servicing and replacement of machinery as and when necessary. Also, at the bottom of the escalator, i.e. in the lower floor, a pit is required to accommodate the truss. The minimum plan dimensions of the pit are approximately $1.5m \times 3m$. The dimension $3.5m$ at the top chamber and the $1.5m$ at the bottom pit are those measured beyond the ends of the escalator truss. Provision of such cut-outs in the floor slabs and their structural implications need to be addressed as appropriate.

2.3 Provisions of safety

In the planning of a metro system, particularly over its underground sections, it is vital to consider all the safety and emergency features that will be required. A continuous passage is generally provided through every carriage of the train, with centre exit doors in the cabs at each end. However, detrainment of passengers through these means in an emergency can be laborious and sometimes a precarious undertaking especially if the damage sustained by the carriages makes it difficult.

In the interests of safety, appropriate alternative provision must be made to detrain the passengers in the event of a serious breakdown or accident, and to enable them to walk along the tunnels to adjacent stations in reasonable safety. In twin-track tunnels, this presents no serious problems; however, in single-track tunnels, it may be necessary to incorporate a narrow 'catwalk' at a level that corresponds with the floor level of the train on one side throughout the route. This provision will, naturally, add to the width and, of course, to the cost of the tunnel structure. It may have an impact, albeit marginal, on the planning of the station structure as well.

The design of the line must also make provisions for some strategically located emergency crossovers between the running tracks to permit trains to bypass the site of a breakdown or, more usually, to reverse on either side of it. Such crossovers must be so located as to enable the curtailed service to reach the nearest station where there are reasonable facilities to disperse the passengers. It is also important not to have too many emergency crossovers of this kind. There must be clear and well-rehearsed arrangements for their use; otherwise there may be confusion at the very time when disciplined emergency operation is vital.

The aforementioned notwithstanding, in the case of major cut-and-cover structures, such as metro stations, the likelihood of a hazard resulting from the inadequacy of design or construction procedures can have potentially serious implications. It is therefore important to carry out, in accordance with the requirements of *CDM* (Construction, Design and Management) Regulations, 1994, UK, or similar, an appropriate and detailed risk assessment. For the assessment to be acceptable, the resulting Risk Assessment Code (*RAC*) should be sufficiently low (i.e. ≤ 3).

2.4 Make-up of size

The spatial considerations, which together define the size and level of a typical cut-and-cover metro station structure, are:

- Overall depth of structure
- Overall width of structure
- Overall length of structure
- Depression below ground.

Overall depth of structure

The overall depth of a typical metro station structure is made up from the following:

- Combined overall thickness (including finishes) of the horizontal elements of the structure. This would, typically, include the thicknesses of the roof, the concourse, the platform(s) and the track slab(s);
- The nature of the subgrade, whether rock or soil. This can influence the thickness of the (lower) track slab, which also forms the base slab of the structure. For example, where the subgrade is rock, a nominal thickness of the base slab may suffice;
- Allowances for voids below the soffits of structural slabs to accommodate the services and thicknesses of the false ceilings;
- Vertical planning clearances. These depend on the required minimum headrooms stipulated by the appropriate planning authority of the country;
- Accommodation for the ventilation and air-conditioning ducts;
- Accommodation for an overhead conductor (wires and pantographs) for motive power distribution if preferred in place of the third rail; and
- Required number of levels of platform/track slabs.

The requirement for the trains and stations to be air-conditioned will impose the need for conducting the supply of conditioned and exhaust air from and to the plant rooms. This will necessitate the provision of longitudinal ducts over the tracks to deal with the emission of heat from the trains.

The depth below the underside of the roof slab to the rail level is generally around 9·5m.

Overall width of structure

The overall width of a typical metro station structure is largely determined by the need to accommodate the following:

- Thickness of the longitudinal sidewalls forming part of the station perimeter;
- Thickness of the intermediate support structural elements such as columns;
- Type and layout of the running tunnels influencing the layout of the tracks within the station;
- Planning clearances incorporating the widths of platforms, escalators, stairs, etc.;
- Setting-out of and construction tolerances on the longitudinal sidewalls;
- Thickness of inner-skin walls, where required to mask the longitudinal sidewalls, and finishes on columns;
- Provision, where required, for drainage channels between the longitudinal side and the masking inner-skin walls; and
- Track widths taking into account kinematic envelope and the required clearances.

The clear internal width between the longitudinal walls is usually in the range of 18–22m. The variation is essentially due to the number of escalators needed to serve the passengers; these can change from station to station, depending on whether the station is lightly or heavily trafficked.

If the running tunnels are twin-bored, practicalities of construction require these to be a certain minimum distance apart. Such a necessity allows the extent of this clearance to be so adjusted as to incorporate the requisite width for the formation of a central island platform within the station structure. Bored running tunnels and island platform stations thus form natural extensions of each other.

Figure 2.2 shows a typical cross-section through an island-platform station structure.

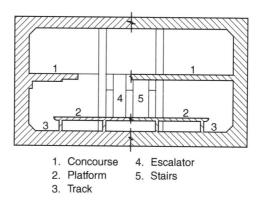

1. Concourse   4. Escalator
2. Platform   5. Stairs
3. Track

Figure 2.2 Typical Cross-Section Island-Platform Station
(adapted from BRTA, draft pre-design report, Vol. 1, 1982)

The right-hand half represents the section through the full (unvoided) concourse whereas the left-hand half through the concourse openings.

In the case of running tunnels constructed by the cut-and-cover method, on the other hand, to ensure economy of construction and to avoid unnecessary expense, the overall width of the running tunnel box structure needs to be kept to a minimum. This requires

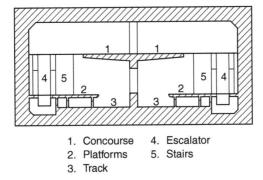

1. Concourse   4. Escalator
2. Platforms   5. Stairs
3. Track

Figure 2.3 Typical Cross-Section Side Platform Station
(adapted from BRTA, draft pre-design report, Vol. 1, 1982)

the tracks to be located side-by-side within a common box structure. Furthermore, it is also sensible to continue the same central side-by-side alignment of the tracks into the station structure as well. This then, logically, leads to the location of the platforms on either side of the twin tracks. It is therefore clear that bored running tunnels lend themselves, naturally, to the provision of island platform stations just as cut-and-cover tunnels do to the provision of side platform stations. These configurations are shown in Figures 2.2 and 2.3, respectively.

Train size is a key parameter that determines the overall planning of a metro system and is critical for design optimization. The cross-section of the train must be known before the station cross-section can be finalized.

Overall length of structure

The overall length of a typical metro station structure is determined by the following considerations:

- Thickness of transverse walls defining the two ends of the station perimeter;
- Design and accommodation of the ventilation and air treatment systems;
- Length of platform. This is, in turn, dependent on the length of the train; and
- Provision of space for the plant room(s) to accommodate electrical switchgear, transformers, emergency generators, signalling equipment, etc.

The length of the station platform constitutes a key parameter making up the length of a typical station structure, and the single major factor determining the platform length is the length of the train. This is, in turn, based on the required passenger design capacity of the metro system, which is determined by estimating the heaviest flow of passengers likely to use the busiest section of the line during the design-year peak hour. In principle, the system needs to be designed to accommodate this flow with trains of appropriate size operating at the necessary frequency. However, platform lengths turn out to be typically in the range of 140–180*m*.

In the case of cut-and-cover construction, the overall length of the station at platform level effectively determines the size of the concourse level. Experience has shown that this provides adequate space for the anticipated passenger circulation, ticketing requirements, nonpublic office accommodation, etc.

The extent of air-conditioning and/or mechanical ventilation needed within the station to provide the desired environment is one of the main factors that determine the space requirements for plant and equipment. The size and the routeing of the air ducts can have a critical effect on the design of the public areas, while the space required to house the mechanical ventilation plant and cooling equipment can significantly influence the overall length of the station box.

A carefully thought out and well-designed ventilation system does not always have to add to the station's subsurface volume or the size of the running tunnel. However, in the layout and design of a ventilation system, it is vitally important to recognize the life-preservation and passenger-safety roles of the ventilation system. Notwithstanding this, the ventilation system designer should, like other discipline specialists, also be prepared to be flexible enough in designing that system in a manner that will help reach mutual accommodation with the requirements of other disciplines.

Depression (*of structure*) *below ground*

Deeper the structure, greater also is the depth of excavation. Increasing the depth of excavation means a commensurate increase in the:

- Hydraulic gradient
- Associated technical difficulties
- Ground movements due to excavation
- Zone of influence of groundwater lowering
- Ground movements caused by groundwater lowering.

In view of this, purely from the structural point of view, every attempt must be made to locate the structure as close to the surface as possible. However, in reality, the extent of the minimum depression of the structure below the ground level is variable and is mostly determined by other site-specific conditions. The depth below ground at which the roof of a metro station structure can be located is generally governed by the following considerations:

- The need to accommodate the existing and/or the future services and utilities
- The need to clear the existing obstructions that have to be maintained
- The desirability of avoiding ponding of water on the roof slab
- Requirements for the civil defence, where appropriate.

Generally, a $2m$ cover of soil over roof slab of the station structure provides an adequate minimum depth to accommodate pavement construction and small services. However, in the case of larger services, such as trunk sewers, if permanent and expensive diversions are to be avoided, greater depths can become necessary.

Where an existing feature such as an underpass or a canal cuts across above a metro station structure, depending on the extent of its depth below the surface, it could easily constitute the governing criterion for the extent of depression of the station box structure below the surface.

'Waterproofing' of the roof slab notwithstanding, some engineers generally favour incorporating falls in the top surface of the slab during its construction in the belief that it will discourage the ponding of water on top and assist its flow away from the structure. This is usually achieved by the formation of a longitudinal ridge along the centre line of the roof slab thereby creating cross-falls. Where such an approach is adopted, the thickness of the soil backfill cover on top of the station roof necessary to accommodate the services and utilities crossing over needs to be adjusted upwards to incorporate the provision of such drainage falls also. This approach, it is argued, provides the first line of defence against the potential seepage of water through the roof slab. Nevertheless, it must be clearly understood that where the roof is permanently below the water table, it is unlikely that any tangible advantage can be gained by the formation of cross-falls in which case these can be dispensed with. Such an approach could, conceivably, be useful but only if the top surface of the roof were to be above the ambient water table; however, even then its effectiveness cannot always be guaranteed.

Where the requirements of civil defence dictate the placement of a shield in the form of a 'burster slab' over the station roof sandwiching a layer of soil in between, the thickness of such a slab together with the sandwiched soil layer must also be allowed for in establishing the extent of the depression of the structure necessary below the ground level.

2.5 Typical size threshold

A threshold for the size of a typical, single-level track, metro station structure can be developed by assigning a range of commonly adopted values to the various parameters making up the overall size as discussed in the preceding section. For illustrative purposes, this principle has been adopted in the following exercise:

- Depth

    Depth below road surface: $2 \cdot 0 – 3 \cdot 5m$
    Thickness of roof slab: $1 \cdot 0 – 1 \cdot 5m$
    Headroom – 'concourse to roof': $4 \cdot 5 – 4 \cdot 7m$
    Thickness of concourse slab: $0 \cdot 8 – 1 \cdot 0m$
    Headroom – 'track to concourse': $4 \cdot 7 – 5 \cdot 3m$
    Thickness of track slab: $1 \cdot 0 – 1 \cdot 5m$
    Based on the above,
    Depth of the box structure: $12 \cdot 0 – 14 \cdot 0m$
    Overall depth of structure: $14 \cdot 0 – 17 \cdot 5m$

- Width

    Longitudinal wall thickness: $1 \cdot 0 – 1 \cdot 5m$
    Track width: $4 \cdot 0m$
    Platform widths including escalators and stairs: $14 \cdot 0 – 15 \cdot 0m$
    Thus, overall width of structure: $24 \cdot 0 – 26 \cdot 0m$

- Length

    End wall thickness: $1 \cdot 0 – 1 \cdot 5m$
    Platform length: $140 – 180m$
    Plant room accommodation: $100 – 117m$
    Accordingly, overall length of structure: $242 – 300m$.

The foregoing exercise places the typical one-level-track, cut-and-cover metro station structure in the overall range of $24m$ (W) $\times$ $14m$ (D) $\times$ $242m$ (L) to $26m$ $\times$ $17{\cdot}5m \times 300m$. However, stations much narrower than the $24m$ (i.e. of the order of $18m$) and much longer than the $300m$ (i.e. of the order of $500m$, or even longer) are not uncommon. Requirement for such unusually long station structures is generally dictated by commercial rather than technical considerations, such as the need to incorporate underground shopping malls, food courts, etc.

2.6 Route location and alignment

The main aim of the alignment is to link the successive stations on the route in a manner that minimizes the extent of construction and related problems, including any adverse environmental impact, performs satisfactorily in operational terms and meets the defined alignment standards, all within budget. It may not be possible to achieve all this system-wide; in certain locations, meeting all these aims may result in conflict. In such cases, it is not unusual to examine an alternative alignment and, if not feasible, to settle for a design compromise.

Horizontal alignment

The horizontal alignment is broadly determined by attempting to follow the broad corridors of greatest traffic demand, actual or potential, and as direct a route as possible that gives the least length of construction necessary to meet the traffic requirements. A direct alignment would meet the objective of passenger transportation most ideally. Since it would avoid severity of curvature and the problems related thereto, it would also permit higher train speeds and achieve shorter travel time. However, in practice, such an alignment is often not achievable. This is so because of the existence, all too often, of high-rise structures on deep foundations in heavily built-up urban areas. Acquisition and demolition of such properties can involve high land-take costs. But, more importantly, it can also give rise to other socioeconomic problems closely associated with the disturbance, movement and the resettlement of a large mass of people. The detailed planning must therefore take into consideration not only the optimum station locations, the requirements of junction layouts and the related factors, but also, particularly, the need to minimize disturbance to the existing structures and the related, large-scale, mass movement of people, since an alignment that cuts across a heavily built-up urban environment can prove to be prohibitively expensive. In view of this, the horizontal alignment corresponds, more often than not, to the alignment of the major roads on the surface.

Vertical alignment

The principal aim of achieving a good vertical alignment is to locate the metro system, consistent with the required operational standards, as close to the surface as possible. This is primarily because a shallow station avoids most problems commonly associated with deeper excavation and construction and is relatively quicker and cheaper to build. It also offers:

- Quick and convenient access to the platforms for the passengers;
- Shorter travel of facilities such as lifts and escalators, making the system less expensive both in terms of capital cost as well as running and maintenance costs.

The vertical alignment of the track is influenced by the following factors:

- Ground conditions;
- Relative levels of other stations;
- Construction methods used for running tunnels;
- Levels and locations of other crossings or adjacent tracks;
- Gradients introduced to assist acceleration and deceleration of trains;
- Gradients required to facilitate the drainage and collection of seepage water;
- Need to clear existing or prospective obstacles, such as rivers, underpasses, foundations, etc.; and
- Need to provide accommodation for the existing and future services and utilities, such as water mains, deep sewers, etc., above the station roof.

It is often possible to construct bored running tunnels in between stations at slightly greater depths, i.e. depressing them in relation to and in between the adjacent stations. Thus the tracks within the stations are maintained at levels relatively higher than those of the running tunnels in between. Such an alignment configuration is known as the 'hump profile'. This enables the use of gradients normally not exceeding 3·0 per cent to be used both to achieve acceleration of the trains during the downhill departure from a station and deceleration during the later stage of their uphill approach to the next station. The use of the hump profile can result in significant savings in the electrical power.

However, in the case of running tunnels constructed by the cut-and-cover method, the cost increases rapidly with depth. In such cases, the use of hump profile may not prove to be cost-effective and therefore, generally, flat grade of 0·3 per cent, adequate for drainage only, is provided, and hump profile is avoided.

It should be recognized that, in certain locations, both horizontal and vertical alignments can be seriously affected by constraints such as those presented by the existence of foundation piles to buildings and elevated roads, especially if it is not feasible or desirable to demolish such structures, and by the access routes and ventilation shafts for the stations. In such instances, it may be necessary to undertake expensive underpinning works to enable the alignment to be achieved to reasonable standards. To avoid escalation of costs, the alignment of a cut-and-cover metro system below the surface therefore follows, for most of its length, the route of the existing highways on the surface above. With such an alignment, however, the construction works, inevitably, present problems for the road traffic.

In addition to the alignment proper, it is also necessary to establish a corridor, commonly referred to as the metro corridor, to protect it against encroachment by other new works. This protection is generally given by legal powers granted to the metro authority. Within the boundaries of the metro corridor, areas must be earmarked as temporary work sites for the duration of construction. Areas must also be identified as permanent sites for the main metro station structures and other associated structures, such as entrances, ventilation shafts, traction substations, escape stairs, etc.

The construction of the cut-and-cover structures associated with the metro system would also require large quantities of plant and materials to be transported by road to the various sites, while the spoil from the excavations would have to be removed to designated tips. The additional traffic generated by the construction activity itself would also add to the problems. At each construction site, traffic circulation would have to be temporarily reorganized. If possible, through traffic would have to be diverted away from the site and corridors of suitable access to premises provided for the local traffic and the essential services.

2.7 Traffic management

The aim of traffic management measures is to relieve, wherever possible, or minimize the (short-term) disruption to normal traffic likely to be caused by the construction of the metro works. The organization of traffic during construction activities should be phased into and coordinated with the long-term strategic traffic plans, developed to cope with the disruption to traffic caused by the construction activity and its attendant works traffic. The traffic management schemes to be adopted should be compatible with the transport and traffic policies of the relevant regulatory authorities. The traffic management measures would need to cope, in safety, with all aspects of traffic, including those generated from:

- Goods vehicles
- Public transport
- Essential services
- Pedestrian movement
- Local and through private traffic.

In addition to the foregoing, the traffic generated by the construction works activity would also have to be accommodated.

Traffic management measures notwithstanding, there is no doubt that, while the work is in progress, cut-and-cover construction activity can have a blighting effect on the shops and businesses in the streets concerned even though pedestrian access to such locations is maintained.

2.8 Surveys

In order to establish a firm alignment for the planned metro route, a series of topographical surveys have to be carried out for alternative routes so that maps can be drawn which are accurate enough to allow a comparative study to be made and the preferred route to be identified. It is also usual to employ aerial photography using stereo pictures for these surveys. Secondary survey stations are also set up close to the preferred route alignment so as to pick details, such as pedestrian underpasses, etc. which may not show up in the aerial survey. Besides, it also helps to pinpoint any potential areas of conflict between the metro and other proposed new developments.

It is absolutely essential to establish the precise location of the services such as sewers, water and gas mains, telecommunication and power cables, etc. This information is generally available and obtained from the managing authorities of these utilities and is then plotted on the metro survey maps. These maps are then examined to identify any potential conflict points. It is often necessary to make trial holes to locate the precise positions of the mains and sewers so that any diversions required can be properly planned. Trial holes may also be needed to confirm the types and exact locations of foundations of buildings and basements along the metro route where such information is either suspect or not available.

Preparatory to undertaking any construction activity, it is advisable to carry out, in collaboration with the owners or their agents, condition surveys of the existing structures which fall within the zone of influence of the metro works. Agreed records of these surveys generally form the bases of any subsequent restitution and settlement of disputes that may arise in the event of any damage being sustained by the existing structures as a result of the metro construction activity.

2.9 Site investigation

Sound understanding of the response of the ground during the groundwater draw-down and the various stages of excavation, and of the soil–structure interaction during the subsequent stages of construction of the structure is vital for the proper design of the structure and its successful construction. It is prudent to collect, through detailed desk studies, all the available information from previous borings and deep excavations associated with any previous construction activity in the area. However, a properly planned and thorough site investigation must be carried out to identify the strata, establish the ambient water table, classify the soil from borehole samples and identify any variations in the ground conditions along the route. Information with regard to the potential of long-term variations in the groundwater level must also be gathered so as to establish the worst credible level over the operational lifetime of the structure. This information is needed not only to refine the gradient profile, evaluate construction methods and carry out cost estimates, but also to obtain appropriate parameters for carrying out the design of the structure.

Most parts of the world are prone to the occurrence of earthquakes albeit to varying degrees. Enough evidence is available from different parts of the world to confirm the extreme seriousness of the scale and intensity of devastation to life and property that seismic activity is capable of unleashing. In the earthquake-prone areas of the world, therefore, it is also necessary to consider the risk of earthquakes and the metro structures have to be designed to withstand the appropriate level of seismic and related pressures.

3   Structural Form

3.1 Introduction

In the course of design development, the process leading to the evolution of the ideal structural form for a cut-and-cover metro station involves inputs from a number of various disciplines such as architectural, planning, structural, geotechnical, alignment-related, environmental system, etc. To begin with, each discipline within the design team identifies its own constraints in the light of which it tries to seek the best accommodation possible for its own requirements. In order to become aware of and appreciate their inter-disciplinary impact, constraints from every discipline are assembled and made available to all the members of the design team. An attempt is then made to coordinate and rationalize, as far as possible, the respective constraints in order to establish an integrated conceptual threshold for the form that satisfies the various core requirements such as, functional, planning, aesthetic, structural, constructional, etc.

During the conceptual design stage, the thought processes of the different disciplines are at work simultaneously, overlapping, influencing and modifying each other's thinking and decision making up to an extent and in such a manner as to help arrive at a commonly acceptable solution. Like the other professionals in the design team, the structural and the geotechnical engineers too contribute their own knowledge, experience and skills to this process. However, achieving a consensus can, at times, be a challenging process.

The various preferences may at times appear to lead to mutually contradictory requirements rendering, on the face of it, a particular proposed solution incompatible. But, given a measure of understanding all-round, common ground can generally be found. However, in a complex situation, the choice of form may be dictated by the most dominant requirement or constraint based upon a clearly defined priority. Nevertheless, for a rational approach, a sympathetic consideration by the structural engineer of the aspirations of other disciplines is as important as a measure of understanding by the other disciplines of the difficulties peculiar to the design and associated with the construction of the specific form of the metro structure as perceived by the structural engineer.

However, it must be recognized that in conjunction with providing comfort and ensuring reliability of service, the most ideal form will be the one that aims at the harmonious resolution of the various requirements, viz., aesthetic, ease of construction, maintenance, safety, cost, etc. as discussed in the following section.

3.2 Guiding considerations

Form of a cut-and-cover metro structure is generally influenced by the following aspects:

- Functionality
- Structural principles
- Ease of construction
- Aesthetic appeal
- Ease of maintenance
- Consideration of cost
- Other considerations?

Functionality

From a purely functional standpoint, a structure enclosed by a perimeter wall incorporating intermediate slabs as horizontal support elements and columns as intermediate vertical support elements, forming a linear rectangular box, designed to withstand the most onerous combinations of all the gravity loads and the lateral pressures from the surrounding ground, and providing, internally, the minimum acceptable standards of horizontal and vertical planning clearances, would offer the simplest structural form. It would also represent the threshold for the cheapest structural solution.

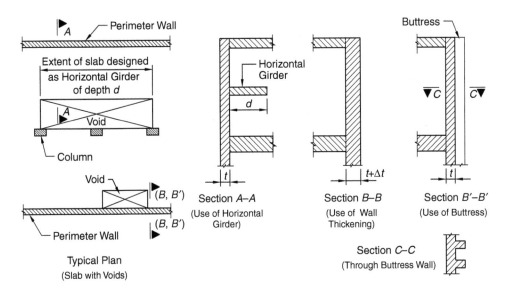

Figure 3.1 Typical Plan and Cross-Sections

In contributing to the structural form, it should be recognized that the horizontal slabs besides having to carry the required gravity loads also act as permanent struts maintaining the opposite perimeter walls the requisite distance apart. Nevertheless, existence of large voids and discontinuities in the floor slabs required to accommodate the escalators and stairs resulting in the consequent loss of strut action over these areas cannot be avoided. However, such voids should be closely examined to ensure that the loss of strut action does not represent also the complete loss of lateral support to the walls and that the loss is compensated in some way.

Consideration may need to be given to the prospect of designing the intact strip of width of the slab between the void and the perimeter wall as a horizontal girder over the extent of the void as shown in Figure 3.1. If needed, the thickness of the wall can be combined with the width of the strip to form an increased girder depth. However, if the void extends as far as the face of the wall itself such that no strip of slab is available to provide a horizontal girder support, the wall would then be required to span over a greater height vertically. In that case, the resulting implications would need to be examined carefully. If required, the wall may need to be strengthened by thickening it or by the inclusion of external buttresses at required centres, or both, over the extent necessary. These options are illustrated in Sections *B–B* and *B'–B'*, respectively.

In certain cases, it may also be possible to use castellated plan shapes of wall panels, as discussed in Chapter 5.

Structural principles

The reasons for the choice of the structural form as preferred by the engineer may, at times, appear to be at variance with those based on the aesthetic preferences of the architect or somewhat out of accord with the expectations of the other disciplines. For example, in a bid to maximize the usable plan area and achieve an open aspect within the metro station structure, the architect may try to do away with the intermediate column supports altogether or, at least, show preference for far fewer and slender columns spaced longer distances apart. To the engineer, on the other hand, columns are important structural elements which, in the event of a major accident, can help postpone complete collapse of the structure and therefore serve as important life savers. Besides, sensibly spaced columns can ensure sensible thicknesses for the floor slabs and also provide a direct and an efficient path for the vertical loads to the subgrade.

To illustrate this point further, compare the case of a typical floor slab with and without a central line of support. With the central line of support removed, the design moments would increase four-fold, and even if the slab thickness were to be doubled, the amount of reinforcement would also be nearly doubled. This would not only, virtually, double the cost of the slab, it would also add to the overall cost in many other ways. For example, increase in the thicknesses of the floor slabs would subscribe to a commensurate increase in the overall depth of the structure and the volume of excavation. This would, in turn, increase the magnitude of lateral loads on the perimeter walls. In the case of a partially submerged structure, it could also increase the buoyancy and affect the potential of the structure to float. In view of these, the engineer may be inclined to aim for reasonably robust column sections at what might be construed, structurally, as a 'sensible' spacing in order to minimize their slenderness (L/r) and maximize efficiency. In order to alleviate, to the extent possible, the problems likely to result from the effects of differential axial deformation between the discrete intermediate columns and the continuous perimeter walls, avoid waste and achieve cost-effectiveness, the engineer could feel perfectly justified in pursuing this approach.

The architect may prefer to have one large uninterrupted void within the concourse slab so as to make it possible to provide unobstructed visual communication between the concourse and the movement of trains approaching or leaving the platforms. The engineer, on the other hand, may not favour a void of large uninterrupted length but may feel obliged to retain at least some band widths of the slab to serve as discrete struts so as to compensate for the loss of strut action over the extent of the void between the opposite walls. The Electrical & Mechanical (*E & M*) discipline may prefer the environmental control ducting to be square or round and without major bends to ensure an efficient flow; the architect may prefer these to be rectangular and shallow so as to be neatly tucked away and concealed within the space above the false ceiling; the structural engineer would, ideally, like to see these routed under the beam soffits in order to obviate the need to provide sizable holes through the webs of the beams, and so on. All these seemingly contradictory issues need to be resolved satisfactorily.

It is as well to recognize that, owing to their confinement underground, opportunities for the visual appreciation of the form of cut-and-cover structures are rather limited. However, for the same reason, their construction too is often complex, difficult and cost intensive. Consequently, the efficient use of the space and the sound construction of the structure within the site-specific constraints, invariably present themselves as the respective principal concerns of the architect and the engineer. Therefore, in the context of cut-and-cover metro station structures particularly, the assertion that whereas the architect designs the space the engineer the structure, is largely true. The key to the success, however, lies in harmoniously blending the contributions of the two disciplines in such a manner as to ensure that the space and the structure complement each other to the best advantage of the metro system as a whole.

Ease of construction

In the planning and design of cut-and-cover metro structures, the use of complex geometrical forms is generally avoided. This is because the construction of such inherently large underground structures can be significantly cost intensive and is often accompanied by major constraints and complexities. The severity of these complexities is a function of a number of factors such as, the type of the soil and the groundwater conditions, permeability of the ground mass, structural environment and the sensitivity of the structures in the vicinity, method and sequence of construction, and any other specific constraint that may be prevalent on site. While these factors invariably vary from site to site, changes in some of these even on the same site from one part to the other or one stage of construction to the other also cannot be ruled out. This makes the precise control of all such factors rather difficult.

Furthermore, complex shapes can subscribe to difficulties in construction with adverse programme and cost implications. They can also lead to serviceability and durability problems. In view of this, a structural form that is simple and can be associated, inherently, with ease of construction is, from an engineering viewpoint, the most preferable.

Aesthetic appeal

The ground reality that the users of a metro are unlikely, ultimately, to measure its worth by the extent of the complexities faced in the design, difficulties overcome during construction or indeed the size of the cost involved, cannot be ignored. Such statistics end up becoming the exclusive preserve of only the connoisseurs and the experts. In the course of time, when all these aspects, irrespective of their size and severity, are no longer topical or relevant and are therefore forgotten by the common masses, usefulness and worth of the underground metro system will continue to be gauged, principally, by the efficiency of its function.

However, without the benefit of a pleasing and inviting station environment, it may not be possible to rule out user reluctance of the facility let alone encourage its active usership. It is therefore important that, in the planning and design of metro structures, aesthetic elegance, particularly of the station structures, is not sacrificed to an overly structural conservatism. By the same token, structural soundness must not be bartered for an exclusive and fanatical pursuit, at any cost, of a pleasing appearance either.

Structural constraints, imagined rather than real, which may be perceived as unnecessarily stifling creative expression of, or imposing unreasonable restrictions on, the architectural freedom may, *ipso facto*, make their acceptance as difficult as an architectural treatment, which may be overambitious and, from a structural viewpoint, bordering on the dangerous. Neither of these approaches can be seriously entertained. However, one must not lose sight of the possibility either that aesthetic elegance and engineering excellence need not, necessarily, be mutually exclusive. The aim therefore must be to come up with a form that combines within it, as far as the budgetary constraints will allow it, both these qualities in a complementary manner, without compromising the primary considerations of function and safety.

It must also be recognized that since the consequences of a failure of a metro structure in terms of loss of life, disruption, cost, etc., can be potentially catastrophic, it is not advisable, given the potential variability of the various parameters, to stretch the design and strength of the structure to their very limits; it is prudent to build-in reasonable safeguards and margins in these to ensure a desired assurance of safety and comfort.

Ease of maintenance

In a typical cut-and-cover metro structure, by virtue of its confinement under ground, the outside perimeter is expected to be in constant contact with the surrounding ground and groundwater over its lifetime. As a result, the outside areas of the structure are not going to be readily accessible and, consequently, gaining access to such areas for the purposes of repairs can be prohibitively expensive. Cut-and-cover metro structures should therefore be robust, durable and designed to warrant minimal maintenance.

In choosing the form of a metro structure as also the method and sequence of its construction, every attempt should be made to ensure that, as far as possible, the need to reach inaccessible areas is eliminated, potentially major maintenance costs are avoided and that other life-cycle costs too are kept to the absolute minimum. This can be achieved by following a method and sequence of construction which is simple, tried and tested and one that can be soundly executed, maintaining close control on the quality of materials and workmanship and building-in, where advisable, appropriate extra sacrificial safe-guards.

Consideration of cost

It is a well-known fact that the construction costs of cut-and-cover structures increase significantly with depth. It is not unusual or surprising, therefore, that cut-and-cover structures in general and, because of their size and complexities of construction, metro station structures in particular, tend to be extremely cost intensive.

Costs of metro structures can vary from country to country. However, depending upon their size, depth below ground level, the method and sequence of construction and the type of ground conditions encountered, the costs of cut-and-cover metro station structures can easily range from £30M to over £100M and those of the running tunnels between £20M and £100M sterling per *km* run. Therefore, putting right any inherent weakness in the structural form can carry, potentially, major cost implications.

While achieving a structural form acceptable both architecturally and in terms of practical engineering is not impossible, it may not always be easy. However, in the event of a problem, invariably, the overriding economic considerations, consistent with safety, are likely to influence the decision leading to the eventual choice of the method and the sequence of construction.

Other considerations

As discussed in the previous chapter, the type of the running tunnel, whether bored or cut-and-cover, can dictate whether the station structure will follow the island or the side platform type of layout and, therefore, influence the form of the structure accordingly.

Also, metro station structures generally tend to be long linear structures; therefore, significant variations in the subsoil conditions (e.g. from rock to soft ground) demanding the use of radically different foundation treatments over the length of the same structure cannot be ruled out. While this cannot be helped, it may also become necessary to introduce appropriate transitional structural systems that will help maintain the compatibility of the different forms of foundations adopted across the interfaces. This may, in turn, also influence the form of the structure.

3.3 Make-up of form

The choice of form for cut-and-cover metro station structures is rather limited. However, in the light of the foregoing discussion, it is clear that, a rectilinear box structure incorporating either a side or an island platform with the concourse directly above the platform/tracks as shown in Figures 2.2 and 2.3 in the previous chapter represents, typically, the most simple and commonly adopted cross-sections for cut-and-cover metro station structures. Notwithstanding this, sometimes commercial

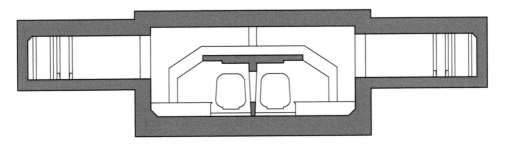

Figure 3.2 Typical Shallow Wide Station Cross-Section with Link Bridge
(adapted from draft pre-design report, BRTA, Vol. 5, Nov. 1981)

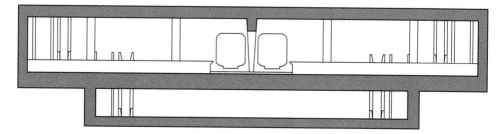

Figure 3.3 Typical Shallow Wide Station Cross-Section with Underpass
(adapted from draft pre-design report, BRTA, Vol. 5, Nov. 1981)

interests may dictate, additionally, the provision of shopping mall, food courts, leisure facilities, etc. to be integrated with the station. This can be easily achieved by incorporating an extra floor level above the concourse dedicated for such use. Naturally, this would make the structure deeper and, inevitably, add to its cost. Such narrow and deep stations are commonly adopted in areas where, on the basis of socioeconomic reasons, wider shallower station structures demanding extensive acquisition of land and large-scale demolition of the existing structures are unacceptable.

On the other hand, where the prevailing difficulty of ground conditions makes it absolutely essential to minimize the depth of the station structure, and where it is feasible to expand laterally, it is possible to adopt a shallow but wider structure. Under such conditions, it may be possible to split the concourse into two halves positioning each half clear of and on either side of the main structure which incorporates the platforms and the tracks. The connectivity between the two halves of the concourse, the platforms and the tracks can be provided by means of a link bridge above. A typical cross-section of such a configuration is shown in Figure 3.2.

Alternatively, for a similar overall depth of the structure as outlined in the foregoing section, it is possible, if so required, to make the rail level even shallower by accommodating the concourse, the platforms and the tracks at one level with an underpass below (instead of the bridge above) providing the link across the platforms as shown in Figure 3.3.

Such wider shallower stations as described above may be feasible in suburban areas where land is available and where such structures prove to be cost-effective.

It is also possible to construct a station structure, where appropriate, with an arched roof similar to that shown in Figure 3.4. Besides enhancing the aesthetics of the concourse, the use of such a form enables the benefit of arch action to be made use of in the design. A number of stations have already been constructed using this principle and are in operation world-wide. Some of the key features of such a structural form are as follows:

- Appropriate proprietary systems can be used to provide temporary lateral ground support during excavation for and construction of the structure.
- Construction can be carried out top-down or bottom-up as dictated by design and site-specific constraints.

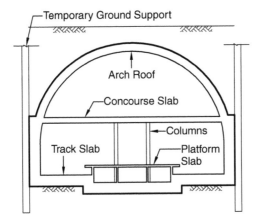

Figure 3.4 Typical Cross-Section with Arch Roof

- No intermediate vertical supports are needed for the arch thus providing an open aspect at the concourse level. However, in order to provide the requisite minimum rise for the arch at the critical locations, the overall structure can become deeper with the obvious cost implications.
- Owing to the arch action of the roof, the requirement of reinforcement in the arch is minimized.
- Concourse slab can be designed to act as a tie to contain the horizontal outward thrust from the arch at its springings.
- Any net uplift forces acting on the base of the structure would be shared between the base/track slab and the concourse slab but not the arched roof.

Note: The thicknesses of the member elements of the structure are, naturally, a function of design and, as such, the relative thicknesses implied in the elements of Figure 3.4 should be treated as indicative only.

3.4 Canary Wharf Station: a form with a difference

Canary Wharf Station, the line diagrams of which adorn the front and the back covers of the book, is part of the Jubilee Line Extension, London. It is built on part of the redundant West India Docks and rests within the basin of a former dock. It measures 290*m* long, 27*m* deep, 36·5*m* wide above and 28*m* wide below the Ticket Hall level. The station not only typifies elegance in architecture but also represents an outstanding example of structural form engineered under difficult ground conditions and severe constraints. Typical cross-section of the station structure is shown in Figure 3.5.

   The geology at the site comprises Thames Gravel overlying Woolwich and Redding Beds over Thanet Sands on top of Chalk. Of the two main aquifers on site, the upper in the Thames Gravel is affected by the River Thames and the impounded dock water level of +104*m* Project Datum (PD), while the lower aquifer is confined in Thanet Sands with a piezometric head of +97*m*PD.

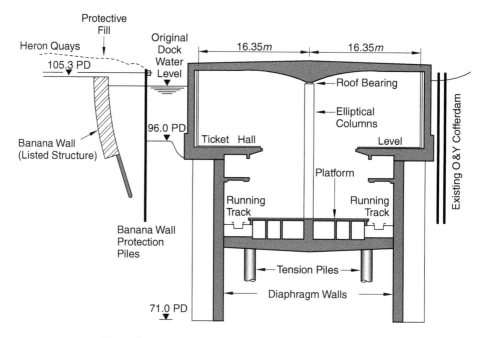

Figure 3.5 Cross-Section Canary Wharf Station London
(adapted from Drake *et al.*, 1999)

Salient features of the design of the station structure (Drake *et al.*, 1999) are as follows:

- The permanent station box structure below the Ticket Hall level is made up of 148 diaphragm wall panels 25*m* deep. In order to withstand the significantly high lateral forces resulting from the high water table and earth pressures acting over the depth of the structure, diaphragm wall panels for the station perimeter comprise 1·2*m*-thick T-sections.
- Over and above the gravity load provided by the mass of the structure and some effective overburden on top, a total number of 163 tension piles reinforced with steel stanchions have been installed into sockets 9–10*m* into the chalk to help counteract the potential uplift force of 200,000 *tonnes* due to flotation.
- The base slab varies up to a maximum of 3*m* in thickness. Besides withstanding the maximum contact pressures vertically, it also acts as the permanent lower strut horizontally between the longitudinal diaphragm walls. Ticket Hall slab provides the permanent strut action at the upper level.
- To accommodate a small amount of rotation and elastic elongation of piles under full hydrostatic uplift pressures at the interface with the base slab, the diaphragm walls have been trimmed and rendered to form a smooth surface and finished with mastic asphalt to offer some flexibility while still permitting the base slab to act as a horizontal strut.
- Spherical pot-bearings, with a typical vertical working load capacity of 2700*t* each, have been used at the top of the intermediate columns. These permit lateral movement in east-west direction and spherical rotation to allow deflection

movements with directional restraint in the north-south direction and, at the same time, prevent moment transfer and development of undesirable stress-resultants in the various members.

- Owing to the limitation of the available working area on site, site offices and the welfare facilities were mounted on barges moored alongside the quay. Also, in order to minimize the impact of construction traffic on the local roads, transport of all the bulk materials for construction and removal of spoil from excavation were transported to and from site by barge.

4 Construction Overview

4.1 Introduction

Cut-and-cover metro structures are enormously large structures that cannot be simply wished into place. Construction of such structures is generally complex; it takes place in constrained, often difficult and variable ground conditions and, invariably, needs to be carried out in carefully phased stages. Besides, it can extend over a relatively long period, over which it can remain exposed and, potentially, susceptible to the influence of the elements.

Furthermore, in respect of cut-and-cover structures, there also seems to be a conspicuous absence of a unified approach to design. This is largely due to the fact that the design of such structures is closely interlinked with the sequence of their construction which, for a given structure, may not be unique but, conceivably, embrace a variety of possible different permutations and combinations involving their own specific complexities.

The sheer number of potential variables involved, including the number of possible construction options on offer, can sometimes make it difficult for the designers to easily home in on the most appropriate solution. Inevitably, therefore, the designers of cut-and-cover structures can be seen to place a lot of reliance on past experience. However, in selecting a 'method and sequence' of construction that is most suited to a given set of conditions and constraints, it is often helpful to carry out a comparative study of the merits and demerits of the various available options, including examining their limitations and design and cost implications.

The surrounding ground, because of its intimate contact with the cut-and-cover structure, generates lateral pressures constituting the principal load acting on it; however, it also provides all round support and stability to the structure. The magnitude of the soil pressures generated and the extent of the soil support available in terms of the reactive pressures mobilized at any given time are, therefore, closely interrelated. Accordingly, as the geometry and the boundary conditions of the structure change from one stage of construction to the next, so also will the extent of the soil support available depending upon the extent of excavation and the state of completion of the structure reached at that stage.

If, for a chosen method and sequence of construction, the changes in the stages of excavation and the geometry of the structure are significant, so also are likely to be the changes in the loadings and the boundary conditions. Since this invariably happens to be the case, the influence of the chosen method and the sequence of construction on the design of a cut-and-cover metro structure can often turn out to be very significant. In achieving a satisfactory design solution, therefore, it is important to develop a clear understanding about the preferred method and sequence in which the excavation is to be carried out and the structure to be constructed. It is equally important also to have a sound appreciation of what implications these can have on design. In view of this, what follows presents a brief overview of the construction process.

4.2 Construction process

Construction of a typical cut-and-cover metro station structure involves four principal operations. These are:

- General excavation and groundwater control
- Ground support as the excavation progresses
- Construction of the principal station structure
- Replacement and reinstatement of the backfill.

First, a hole is made in the ground that is large enough to accommodate the structure and deep enough to found it at the required depth. Furthermore, for the excavation and the construction activities to proceed in the dry, dewatering is commenced ahead of the excavation so as to maintain a lead-in depth of at least 2–3m.

The response of the ground to the lowering of the groundwater level can be of special importance in the selection of the appropriate construction method. It can also form the basis for the assessment of the influence of the construction process on the surrounding ground and structures. Towards this end, carrying out some full-scale deep-well groundwater lowering tests before the commencement of the detail design work and the results obtained from it could furnish immensely useful information and is therefore always worth considering.

In addition, to ensure that the construction activity can be carried out safely, the faces of the cut must be adequately 'stabilized' or properly 'retained'. Construction of the structure itself, the subsequent replacement of the backfill and the reinstatement of the features on the surface can then follow in accordance with the recognized principles and accepted practices.

Excavation for and construction of the structure may be carried out inside:

- An open cut with 'stabilized side slopes'
- An open cut with 'vertically stabilized sides' or
- A covered cut with 'vertically retained sides'.

Open cut: 'stabilized side slopes'

The extent of the hole to be made in the ground will depend, primarily, on the size (plan area and depth) of the structure to be constructed. The depth of the cut, even for the shallowest of metro station structures, can easily be of the order of 14m or more. The sides of the cut for an excavation depth of such a magnitude if extended vertically downwards from the surface without any form of ground support mechanism would be unstable and highly unsafe unless, of course, the ground comprises rock or is rock-like in which case the faces of the cut could, in all probability, be self-supporting. With the progress of the excavation, therefore, the sides of the cut would need to be secured against slips and failures. One way of ensuring this would be by profiling the sides to form stabilized slopes that are self-supporting. However, this would be possible only if a sufficient width of the corridor, necessary to accommodate such slopes, were to be available around the perimeter of the structure.

The depth to the formation of the base slab of the structure below the existing ground level, the shear strength of the soil and the groundwater conditions are likely to influence what side slopes can be safely worked to and therefore what minimum corridor width is necessary. On the other hand, the proximity of the existing structures and the subsurface structural environment are likely to dictate what width might be available for construction. Clearly, deeper the cut, greater would be the width of the corridor necessary to accommodate it.

Depending on the type of the ground to be excavated and its drainage characteristics, it may be possible to stabilize the sides at slopes anywhere within the range (horizontal to vertical) of less than '1 in 1' to greater than '1·5 in 1'. To get an idea of the scale of things, assume that an overall depth of excavation for a single-track metro station structure is to be of the order of, say, 15*m*. Using a side slope of 1 in 1 and allowing for a minimum construction width of 0·5*m* at the base, the width of the corridor around the perimeter of the structure necessary to enable the formation of stabilized sides would be of the order of 15·5*m*. Furthermore, over the 15*m* depth, it would be sensible from the logistics point of view to provide also at least one level of terrace at mid-depth of the cut. Assumption of a 3*m* width for such a terrace would then push the necessary corridor width around the perimeter of the structure up to 18·5*m*. Assuming an overall width of the structure itself to be of the order of, say, 26*m*, the required overall width of the excavation at the surface to accommodate such stabilized side slopes as shown in Figure 4.1, would then add up to 63*m* – nearly two and a half times the width of the structure!

For the length of the structure of, say, 250*m*, the provision of stabilized side slopes would, likewise, increase the overall length of the excavation on the surface by 37*m* to 287*m* – also a substantial increase in its own right.

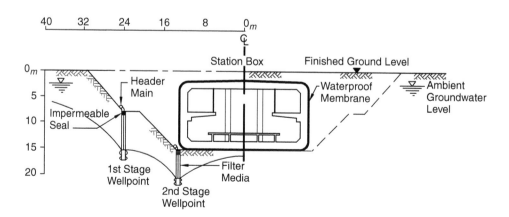

Figure 4.1 Open Cut Construction (within Stabilized Side Slopes)

Excavation of the magnitude necessitated by the construction of the structure as outlined above would, in order to allow it to proceed in the dry, also necessitate lowering of the water table ahead of excavation, as shown in the figure. The effects of such a draw-down are bound to extend to an area stretching well beyond the immediate limits of the excavation activity. Consequently, the effect on, and the settlements of, the surrounding ground and the structures in the vicinity cannot be ruled out and should not be ignored. Furthermore, the stabilized slopes would also need to be made secure against the adverse effects of the elements and, in particular, the potential for slips to occur especially during wet periods at the most vulnerable levels such as the clay-sand interfaces. However, by discouraging the ingress of water, it may be possible to prevent the occurrence of such slips. This can be achieved by promptly and progressively blinding with concrete the side slopes as they become exposed upon excavation.

Excavation within an open cut with stabilized side slopes as outlined above would be self-supporting and would, therefore, not require any ground support measures. In other words, no strutting or anchoring would be necessary. The use of stabilized side slopes would thus present the prospect of carrying out excavation unobstructed and unhindered and, *ipso facto*, also at speed. It would also allow the subsequent construction of the structure to be carried out most expeditiously. As a result, the construction activity would remain exposed to the effects of the elements for only the shortest possible time.

For construction in an open cut with stabilized side slopes, no specialist techniques would be necessary; the structure would be constructed in the conventional manner, base slab upwards, simply and efficiently and with better control on construction tolerances and workmanship. Given such a method of construction, it would also be possible to wrap the entire structure, if the aggressivity of ground and groundwater so warranted, in a waterproof membrane without any problems. Furthermore, in view of its conventional nature, the method and sequence of construction would not impose any unusual boundary conditions. However, replacement and compaction of the backfill in discrete layers around the perimeter of the structure as necessitated by this method would give rise to increased lateral pressures due to compaction forces. Accordingly, such increase in pressures would need to be taken on board in design as appropriate.

In the light of what has been discussed so far, there is little doubt that the construction of a cut-and-cover metro station structure inside an open cut with self-supporting (stabilized) side slopes would offer the most simple design solution; it would probably represent the simplest construction option as well. However, for such a solution to be seriously considered, it is necessary to ensure that, in principle, the following five conditions can be satisfied:

- Either there are no utilities encountered within the construction zone or, if present, these can be temporarily diverted, or permanently rerouted outside the zone of construction activity.
- There are no major thoroughfares within the construction zone. Nevertheless, if there are, the traffic can be satisfactorily diverted elsewhere for the duration of construction.
- Adequate width of corridor for construction activity surrounding the perimeter of the structure is available without involving enormous land-take costs or

necessitating demolition of a large number of existing structures. Demolition on a large scale can lead to significant socioeconomic problems generally associated with the displacement and resettlement of a large mass of people.

- The extent of the anticipated settlements of the surrounding ground and structures in the vicinity as a result of the required amounts of water table draw-down and excavation are not significant and are, as such, unlikely to be of concern.
- The weather is generally favourable and the exposure of the excavation to elements over the period of construction activity does not give cause for concern. However, if any concern does exist, it can be adequately and cheaply addressed.

The nature of the above conditions would lead to the conclusion that the construction of a cut-and-cover metro station structure can be undertaken in an open cut within stabilized side slopes only where there are, ideally, no constraints, e.g. those from services, traffic, existing structures in the vicinity, etc. Most of the time, however, the reality is that the existing structures abutting the likely sites seldom leave enough space for a cut with sloping sides to be accommodated and, as such, locations offering such luxuries are likely to be very rare. Besides, such sites, owing to their inherently poor catchment potential also represent, in practice, equally poor station locations.

There is no doubt that the success of a metro can be gauged by the size of its usership. For maximizing the usership of the metro, inevitably, the most efficient alignment is likely to be the one that seeks out the best catchment areas. In view of this, station structures, invariably, end up being located under busy thoroughfares and in heavily built-up urban areas. However, such locations, *ipso facto*, also present problems related to traffic congestion, constraints imposed by the location of existing services, proximity of existing structures and their foundations, ground movements, etc. which can very often present major concerns to the engineer. Construction of a cut-and-cover metro station structure in an open cut within stabilized side slopes can, therefore, be seriously considered only where such a structure is located, virtually, on a 'green-field' site.

Open cut: 'vertically stabilized sides'

In mitigation of the open cut method, it must be said that, where ground conditions allow it, it may be possible to dispense with the use of stabilized side slopes and avoid the extra land-take and the related costs associated with it. This can be achieved by advancing the sides of the cut vertically (or near vertically) downwards thereby restricting the size of the excavation very closely to that of the structure itself. However, this is possible only where some technique, such as 'soil nailing' or similar, to stabilize the faces of the cut can be successfully adopted. Notwithstanding this, it must be appreciated that the technique would, at best, remove only one of the five conditions listed earlier; for the success of the open cut method, therefore, the remaining four conditions would still remain to be adequately met.

Soil nailing

According to the Department of Transport's (UK, 1994) manual, soil nailing involves the reinforcement of the *in situ* ground (virgin soil or existing fill material) by the insertion of tension-carrying soil nails. Soil nails may be of metallic or polymeric material, grouted into predrilled holes or inserted using a displacement technique. The 'nails' are normally installed at a slightly subhorizontal declination.

Soil nailing is a technique whereby a vertical or near-vertical exposed face of a cut can be stabilized progressively as the excavation is advanced downwards in discrete depths. The ground mass is stabilized by inserting into it through the exposed face, apace with the advancement of excavation downwards, steel reinforcement commonly referred to as 'nails'. The nails are not, ordinarily, prestressed and usually constitute steel rods 20–50*mm* in diameter. However, with the use of a torque wrench and lock nut arrangement, it is easily possible to lock-in, typically, a load of about 10 per cent of the working load (Bruce and Jewell, 1987). It is helpful to remember that, where such an approach is preferred, such nominal post-tensioning would warrant only light steel bearing plates, possibly no bigger than 200 × 200*mm* and 10*mm* thick, and that stiff walers may not be required.

In granular soils, with the use of soil nailing, cut depths greater than 2*m* or less than 0·5*m* are rarely adopted. In overconsolidated cohesive soils, on the other hand, it is known that greater depths have been used safely. In a given ground environment, the maximum depth of cut at each stage is dictated, essentially, by the extent of the acceptable deformability of the soil and the ability of the cut to 'stand up' and maintain its integrity at least for a few hours prior to shotcreting and nailing. However, excavation of the soil, progressively, in depths of 1–1·5*m* is recognized as a commonly accepted practice.

The length of the cut, just like its depth, is also dictated by the considerations of soil deformation. Where deformations have to be minimized, excavation may be carried out piecemeal and by limiting the length to 10*m* or less. It should also be appreciated that a working platform width of at least 6*m* in front of the exposed face is generally required for accommodating the nailing equipment. However, with regard to the width of the excavations (significantly more than the 6*m*) generally to be expected in the construction of metro station structures, this should pose no problems.

The extent of the soil face exposed at a stage is subsequently protected by layer(s) of sprayed concrete (gunite or shotcrete) reinforced with high tensile steel mesh, or by the emplacement of prefabricated concrete units. A face thickness of the order of 50–150*mm* is commonly employed for temporary applications (Bruce and Jewell, 1987) whereas, for permanent works, a thickness of 150–250*mm* is generally sufficient. Sprayed concrete is generally applied in one or more successive layers depending upon the type of the nail used and the sequence of construction adopted. As a rule, spraying must be carried out expeditiously to ensure that the least amount of relaxation of the soil mass takes place and to prevent its loosening at the exposed face. The spraying immediately following the excavation prevents the soil moisture from evaporating. The soil moisture thus retained helps preserve the apparent cohesion in the granular medium. As a result, the natural stability of the ground can be prolonged.

Spraying is often discontinued about 300*mm* above the bottom of the cut. This facilitates the placement of the mesh for the following lower cut; it also makes it possible to achieve an overlapped joint for the sprayed concrete, which is preferable. The nails are subsequently grouted into predrilled holes or driven or drilled into the ground using a percussion or high-energy drilling device. Depending upon the perceived risk of soil instability, the nailing can be carried out either before or after spraying the exposed face with concrete. However, if prefabricated concrete units are used to stabilize the face, the nailing is carried out through predrilled holes in the units and only after their emplacement. The sequence of excavating, spraying and nailing is repeated as the work progresses downwards and is continued until the desired formation level is reached.

The soil mass, after it has been nailed, changes from behaving as a relatively 'loose' medium into a well-defined coherent, co-acting mass which is able to function as a massive gravity retaining 'wall'. With nailing, it also acquires the ability to withstand the lateral earth pressures, forces generated by surcharge loads and, of course, its own dead weight. The nails provide the tensile strength to the soil mass and, by extending these across the potential slip surface, they also increase its shear strength.

A similar technique, but using steel angle sections instead of bars, and known as Hurpinoise, was successfully used to stabilize an 11·6m deep face of a motorway cut east of Paris in France (Cartier and Gigan, 1983). Excavation was carried out in 1·4m deep stages and steel angle sections measuring 50 × 50 × 5mm, 5·5m long, and 60 × 60 × 6mm, 7m long, inclined at 20° to the horizontal were driven at a grid of, generally, 0·7m× 0·7m as illustrated in Figure 4.2. The face was protected by 50–100mm thickness of gunite applied on a welded wire mesh.

Soil nailing is generally ideal for dry ground conditions, i.e. where water table is low. Failing this, the success of the technique depends upon ensuring that there is no adverse build-up of groundwater pressure behind the facing. Where surface water is present, it may be possible to lead it away by installing a drainage ditch. It is also possible to install weep holes or 'horizontal' deep drains to release water immediately behind the facing, taking care to ensure that these are not blocked by impregnation from the sprayed concrete. Deep drains may take the form of slotted tubes, 50mm in diameter which may be at least the same length as, or longer than, the nails but inclined 5–10° upwards from the face.

Alternatively, face drains, placed vertically over the full height of the cut, may be employed to achieve the release of water pressure. It must be recognized, however,

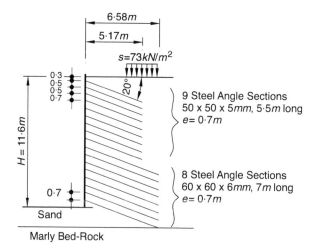

Figure 4.2 Typical Cross-Section, Hurpinoise Wall (adapted from Hulme *et al.*, 1989)

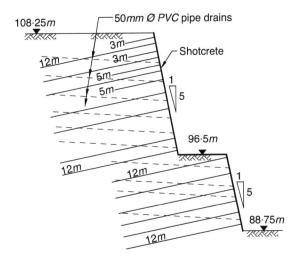

Figure 4.3 Nailed Slopes at Orchard Station, Singapore (adapted from Hulme *et al.*, 1989)

that the consolidation of the ground surrounding the cut as a result of removal of groundwater in this manner cannot be ruled out. Therefore, where ground movements are to be avoided at all costs, soil nailing may not be the most favourable option to pursue.

Figure 4.3 shows nailed slopes at Orchard station in Singapore (Hulme *et al.*, 1989) in completely weathered granite, incorporating the use of 'horizontal' deep drains. During the excavation, an incipient failure was detected and the initial design was amended by lengthening the nails to 12*m*. In conjunction with the nails, perforated PVC pipes, 50*mm* in diameter and 12*m* long were installed at 5 per cent slope upwards to relieve the pressure of water and to ensure stability.

If soil nailing were to be used in soft clays with low frictional resistance, ensuring adequate levels of stability could easily necessitate an unrealistically high frequency and considerable length of nails. Such soils are therefore unsuited to stabilization by this technique generally.

Plate and anchor underpinning

This method can be seen, essentially, as a variant on the 'soil nailing' theme. It is a conventional underpinning method whereby the excavation is progressed vertically downwards in panels of short lengths and depths and reinforced concrete wall sections are cast against these panels. Each individual wall panel is secured against the excavated face by means of a ground anchor before proceeding to excavate for the next panel. The size and thickness of the concrete panel depend upon the anchor load that can be safely tolerated by the soil and can be designed on the principle of a (horizontal) 'base' plate. The technique is applicable in dry, relatively dense or stiff subsoil conditions, which can develop the required anchor loads and allow the excavation of the required panel sizes to proceed without the loss of soil on the cut face. It has been successfully applied in a number of places around the world, i.e. Lisbon, Madrid, Frankfurt, etc.

After the full faces of the excavation all the way down to the formation are stabilized either as slopes or vertical cuts or underpinned in the manner outlined earlier, the construction of the station structure can commence. Upon completion of the construction of the structure, including its enwrapment with waterproof membrane, if so required, the backfill can be replaced. However, in the case of stabilized side slopes, backfill against the walls must be placed uniformly all round the perimeter and compacted in discrete layers in a manner that does not give rise to significant out-of-balance horizontal forces across the structure which could, potentially, cause onerous loading conditions. Finally, backfill on top of the structure can be replaced ensuring that it is adequately compacted, and the surface features reinstated.

Method of underpinning can be used not only to stabilize the ground as explained above but also to install the perimeter wall as an integral part of the structure. In the self-supporting Miocenic ground of Lisbon particularly with low water table, use of this principle in the construction of cut-and-cover structures has been used successfully. As an illustration of the principle, Figure 4.4 shows the possible method of construction of a typical metro station structure wall by this method and the use of temporary anchors.

Discrete wall panels are cast *in situ*, top-downwards, in the underpinning sequence and temporarily secured by means of anchors. Structural continuity of the wall vertically across the joints is achieved with the help of couplers as shown in detail 'A'. Once the structure is completed, the anchors can be destressed. If the structure is not above the water table but is partially or fully submerged, watertightness can be achieved by incorporating external water bars across the construction joints of the perimeter walls.

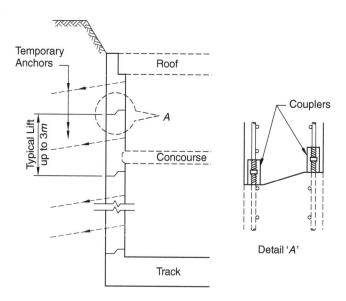

Figure 4.4 Integral Wall Construction by Underpinning

Design considerations: stabilized sides

Having gone through the various stages involved in the open cut method of construction using 'stabilized sides' as outlined thus far, the principal aspects needing to be addressed in design can now be identified as follows:

- Checking the stability of the side slopes at steady-state seepage (defined in Chapter 11) or, if the excavation is advanced as a vertical cut, designing the soil nailing or the concrete panel size and anchor;
- Estimating the settlement of the surrounding ground and structures and the heave of the formation, respectively, following pressure relief due to water table draw-down and excavation;
- Assessment of loadings including ground and groundwater pressures. In the case of stabilized side slopes, the enhancement in the lateral earth pressures due to the compaction of the backfill would also need to be taken on board; in the case of a vertical cut, naturally, there would be no compaction forces involved;
- Checking the stability of the completed structure by investigating its potential to float when the ambient water table is restored; and
- Analyzing the box structure and carrying out the design of its structural elements.

Evidence available from the metro systems around the world suggests that the cut-and-cover station structures generally end up being located in heavily built-up urban environments where the existence of high-rise structures almost abutting the construction works is not all that unusual or uncommon. Under these circumstances, an adequate works corridor outside the immediate perimeter of the structure can be made available but only at a prohibitively high premium. Therefore, for all practical purposes, it can be safely assumed that such corridors are unlikely to be available for construction use. Consequently, in most situations, the use of self-supporting stabilized side slopes can be ruled out. In view of this, consideration would need to be given to some other method of advancing the excavation which would ensure safe verticality of the faces of the cut.

With regard to maintaining the verticality of the faces of excavation, it is useful to appreciate that the techniques of soil nailing and underpinning have their limitations in that they can be applied most economically, generally, in granular soils only. Besides, they also rely on relatively dry ground conditions for practical installation. Therefore, where the type of the ground and groundwater conditions encountered or the presence of existing deep foundations in the vicinity, or both, preclude the use of vertical face stabilization by such techniques, some form of ground 'support' (rather than 'stabilization') system which can actually 'retain' the face vertically needs to be considered.

Covered cut: 'vertically retained sides'

Where traffic can be safely diverted elsewhere and services can be temporarily or permanently rerouted, where adequate construction corridor is not available outside the perimeter of the station structure and soil nailing is not a viable option and where weather conditions are favourable, excavation can be advanced downwards vertically in an 'open' cut with the provision of some form of ground support system and without the necessity for any protective cover on the surface.

However, where the volume and the intensity of the existing traffic on the surface is such that it is not possible to divert it elsewhere without increasing the problems to an unacceptably high level, it is then necessary to install a temporary traffic deck at the surface which will allow continued flow of the bulk of the existing traffic to be maintained. Where the existing services, likewise, cannot be temporarily rerouted but need to be maintained *in situ*, the temporary deck can provide the necessary support from which to hang these. Furthermore, where, because of the possibility of adverse weather conditions, the cut cannot be left open to, and at the mercy of, the elements for any length of time let alone for the duration of the construction which can easily run beyond a year, it would be possible to incorporate additional measures within the temporary traffic deck as necessary so that it can also provide adequate protection against the elements.

Before proceeding further, it would be useful at this stage to briefly recount the stages generally involved in the construction of a typical cut-and-cover metro station structure within a cut with 'vertically retained sides'.

4.3 Stages of construction

The various stages involved in the construction of a typical cut-and-cover metro structure carried out within a cut with 'vertically retained sides' can be identified as follows:

- Installation of the boundary wall
- Construction of the traffic deck
- Groundwater control measures
- Excavation and wall support
- Construction of the structure
- Backfilling and reinstatement.

The choice of the most suitable method of construction is influenced, primarily, by the geological and hydrological conditions, size (depth and width) of excavation and by the location and characteristics of the adjacent structures. A number of secondary influences such as the existence and the disposition of the underground public utilities and the volume and intensity of traffic on the surface also need to be considered.

The various stages of construction listed in the preceding section are briefly discussed next.

Installation of the boundary wall

Excavation for and construction of a typical metro station box structure generally proceeds within a preinstalled perimeter boundary wall. The wall provides the main ground support system as the excavation progresses downwards. Installation of such a wall thus represents the first principal stage in the construction process.

The boundary wall does not always form an integral part of the main station structure. Sometimes, it merely constitutes a temporary outer cofferdam within which the entire box structure, including the permanent perimeter wall of the structure itself, is expected to be constructed. Such a prospect can present itself if, for example, the existing ground and groundwater conditions are highly aggressive and the structure is required to be completely wrapped within a waterproof membrane. Clearly, access for such an encasement around the entire structure for waterproofing purposes can be possible only if it is constructed within an outer cofferdam in the conventional manner. This is illustrated in Figure 4.5.

Alternatively, the boundary wall may not only retain the surrounding ground during excavation but it may also form an integral part of the permanent structure as shown in Figure 4.6. In such a case, owing to the lack of ready access around the perimeter of the structure, waterproofing in terms of an all round encasement with a membrane would not be possible. Slab-wall connection details also turn out to be intolerant of errors and difficult to waterproof. Furthermore, the degree of fixity achievable at these details needs to be carefully quantified and modelled accordingly in design.

In the event that the boundary wall has to act purely as an outer cofferdam, there is then bound to be a duplication of the walls. Where the aggressivity of the existing ground and groundwater conditions dictates such an approach, it may not be possible to avoid such duplication. This will, inevitably, result in the escalation of costs as it is bound to involve, along with the extra structure, additional land take and excavation as well. This will be all the more so if the outer cofferdam is treated as a purely temporary, totally sacrificial and, as such, a nonparticipating component of the permanent structure.

However, if the outer cofferdam is not deemed to be entirely sacrificial and can be accepted as structurally participating, albeit partially, it may be possible to achieve some savings in the thicknesses of both the outer cofferdam and the inner wall, and in the amounts of the reinforcements required. In this way, at least some of the extra costs can be avoided. For instance, it could be argued that, to allow for its deterioration in the long term, the capability of the outer cofferdam could be downgraded and deemed good enough to carry just the long-term lateral 'soil' pressures whereas the permanent perimeter wall of the structure itself need only be designed for the full hydrostatic pressures.

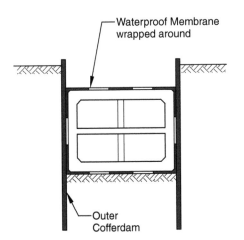

Figure 4.5 Construction within Outer Cofferdam

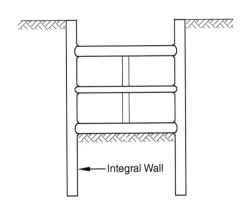

Figure 4.6 Perimeter Wall as an Integral Element of Structure

The logic underlying the aforementioned approach is that, as the excavation progresses downwards, the effects of the short-term earth and the somewhat reduced hydrostatic pressures would, initially, be fully locked into the 'temporary' outer cofferdam. However, subsequently, when the permanent inner wall is in place and, with time, the outer cofferdam suffers some deterioration and there is therefore some relaxation in the previously locked-in effects, there is then bound to be some load shedding on to the inner wall. In the event that this approach is adopted, care should be exercised to ensure that the design errs on the side of safety and that the anticipated load on each wall at any stage is in no way underestimated.

To conclude, comparison of the two concepts for the use of the perimeter wall may be summed up as follows.

As an outer cofferdam:

- Tighter construction tolerances can be achieved.
- Construction of the structure can proceed conventionally.
- Slab-wall junction details present no construction problems.
- Waterproofing of the structure can be achieved relatively easily.
- There is duplication of wall, extra land take and extra excavation.
- Relatively higher overall construction times and costs are involved.
- No specialist techniques for construction of the structure are necessary.

As an integral wall:

- Complete waterproofing is difficult to achieve.
- Slab-wall junction details are generally problematic.
- Relatively bigger construction tolerances are involved.
- Construction of structure can be speedier and economical.
- Certain aspects of construction can become relatively complex.

In view of their particular importance as the key elements in the construction of cut-and-cover metro station structures, different types of wall systems available and tried and tested in the different parts of the world to provide support to the ground as the excavation advances downwards as well as the principal wall support systems commonly employed have been dealt with separately in Chapter 5.

Design considerations

The considerations that can be identified for the design of boundary wall are as follows:

- To ensure that no untoward problems present themselves during the excavation for and the construction of the wall, stability of the trench under the given ground and groundwater conditions needs to be investigated. This, including ways of improving stability where necessary, is covered in detail in Chapters 16 and 17.
- Boundary conditions of the wall and changes in these at different stages of excavation for and construction of the station structure need to be identified and allowed for in the design.

- Type and magnitude of the lateral loads and the redistribution in these at the different stages need to be carefully assessed.
- The effects of carrying out the excavation and construction in stages and, in particular, any carryover of the locked-in effects from the previous into the subsequent stages and their implications on design must be carefully considered.
- Design of wall element must be based on the envelope of stress resultants obtained from a combination of both the stages of excavation and construction as well as the operational lifetime of the structure. It must also reflect the consideration of the locked-in effects.
- Being inaccessible from the outside, maintenance of or remedial measures for the perimeter wall can prove to be prohibitively expensive. In order to minimize its maintenance, therefore, particular attention must be paid to the aspects of durability of the wall.

Construction of temporary traffic deck

The traffic capacity of a network of city streets is generally dictated by the capacity of its at-grade junctions. Because of their inherent catchment potential, it is these very locations that are also often the most favoured sites for the metro stations. However, construction of such structures, owing to the enormity of their sizes, generally, involves large volumes of excavations handling of which can prove to be very disruptive to traffic. To prevent large-scale disruption to the surface traffic during the construction of a cut-and-cover metro station structure, two possibilities exist – the existing traffic is diverted through nearby streets or, where this is not feasible, a temporary deck is installed to maintain the continuity of the existing traffic.

From a purely construction point of view, it would be ideal if unrestricted access for construction purposes were made available by closing the site to and rerouteing all the nonconstruction vehicular traffic as this would also reduce the construction time significantly. However, unless the traffic can be efficiently rerouted, any reduction in the traffic capacity due even to partial closure of a street or a junction could have a significantly adverse effect on the levels of congestion.

In the case of a heavily trafficked street where the existing traffic cannot be diverted elsewhere efficiently or without merely enlarging the zone of disturbance, and where the station structure straddles a busy road junction, it is often necessary to install a temporary deck on the surface to maintain continuity of the existing traffic. This can permit the construction activity to be carried out beneath the deck while allowing bulk of the existing flow of traffic on the surface to be maintained even during the construction period. Where it is so required, construction of such a temporary traffic deck represents the second principal stage in the construction activity.

Given that the services are generally embedded under and follow the alignment of the roads on the surface, it is imperative that, during the installation of the traffic deck or the undertaking of any related excavation activity, the existing services are not damaged. It is therefore important to identify these and, preparatory to the commencement of excavation, to establish their existence and location from the available records maintained by the various utility agencies. However, this may not always be enough; it may, at times, also be necessary to confirm the precise level and disposition of these services through surveys and exploratory works on site. There is

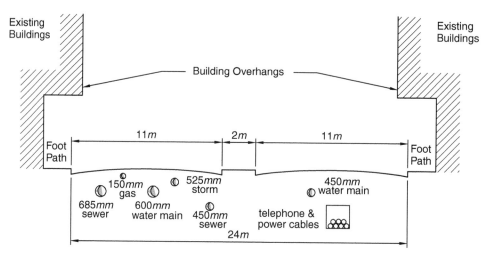

Figure 4.7 Nathan Road, Hong Kong (Building Overhangs above and Services below)
(adapted from Edwards, 1976)

no doubt that the satisfactory identification of all the services can, at best of times, be a time-consuming activity. However, it has to be undertaken and, in fact, rigorously gone through if major delays, costly emergency works and heavy claims are to be avoided.

Depending upon the number and disposition of the existing services encountered, it may be possible to shave off, without any interference with them, a sufficient depth of the surfacing to accommodate the temporary traffic deck within the ground in a manner that maintains the existing road level unaltered. Alternatively, it may be necessary to ramp up the deck, partially or fully, above the existing road level if the potential interference with the existing services is to be avoided at least until such time as the deck is in place and is in a position to provide the temporary support to them during the main excavation. In any case, some interference with the existing traffic during the placement of the temporary deck cannot be avoided. However, to ensure that such interference is kept to the very minimum and for as short a period of time as possible, assembly and installation of the deck can be confined to night-time or other designated off-peak hours.

The temporary traffic deck generally comprises a grillage of primary longitudinal and transverse girders underlying secondary longitudinal girders supporting deck elements on top. The deck elements can be in timber, steel or even precast concrete units. These can be designed and placed as modular panels so that it is possible to gain access, if so required, to any area below by removing the appropriate panel(s).

Preparatory to the installation of the temporary traffic deck structure, it is necessary to have the appropriate vertical supports for it in place. Invariably, longitudinal walls of the station perimeter are used as the two outer supports to the deck. Additionally, depending upon the width of the box structure, at least further one or two intermediate, longitudinal lines of support are also generally necessary. These can be provided in the form of discrete rows of steel stanchions installed at regular centres longitudinally. At the intended locations of the stanchions, pile bores are made from the surface and extended to appropriate depths. Concreting of the bores, however, is confined to the bottom few metres only. The stanchions are then inserted vertically downwards through the empty bores and pushed some distance into the wet concrete at the bottom. After the concrete has set, the hardened concrete cylinders thus formed at the bottom constitute individual foundations for the stanchions. The empty bores above these foundations are then backfilled with sand, pea shingle or lean concrete.

As an example, Figure 4.7 shows a typical cross-section of a cluster of services encountered under Nathan Road in Hong Kong. Constrained headroom condition under the adjoining building overhangs is also clearly in evidence. To illustrate the procedure of installing a traffic deck, a typical stage-by-stage sequence is set out in Figure 4.8. This assumes the two outer longitudinal walls of the structure to be already in place. The deck is then installed in a phased manner. The usual sequence is to confine

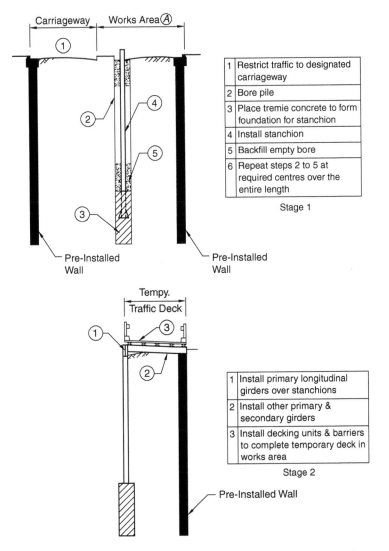

Figure 4.8 Sequence of Construction for Temporary Traffic Deck

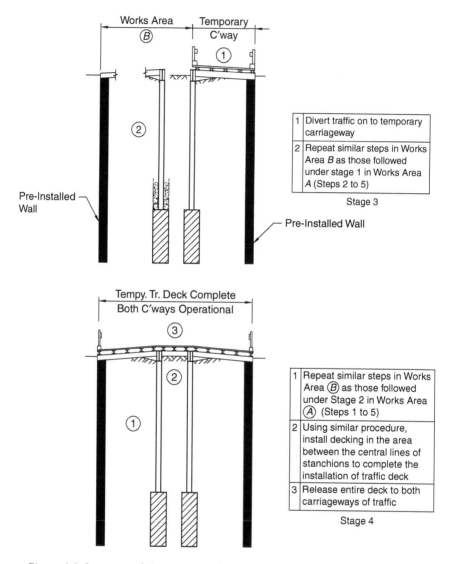

Figure 4.8 Sequence of Construction for Temporary Traffic Deck (*Continued*)

traffic to one half of the existing roadway while the activity of installing the traffic deck is undertaken in the other half. The roles are then reversed; the traffic is diverted on to the part decking in place on the second half while the process of installation of the remainder of the deck is carried out in the 'first' half.

Figure 4.9 shows a somewhat similar temporary traffic deck in place under Nathan Road in Hong Kong.

Location of the temporary support stanchions *vis-à-vis* that of the permanent column supports of the structure can be very important. If the locations of the temporary stanchions can be made to coincide with those of the permanent column supports, it is then possible to encase the stanchions in reinforced concrete so as to form, eventually,

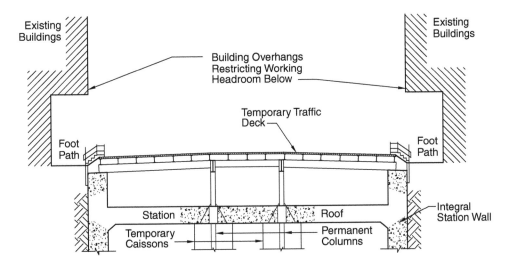

Figure 4.9 Temporary Traffic Deck over Nathan Road, Hong Kong
(also showing Building Overhangs) (adapted from Edwards, 1976)

the permanent column supports as well. Utilizing a support in this manner for both temporary and permanent purposes offers not only cost savings but subscribes to efficiency in design also. Therefore, such a possibility should not be left unexplored.

Design considerations

While the design of the temporary traffic deck can proceed on the basis similar to that of a conventional bridge deck design, appropriate consideration must also be given to the following aspects:

- The effects of differential stiffnesses between the permanent (outer walls) and the discrete temporary (intermediate stanchions) supports on the stress-resultants in the deck must be allowed for in design.
- Impact of the progressive reduction in the axial stiffness of the stanchions on the stress-resultants in the deck elements as the exposed height of the stanchions increases with the downward progression in excavation should be acknowledged and duly allowed for in design.
- If the existing services are to be supported *in situ*, i.e. by means of hangers supported off the deck, additional loads from these must be taken on board in the design of the deck.
- Design of the foundations to the stanchions must also ensure that the local settlements are kept within acceptable limits or, otherwise, allowed for in design.
- After the completion of the construction of the structure, if the presence of hard spots under the base slab resulting from the existence of the stanchion foundations is to be avoided, consideration must be given to the manner in which these foundations can be divorced from the station structure. If this is not practical, implications of the presence of such hard spots on the design of the structure must be duly considered.

- In the case of those of the temporary stanchions which may be encased, eventually, to form permanent column supports, the composite sections need be designed to carry only the extra load from the structure over and above that already locked into the stanchions from the dead weight of the deck. In other words, the spring stiffness of the composite column section should be used for computing the effects of the extra-over loads.
- Allowance may have to be built into the design for the provision of holes in the temporary deck for the removal of spoil from the excavation below the deck.

Groundwater control measures

In order to carry out excavation and construction of the works in the dry and to avoid excessive movements in the surrounding ground and structures, ensuring an appropriate control of groundwater is essential. This represents the third principal stage in the construction activity. In view of their importance in the design and construction of cut-and-cover structures, various measures commonly available for groundwater control and their implications on design are discussed separately in Chapter 6.

Excavation and wall support

Excavation for the construction of the structure represents the fourth principal stage in the construction activity. In the construction of the temporary traffic deck, it is usual to incorporate, at the very least, two access holes large enough to allow the necessary equipment to be lowered into place to carry out excavation and through which to remove the spoil. Initially, a hole is made in the ground from the deck level into which a digger is lowered which then starts to mole its way around below the deck. Muck is either raised to the deck level through the holes and transported away to the tips, or conveyed over belts to designated collection points and thence removed for disposal.

For reasons of safety, the holes in the deck are cordoned off by appropriate fencing. Clearly, in the presence of such barriers on the deck, it is not possible to have the use of the full width of the deck over its entire length for normal vehicular traffic. This cannot be helped. However, it may be possible, in certain circumstances, to organize the collection of spoil and measures for its disposal away from the curtilage of, and without the need to bring it to the surface through, the deck in which case the normal flow of traffic can have the full, unobstructed, use of the entire deck. In any case, execution of an efficient traffic management scheme, which minimizes disruption and ensures that the emergency services have continuous access to all the locations including the construction site, is most essential. The scheme must be prepared in sufficient detail and in full collaboration with the highways authorities, the police and other emergency services; it must incorporate all the necessary signing, signalling and appropriate measures for advance publicity and the timely dissemination of information to the general public.

As the excavation is advanced downwards, owing to the resultant loss of lateral ground support on the inside, the wall, in its bid to retain the ground on the outside, will tend to deflect, under pressure, inwards towards the cut. To limit such movements, it becomes necessary to introduce wall support system as the excavation progresses. The frequency of the supports is dictated not only by the design requirements, such as the stiffness and the strength of the supports themselves, but also by the type of the ground and groundwater conditions, stiffness of the soil mass, the extent of the permissible ground movements and the need to ensure a measure of freedom of movement for the construction activity and equipment. If it is required that the wall movements are to be kept to the absolute minimum, it may even be necessary to preload the supports to precompensate for their elastic deformation when they take on the load. All these aspects, including the related design considerations, are covered in Chapter 5.

Construction of the structure

Construction of the station structure represents the fifth principal stage in the construction process. The station box structures are invariably located in heavily built-up urban areas where space is at a premium. In view of the restricted nature of the available space in such locations, construction of such structures is carried out, almost exclusively, in vertically retained cuts and using the principal perimeter retaining walls, often but not necessarily always, as integral parts of the structure.

Furthermore, in the case of structures below ground, unlike those above ground, because of the intimate contact of the surrounding ground with the structure, soil–structure interaction plays a prominent role in design. As a result, there is also a close interrelationship between the method and sequence of construction of the structure and its design. In other words, the design of cut-and-cover structures cannot be treated in isolation from their construction. The successful design and construction of such structures therefore, particularly in highly constrained urban environment and difficult ground and groundwater conditions, can present some of the greatest challenges faced both by the designer as well as the constructor.

Depending upon the prevailing site-specific constraints, it may be possible to construct a typical cut-and-cover metro station structure by any one of the following means:

- Bottom-Up method
- Top-Down method
- Combined method
- Caisson sinking method.

The principles underlying these methods are briefly discussed hereunder.

Bottom-up method

In the bottom-up method of construction, as the first step, a boundary wall, in the form of either an outer perimeter cofferdam, or one that forms an integral part of the permanent structure, is installed from the surface. Thereafter, and prior to the commencement of the bulk excavation within the perimeter, invariably, a capping beam with appropriate shear stiffness, which connects the wall panels together at the top and helps maintain the alignment of the perimeter wall in tact during subsequent excavation, is also constructed. This is done in order to ensure that there is no differential, out-of-plane, lateral movement between the adjacent panels of the boundary wall when these are exposed to the progressively increasing lateral earth pressures as the excavation within the perimeter is advanced downwards. After the capping beam is in place and has attained the requisite strength, the main excavation within the perimeter can commence.

As the excavation progresses downwards, stages of wall bracings as dictated by the design, in the form of internal horizontal struts or external ground anchors or, where so required, even some combination of both, are installed to provide lateral support to the perimeter walls. This alternating sequence of excavation and bracing is carried on until the required formation is reached and the vertically retained cut of the required size is achieved.

Within the cut thus formed, construction of the structure is then carried out, as the name suggests, bottom slab upwards in the conventional manner. As the construction of the structure advances upwards and those elements of the structure which form the permanent lateral supports are in place and operational, and it is safe to do so, each corresponding stage of temporary horizontal bracing is, sequentially, released and removed. The backfilling and surface reinstatement are finally carried out only after the roof slab and the associated support structure are in place and have attained the requisite strength to sustain the imposed loads.

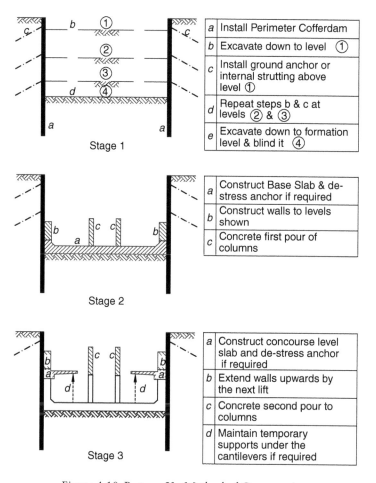

Figure 4.10 Bottom-Up Method of Construction

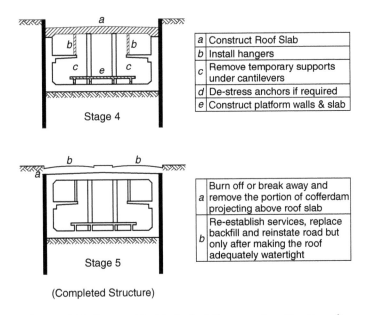

a	Construct Roof Slab
b	Install hangers
c	Remove temporary supports under cantilevers
d	De-stress anchors if required
e	Construct platform walls & slab

Stage 4

a	Burn off or break away and remove the portion of cofferdam projecting above roof slab
b	Re-establish services, replace backfill and reinstate road but only after making the roof adequately watertight

Stage 5

(Completed Structure)

Figure 4.10 Bottom-Up Method of Construction (*Continued*)

A typical sequence of construction utilizing the 'bottom-up' method is illustrated in Figure 4.10, Stages 1–5.

If, as shown in the figure, ground anchors are used as tie-backs for the perimeter wall, then excavation and construction within the perimeter will have a clear unhindered run. However, if the anchors need to be de-stressed later on when no longer necessary, the relationship between the location of the anchors and the time of their destressing *vis-à-vis* the construction of the corresponding floor slab needs to be carefully considered so as to avoid any clash between the different operations.

If, on the other hand, bracing in the form of internal strutting is used to support the outer cofferdam walls, then depending upon their location, permanent walls of the structure may have to be constructed around the struts. In other words, discrete holes may have to be left for and around the struts temporarily which will need to be filled and made good later on when the permanent floor slabs become operational and after the temporary struts have been removed. Careful attention to detail is required to make such in-fill concrete adequately watertight. It is also essential to remember that, at no time should the walls remain without adequate lateral supports, temporary or permanent, at the appropriate locations. This may, at times, even necessitate the introduction of secondary strutting between the inside faces of the permanent walls to enable the removal of the primary strutting between the outer cofferdam walls and the construction of the floor slabs to take place.

In Figure 4.10 particularly, Stage 3 indicates, among other things, temporary vertical supports under the 'cantilever' slabs. In lieu of these, it may be possible to actually design these slabs also as cantilevers carrying their self-weight until such time as the permanent hangers are in place and operational in which case the temporary supports would not be necessary as long as the associated short-term deflections are acceptable.

However, the preferred choice between the two options is likely to be influenced by the consideration of cost.

Bottom-up method of construction is ideally suited to locations where:

- There are no major services existing within the area of construction activity, and those minor services that are, can be easily and cheaply diverted temporarily or rerouted permanently.
- Road traffic can be safely diverted elsewhere without any problems and therefore no temporary traffic deck is required.
- Exposure of excavation to elements does not give cause for concern.
- Anticipated ground and groundwater pressures are not very high, i.e. ground is reasonably strong and water table low.
- Anticipated movements of the surrounding ground and structures are not critical.

Where lateral pressures are not all that high and the extent of the anticipated movements of the surrounding ground and structures are also unlikely to cause concern, it may even be possible to use a relatively flexible ground support system, i.e. a sheet pile cofferdam, in combination with adequate stages of bracings. It may also be possible to reduce pressures on the walls by lowering the water table outside the perimeter but only so long as the resulting settlements of the surrounding ground and structures are not unacceptably high. Where such is not the case, i.e. where lateral pressures are likely to be high and the anticipated movements of the surrounding ground and structures could also cause concern, it may still be possible to use the bottom-up method of construction; however, in that case, it is important to ensure that the movements of the walls are carefully controlled by introducing relatively stiffer wall sections, more stages of supports, preloading of the supports, etc.

To put the bottom-up method of construction and its utility into perspective, it may be helpful to compare its advantages and disadvantages as follows.

Advantages:

- It presents ease of excavation due to an open environment.
- Use of ground anchors, particularly, can present the prospect of an unhindered, and therefore speedy, excavation and construction.
- No provision of any temporary holes within the slabs of the structure is needed for the removal of spoil.
- Being conventional in nature, such a method of construction is unlikely to present any unexpected surprises in design.
- No special measures or details are necessary at column-slab junctions.
- Where the construction is carried out within an outer cofferdam, waterproofing of the entire structure by wrapping it in a membrane is relatively simple.
- Analysis of the structure is relatively simple and essentially conventional.

Disadvantages:

- Extensive falsework is needed for the construction of the heavy suspended floor slabs involving obvious time and cost implications.
- Road reinstatement is not possible, nor can its use be restored, until the very end, i.e. after the completion of the structure and reinstatement of the backfill.

However, if other considerations dictate that a temporary traffic deck has to be used, then this may not be a critical issue.

- The method is not always very effective in controlling ground movements. This can be a serious drawback if control of movements is critical.
- Existence of noise and dust pollution during the period of construction cannot be avoided.

In selecting this method of construction, the aforementioned pros and cons should be carefully considered in the light of the prevalent site-specific constraints.

DESIGN CONSIDERATIONS

Where the ground support system constitutes a temporary outer cofferdam within which the main structure is to be constructed and the cofferdam is supported by internal bracing, the following design considerations can be identified:

- The temporary cofferdam must be designed for all the stages of excavation and bracing.
- Changes in the boundary conditions and the redistribution of the lateral loads at different stages must be recognized and taken into account.
- It should also be appreciated that any movement of the wall inwards towards the excavation is likely to be accompanied by the movement of the surrounding soil mass in sympathy; any attempt thereafter to force the wall back fully to its premovement position may not be successful. Therefore, locked-in effects of stress resultants may have to be taken on board in design as appropriate.
- As the structure is constructed bottom-upwards and each bracing stage is progressively removed, the resulting effects of changes in the boundary conditions of the wall should be carefully examined and allowed for.
- Impact of variation in the bracing loads due to the diurnal effects on the stress resultants in the wall, where appropriate, must not be overlooked.
- The effect, the manner and the timing of load transfer from the temporary bracing on to the permanent structure as the bracing is removed and the structure assumes the bracing function must be taken into account, as appropriate.
- The design of the ground and the wall support systems must be based on the envelope of stress-resultants obtained from all the different stages of excavation and construction.
- Particularly in the case of structures similar to that illustrated in Figure 4.10, given their height, the possibility of the intermediate columns being slender and the implications associated therewith must be examined.
- The effect of elastic shortening of the intermediate columns differentially with respect to that of the continuous perimeter wall on the stress-resultants of the roof and, where applicable, other floor slabs must be duly considered. Alternatively, the effect may be compensated by building-in appropriate provisions to allow for the elastic shortening, i.e. by constructing the columns taller by the amount of their anticipated shortening under the self-weight of the suspended slabs. The same result may also be achieved by jacking up the floor slabs at the column locations by the requisite amounts.

Where ground anchors are used as the wall support system, boundary conditions and the loading cases for design are fewer and simpler. However, in the event that the anchors are required to be destressed as the construction of the structure progresses, it is important to ensure that the sequence of construction is so chosen as to make this possible and allow the effects of load transfer on to the structure to be taken on board.

The cross-section chosen in Figure 4.10 is indicative only to illustrate the principle. However, even so, the concourse level floor as shown can be seen to be represented by two tapered cantilevers – the distance between the cantilever ends being indicative of the width of the void in the floor slab. Because of the discontinuity in the concourse slab due to the presence of this void, the transverse strut action over the extent of the void, that would otherwise be available if the slab were to be continuous wall-to-wall, is absent. It is therefore imperative to examine whether the tapered cantilever floor slabs at this level are capable of acting also as deep horizontal girders providing adequate lateral supports to the walls over the extent of the void. If that does not turn out to be the case, then one of the following two options may be looked at:

- Removing the taper and making the cantilevers of uniform thickness as necessary taking care to ensure that the increase in the thickness and the deflection are acceptable and do not impinge upon the vertical planning clearances.
- In the event of the extent of the necessary increase in the thickness not being viable, mobilizing some strut action by incorporating discrete bandwidths of the slab transversely across the void. This would split one long void into two or more relatively shorter ones.

Top-down method

Top-down method of construction refers, essentially, to the sequence in which the horizontal elements of the structure, principally the floor slabs, are constructed after the perimeter wall is in place. First of all excavation is taken down far enough to enable the topmost (i.e. the roof) slab to be constructed; it is then followed by the excavation down to the underside and the construction of the next slab below and so on, with the lowermost (i.e. the track or the base) slab being constructed last of all. This method of construction is preferred where it is essential to minimize ground movements. It relies upon each permanent floor element, as it is cast into place and becomes operational, to function also as a continuous temporary strut before the excavation progresses downwards to the next floor level below.

In the top-down method also, as in the case of the bottom-up method of construction discussed previously, the installation of the perimeter ground support system together with the capping beam on top represents the first stage of construction. Excavation within the perimeter is then taken down just far enough to enable construction of the topmost (i.e. roof) slab to be carried out either directly on the blinded formation without the need for any falsework, or on shuttering placed on the blinded formation. Discrete temporary holes are formed within the roof slab in strategic locations to facilitate the subsequent excavation related activities through and below these. The holes allow excavation and related activity below the slab to be carried out after lowering through them the necessary plant and equipment; they also provide avenues for the removal of the spoil.

It is important to recognize here that, prior to carrying out the excavation *en masse* below the slab level and while the slab is still ground-bearing, intermediate vertical supports, either temporary or permanent, as required by design, are in place. This is to ensure that when the excavation is carried out and the slab becomes suspended, it is supported adequately without suffering any distress. These supports, invariably, take the shape of 'plunge-down' stanchions installation of which follows a similar procedure as outlined previously for the intermediate vertical supports for the temporary traffic deck. The excavation is then progressed to the formation level of the next lower slab, which is then constructed in a similar fashion as the slab above.

During the construction of the floor slabs, care must be exercised to ensure that the support stanchions are adequately shear connected and anchored into them at each level. Each activity, i.e. excavation, shear-connection of stanchions and construction of the floor slab, is progressed to each lower level in this manner sequentially until the base slab is constructed. Thereafter, the permanent intermediate column supports are installed floor-by-floor from base slab upwards. Once the permanent supports are in place and operational, the temporary stanchion supports can be removed. Alternatively, and preferably, the stanchions should be so located as to make it possible to encase these in structural concrete and be incorporated as integral parts of the permanent supports.

One of the significant features of the top-down method of construction is that it is not necessary to wait until the completion of the construction of the entire structure for the backfill on the rooftop to be replaced. Backfilling can take place much earlier, i.e. soon after the construction of the roof slab and upon its attaining sufficient strength, and even while excavation and construction activities below the roof level are in progress. This enables early re-establishment of utilities and amenities on top, and release of surface area for the reinstatement of the road and, thereafter, to traffic.

A typical sequence of construction utilizing the top-down method is illustrated in Figure 4.11, Stages 1–6.

It is often found that placing of concrete especially in the top reaches of the columns below when the slab above is already in place is not straight-forward. A possible method of achieving this adequately for a typical 'column-roof slab' interface is shown in Figure 4.12. Concrete is poured through a central hole in the slab directly above the column. Provision of air vents is also made to ensure that there is no entrapment of air which could call to question the integrity of the poured concrete. Continuity of reinforcement between the column and the slab is achieved through the use of couplers accommodated within the downward projection of the slab locally forming an inverted kicker for the column. Note that the reinforcement shown in the figure is indicative only.

In the top-down method of construction, the horizontal structural elements, i.e. the floor slabs, also function as struts during construction. The inward movement of the perimeter walls is, accordingly, limited to the extent of the deformation, i.e. shortening and vertical deflection, of such elements under the action of the lateral earth pressures and self-weight, respectively. The movement will also be inhibited somewhat by the inherent stiffness of the walls themselves. Although the deformation and the associated ground movement may be small, yet it is conceivable that there may be instances where, owing to the existence of highly sensitive structures in the immediate vicinity, 'zero' tolerance to movements is called for.

One way of overcoming the potential problem of deformation of the support elements associated with supporting the walls horizontally would be to install horizontal struts within discrete trenches and, to forestall any inward movement, preload them to the extent required, before the bulk excavation is commenced.

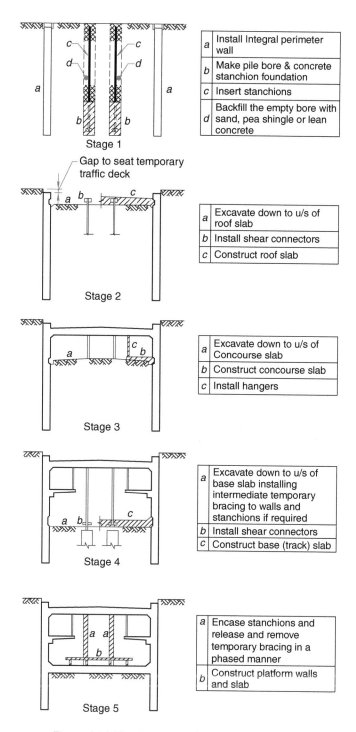

Figure 4.11 Top-Down Method of Construction

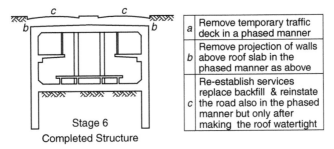

	a	Remove temporary traffic deck in a phased manner
	b	Remove projection of walls above roof slab in the phased manner as above
	c	Re-establish services replace backfill & reinstate the road also in the phased manner but only after making the roof watertight

Stage 6
Completed Structure

Note:- Temporary traffic deck not shown for clarity

Figure 4.11 Top-Down Method of Construction (*Continued*)

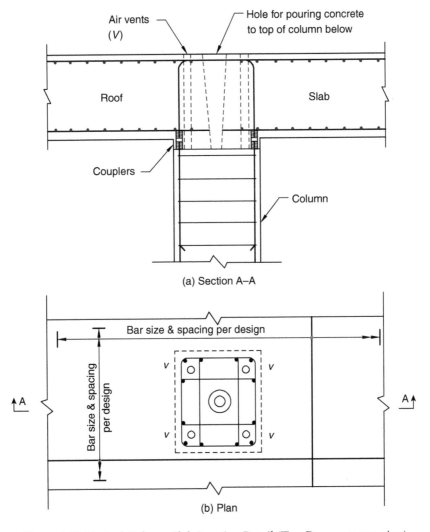

(a) Section A–A

(b) Plan

Figure 4.12 Typical Column-Slab Junction Detail (Top-Down construction)

However, in that case, the struts would need to be positioned, ideally, within the thickness of the floor slabs so that they could be concreted-in, eventually, as part of the permanent floor slabs as shown in Figure 4.13. In this way, the movements would, in theory, be completely eliminated. In practice, however, achieving precise correspondence of the jacking load and the anticipated movement may not always be possible and so, expecting 100 per cent elimination of ground movements may not be realistic. It may not be entirely necessary either; a more pragmatic view of the acceptable movements should, therefore, be taken.

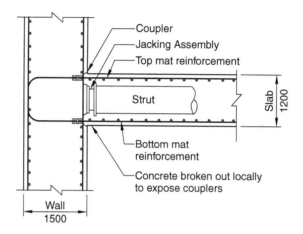

Figure 4.13 Pre-Loaded Strut Incorporated in Floor Slab

In the top-down method of construction, it is necessary to integrate the stanchions into the floor slabs at their respective levels. This can be achieved by connecting to the stanchions at these levels appropriate shear connectors which can be concreted-in within the structural slabs.

Top-down method of construction is ideally suited to locations where:

- Existing services within the area of construction activity cannot be rerouted and need to be temporarily supported *in situ*.
- Road traffic cannot be satisfactorily diverted elsewhere and so temporary traffic deck is required.
- Anticipated lateral pressures are likely to be high, as may be the case in soft ground with high water table.
- Ground movements, including settlements and heave, are critical and need to be tightly controlled.
- Structures in the vicinity are sensitive and intolerant of movement.
- It is essential to reinstate the road and release it to traffic as early as possible.
- Noise and dust pollution have to be minimized.
- Construction works have to be shielded from the severity of weather conditions.

In adopting the top-down method of construction at a site, it is important to critically examine all the constraints outlined above for their relevance to the given site.

It is equally important also to appreciate the scope and be aware of the limitations of the method. Towards this end, it may be helpful to compare the advantages and disadvantages associated with this method as follows.

Advantages:

- It can, potentially, provide better control of ground movements.
- Permanent floor slabs provide continuous strut action that is much superior to that of the isolated struts at discrete locations and the need for additional bracing is significantly reduced with obvious cost savings.
- The need for extensive falsework is eliminated; again, this can lead to significant savings in time and cost.
- Substantial part of backfill can be replaced much earlier. This helps to isolate the environment from the effects of subsequent excavation. The top surface is freed and any construction activity on or aboveground can be undertaken concurrently with the construction activity belowground.
- By working under the cover of the roof slab, the adverse effects of construction under exposure to severe weather conditions are largely eliminated.
- Noise and dust pollution are significantly reduced.

Disadvantages:

- A number of temporary 'muck-out' holes need to be provided through all but the base and the platform slabs. This also necessitates the provision of some form of falsework for the eventual filling up and making good of the holes.
- Difficulty of earth moving and a slowing-down in the rate of excavation can be faced as compared to the relative freedom and speed associated with open and conventional excavation.
- Confinement of area is likely to impose limitations on access for construction activities. As a result, there is likely to be some escalation of time and costs.
- As it cannot be wrapped in a membrane, waterproofing of the box structure can be problematic.
- Construction of the permanent columns especially at their junctions with the already-constructed slab above will demand special attention to detail.
- Where the cofferdam is nonintegral, concreting of the permanent walls right up against the slab soffits above requires special consideration.
- Invariably, the stages of construction have a dominant influence on the geometry of the structure, the boundary conditions and the loading. Implications of all these factors need to be thoroughly appreciated and taken into account in design as appropriate.

The obvious technical superiority of the top-down method of construction, particularly where ground movements have to be tightly controlled, notwithstanding, its use should not be regarded as the *sine qua non* for all types of ground and groundwater conditions encountered. Before coming to a decision as to the most suitable method of construction to be adopted, all the site-specific constraints should be carefully considered and every situation examined on its merits.

DESIGN CONSIDERATIONS

Complexities of the top-down sequence of construction can be varied and many. As such, it is not practical to list every conceivable aspect that might present itself for design while using this method of construction. However, the main considerations for design can be broadly identified as follows:

- The difference in the geometry and boundary conditions between construction within an outer cofferdam and using integral walls and their effect on design must be recognized and taken into account as appropriate.
- If the construction is carried out within a temporary outer cofferdam, the cofferdam must be designed for the envelope of stress resultants obtained from all the different stages of excavation and bracing provided by the permanent floor slabs taking into account the appropriate geometry and the relevant boundary conditions in each case. It is important to recognize the difference between the respective nature and location of the 'softer' boundary conditions when the formation level for a particular floor level is reached and the 'harder' boundary conditions after the floor slab is in place and operational as the excavation is advanced downwards, and make allowances for their respective effects in design accordingly.
- Elastic properties of the floor elements should be taken into account to allow for their axial compression and their impact on the stress-resultants of the walls.
- The effects of soil–structure interaction as the wall deflects under lateral pressure from the surrounding ground and the soil moves in sympathy to harmonize with the deflected shape of the wall should be taken into account. The wall may be unlikely, thereafter, to return fully to its predeflected profile. This possibility should be recognized.
- Locked-in effects of stress-resultants should be taken on board in design.
- Design of the stanchion supports should proceed on similar lines as those for the temporary traffic deck discussed earlier taking on board, in this case, the relatively heavier loading due to the self-weight of the structure and, possibly, the backfill on top.
- The effect of the differential stiffness between the vertical support elements, i.e. the discrete intermediate stanchions and the continuous perimeter walls, and their implications on the design particularly of the floor elements must be considered.
- If the structure is to be constructed within, and its floor slabs are initially supported off, a temporary outer cofferdam, the design of the floor elements must recognize and reflect the commensurate increase in the lengths of the spans during the construction stages.
- The locked-in effects under the self-weight as the floor elements are free to take up the deflected shapes as the soil support below is removed through excavation must not be overlooked.
- If the locations of the temporary stanchions do not coincide with those of the permanent column supports, the effects of transfer of loads from the former to the latter when the permanent column supports are in place and operational and the temporary supports are removed must be duly taken into consideration.
- If the temporary stanchions are so located as to coincide with the permanent column supports, adequate shear connection details with the floor slabs must

be devised. Consideration must also be given to the implication of loads from the self-weight of the structure and the backfill on top being locked into the stanchions and the need, thereafter, to design the composite column supports to carry only the extra-over loads.

- Where it may not be practical to achieve a discrete separation between the stanchion foundation below and the base slab of the structure, the implications of the hard spots provided by the existence of such foundations on the design of the structure must be duly taken into account.

The principles illustrating the implications of differential support systems, method and sequence of construction and changes in the boundary conditions, etc., on the design of a typical cut-and-cover metro station structure have been addressed through a number of solved examples in Chapter 26.

Combined method

Under certain circumstances, it may neither be practical nor advisable to follow, completely and exclusively, either the bottom-up or the top-down method for the construction of an entire cut-and-cover metro structure. However, it may well be that, in such a case, a hybrid approach which combines within it principles of both these methods of construction presents itself as the most sensible option.

Consider, for instance, a geological formation comprising loose soil accompanied by high water table overlying a very hard stratum. Consider also the possibility of having to construct, under such ground conditions, a station box structure which falls partly in the loose soil above and partly in the hard stratum below as shown schematically in Figure 4.14. Let us further assume that important and sensitive buildings of national and historic interest exist in the immediate vicinity of the structure, which demands that the ground movements are restricted to the absolute minimum.

The ground most susceptible to movement is likely to be the overlying loose layer of soil. To minimize movements within this stratum, it would be sensible to construct the upper half of the structure that falls within it by the top-down method.

However, it is more than likely that the need to remove the requisite mass of the hard stratum below, may be even with the use of small charges under controlled and confined conditions, would preclude the use of top-down method for the construction of the lower half to be carried out in it. Let us also assume that the vertical cut within the hard stratum is self-supporting without the need for any bracing or can be made secure by the use of some technique such as soil nailing, etc., then the construction of the lower

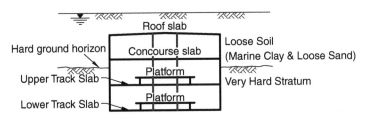

Figure 4.14 Schematic of a Metro Station Structure

half can safely follow the bottom-up method. A potential sequence of construction following such a combined method is illustrated in Figure 4.15, Stages 1–7.

Being a hybrid method, it will combine within it, to a large extent, the aspects of design considerations of, as well as the advantages and the disadvantages inherent in, both the top-down and the bottom-up methods of construction. However, connection details particularly at the interfaces between the two methods such as the junction between the top-down and the bottom-up parts of the perimeter walls, and certain aspects of design, peculiar to this method, warrant careful consideration. For example, to achieve structural continuity across the interface between the already-constructed top-down and yet-to-be-constructed bottom-up portions of the perimeter walls, it would be necessary to have couplers in place at the bottom of the former at the time of its construction into which extension bars from the latter could be screwed-in preparatory to the placement of concrete.

It is also important to appreciate that, initially, the top-down portion of the perimeter wall would be directly supported on the hard stratum. The removal of the support hitherto provided by this stratum and the construction of the 'bottom-up' portion of the wall would then be carried out in a carefully phased, piece-meal sequence. When this

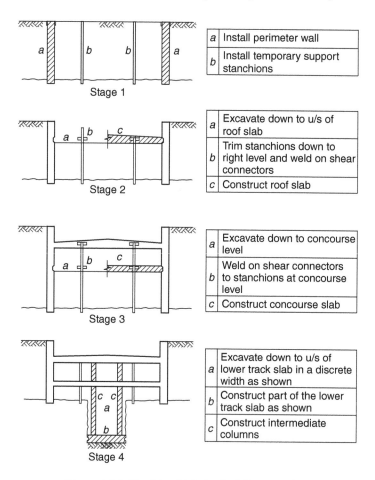

Figure 4.15 Combined Method of Construction

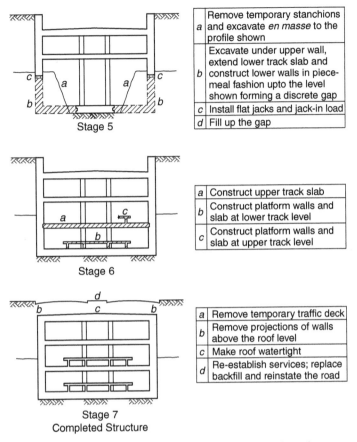

a	Remove temporary stanchions and excavate *en masse* to the profile shown
b	Excavate under upper wall, extend lower track slab and construct lower walls in piece-meal fashion upto the level shown forming a discrete gap
c	Install flat jacks and jack-in load
d	Fill up the gap

Stage 5

a	Construct upper track slab
b	Construct platform walls and slab at lower track level
c	Construct platform walls and slab at upper track level

Stage 6

a	Remove temporary traffic deck
b	Remove projections of walls above the roof level
c	Make roof watertight
d	Re-establish services; replace backfill and reinstate the road

Stage 7
Completed Structure

Note:- All the above operations need to be carried out in a phased manner.

Figure 4.15 Combined Method of Construction (*Continued*)

element of the wall becomes operational and is ready to support the weight of the structure above, there would be load transfer on to it via the previously constructed wall above. As the lower (new) part of the wall takes on this load and undergoes vertical compression under it, as it is bound to, the upper (older) portion of the wall would, correspondingly, be free to experience a downward movement of a magnitude commensurate with and as permitted by the flexural and shear stiffnesses of the slab elements above. This could cause extra loads to be shed on to the intermediate support columns and, potentially, adversely affect the stress-resultants in the upper floors as well.

One way of avoiding the adverse consequences of load transfer in the manner described above would be to stop placement of concrete some distance below (rather than extending right up to) the bottom of the upper portion of the wall so as to create a discrete gap deep enough to allow flat jacks to be inserted. By loading the jacks, thereby precompressing the lower wall it would then be possible to transfer the load on to it without the danger of the upper wall experiencing any downward movement or the stress-resultants of the upper floors any adverse consequences. The jacks would need to be so spaced as to avoid adverse localized concentrations of stresses in the walls. All that would remain to be done thereafter would be to pack the gap discretely, allow the jacks to be released and removed and the gaps filled with concrete appropriately. Alternatively, the jacks could be left in place, if so desired, and the gaps concreted.

Notwithstanding the merits of the discussion so far, it must be recognized that the effects of load transfer can be completely neutralized only if the structure, in reality, conforms to the boundary conditions as assumed and behaves precisely as theoretically anticipated, and the jacking load and the actual load transferred from the upper to the lower half of the wall are identical. Achieving such an exact correspondence in practice cannot be guaranteed. It would therefore be sensible to examine whether and to what extent the design as it stands is tolerant of any variations albeit allowing for certain amount of overstress. If it is not, it would be prudent to build into the design of the structure allowances for reasonable variations to avoid any unpleasant surprises later on.

Consideration would also need to be given to the implications of transfer of load from the temporary stanchions to the permanent columns on the stress-resultants in the upper floors. It would also be necessary, particularly in the case of the chosen example, to examine the slenderness of the permanent columns over the time the upper track slab remains to be constructed and to establish whether any temporary bracing might be required. A typical example of a hybrid construction of a deep, multilevel metro station is illustrated in Figure 4.22.

Caisson sinking method

The method of sinking a pneumatic caisson involves assembling the cutting edge on ground, building on it the caisson structure and then pushing it down to its founding level under its own weight. The process of sinking is aided by excavating and undercutting the underlying soil. The area below the base slab of the caisson structure and enclosed by the perimeter of the cutting-edge constitutes what is known as the work-chamber. It is so named because all the activities associated with the sink, i.e. water jetting and flushing operations for the removal of the spoil and controlling and achieving the sink, are carried out under confined conditions from this chamber. The work-chamber is accessible through an airlock and the spoil is either transported through a material lock or mixed with water and pumped out through pipes.

In order to achieve dry working conditions within it, compressed air pressure in the work-chamber is so maintained as to balance the pressure of the groundwater and prevent its ingress into the chamber. The amount of the pneumatic pressure required at any given time of the sink is given by $\gamma_w \cdot H$, where γ_w is the unit weight of water and H the corresponding differential hydrostatic head at the formation level of the chamber. As the head differential increases with the progress of the sink, the necessary pneumatic pressure too needs to be increased accordingly. However, in the context of the depth of metro structures encountered generally, the pressure is unlikely to exceed around 1·5 bars.

In soft ground with high permeability where water table is also high, the existing structures in the vicinity of the metro works can be highly susceptible to settlements. Where the existing structures are of historical or national importance, or, are otherwise sensitive to ground movements, lowering of water table would not be feasible; if anything, control of movements would be crucial and could constitute the governing criterion. However, achieving adequate control of groundwater and ground movements during the construction of underground metro structures under such difficult conditions, irrespective of whether bottom-up or top-down sequence of construction is adopted, can be extremely difficult. Under these circumstances, the technique of pneumatic caisson sinking under compressed air can present a viable alternative provided that:

- The ground at the surface is strong enough to safely sustain the deadweight of the caisson unit including that of the cutting-edge without causing any uneven settlements.
- The soil is amenable to excavation by the use of water jets and, when mixed with water, can be easily pumped through pipes.
- Adequate founding level is available at the depth required for the structure.
- Sufficient and easily accessible working space is available around the units.
- Where the structure is located under a thoroughfare, the site can be closed to nonconstruction vehicular traffic which can be diverted elsewhere.
- Utilities can be permanently rerouted or temporarily diverted during the period of construction.

In respect of the first proviso, if the ground is not strong enough to sustain the deadweight of the caisson unit, it may be necessary to assemble the unit on a purpose-made concrete strip large and strong enough to spread the load more tolerably. After the unit has been completed full height and has attained the required strength, the strip would be broken up preparatory to the commencement of the sink.

With regard to the third proviso, invariably, owing to the cellular geometry of the structure, generally more weight is removed in terms of the soil excavated than that replaced by the structure; accordingly, the formation is likely to be subject to a net pressure relief. As such, adequacy of the founding level is unlikely to be a critical issue as far as the deadweight of the structure on its own is concerned. However, it can become an issue if the extent of vibration due to the dynamic loading is likely to lead to settlements or the settlement of the ground is ongoing in which case adequacy of the founding level can be important.

The method of caisson sinking under compressed air is a specialized technique. It was successfully used during the construction, in the 1970s, of the Metro Eastline in Amsterdam. On that project, caisson units ranging from 40 to 70*m* in length and 10 to 14*m* in depth were used for different metro structures (Brink, 1990).

Figure 4.16, Stages 1–6, outlines, in a simplified way, a typical sequence of construction and sinking of a caisson structure.

In Stage 1, preparatory to assembling the cutting-edge, ground is prepared by excavating the top 2–3*m* to remove the road pavement, divert utilities, etc. or as deep as is necessary to remove any other man-made obstructions, in order to get down to the virgin ground. The excavated area is then filled with sand which is compacted in layers and the original level restored. In the case of compressible subsoil, it may be advisable to allow the sand to settle for a few months to ensure a more stable base for the construction of the caisson. However, this may not always be feasible, for instance, if there is a time constraint, in which case the sand-fill may be preloaded by raising the

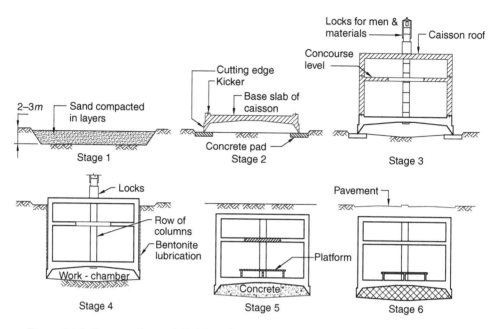

Figure 4.16 Construction and Sinking of a Typical Caisson (adapted from Brink, 1990)

level of the fill sufficiently above the surface to allow the required settlement to take place before regrading it down to the required surface level.

In Stage 2, concrete strip is constructed on levelled sand to form a continuous pad under the cutting-edge perimeter. The steel cutting-edge of the required size is assembled on the pad strip, reinforcement and formwork are fixed into place and the base slab of the caisson and the wall up to the 'kicker' (casting joint) level are concreted. Clearly, the size and thickness of the pad should be such that it can sustain the weight of the completed caisson without permitting any uneven settlement.

Construction of the caisson is then progressed upwards incorporating locks for entry and exit of men and materials. Temporary bulkheads are also installed to plug each end of the unit. Compressed air and flushing equipment are installed in the caisson and the pumps, pipelines and the necessary control equipment for enabling the sink and the related activities to be carried out are fixed into place. All this is done in Stage 3.

Stage 4 represents a typical intermediate stage during the caisson sink. Stage 5 is indicative of the completion of the sink and reaching the required founding level. The work-chamber is then concreted after which the equipment can be dismantled and removed and the works interior to the caisson completed. Stage 6, finally, shows the completed structure.

For the purposes of the sink, full weight and the finished geometry of the in-service caisson structure are not normally required; sinking weight need only be slightly greater than the force of buoyancy or of contact resistance whichever is the greater. The construction of the caisson during the sink is therefore limited to that extent of the structure which is sufficient to overcome this force and, may be, incorporate some marginally spare capacity. The construction of the remainder of the structure internally is then progressed to completion later on after the sink is complete.

However, given the sizeable area of the soil–structure interface around the caisson perimeter, the possibility that a significant amount of residual friction would also need to be overcome to achieve the sink cannot be ruled out. If the extent of this friction is likely to present a problem, it can be minimized with the use of bentonite as a lubricant. For this purpose, the outside face of the cutting-edge is normally set 30–40*mm* proud of the outside face of the caisson walls. This helps create a recess around the perimeter as the caisson sinks. With the injection of bentonite into it, a lubricating annulus surrounding the caisson is formed.

After the founding level is reached and the work-chamber is flushed clean, it is filled with concrete. This helps distribute the loads on the subgrade more evenly. Besides, if the weight of this concrete is also required to make the in-service structure safe against flotation, it would be necessary, in that case, to integrate it with the structure. However, this would call for the prior provision of couplers into the underside of the base slab of the structure in Stage 1 to enable screwing-in dowel bars into these preparatory to filling the chamber with concrete in Stage 5.

Clearly, for completing the full length of a typical station structure, a number of caisson units would need to be sunk. Longitudinal section through one such typical unit is shown in Figure 4.17.

To avoid problems during the sink, clearance between the units of 300–1000*mm* (an average of 650*mm*) is aimed at (Brink, 1990). These *in situ* gaps end up representing the width of the joints between the units. The soil in and around these gaps is frozen by the use of liquid nitrogen forming watertight joints for the adjacent caissons. After the freezing has extended a sufficient distance into the soil surrounding the joint and the units, the temporary bulkheads are demolished and the gap between the units cleared off the frozen soil preparatory to the construction of the joint. Section through a typical joint as used in the metro tunnel in Amsterdam is shown in Figure 4.18.

DESIGN CONSIDERATIONS

Pneumatic caisson sinking can present itself as an ideal technique for the construction of an underground station structure especially in poor ground conditions and high

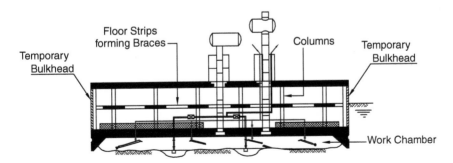

Figure 4.17 Longitudinal Section Through Caisson during Sinking
(adapted from Brink, 1990)

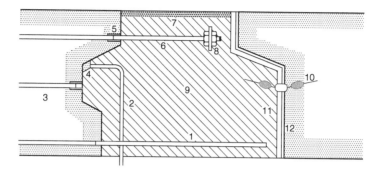

Figure 4.18 Design of Typical Connection-Joint Between Two Caissons (adapted from Brink, 1990) (1 Screw bar 14*mm* Ø; 2 grouting tube; 3 reserve connection; 4 grouting duct filled with styropor; 5 screwed fitting; 6 screw bar 24*mm* Ø; 7 insulation; 8 end anchorage; 9 joint concrete; 10 joint dealing with joint sheet; 11 lining with foam rubber 20*mm*; 12 triplex d=5*mm*)

water table. During the sink, the surrounding ground will not only impose loads on the caisson structure but also provide all-round support to it. However, the quality and the measure of the support that can be expected particularly in the event of an erratic sink will be reflected in the extent of deviation from verticality suffered by the structure. If excessive, it can have a significant impact on the design of the structure. In the light of these, some of the main considerations for design are listed as follows:

- It is important to recognize that, for reasons discussed next, the geometry of a caisson structure and its boundary conditions during its construction and sink could be different from those of its in-service condition. The loadings could prove to be more onerous particularly during the first two stages. However, the design must ensure that the structure is capable of withstanding the respective loads at all these stages adequately.

- During the course of sinking a caisson, it is imperative to have a measure of control in achieving a uniform descent. It is equally important to achieve a balance between the weight of the structure and the desired rate of sink. In other words, given the softness of the ground, if the caisson is too heavy, it may prove difficult to control its descent causing problems beyond the stipulated tolerances. Accordingly, the aim, generally, is to achieve a structure with self-weight just sufficient to overcome buoyancy or the contact pressure whichever is the greater. One possible way of achieving this would be by limiting the extent of concrete placement of, say, the intermediate (concourse) slab to strips along the longitudinal walls with discrete transverse strut strips across as indicated in Figures 4.16 and 4.17. However, this on its own may not always be enough if, for instance, the residual soil–structure interface friction even after using bentonite lubrication proves difficult to overcome. In that case, either the concrete pour would need to be extended to provide the desired sinking weight or, if the additional need is only marginal, temporary kentledge in the form of sand bags or limited flooding of the caisson with water could be used. This then defines the threshold for the minimum weight of the structure for the sink.

- The 'lightweight' structure defined thus will be exposed to, and has to be able to withstand, the soil and groundwater pressures which increase progressively during the stages of the sink. Furthermore, in the case of soft ground combined with high water table, these pressures could be significantly high. It may therefore be necessary to install temporary or permanent stiffening walls to safely accommodate the forces generated during the construction of the caisson as well as during the sinking process.

- If the caisson maintains a truly vertical descent, the lateral forces can, in theory, be reasonably assumed to be those at rest. However, the possibility of maintaining such a descent cannot be guaranteed. It is important to recognize that any deviations beyond the acceptable tolerances could prove to be extremely difficult or prohibitively expensive to rectify. It is therefore imperative to ensure a controlled descent within the permissible tolerances. To achieve this, uneven loading of the 'lightweight' caisson structure must be avoided. However, in the event of any deviation in the verticality during the sink, it is, ironically, the uneven loading itself in the form of strategically disposed kentledge, albeit coupled with selective undercutting of soil below, that is relied upon to help correct it.

- The above notwithstanding, even if the acceptable tolerance in the verticality of the caisson in the order of, say, $\pm50mm$ is achieved, the lateral pressure on the side leaning into the soil is bound to rise above whereas that on the side opposite drop below their at-rest values. It is therefore advisable also to consider a differential lateral load case across the 'lightweight' structure whereby the pressure on one side is less and that on the other correspondingly more than the at-rest value.

- Out-of-verticality can also impose concentration of stresses particularly in the vulnerable sections of the walls, i.e. above the 'kickers' (casting joints). To alleviate the generation of cracks potentially due to this problem, the amount of horizontal reinforcement as a percentage of the cross-sectional area in the lower third (Brink, 1990) of the walls is increased to 0·5 per cent from 0·3 per cent elsewhere. Besides, it is disposed in the outside faces in the form of very-small-diameter, closely spaced bars.

- Economic and practical considerations rule out the possibility of wrapping the caisson structure in a waterproof membrane. Since it is likely to be in contact with groundwater throughout its operational lifetime, achieving watertightness of the structure therefore assumes major importance. This can be realized by aiming for 'impervious' concrete. Based on the experience of the construction of the Metro Eastline tunnels in Amsterdam, Brink (1990) recommends the following measures:

  - Use of formwork without (tie) rods passing through the concrete;
  - Limiting the number of casting joints ('kickers');
  - Adapting the composition of concrete (by special attention to graduation of particle sizes including addition of glacial fine sand, use of appropriate admixtures and aiming for low heat of hydration during setting of concrete);
  - Use of crack-distributing reinforcement (as outlined earlier);
  - Reduction and control of setting temperature (reducing the temperature differential and gradients by reducing initial temperature of concrete and by its cooling during its setting);

- Subsequent treatment of the concrete (to minimize differential shrinkage between the horizontal and vertical structural components by maintaining the moisture in the floors by spraying, and protecting the walls against drying using a curing compound or covering with tarpaulin as a protection against overcooling as appropriate).

Backfilling and reinstatement

Backfilling on the roof and reinstatement of the surface features represents the sixth and the last principal stage in the construction activity. The specific items of work associated with this stage of construction can generally be listed as follows:

- Waterproofing of the roof slab;
- Protection of the waterproofing, may be, by the placement of a layer of concrete cover over it;
- Replacement and compaction of backfill in layers;
- Reinstatement of services such as: Gas, Drainage, Water Supply, Electrical Power, Telecommunications, etc.; and
- Replacement of the subgrade and the base and the reinstatement of the road surfacing.

It is worth recalling here that this stage of construction, i.e. backfilling on roof and the reinstatement of surface features, does not have to be delayed until the very end of construction of the structure; it can be carried out much earlier. The only prerequisite is that the roof slab must be in place, have achieved the requisite strength and be appropriately supported. In the case of top-down construction as discussed previously, unlike that of bottom-up construction and the use of caisson sinking technique, this is not difficult to achieve.

4.4 Construction sequences: some illustrations

Following the discussions in the previous sections it is abundantly clear that, in order to establish the sequence most appropriate for the construction of a particular cut-and-cover structure in a given environment, due cognizance must be taken of a number of factors including the following:

- Type of ground and groundwater conditions
- Type of soil, its strength, stiffness and permeability
- Proximity to site of the existing structures in the vicinity
- Type of construction and age of the existing structures
- Extent of control necessary over ground movements
- Nature of the subsurface structural environment
- Constraints imposed by environmental issues
- Size of structure and depth of excavation
- Control and management of traffic
- Constraints of time and cost
- Any other constraints.

In view of the likelihood that the relevance and criticality of the various factors listed on the previous page can vary from site to site, it is not possible to devise a unique, 'cover-all' sequence of construction for all eventualities. Each situation presenting itself must be considered on its merits and, in the light of the aforementioned factors, to the extent as appropriate. However, for illustrative purposes only, a number of construction sequences have been outlined in Figures 4.19–4.27 incorporating, in each case, some special features.

Top-down sequence: use of hangers

Stages 1–8 outlined in Figure 4.19 represent a top-down sequence of construction. It shows the construction of the *6m* wide longitudinal strips of the roof slab and temporary retaining walls within strutted sheet pile excavation. During this operation,

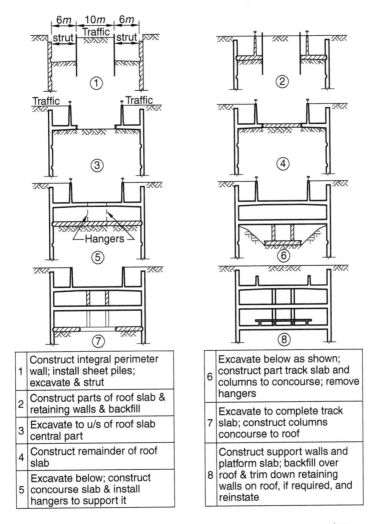

1	Construct integral perimeter wall; install sheet piles; excavate & strut
2	Construct parts of roof slab & retaining walls & backfill
3	Excavate to u/s of roof slab central part
4	Construct remainder of roof slab
5	Excavate below; construct concourse slab & install hangers to support it
6	Excavate below as shown; construct part track slab and columns to concourse; remove hangers
7	Excavate to complete track slab; construct columns concourse to roof
8	Construct support walls and platform slab; backfill over roof & trim down retaining walls on roof, if required, and reinstate

Figure 4.19 Top-Down Construction Sequence, Stages 1–8 (Use of Hangers)
(adapted from Coulson and Stubbings, 1980)

the traffic is confined atop the 10*m* wide ground retained by the sheet piles. After the permanent backfill is replaced over the longitudinal strips of the roof slab, traffic is diverted on to these areas and the central dumpling is excavated deep enough to enable remainder of the roof slab to be constructed. 'Muck-out' holes of appropriate size and spacing are accommodated within the central 10*m* of the roof slab; the area above the roof slab between the temporary retaining walls is used to progress excavation, remove the spoil and facilitate the construction of the remainder of the structure in the top-down sequence. This area is therefore not backfilled until

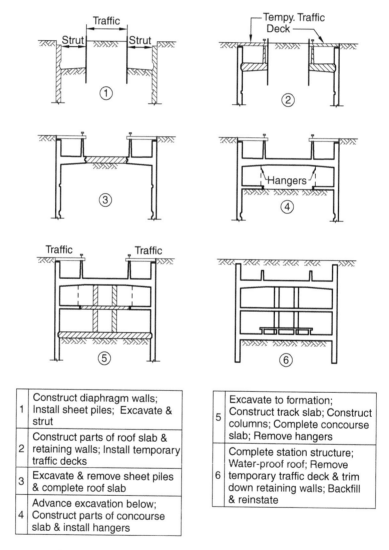

1	Construct diaphragm walls; Install sheet piles; Excavate & strut		5	Excavate to formation; Construct track slab; Construct columns; Complete concourse slab; Remove hangers
2	Construct parts of roof slab & retaining walls; Install temporary traffic decks		6	Complete station structure; Water-proof roof; Remove temporary traffic deck & trim down retaining walls; Backfill & reinstate
3	Excavate & remove sheet piles & complete roof slab			
4	Advance excavation below; Construct parts of concourse slab & install hangers			

Figure 4.20 Hybrid Construction Sequence, Stages 1 to 6 (Use of Partial Traffic Deck and Hangers)

after the completion of the track slab as a result of which, and during which time, the roof slab retains some spare carrying capacity. This reserve capacity during construction, if adequate, may enable the concourse slab to be temporarily supported off hangers from the roof slab until such time as the permanent intermediate column supports below are in place and become operational after which the hangers can be removed.

The key to the success of this scheme rests in the ability of the roof slab to carry, without any distress, the backfill and the traffic load on the side strips, allow the concourse slab to safely hang off it and to be able to span from wall-to-wall without the benefit of the intermediate column supports being in place. Where this is not possible without major modifications to the design of the roof slab, the sequence of construction itself may need to be modified in which case the construction sequence such as that presented in Figure 4.20 may provide the answer.

Hybrid sequence: use of traffic deck and hangers

The sequence outlined in Figure 4.20, Stages 1–6, is a hybrid scheme combining within it the aspects of top-down as well as bottom-up methods of construction.

In the top-down part, the sequence is similar to that of Figure 4.19 insofar as the construction of the temporary retaining walls above the roof, temporary confinement of the traffic to the longitudinal strips on the sides and the use of temporary hangers are concerned. However, there the similarity ends. To ensure that there are no major modifications needed to the design of the roof slab, the extent of the temporary loading on it is alleviated by ensuring that the longitudinal side strips are not backfilled until after the completion of the structure; instead, temporary decks are installed to carry the traffic. Furthermore, by limiting the top-down sequence of construction to only the discrete longitudinal strips of concourse slab along the sidewalls, the load transferred to the roof slab via the hangers is much reduced. The track slab and the remainder of the concourse slab are subsequently constructed in the bottom-up sequence. For the success of this scheme, the discrete longitudinal strips of concourse slab must be able to act as horizontal girders wide enough to provide adequate lateral support to the longitudinal walls at the concourse level.

Top-down sequence: early replacement of road

Generally, metro stations are located in heavily built-up, constrained urban areas below heavily trafficked thoroughfares. Reinstatement of these thoroughfares as early as possible often represents a key requirement. This, invariably, dictates a top-down sequence of construction and demands replacement of the backfill and reinstatement of the road on the surface soon after the roof slab is in place and able to sustain the load.

One such scheme is outlined in Figure 4.21, Stages 1–6. The other special feature of this scheme is the location of the temporary stanchions which, eventually, when encased in structural concrete, also form the permanent column supports. This ensures that the stanchion loads remain locked-in and, as a result, there are no load transfer and related problems to be dealt with.

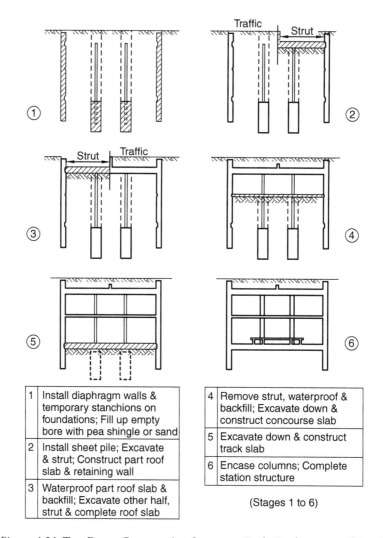

1	Install diaphragm walls & temporary stanchions on foundations; Fill up empty bore with pea shingle or sand
2	Install sheet pile; Excavate & strut; Construct part roof slab & retaining wall
3	Waterproof part roof slab & backfill; Excavate other half, strut & complete roof slab

4	Remove strut, waterproof & backfill; Excavate down & construct concourse slab
5	Excavate down & construct track slab
6	Encase columns; Complete station structure

(Stages 1 to 6)

Figure 4.21 Top-Down Construction Sequence (Early Replacement of Road)

Hybrid sequence: variable cross-section

Figure 4.22, Stages 1–7, outlines a sequence of construction for a large and variable width station cross-section incorporating both the bottom-up as well as the top-down methods of construction.

The scheme also demonstrates the method of harmoniously blending two different concepts of the ground support systems. The wider top half of the structure is constructed within a temporary sheet pile cofferdam in the bottom-up sequence and carries a temporary traffic deck on top. The narrower bottom half is constructed using integral perimeter diaphragm walls and temporary intermediate 'plunge-down' stanchions on bored pile foundations in the top-down sequence. The unique feature of this scheme is that, once the enabling works are in place, both the construction of

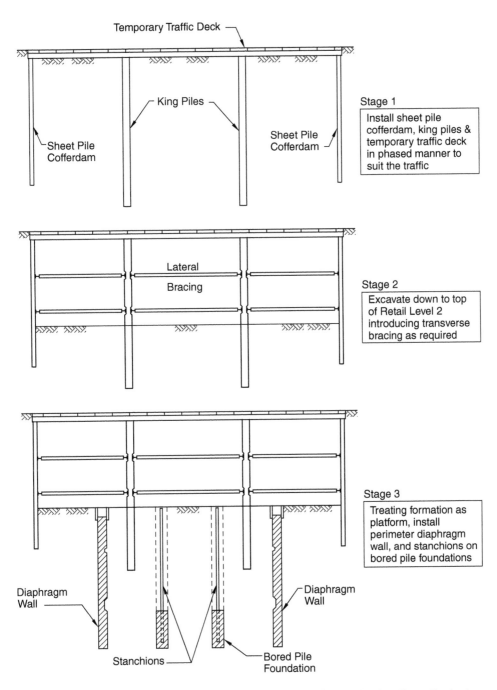

Temporary Traffic Deck

King Piles

Sheet Pile Cofferdam

Sheet Pile Cofferdam

Stage 1

Install sheet pile cofferdam, king piles & temporary traffic deck in phased manner to suit the traffic

Lateral Bracing

Stage 2

Excavate down to top of Retail Level 2 introducing transverse bracing as required

Stage 3

Treating formation as platform, install perimeter diaphragm wall, and stanchions on bored pile foundations

Diaphragm Wall

Diaphragm Wall

Stanchions

Bored Pile Foundation

Figure 4.22 Hybrid Construction Sequence (Variable and Large Station Cross-Section)

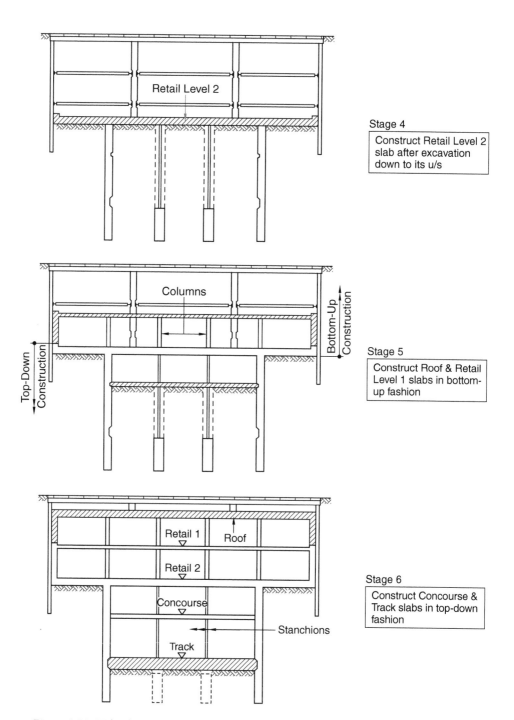

Figure 4.22 Hybrid Construction Sequence (Variable and Large Station Cross-Section) (*Continued*)

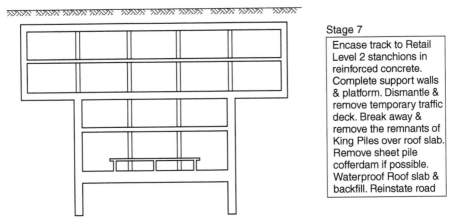

Stage 7

Encase track to Retail Level 2 stanchions in reinforced concrete. Complete support walls & platform. Dismantle & remove temporary traffic deck. Break away & remove the remnants of King Piles over roof slab. Remove sheet pile cofferdam if possible. Waterproof Roof slab & backfill. Reinstate road

Note:- Dismantling of traffic deck, waterproofing of roof slab & backfilling to be carried out in a phased manner to suit traffic.

Figure 4.22 Hybrid Construction Sequence (Variable and Large Station Cross-Section)
(*Continued*)

the top half in the bottom-up sequence and the bottom half in the top-down sequence can proceed concurrently.

Hybrid sequence: use of jet grouting

Consider the construction of a cut-and-cover station structure located directly under a major thoroughfare and in the proximity of existing high-rise structures which are

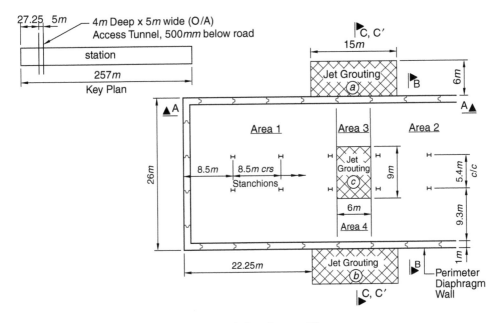

Figure 4.23 Part Layout Plan

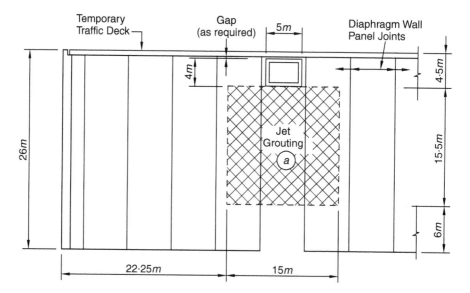

Figure 4.24 Elevation A–A

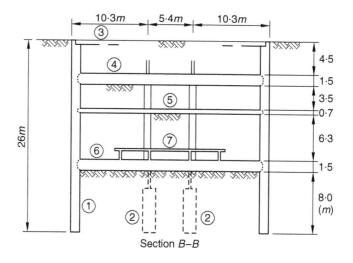

Section *B–B*

Figure 4.25 Top-Down Construction Areas 1 and 2

sensitive to ground movements. It is stipulated that the existing traffic has to be retained without effecting any diversions. There is an additional constraint presented by an existing 5*m* wide × 4*m* deep tunnel running at right angles to the station providing one-way access to an underground car park as shown in the Key Plan of Figure 4.23. It is also stipulated that the tunnel has to be retained without demolition. However, if it helps the construction process and the design so dictates, it is permissible to discontinue the traffic within it for a specified period. Geotechnical information available indicates the soil to be essentially silty. Back analysis from the available

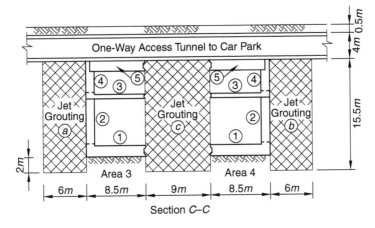

Figure 4.26 Bottom-Up Construction Areas 3 and 4

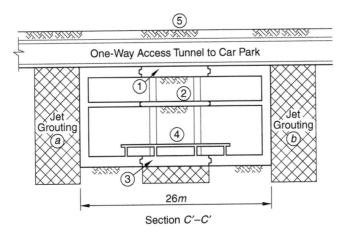

Figure 4.27 Top-Down Construction Between Areas 3 and 4

details of the tunnel suggests that the box structure, when suspended, is capable of carrying its self-weight, safely and without giving any cause for concern, but only over a span of 10*m* longitudinally. Transverse reinforcement in the base slab is also sufficient to carry its own weight, if suspended, wall-to-wall (i.e. across the 5*m* span).

Given the sensitive nature of the existing structures in the vicinity demanding control of ground movements, top-down sequence of construction, as indicated in Figure 4.25, would appear to be the likely choice. For most of the length of the structure, this would be the case. However, in view of the specific constraint presented by it, it is not possible to install any form of a prebored type of proprietary wall from the surface directly under and over the width of the tunnel. Construction of the part of the station box structure directly under the tunnel thus presents a problem which would need to be addressed.

It is conceded that, there could possibly be more than one acceptable method and sequence for the construction of the structure under the tunnel with, inevitably, different cost and construction implications attached to each. However, it is important, given the extremely constrained conditions prevalent on the site, to select a method that does not, in any way, fall foul of the site-specific constraints. To illustrate the principle to be adopted, Figures 4.23–4.27 outline one possible method which can be executed incorporating appropriate safeguards and a reasonable degree of control at every stage of construction. Figure 4.23 shows the layout plan outlining the proposed scheme.

Rationale

As mentioned in the preceding section, given the site-specific constraint, it is not possible to install a proprietary wall from the surface over the width of the existing tunnel. Therefore, the prospect of having to install the retaining walls locally under these areas, *in situ*, presents itself. However, that would require the ground below the tunnel corresponding to the depth of the station box structure, i.e. 13·5*m* height, to be excavated as part of the enabling works preparatory to the construction of the retaining wall. Safe retention of the ground over such a depth within a confined environment, clearly, presents a problem. However, it would be possible to overcome this problem by stabilizing sufficient blocks of the ground immediately outside the station perimeter (represented by areas *a* and *b* in Figures 4.23 and 4.24) through jet grouting so as to enable the excavation for and construction of the sections of the perimeter wall to be safely proceeded with. The *in situ* construction of the wall sections under the tunnel in this manner would lend itself to the conventional bottom-up sequence of construction of parts of the station structure in these areas as indicated in Figure 4.26.

The overall width of the station box structure is 26*m*. The constraint that the existing tunnel box structure is good enough to carry its self-weight over a span of 10*m* only, therefore, obviates the possibility of considering the removal of the existing ground support under the entire length of the access tunnel. However, the problem can be overcome by creating an additional jet-grouted block (zone *c*), as shown in Figure 4.23, to provide a temporary central support of a width that would limit the suspended span of the tunnel, when phased excavation below it is carried out, to less than the permissible 10*m*. Had the safe span of the tunnel been less than 10*m*, say, 4*m*, same principle could still be adopted; however, in this case, three (instead of the one) alternating zones, 3·7*m* long each, of additional jet-grouted blocks could be created. These blocks would be symmetrically disposed along the length of the tunnel and the structure constructed in seven smaller bites instead of the three bigger ones as proposed in the present case.

To ensure that the loads in excess of those which the tunnel box, when suspended, can safely carry, are avoided, use of the tunnel for vehicular traffic would have to be discontinued for a specified period. However, it would be possible, if necessary, to install another temporary traffic deck inside the tunnel located a discrete gap above, and discretely supported off, the base slab to allow vehicular traffic within it. The gap would ensure that the traffic load is not passed on to the suspended part of the base slab of the tunnel. Furthermore, by maintaining a discrete gap between the main outside (surface) temporary traffic deck and the tunnel roof, as indicated in Figure 4.24, the impact of surface traffic loading would also be excluded from impacting upon the station structure during its construction under the tunnel.

Finally, excavation and removal of the jet grouted zone *c* and completion of the remainder of the station box structure in this area can follow the top-down sequence as indicated in Figure 4.27.

The proposal

In view of the foregoing rationale, it is possible to carry out the scheme as follows:

- Install the perimeter diaphragm wall taking it as close as possible to and on either side of the tunnel as shown in the Figures 4.23 and 4.24.
- Install two rows of 'plunge-down' stanchions into prebored piled foundations at centres as indicated in Figure 4.23 to provide intermediate supports for the temporary traffic deck as well as to double up, when encased, as permanent column supports for the station structure.
- Jet grout zones *a*, *b* and *c* each to a depth of 15·5m below the underside of the tunnel box structure as shown in Figures 4.23 and 4.26.
- Install temporary traffic deck over the entire area maintaining a discrete gap above the tunnel structure, as shown in Figure 4.24, to ensure that no traffic load is transmitted on to it.
- Construct parts of the station structure in Areas 1 and 2, adopting top-down sequence of construction as identified by Stages 1–7 in Figure 4.25; Stage 3 represents the installation of the temporary traffic deck.
- In Areas 3 and 4, follow bottom-up sequence of construction as identified by Stages 1–5 in Figure 4.26, resin anchoring dowel bars into adjacent wall panels to ensure continuity.
- Excavate and remove the soil in the jet grouted zone *c* and complete the station construction in this area (i.e. between Areas 3 and 4) following the top-down sequence of construction as identified by Stages 1–5 in Figure 4.27. Stage 5 represents the dismantling and removal of the entire temporary traffic deck, replacement of the backfill and the reinstatement of the road.

It is conceded that the fixing of reinforcement and concreting of the station roof slab within the constrained space immediately below the tunnel structure is unlikely to be easy, however, given the extent of its thickness, certainly not impossible; accordingly, careful attention needs to be given to working out the logistic details. For example, it may be prudent to leave discrete sections of the station roof slab either side of the tunnel to be constructed after the construction of the section directly under the tunnel in order to maintain helpful access from the sides for construction purposes.

5 Ground and Wall Support

5.1 Introduction

In the case of a typical cut-and-cover metro station structure constructed in an open cut within stabilized side slopes, vertically stabilized sides or using the technique of underpinning, as discussed in the previous chapter, excavation and side stabilization proceed in tandem. However, in the case where the sides are 'retained' vertically, it is essential that the boundary wall intended to support the ground outside the perimeter is in place and capable of functioning as such prior to the commencement of any excavation within the perimeter.

As also discussed in the previous chapter, the boundary wall may form a 'temporary' outer cofferdam within which the main structure is constructed, or it may represent the main wall thereby constituting an integral part of the permanent structure. However, it must be understood that, in the former case, even though treated as 'temporary', it may neither be practical nor possible to extract the cofferdam from the ground when no longer necessary and may, therefore, have to be left in place permanently. Nevertheless, its contribution in supporting the surrounding ground is assumed, strictly, as short term, i.e. limited to the stages of construction only; for the long term, the wall is treated as sacrificial and redundant and any permanent contribution, in respect of strength and stability, it might be capable of making to the main structure, is disregarded. It is for this reason and in this context alone that the outer cofferdam is referred to as 'temporary'.

As the excavation advances downwards exposing, progressively, more and more of the wall, it becomes necessary to provide lateral support to it by installing stages of bracings in accordance with the requirements of design. Without the provision of this support to the wall, the wall, in turn, would not be in a position to provide the necessary support to the ground. It would therefore seem sensible not to treat the wall support system as an independent structural system but to consider it, essentially, as an extension of the ground support system as a whole. It is for this reason that, along with the methods of ground support, the wall support systems too have been covered in this chapter.

5.2 Factors influencing choice

It is important to have a clear understanding of the scope of application, i.e. the relevance and limitations, of the various types of the ground support systems available so that the one best suited to a given set of conditions and constraints can be easily identified. The choice of a ground support system and the method of installing it are largely dictated by the following considerations:

- Nature of the ground and the groundwater conditions
- Proximity of existing structures to the excavation
- Necessity to control the ground movements
- Method and sequence of construction
- Necessity to control groundwater
- Required excavation depth
- Any other constraints.

In deciding upon the type of the ground support system that is best suited under the prevailing ground conditions, it is important to examine, among other things, the geology of the site carefully. Whether the ground is soft or hard and containing boulders may, in respect of the ease or the difficulty of their installation, either favour or preclude the use of certain types of the support systems available. Similarly, softer, looser soils with high water table likely to generate high ground pressures may warrant stiffer and stronger wall sections capable of withstanding such pressures and may, therefore, exclude inherently flexible and less strong wall systems. Where the ground may be firm and stiff and the water table low, ground pressures are also likely to be relatively low. Under such conditions, a relatively flexible ground support system might, on the face of it, appear to be adequate. However, this may not always prove to be the case if, for instance, installation of such a system in hard ground is difficult and presents severe problems. All these factors need to be taken into account in choosing the most appropriate ground support system.

It is only to be expected that different types of ground support systems may require different types of plant and equipment for their installation. They may also necessitate different minimum operational clearances from the nearest obstructions. If such clearances are not available, the use of certain types of plant and equipment and, failing that, even certain types of ground support systems themselves, may have to be ruled out.

Sometimes, the existing structures abutting the site of works may carry cantilever balconies which might overhang the perimeter boundary wall of the metro structure to be constructed. These can present problems of headroom under the balconies for certain types of plant and equipment and thus impose additional constraints. In such a case, the plant may have to be modified to suit the headroom, and if this is not feasible, alternative systems of wall construction may have to be considered.

In terms of design also, if the foundations to the existing structures in the vicinity are heavily loaded and in close proximity to the construction perimeter, they are likely to generate high lateral surcharge loads. Such a subsurface structural environment can have a significant influence on the size and stiffness of the wall sections required and, consequently, also on the choice of the ground support system to be adopted.

In the course of the groundwater lowering, the excavation and the construction of cut-and-cover structures, the effect of the resulting ground movements on the structures in the vicinity, if excessive, can often become a cause for great concern to the engineer. Where the movements need to be restricted to within certain specified limits, it is important to ensure that the ground and the wall support systems chosen are able to achieve this. This requires an estimate of the anticipated ground movements to be carried out and examined prior to the adoption, and as a confirmation of the suitability, of the chosen system.

Method and sequence of construction of a cut-and-cover structure can have a significant impact on the choice of the ground and wall support systems adopted. For instance, where the type of ground and groundwater conditions prevailing, the type of the existing structural environment and the need to limit movements and minimize the stages of bracing, dictate the use of top-down construction, adoption of stiff wall sections as integral elements of the structure may be absolutely necessary. Likewise, in the case of the ground support system functioning as a 'temporary' outer cofferdam, and where ground and groundwater conditions, etc. so permit, the use of bottom-up method of construction may be amenable to the adoption of less stiff ground support system with more frequent stages of bracing in order to keep the costs down to reasonable proportions.

To enable the excavation to proceed in the dry, it is usual to undertake dewatering within the perimeter in advance of the excavation. This is generally so timed as to provide at least 2–3m of lead-in depth over excavation. If the boundary wall is reasonably watertight and can be adequately toed into an impermeable stratum providing an efficient water cut-off such that the lowering of water level within the perimeter does not cause any draw-down outside the perimeter, the wall sections would then need to be designed for full hydrostatic pressure differentials. This may accordingly necessitate the use of thicker wall sections.

On the other hand, if an appropriate thickness of an impermeable stratum for toeing into is not available at a reasonable depth and dewatering within the perimeter results in the draw-down of the water table outside the perimeter also, leading to a steady-state of seepage, then the lateral pressures on the perimeter will be much less requiring the use of relatively thinner wall sections. However, in that case, it is also important not to overlook the extent of the resultant movements in the surrounding ground and structures accompanying the water table draw-down outside the perimeter especially if they are likely to be unacceptably severe. Implications of all these aspects on both the design and the construction need to be taken into account in selecting the most appropriate ground support system.

Different types of ground and soil strengths will warrant different depths of embedment of the boundary walls below the formation level as a proportion of the depth of excavation. Depth of excavation can therefore influence the overall depth of the wall required. If the depth required turns out to be significantly large, for instance, as would be the case for a very deep structure, it may preclude the use of certain types of ground support systems whose inherent capability falls short of such a requirement and therefore restrict the choice somewhat.

Besides the factors discussed above, the choice of the ground and wall support systems is also dependent upon many other factors including environmental constraints such as the need to contain noise and vibration within permissible limits and to minimize interference with the existing traffic; local expertise and the availability of plant and technology; consequence and the risk balance and the required design life of the system; whether and to what extent any leakage through the joints in the wall panels is acceptable; and inevitably, the considerations of maintenance and cost.

The list of the factors influencing the choice of ground support systems discussed here is by no means exhaustive. However, these factors need to be gone into thoroughly as the very minimum before one can expect to home in on the most cost-effective choice. Furthermore, it is also imperative to ensure that any other constraints, which may be peculiar and specific to a given site and likely to influence the choice, are not overlooked.

5.3 Design criteria

Installation of the ground support system, excavation and the placement of the stages of wall bracings (i.e. wall support system) together constitute a very important phase in the construction activity of a typical cut-and-cover metro structure. In view of the constrained environment within which these operations are often required to be carried out and recognizing the potential enormity of contractual and cost implications in the event that site-specific constraints are not properly understood, fully appreciated and adequately addressed, the design and construction of the chosen ground and wall support systems must satisfy the following criteria. They must:

- Not compromise the required aspects of safety;
- Conform to the required standard of water exclusion;
- Recognize and respect the prevailing site-specific constraints;
- Offer temporary and/or permanent ground support in the manner required;
- Follow sound principles of constructability using recognized and proven techniques;
- Ensure that adequate measures of control can be exercised at every stage of construction;
- Avoid measures likely to carry persistent and long-term maintenance and cost implications; and
- Offer appropriate strength and stiffness in conformity with the dictates of sound engineering principles and design.

It must be recognized, however, that it may not always be possible to achieve an ideal ground support system that satisfies all the design criteria fully. Since it is dependent upon many factors including, among others, ground and groundwater conditions for which it may not always be possible to quantify the parameters precisely, the possibility that the choice may have to rely on a sensible compromise should not be ruled out.

5.4 Ground support systems

A typical list of the ground support systems used around the world over the years includes:

- Sheet Pile walls
- Soldier Pile or Berlin walls
- Soldier Pile Tremie Concrete walls
- King Piles and Jack Arch walls
- Contiguous Bored Pile walls
- Secant Pile walls

- Diaphragm walls
- Hand-Dug Caissons.

The principal functions of a typical ground support system are essentially fourfold; these are to:

- Support the surrounding ground adequately and safely;
- Enable excavation and construction to proceed safely in the dry;
- Inhibit ingress of water from the surrounding soil into the excavation;
- Restrict the movements of the surrounding ground to within acceptable limits.

It is not necessary that in all the situations and circumstances the same degree of control with regard to all these functions need be called for. Besides, the ground support systems listed in the preceding paragraph, because of the inherent limitations peculiar to each, may not be able, and cannot therefore be expected, to perform all the functions to the same degree of success. This then helps narrow down the field by identifying the systems that might be feasible and more appropriate under a given set of conditions and constraints and excluding those that might not.

The principal ground support systems listed above are briefly discussed as follows.

Sheet Pile walls

This type of ground support system comprises continuously driven, interlocking standard steel sheet pile sections to the required toe level on the assumption that no obstructions likely to inhibit the driving will be met. With such a system, it is possible to achieve a tolerance of 1 in 75 in the verticality. If obstructions, natural or man-made, are present, they need to be removed or appropriately dealt with before the piling can proceed.

Provided that no 'de-clutching' occurs, sheet piling can provide a continuous steel cofferdam. With the progress of excavation downwards, waling beams and stages of bracings are installed at the levels dictated by design and as favoured by the stages and sequence of construction. Upon reaching the formation level, the construction of the main structure then follows in the bottom-up sequence.

Application

Sheet piles are relatively flexible and, inherently, lack sufficiency of flexural and shear stiffness. They are, characteristically, poor performers in controlling deflection and inhibiting soil movements. Besides, ingress of water at any openings resulting from declutching of the sheet piles which become exposed on excavation cannot be ruled out. In comparison with the other more stiff forms of ground support systems, sheet pile walls require greater frequency, i.e. more stages, of bracings. In fact, large deflections of the walls and the associated ground movements can be expected if:

- Pile embedment is inadequate.
- The spacing between successive stages of bracings is large.
- The first bracing level is too far below the ground level making the initial top cantilever section too long.

In the light of these reasons, the use of sheet pile walls in cut-and-cover construction has, traditionally, come to be associated with and restricted to mostly temporary ground support only. In the context of metro station structures, therefore, sheet piling would be employed only to form a 'temporary' outer cofferdam within which the main structure would be constructed. Even so, where it is required to limit ground movements, it becomes necessary, very often, to preload the bracings.

Driving of sheet piles through hard ground, such as hard boulder clay, can be particularly difficult and problematic. Where man-made or natural obstructions are likely to cause difficulties and impede progress seriously, preboring followed by backfilling with more amenable material, such as sand, through which driving the sheet piles can then be relatively easy, might become necessary. However, considerations of cost may limit the use of such an approach to exceptional circumstances only.

The success in the driving of sheet piles is also dependent upon how easily the frictional resistance of the soil, given the large surface area of the soil-pile interface, can be overcome. Pile driving through heavily overconsolidated London Clay, for instance, to any significant depth can often be difficult; whereas, by comparison, driving through soft alluvial clays should pose no serious problems.

In general, the aspects which render the use of sheet piling as a ground support system in the construction of cut-and-cover structures less attractive can be summarized as follows:

- Noise and vibration in driving;
- Greater frequency of wall bracings;
- Inability to inhibit ground movements;
- Extending the length of the piles problematic;
- Lack of certainty in achieving groundwater cut-off;
- Ingress of water through the pile clutches and split clutches;
- Groundwater draw-down subscribing to further ground movements;
- Possibility of even further ground movements due to extraction of sheet piles;
- Need to account for the implications of load transfer from the temporary to the permanent structure;
- Difficulties likely to be faced in the extraction of sheet piles leading to loss of reuse and the consequent increase in the costs.

The aforementioned aspects notwithstanding, in the less urbanized areas where the environmental restrictions are moderate, depth to formation does not exceed 15–16*m*, existing structures are some distance away, adequate working space is available and the subsurface ground and groundwater conditions are amenable, the use of sheet piling as a ground support system in the construction of cut-and-cover structures can prove to be economical. Successful construction within sheet pile cofferdams of a number of station structures for the various metros around the world bears testimony to this fact.

However, persistent pressures from the public and strengthening of environmental controls over the years demanding exclusion of noise and vibration commonly associated with the driving of sheet piles have increasingly shifted the choice of ground support systems towards the use of techniques which are relatively silent and do not subscribe to any vibrations. The stiffer, more robust, systems which use such techniques and which favour combining temporary with permanent retention of ground in the construction of cut-and-cover structures are, therefore, meeting with far greater acceptance in preference over the use of sheet piling.

The advancement in technology, however, has now made it possible to allow the sheet pile installation, with the use of high-frequency vibratory equipment in the case of granular soils and hydraulic pile hammers in the case of cohesive soils, to proceed with minimal noise levels. The use of acoustic covers (Puller, 1996) to pile frames has also allowed sheet piles to be installed in a built-up urban area relatively recently where it would not have been possible previously. However, as long as the vibrations caused by such methods continue to be at a level that may not be acceptable, particularly in stiff cohesive soils or in the case of sheet piling toed into soft rocks or dense soils, resistance to the use of sheet piles will continue. Also, in the vicinity of buildings housing medical facilities etc., which can be sensitive to disturbance by vibration, use of sheet piling technique may not be appropriate and it may be necessary to look at alternative methods of ground support.

It is worth mentioning here that it is possible to cut out vibration by combining sheet pile insertion with slurry trench technique. The sheet piles can be pitched into cementitious self-hardening slurry-filled trenches and the pile toes concreted up to formation level through a pair of tremie pipes. This would certainly exclude noise and vibration totally. However, feasibility or otherwise of such an approach under a given set of constraints would also need to be examined on the considerations of cost.

Design considerations

In the construction of cut-and-cover metro structures, sheet pile walls are used to support the ground temporarily. For the design of such a wall, therefore, the following aspects should be considered:

- Strength and stiffness of the sheet pile sections chosen should be those commensurate with the desired vertical frequency of lateral bracing and the permissible movements of the wall.
- A sheet pile wall need not be designed for earth-at-rest but pressures which are commensurate with the anticipated wall movements. These may lie somewhere between the at-rest and the active values.
- Where the hydrostatic pressure represents the predominant component of the lateral loads, variation in the loads due to any variation in the earth pressure coefficient may not be significant. Nevertheless, appropriate sensitivity study may need to be carried out to confirm this.
- Due account should be taken of the changes in the boundary conditions and the build-up and redistribution of the lateral loadings during the different stages of excavation and construction.
- Estimate of the anticipated wall movements should be carried out taking into account the frequency and the stiffness of the wall bracing and, where appropriate, the magnitude of the preload.

Soldier Pile or Berlin walls

Like the sheet pile wall, soldier pile wall also represents a 'temporary' form of ground support system. This system is made up of a combination of vertical soldier piles

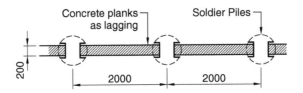

Figure 5.1 Soldier Pile or Berlin Wall

installed at discrete centres with horizontal laggings slotted in between as shown in Figure 5.1. Since this system was originally used in the construction of the Berlin subway, it also came to be known as the Berlin wall method.

The technique

The method consists of boring *600mm* diameter holes (or larger if needed), typically at 2–3*m* centres, along the centreline of the wall. Steel joists, twin steel channel sections or H-beams, representing the principal structural element of the wall, are inserted into the holes extending well below formation level forming vertical 'soldier' piles, which are then encased in no-fines concrete. As the bulk excavation progresses downwards, horizontal lagging elements in timber or precast concrete with sufficient strength to span horizontally between the soldier piles are wedged in between the piles. An example of this is a *24m* deep excavation carried out successfully in silty sand varying to sandy silt at the Raschplatz subway station in Hanover (Puller, 1996). The cut was dewatered by deep wells and retained with the help of an anchored soldier pile wall using horizontal timber laggings.

Clearly, the laggings can be taken down only as far as the formation. In addition to conforming to the design requirements, the size of the typical lagging units, particularly those in concrete, is also dictated by the type of the available equipment and its lifting capacity.

The soldier pile wall is secured by directly anchoring into the ground each soldier pile individually thus obviating the need for any waling beams. In this case, the magnitude of the load on each anchor corresponds to and is a function of the horizontal spacing, centre-to-centre, of the soldier piles and the vertical spacing of the anchors.

In the case of ground that is strong enough to carry higher anchor loads, anchors need not be installed at every soldier pile position but can be located at less frequent intervals commensurate with such higher carrying capacity of the ground. In that case, however, waling beams will be required to carry the thrust from the soldier piles into the anchors.

Application

Particular attention needs to be drawn to the risk of failure under the vertical component of the anchor force upon its transfer to the foot of the soldier pile (Puller, 1996). Where the anchor is installed at a steep angle and, consequently, the vertical component of the anchor load is high, the extent of vertical settlements of the soldier piles could be serious and merit special consideration. However, it may be possible to alleviate the problem or even forestall such a possibility by creating discrete concrete foundations of the size necessary for the soldier piles preparatory to their installation on similar lines as those proposed for the intermediate stanchions under the temporary traffic deck as discussed in Chapter 4.

Soldier pile wall is not likely to provide a sufficient enough cut-off unless the lagging is taken several metres below the formation level. This, however, is not practical. Such a wall does not provide a watertight cofferdam *per se* and dewatering within such a cofferdam causes a draw-down in the water table outside the cofferdam as well. This can, potentially, lead to the settlements of the surrounding ground and the structures in the vicinity. This, together with the potential of blow-up and heave at the formation, is the greatest drawback associated with this system. This technique is therefore feasible only in relatively dry ground conditions. However, even if the conditions are not dry but as long as the ground is at least amenable to dewatering and the extent of the resulting ground movement is unlikely to be of concern, the technique can still be competitively used provided that the ground permits economic anchoring.

In the context of the size of excavation generally associated with metro station structures, soldier pile wall type of ground support system requires a greater frequency of the stages of bracings. The success of this system also depends on the achievement of uniform inter-face contact between the lagging and the excavated soil face. Poor contact due to loss of soil during the excavation and the installation of the lagging can induce ground movements. In the event of nonuniform soil-lagging contact, it may be advisable to restore uniformity of contact, where this is possible, by adequately grouting the gaps.

As a variation on the aforementioned theme but using the same principle, steel I or H-sections can be 'driven' into the ground, but at somewhat closer spacing, to act as soldier piles as well as guides for driving steel sheet piles abutting the flanges of the soldier piles which act as laggings. This arrangement is shown in Figure 5.2. Depending upon the clutch configuration and the type of soil through which the piles are driven, it may be possible to obtain reasonably watertight conditions.

In conclusion, this system of ground support, mainly used for bottom-up sequence of construction of the structure, can prove to be very economic under favourable ground conditions. Also, a tolerance in the verticality of 1 in 100 can be achieved. It has, for instance, found great favour in metro related works in the residual soils of Singapore. Its success is conditional, however, upon being able to place the lagging with ease and without ravelling or loss of ground. It is therefore not feasible in soft and loose soils particularly under high water table. With predrilling, it can also be adopted in boulder clays. However, it is unlikely to be successful in loose granular or very soft soils where local collapse, particularly below the water table, could render the placement of lagging extremely difficult.

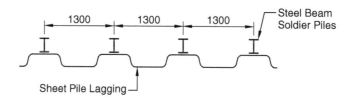

Figure 5.2 Soldier Pile–Sheet Pile Wall

Design considerations

Soldier pile wall also represents a temporary system of ground support. For its design, the following aspects, based on similar lines as those discussed for a typical sheet pile wall, should be considered:

- Strength and stiffness of the typical soldier pile section chosen should be that commensurate with the desired vertical spacing of anchors or bracings and the permissible movements of the pile.
- Soldier pile wall need not be designed for earth-at-rest but pressures which are commensurate with the anticipated pile movements. These may lie somewhere between the at-rest and the active values.
- Laggings should be designed to span between the soldier piles.
- Due account should be taken of the changes in the boundary conditions and the build-up and redistribution of the lateral loadings during the various stages of excavation and construction.
- Estimate of the anticipated wall movements should be carried out taking into account the frequency of the anchors or the bracings and, where appropriate, the magnitude of the preload in them.

Soldier Pile Tremie Concrete (SPTC) walls

The technique

This is essentially a variation on the soldier pile wall theme discussed above. According to the *SPTC* procedure, the wide-flange soldier piles are inserted in predrilled slurry-filled boreholes spaced at regular intervals. The diameter of the hole is generally equal to or slightly less than the depth of the soldier pile so that when the pile is inserted into place the flanges are in intimate contact with the soil. Using the flanges of the two adjacent piles as guides, a slot to the required depth is excavated between them by means of a special grab while, simultaneously, being filled with slurry. The slurry-filled slot is then filled with tremie concrete forming a finished wall panel of mass concrete as shown in Figure 5.3. The significant advantage of this method over that of the lagging method is that it is possible to extend the wall panels well below the formation level.

The bore holes eliminate the need for driving the piles and also help achieve their verticality. The piles are spaced at about twice the thickness of the wall. This allows

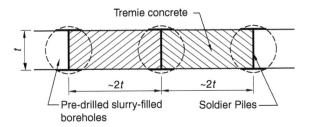

Figure 5.3 Soldier Pile Tremie Concrete Wall

a 45° dispersion of the lateral load across the panel thickness thereby obviating the need for any reinforcement in the wall. The wall has to be in place before any excavation can be commenced.

Application

SPTC wall can be modified for use also in the top-down sequence of construction of the main metro structure.

Mesh-reinforced gunite has been successfully used in place of the mass concrete used in the *SPTC* method to retain the subsoil between anchored vertical soldier pile steel beams. The absence of groundwater, however, is a pre-requisite for the successful use of 'gunite over mesh' technique.

For fairly heavy *SPTC* walls, 900*mm* deep soldier pile sections at under 2*m* centres together with a commensurate thickness of the concrete panel have been successfully used. Such a combination of structural steel and uncracked concrete easily represents an enhancement in the wall stiffness of around two orders of magnitude over the heaviest rolled sheet pile sections. Such walls may alleviate the problem of ground movements, but they in themselves do not permit deep excavations without the use of bracing at vertical intervals of less than about twice those that could be tolerated with more flexible sheeting. It is for this reason that such walls are regarded only as semirigid.

Design considerations

Design considerations for a typical *SPTC* wall are similar to those for the soldier pile wall outlined earlier. The correlation between the spacing of the soldier piles and the thickness of the tremie wall between them is determined by the need to avoid reinforcement in the wall. This is achieved by limiting, in plan, the dispersion of lateral load across the thickness of the wall to 45°.

King Piles and Jack Arch walls

The technique and application

This method of ground support also involves installation of steel universal beam sections (around 900*mm* deep) forming king piles in prebored shafts for which a tolerance in the verticality of 1 in 100 can be achieved. The shafts and the king piles are generally spaced at 2·5–3*m* centres. Semipermanent jack arches and temporary arches facing each other, located between and linking the king piles are then constructed in the underpinning fashion. Spanning between the king piles and relying upon their inherent geometry for strength, both the arches are usually constructed unreinforced. The arches are generally up to 300*mm* thick and the space enclosed by them is maintained sufficiently wide to allow the excavation of the soil within it and the construction of the arches to be carried out safely. The inner arches provide temporary support to the soil until the completion of the main excavation. Unlike that of the soldier pile wall system, the arches in this system can be extended, if so required, below the formation level also.

With the progress of the bulk excavation and the construction of the metro structure, the inner arches, i.e. on the excavation side, are also broken out and removed and it is for this reason that they are labelled as temporary. However, the outer jack arches, i.e. on the opposite (soil) side, are left in place as laggings. The method is illustrated in Figure 5.4(*a*). The ground support system thus formed, comprising the king piles and the outer jack arches, is then incorporated into the permanent wall structure. The outer jack arches form the permanent back shutter and a new shutter is required to form the inner face of the wall. This system of ground support has been successfully used in the bottom-up construction of a number of metro stations in Hong Kong.

Figures 5.4(*b* and *c*) show variations on the same theme; one using structural steel as before and the other reinforced concrete. In both the variations, unlike those of Figure 5.4(*a*), the inner arches are located further inwards as shown in the figures (*b* and *c*) and the voids between the arches are then filled with reinforced concrete thus forming a secant type wall. These variants can be adapted also to suit the top-down sequence of construction of the main metro structure.

Incidentally, the king piles, generally 1500*mm* or more in diameter, can also be constructed, where appropriate, by the hand-dug caisson method. This method is described later on.

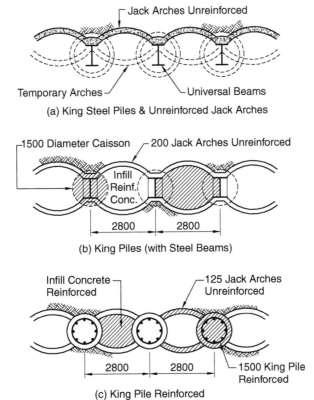

Figure 5.4 King Pile and Jack Arch Walls

Design considerations

'King Pile and Jack Arch' wall is generally used for a temporary system of ground support but can be adapted to represent a permanent integral wall also. For its design, the following aspects should be considered:

- Strength and stiffness of the typical king pile section chosen should be those commensurate with the desired vertical spacing of bracing and the permissible movements of the pile.
- As for the soldier piles discussed before, the king piles also need not be designed for earth-at-rest but pressures which are commensurate with the anticipated pile movements. These may lie somewhere between the at-rest and the active values. However, if the piles form part of the permanent integral wall, then the king pile should also be designed for earth-at-rest pressures to reflect the long-term loading conditions likely to obtain for the completed permanent structure.
- Due account should be taken of the changes in the boundary conditions and the build-up and redistribution of the lateral loadings during the various stages of excavation and construction.
- Estimate of the anticipated wall movements should be carried out taking into account the frequency of the bracings and the magnitude of the preload in them.
- In the interests of safety, the thickness of the outer jack and the inner temporary unreinforced concrete arches should be designed for earth-at-rest pressures.
- In the design of the reinforced concrete infill between the arches, any contribution from the arches is generally disregarded.

Contiguous Bored Pile walls

Where the ground and groundwater conditions are amenable, the use of low-cost bored piling techniques, more particularly, using the continuous flight auger (*CFA*) rigs to drill unconnected piles sequentially, can provide an economical wall for supporting excavations to medium depths especially when temporary casings are not required and tolerance in the verticality of 1 in 100 can be achieved. However, recent developments in the auger rig technology have made it possible to develop enough torque to enable construction of walls with casings even to larger depths. Contiguous bored-pile walls can be used for temporary and, in the case of shallow excavations, for permanent ground retention also.

The technique

A typical contiguous bored pile wall, as illustrated in Figure 5.5(*a*), comprises piles installed along the required wall alignment at centres given by the diameter of the pile plus a margin of about 1 per cent of the pile length, subject to a minimum of 150*mm* (Fernie, 1990), to allow tolerances for construction. The tolerances are generally rounded up to the nearest 50 or 100*mm*. In other words, for pile lengths less than or equal to 15*m*, the allowance would be 150*mm*; for a 16*m* long pile, it would work out to 160*mm* plus 40*mm* for rounding up to the nearest 100*mm*,

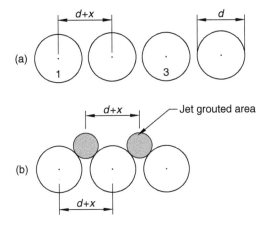

Figure 5.5 Contiguous Pile Walls (adapted from Fernie, 1990)

taking it to 200*mm*; for a 23*m*-long pile, it would work out to 230*mm* plus 20*mm* for rounding up to the nearest 50*mm*, amounting to a total allowance of 250*mm*; and so on.

CFA injected piles are available, nominally, in the range of 300–750*mm*. In this system, a hollow-stemmed continuous auger advances the hole to the required depth, the powered auger and the flighted casing acting as the ground support. When the required depth is reached, the end 'bung' is blown and grout or high-slump concrete is pumped into the hole through the auger stem as the auger is withdrawn.

CFA piles do not require top casings during the concreting process. However, the depth of wall constructed by this method is limited by the length of the reinforcement cage that can be pushed through the concrete. Maximum lengths of up to 12*m* are commonly quoted, although the installation of much greater lengths, up to 20*m*, may be achieved with the aid of an *H*-pile mandrel to which the prefabricated cage is attached. A vibrator is then used to drive, and extract, the mandrel. Alternatively, permanent structural steel sections may be inserted to reinforce the piles to lengths in excess of 12*m* (Puller, 1996).

Application

In the case of rotary piles, powered cutting augers are used to advance the piles to the required depth. They are used mainly in clays and soft rocks. Although, with the support from casing or stabilizing slurry, they can be used in granular materials also, their economic viability drops if large volumes of granular material need to be piled. CFA piles, on the other hand, are ideally suited to mixed ground conditions. However, their efficiency can drop significantly in ground with significant obstructions.

The main advantages of contiguous bored pile wall installed with the use of *CFA* rigs are:

• Absence of noise and vibration;
• Speed of installation and at low cost;

- Not affected by the presence of groundwater;
- Ability to perform under low headroom conditions;
- Ability to operate within a wide range of soil conditions;
- Ability to operate within small clearances from obstructions.

The range of soil conditions in which *CFA* piles can be installed is: $c' = 0, \phi'$ soils with few exceptions; $c', \phi' = 0$ soils except for penetrations in excess of about 8m into hard clay; soft rocks, e.g. soft marls and chalk (Puller, 1996).

It is also worth noting that, with the introduction of quality control measures, such as electronic sensors with visual cab displays to monitor and record auger depth, rate of concrete flow, torque, penetration and withdrawal rates, concrete pressure, etc., the performance of *CFA* rigs has vastly improved making the bored pile technique, in certain circumstances, very attractive.

However, the main disadvantages are:

- Limitations of operating depths;
- Risk of groundwater ingress between piles;
- Escalation in cost where temporary casing required;
- Low structural efficiency on account of its circular cross-section;
- Unsuitability in soft soils where undrained shear strength is less than $10kN/m^2$, in weak organic soils due to the potential of wall bulging and in hard rocks; can also be problematic in hard mudstones (Puller, 1996).

In granular soils particularly, contiguous bored pile walls suffer badly from the ingress of groundwater and, along with it, potentially, the loss of soil as well. However, by jet-grouting the area immediately behind and between the piles, prior to excavation, as shown in Figure 5.5(*b*), it is possible to achieve, potentially, a good water seal and prevent the loss of soil.

The development of mini rotary rigs has made it possible for small-diameter (up to 300*mm*) contiguous pile walls to be constructed within a clearance of as little as 100*mm* from the existing structures.

Variation on the technique

Where the driving of sheet piles or steel *H*-piles is not feasible due to the presence of boulders or other ground conditions, smaller-diameter bored piling can be used to form a temporary cofferdam within which the main structure can be constructed in the bottom-up sequence. One such system is the *PIP* (Pakt-in-Place) system. The *PIP* pile, in the diameter range of 300–550*mm*, is bored with the aid of a continuous flight hollow shaft auger to the required design depth or refusal. The auger is augmented by a chisel for breaking through if boulders are anticipated. Cement-sand mortar is injected through the auger stem pushing soil cuttings out as the auger is withdrawn. A prefabricated reinforcement cage or an *H*-beam is then lowered into the mortar filled shaft to form a completed reinforced *PIP* pile.

To form a contiguous pile wall, *PIP* piles (*A* piles) are installed in this manner in alternate pile positions. The infill piles (*B* piles) between the *A* piles are then formed in the same way as the *A* piles, except that during the auger withdrawal and mortar placement, a high-pressure plunger pump is used for injecting cement paste through the pile. This provides a vertical mortar cut-off between the *B* and the two adjacent *A* piles. After the temporary cofferdam is completed in this fashion, excavation is carried out and bracing installed all the way down to the formation level. The permanent wall of the structure is then constructed using the temporary cofferdam as the permanent back shutter.

In the case of deeper excavations, generally, large-diameter bored piles, in the range of 800*mm* to over 2*m*, are employed using large-diameter bored piling equipment and, if so required, with cutting tools to suit even harder ground conditions.

With the exception of the laggings, which are absent in a contiguous bored pile wall, the rest of the design considerations are similar to those of the soldier pile wall outlined previously.

Secant Pile walls

The main drawback with contiguous bored pile walls is the existence of gaps between the piles; these are responsible for the inherent weakness of the walls. The gaps not only impose a significant limitation on the overall strength and stiffness of the wall making it structurally less efficient, they also make it difficult to achieve watertightness. These aspects render it unsuitable, on its own, as a permanent wall in cut-and-cover construction. Avoiding the gaps by closing the pile spacing such that the adjacent piles intersect can, therefore, significantly enhance strength, stiffness and watertightness. Bored piles interlocking in this fashion constitute what is known as a secant bored pile or, simply, a secant pile wall.

The technique

Typically, 'primary' piles are installed along the alignment of the wall at centres slightly less than twice the nominal pile diameter, using large-diameter bored piling equipment augmented with tools to suit the particular ground conditions. 'Secondary' piles, of the same diameter as the primary piles, are then bored in between, the boring equipment cutting a secant section from each of the primary piles on either side. The secondary pile, also known as the intersecting pile, is formed in a temporary casing usually rotated by an oscillating platform and guided by the primary piles. It is important that the secondary pile cuts through before the primary piles have achieved much of their strength. Therefore, to facilitate construction, a slower rate of strength-gain is specified for the primary piles.

The final wall is thus made up of a line of interlocked alternating primary and secondary piles. For such piles, installation depths of up to 40*m* have been achieved with the use of high-torque *CFA* rigs in London and elsewhere; low-torque *CFA* methods being limited to about 20*m* depth. Each pile is provided with reinforcement, as dictated by the design requirements, in the form of prefabricated conventional cages or steel joists. Such a wall, where both the primary as well as the secondary piles are reinforced and are required to contribute towards its strength and stiffness, is referred to as a 'hard-hard' secant pile wall. For such a wall, a tolerance in verticality of up to 1 in 200 can be achieved. A typical example of a 'hard-hard' secant pile wall is illustrated in Figure 5.6(*a*).

Provision can also be made for the incorporation of horizontal pullout starter bars for connection with the floor slabs within the box-outs formed in the piles at

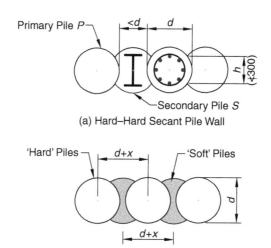

Figure 5.6 Secant Pile Walls (adapted from Fernie, 1990)

the required levels. Excavation is then made and the floors cast top-down providing permanent lateral supports and giving continuity to the wall vertically. However, the use of secant pile walls is not exclusive to top-down construction; they can be used in bottom-up construction also.

The spacing, centre-to-centre, of the primary piles is influenced to a large extent by the need to achieve the required degree of water seal, and to that extent, it is essentially a matter of judgment. To a lesser degree, however, it is also influenced by the requirements of design. Greater the 'overlap' between the primary and the secondary piles, greater is the minimum thickness (dimension h) of the wall and better, potentially, also the seal. For pile diameters in the range of 900–1500*mm*, overlap range of 50–150*mm* yields the minimum wall thickness in the range, nominally, of 430–935*mm*.

In the case of shallow excavations and ground conditions where watertightness of the wall, rather than its strength, is the critical issue, it may be possible to adopt a 'hard-soft' version of secant pile wall. This comprises primary 'soft' piles alternating with secondary 'hard' piles forming an interlocking 'hard-soft' wall as illustrated in Figure 5.6(*b*).

The primary piles are spaced at the same centres as the piles in a contiguous bored pile wall and are, as such, closer pitched than those of the hard-hard version. The primary piles are not reinforced and are cast using bentonite-cement mix, *PFA*-cement mix or a silica-fume mix to form piles of lower strength so that they are 'soft' enough to allow drilling for the secondary piles through them using bits or augers rather than heavy oscillating casing (Fernie, 1990). The secondary piles are reinforced and provide the main strength for the wall. Tolerance in the verticality of up to 1 in 125 can be achieved for the hard-soft secant piles.

Because of the close pitching of the piles, the hard secondary piles cut substantial chunks out leaving very little of the soft primary piles in place. As a result, the hard-soft secant pile wall is likely to be only marginally better in strength and can, therefore, be looked upon, essentially, as only a watertight contiguous bored pile wall. No further reference will, therefore, be made to hard-soft secant pile wall and what follows is confined to the hard-hard version only.

Application

With the advancement in the equipment and the drilling technique, piles with diameters in the range of 600–1500*mm* are now available and can be installed even with a clearance of only 100*mm* from the nearest obstruction (Puller, 1996). Furthermore, depending upon the soil conditions, tolerance in verticality in the range of 1:200–1:300 can also be achieved. However, it must be noted that, to achieve such high tolerances as well as for the accurate location of the piles, guide walls are needed. Construction of the guide walls, inevitably, carries with it a cost implication and this must not be overlooked when comparing the feasibility of the available options for the ground support system.

The circular cross-section of a typical pile bore is, inherently, better able to develop horizontal arch action involving lesser strain in the surrounding soil than would be the case with a long, rectangular cross-section of, say, a diaphragm wall panel. Besides, the secant pile wall technique carries its own casing as part of its standard installation procedure making it suitable in most ground conditions. Furthermore, in poor ground conditions, particularly, the size of the excavation, the time over which it remains open and the expeditious completion of the wall unit can be of the essence. Under such conditions, the circular cross-section of a bored pile in a secant pile wall can steal a march over other types of walls employing linear panels entailing greater amounts of excavation which are likely to remain open over relatively longer periods of time.

The main advantages and disadvantages of the 'hard-hard' secant pile walls may now be summarized as follows.

Advantages:

- Substantial watertightness;
- Absence of noise and vibration;
- Can be concreted under slurry, if needed;
- Construction in all ground conditions possible;
- Pile bore better able to stand in poor subsoil conditions;
- Possible to install the piles at a small inclination to the vertical;
- Ability to work to small clearances and achieve exceptional tolerances;
- Ability to cope with most obstructions even without any need for chiselling;
- Ground movements accompanying the construction of piles much smaller than those of diaphragm wall panels.

Disadvantages:

- Not suitable for depths in excess of 40*m*;
- Less efficient structurally than diaphragm walls;
- More joints and, potentially, as many locations of leakage;
- Strength for strength, more reinforcement needed in secant piles than in diaphragm walls.

Design considerations

In the design of a typical secant pile wall, consideration should be given to the following aspects:

- Strength and stiffness of the typical pile section chosen should be that commensurate with the desired vertical spacing of bracing and the permissible movements of the wall. Due consideration should also be given to the implications likely to result from the sequence, bottom-up or top-down, in which the permanent structure is to be constructed.
- Pile sections should be designed for pressures which are commensurate with the anticipated pile movements. These may lie somewhere between the at-rest and the active values.
- Due account should be taken of the changes in the boundary conditions and the build-up and redistribution of the lateral loadings during the various stages of excavation and construction.
- Where it is used as part of the permanent integral wall, the pile section should also be designed for earth-at-rest pressures to reflect consideration of the log-term boundary conditions for the completed permanent structure.
- Design should be based on the envelope of stress resultants covering the structural geometry, boundary conditions and load cases for all stages of excavation and construction, including the permanent, in-service, case.
- Estimate of the wall movements anticipated during the various stages of excavation and construction should be carried out taking into account the frequency of the bracings and if applicable, the magnitude of the preload in them.

Diaphragm walls

The use of slurry trench walls as structural diaphragm walls in the construction of cut-and-cover metro structures started in the late 1950s. They are continuous concrete walls built below the ground level before the commencement of excavation. Diaphragm walls were used, for the first time, in the top-down sequence in the construction of the Milan Metro. It is for this reason that such a sequence of wall construction also came to be known as the Milan method. Over the last half a century, the use and popularity of diaphragm walls have grown to such an extent that they have come to be recognized, worldwide, as important elements in the cut-and-cover construction scene especially where control of ground movements is critical. They are most versatile and can be adapted to serve a variety of functions. They can be used in almost any soil and, virtually, to any depth, limited only by the capability of the equipment. However, as compared to other ground support systems, diaphragm walls can be relatively expensive.

The technique

Installation of a diaphragm wall involves the construction to the required depth of a series of wall panels of predetermined thickness and length. The thickness is determined by the requirements of the design in respect of both the temporary and the permanent loadings and boundary conditions. However, the plan length is dictated by

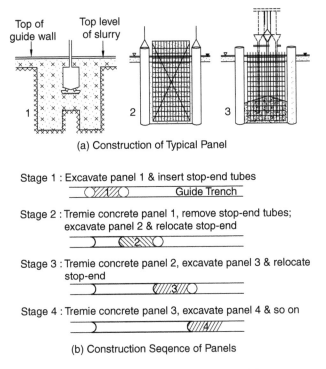

(a) Construction of Typical Panel

Stage 1 : Excavate panel 1 & insert stop-end tubes

Guide Trench

Stage 2 : Tremie concrete panel 1, remove stop-end tubes;
excavate panel 2 & relocate stop-end

Stage 3 : Tremie concrete panel 2, excavate panel 3 & relocate
stop-end

Stage 4 : Tremie concrete panel 3, excavate panel 4 & so on

(b) Construction Seqence of Panels

Figure 5.7 Construction of Diaphragm Wall (adapted from Haws, 1990)

the aspects related to the practicalities of construction such as ground and groundwater conditions, stability of the trench, extent of the concrete pour before any significant stiffening or setting occurs, etc. and the need to minimize the movements of the surrounding ground and the structures. The steps involved in the construction of a typical diaphragm wall are illustrated in Figure 5.7.

First of all, guide walls, generally up to 500*mm* thick and 1*m* deep and reinforced with structural mesh, are constructed parallel to and about 25*mm* outside each face of the intended location of the wall. Where the ground conditions (or the structural environment) so dictate, any other suitable configuration of the guide walls can also be adopted. The distance face to face of the guide walls is thus about 50*mm* wider than the nominal wall thickness. The reason for providing this margin is to ensure that there is little prospect of the walls being hit by the grab on its travel up and down the trench. The principal function of these walls is to guide the excavating grab into the required position to excavate the trench. The outside face of the guide wall is usually cast against the soil whereas the inside face is formed. The wall panels are numbered and their lengths marked on the top surface of the guide walls.

Next, the extent of the ground between the guide walls and as identified by the markings on top and along the guide walls as a panel length is then excavated by the use of clamshell or a grab attached to a Kelly bar. As the excavation progresses downwards, the trench is simultaneously filled with slurry. The slurry is usually a thixotropic suspension of bentonite, which is used to stabilize the sides of the trench. The characteristic of the thixotropic gel is that, when stirred, it becomes fluid again.

By the time the trench excavation is complete and reaches the desired level, the stabilizing fluid or the slurry would have simultaneously filled the full height of the trench panel. Thereafter, two circular steel tubes are inserted vertically, one at each end, into the slurry-filled trench, although elements of noncircular cross-section, such as channel sections, etc. have also been used. These form the stop-ends for the construction of the first panel.

Reinforcement for the full panel height is prefabricated as a cage with concrete rings fixed on all four sides to act as spacers, which help maintain the right cover to the reinforcement all-round. The cage is then lowered into position through the slurry filled trench and suspended at the right level on crossheads placed across the guide walls. Finally, ready-mixed concrete is poured through a tremie pipe progressively discharging from bottom upwards ensuring that the mouth of the pipe is maintained immersed under concrete all the time. In longer panels, two tremie pipes may be required.

Concrete, after pouring it in a diaphragm wall panel, is not vibrated and its unimpeded flow around the reinforcement relies upon its workability and gravity. It is, therefore, advisable to avoid congestion of reinforcement as much as possible. Also, to ensure that the required workability is maintained for sufficient time without any diminution in strength, high slump (150–200*mm*) concrete is used. This is generally achieved by the use of plasticizers in concrete. By the time the stabilizing fluid is completely displaced by the concrete and collected for cleaning and recycling or disposal, the first primary diaphragm wall panel is in place.

After the concrete has gained sufficient strength to stand vertically, the two stop-end tubes are withdrawn and installed in the next alternate location for the construction of the second primary panel to take place. The secondary panel in-between can then be excavated and constructed against the curved smooth concrete stop-ends provided by the ends of the two primary panels. The entire perimeter wall can thus be completed by the alternate installation of the primary and the secondary panels. However, alternatively and with care, panels can also be installed sequentially with the use of one stop-end tube and against the concrete stop-end of the previous in-place panel. It is this latter alternative that is now commonly adopted and is the one illustrated in the figure.

It has to be emphasized that, of necessity, the construction of a typical panel has to be a continuous process, without any stoppage, right from the excavation, through the installation of the steel cage, concreting and up to the withdrawal of the stop-ends. Any disruption at any stage in the sequence can lead to a poorly constituted panel which can be difficult and expensive to rectify.

Recent developments

While excavation by clamshell or grab continues to hold its place as a commonly adopted method in certain soil conditions on small sites, developments in the mechanical excavation equipment have revolutionized the use of diaphragm walling technique. With the use of modern and sophisticated machines employing modern reverse circulation techniques, it is now possible to cut through most types of ground and achieve extremely high tolerances in the verticality.

Hydrofraise machines are capable of cutting through rock with a compressive strength of the order of even $200MN/m^2$ and, in soils, of achieving depths in excess of $100m$ (Puller, 1996). Trench-cutter reverse circulation rigs can, with the use of rock roller bits on the vertical cutter wheels, cut through moderately strong and strong rocks. Furthermore, the compact low-headroom, crawler-mounted version, 4·1*m* wide and 4·7*m* long, able to operate even in headroom only just over 6*m* and, as such, particularly useful in heavily built-up urban environment, can achieve maximum cutting depths of up to 55*m*. Notwithstanding this, the fact that with the use of Hydrofraise and Trench-cutter equipment larger quantities of slurry are required for circulation must not be overlooked as this can have significant cost implications.

However, drawing on the experience gained from the soil drilling industry (Puller, 1996), polymeric and the mixture of polymeric and bentonite slurries have been successfully used on large-diaphragm wall jobs where reasonable economies have been achieved.

Other developments include:

- Improved cage handling methods;
- Reusable precast concrete guide walls;
- Fabricated permanent stop-ends which allow the transfer of shear and tensile forces through the panel joints;
- Stop-end fabrications which allow water bar installation in panel joints without requiring extraction during or immediately after concreting.

With the help of all the developments available, it is now possible to achieve high quality and substantially watertight diaphragm wall panels. Nevertheless, there is a growing realism among the engineering fraternity that acknowledges the possibility that the incorporation of water bars may not, *per se*, automatically provide a watertight wall. It is, therefore, considered prudent to incorporate the provision of a drained cavity construction enclosed by an inner leaf of blockwork lining wall in the planned width of the structure even if, in the end, it may turn out to be redundant.

Application

The use of diaphragm wall as a means of ground support is preferred where:

- Achieving watertightness is essential.
- Requirement of high tolerances is essential.
- Top-down sequence of construction is required.
- Limited working space is available outside the wall.
- Noise and vibration have to be kept to the very minimum.
- Lateral pressures are not high enough to threaten trench stability.
- Excavation is deep enough and wall depth in excess of 40*m* is required.
- Deep cut-offs are required to alleviate the potential problems of 'blow-up'.
- Speed of construction does not favour the use of extensive stages of bracings.
- Deep cut-offs are required to prevent excessive draw-down outside the perimeter.
- The wall is also required to transmit significant vertical loads down to the subgrade.
- Slurry in the trench can be maintained at a level high enough to ensure the trench stability.
- Dilution of the stabilizing fluid by the presence of subsurface water can be controlled.

- High stiffness and strength of the wall are required to restrict movements of the surrounding ground and structures.
- It is necessary, for cost-effectiveness, to combine the temporary with the permanent retention of the ground by using the wall as an integral part of the permanent structure.

The aspects listed above also represent, by implication, the advantages of the use of diaphragm walls. It is also important to remember that, in comparison with other ground support elements, strength-for-strength, it is the most efficient wall element.

By comparison, the principal disadvantages are very few. These are:

- Roughness of the face of wall;
- The risk of loss or spillage of the slurry;
- Large storage and working areas necessary;
- Large cranes needed to handle reinforcement cages;
- Relatively high cost of cleaning and disposal of the slurry;
- Intolerant of disruption during the cycle of panel construction;
- Defects resulting from poor workmanship very expensive to rectify.

For walls deeper than 25m, besides diaphragm wall, cased secant pile wall constructed with the use of high-torque rigs is also available as the likely ground support system. However, based on the current state of technology, for depths in excess of 40m, diaphragm wall stands alone as the single, most suitable method. It is also clear from the comparison that the principal advantages far outweigh the disadvantages inherent in the use of diaphragm walls as a ground support system. In fact, the aforementioned disadvantages constitute, potentially, a very small price to pay for the quality and the extent of the benefits to be derived by using the diaphragm wall technique.

Design considerations

The considerations outlined for secant pile walls above are applicable, in principle, to the design of diaphragm walls also. However, consideration should be given, additionally, to the following aspects:

- Formed, trench-side faces of the guide walls must be cast within a tolerance of 1:200. Although verticality of the wall panels of 1:100 or better is achievable with the use of clamshell or grab, tolerances of better than 1:300 and up to 1:400 can be expected with the use of modern equipment such as the Fraise or the Trench-cutter machines. Positional tolerance of the pullout bars depends upon the integrity of their fixing to the cage; however, tolerances of ±50mm are commonly achieved.
- Guide walls must be sufficiently robust, and adequately braced if need be, to avoid movement due to loads likely to be imposed by the construction activity, buffeting from reinforcement cages and the reactions from the stop-end jacking systems.
- Panel thicknesses ranging from 600 to 1500mm are available. Within this range, the wall thickness is dictated, exclusively, by the requirements of design.

However, the panel length is dependent upon the practicalities of construction, such as the capability of the grab, the type of ground conditions encountered and the safe size of the concrete pour. Grab bites vary between 2·3 and 2·8m in length. Accordingly, the panel lengths can vary from a minimum of one grab bite to a number of bites with smaller widths in between them. However, in poor ground conditions with high water table, it may be necessary to minimize the panel length to ensure trench stability. Design aspects of trench stability have been covered in detail in Chapters 16 and 17.

- The minimum differential slurry head of 1m over that of the groundwater level is considered as the norm for trench stability. However, where groundwater flows in permeable strata are expected, a higher differential, of the order of 1·5m, may be advisable.
- To ensure good quality concrete particularly around the reinforcement, congestion resulting from closely spaced bars should be avoided. Good results are generally achieved by avoiding the spacing of vertical bars to less than 100mm and the horizontal bars to less than 200mm.
- Box-outs should be such as can be cleared and cleaned relatively easily without damaging the reinforcement. However, easy removal of the former is not always possible. This practice has now given way to the use of couplers which, though comparatively expensive, results in a better job. In the fixing of the couplers to the reinforcement cage, due consideration must be given to their diameter *vis-à-vis* the size of the bars to be coupled and the extent of concrete cover required to both the couplers and the bars. This should be reflected in design accordingly.
- The joints between the panels were, hitherto, geared to transmitting only shear. However, with the inclusion of jacks in the concreted panel near the joint and the ability to activate these to thrust bars horizontally into the adjacent panel as soon as it is concreted, structurally continuous diaphragm wall technique has now entered the portals of reality. The feasibility for the use of such a measure is generally determined by the considerations of cost.
- Where, for a typical slab-wall junction, a 'pinned' connection is assumed in the design, engineers have been known to go to a lot of trouble in producing a compliant 'hinge' detail. However, it is worth remembering that, given the confined nature of a cut-and-cover box structure, as the surrounding ground moves in, it is likely to subject the structure, including the slab-wall junction, to the action of a significant axial compression. In view of this, the suspicion that the hinge may not function entirely as intended need not be unfounded. Therefore, the practicalities of going to great lengths in producing elaborate details the functioning of which cannot be guaranteed, must seriously call to question the need to assume such a boundary condition in design. It may be sensible to settle for partial fixity in design if, in reality, the development of full fixity cannot be fully guaranteed.

Other types of diaphragm walls

Wide ranging application of the use of the slurry trench technique and demanding design requirements have led to the introduction of a number of innovations in the plan shape of the diaphragm wall panels. For example, it is possible to construct the panels as *T*, *Z*, *H*-sections, etc.

In the case of cut-and-cover metro station structures, existence of voids in a floor slab over large lengths directly against the walls is not uncommon. In such cases, the lateral support to the wall from the strut action of the floor over the extent of the void may be nonexistent. As a result, the walls at such locations are required to span larger heights vertically. Coupled with this, where water table is high and the ground pressures are significant, normal rectangular panel section, even the thickest one available, may not always prove to be adequate. Under these circumstances, and provided that the necessary space is available, buttressing effect from the stem on the soil-side, in terms of enhanced stiffness and strength, may make the use of the T-section necessary.

Alternatively, it may be possible to make use of the shear joints between individual Z-sections to form a cost-effective diaphragm wall of a castellated plan shape as shown in Figure 5.8(a) provided, again, that the space required to accommodate such a configuration is available. The enhancement in the elastic modulus and, by implication, also in the strength and stiffness can be deduced from figure (b). This principle was used in the cut-and-cover construction of a part of the perimeter wall of Milan metro.

Where vast corridor of space is available around the perimeter, it may be possible to install a cellular diaphragm wall formed from H-sections butting against each other. However, in the context of the construction of cut-and-cover metro station structures which are generally located in heavily built-up urban areas, enough space to accommodate such a configuration is rarely available. Therefore, no further reference will be made to this type of wall.

Precast diaphragm walls can lay claim to a number of advantages in terms of the improved quality of concrete, accuracy in the placement of reinforcement, finer tolerances, saving in the section thickness, avoidance of site concreting operations and

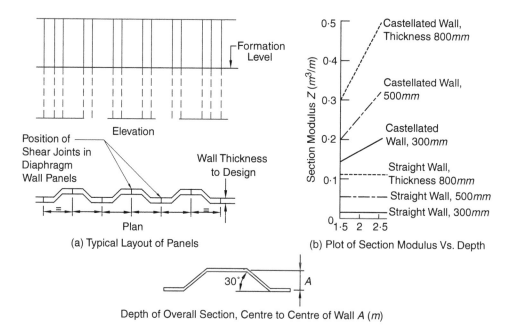

Figure 5.8 Diaphragm Wall, Castellated Shape (adapted from Puller, 1996)

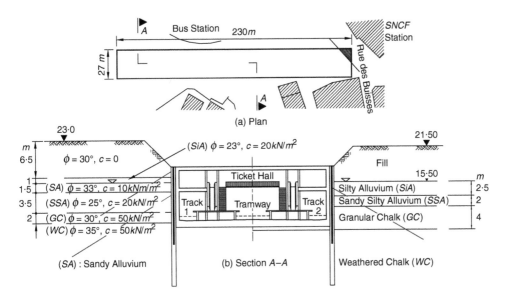

Figure 5.9 Gare de Lille Metro Station (adapted from Puller, 1996)

stop-end extraction, etc. The technique using precast panels was first introduced in 1970 and, by the late 1970s, with its help a number of metro station structures were constructed in France. The use of the technique is illustrated in Figure 5.9.

However, over the years, the popularity of the precast diaphragm wall technique appears to have waned. Besides the considerations of cost, possibly, the main reasons responsible for the decline in its use are:

- Limit on the depth of the panels;
- Limit on the lifting weight and the size of the panels;
- Consideration of flotation imposing preference for thicker wall sections;
- Significant improvements and innovations in the *in situ* slurry wall technique;
- Limit on the panel thickness necessitating additional *in situ* wall to make up the deficiency in the required strength and stiffness.

Notwithstanding the above, the possibility that, under certain circumstances, it might be appropriate and preferable to use precast panels, for instance, as lagging in combination with soldier piles should not be ruled out.

Hand-Dug Caissons (HDC)

In the context of cut-and-cover metros, a hole, generally up to 3*m* in diameter, extended downwards in stages by the underpinning method is described in some parts of the world as a hand-dug caisson and is commonly abbreviated as *HDC*. Although smaller holes have been formed over relatively shallow depths, a minimum diameter of 1200*mm* provides a reasonable working space for relatively deeper metro works. However, for columns in non-metro structures, caissons as large as 9*m* in diameter have also been constructed with this technique. The technique involves manual excavation, is well suited to overcoming obstructions and requires headroom of no more than 2–3*m*. It has been successfully used in the construction of perimeter walls providing the main ground support and isolated columns on a number of cut-and-cover metro station structures in Hong Kong.

The technique

In the construction of a hand-dug caisson, generally, a two-man team, comprising a miner and a winch operator, is involved. An *A*-frame, supporting a winch and a bucket for the removal of the spoil, is installed on the surface directly over the excavation. The miner carries out hand excavation in stages not exceeding 1*m* in depth using pick and shovel or pneumatic spades. However, depending upon the ground conditions and if warranted by safety, the depth can be even less. The winch operator winches up the bucket for the disposal of the spoil.

Each stage of excavation is lined with an *in situ* concrete ring before proceeding to the next stage below. The ring is concreted within a tapered steel shutter giving a variable wall thickness with a minimum of no less than 75*mm* at the bottom. In the case of rings much larger in diameter than 1200*mm*, significantly greater minimum thickness than the 75*mm* may be required as dictated by design. The shutter is suitably braced and designed for ease of striking. After the excavation for the next stage, the shutter is duly lowered into place for concreting the next ring. The sequence is repeated stage-by-stage until the required depth is reached. A section through a typical *HDC* is shown in Figure 5.10.

At the completion of each stage, a sump is formed at the bottom of excavation for the collection of water seeping into the caisson and its removal by a submersible pump. Also, for ensuring safe, nontoxic environment for working in the caisson, air is supplied via light plastic tubing directly to the bottom of the caisson using electric fan blowers.

After the caisson is constructed to its full depth, the base is duly blinded. Then, as dictated by design, either the structural steel section is installed or the conventional reinforcement fixed into place in the empty bore. Thereafter, with the pouring of the concrete, a king pile is formed.

Application

Just as in the case of king piles, the arches between the hand-dug caissons also can be constructed in the underpinning way. Then, with or without the infill reinforced concrete between them and together with the king piles they can represent the ground support systems as discussed previously. This technique has been used in the construction of the Diamond Hill and the Choi Hung Stations of the Hong Kong Mass Transit Railway as illustrated in Figure 5.11.

In the case of the Diamond Hill Station, the combination of the *HDC* shafts reinforced with 911 × 418 × 342*kg/m* universal beams, Grade 43A, and 300*mm* thick outer jack arches represented the ground support system. However, at the Choi Hung Station, the conventionally reinforced shafts and the reinforced infill panels butted against each other forming a secant wall. In both the cases, the king piles were 1·5*m* in diameter and located at 2·8*m* centres.

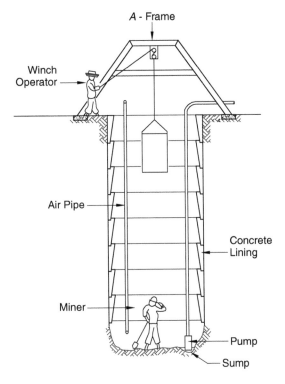

Figure 5.10 Section through a Typical HDC

In the case of Argyle Station also, where lack of headroom under the overhead canopies precluded the use of the Benoto piling rig, 2·2*m* diameter *HDC*s were constructed to form a secant wall.

The *HDC* technique has been successfully and extensively used in Hong Kong. Under the right circumstances, it can continue to prove to be an extremely useful technique in the ground support of cut-and-cover metro structures. Some of its principal advantages may be listed as follows:

- Economy of workspace
- No restrictions of headroom
- Absence of noise and vibration
- Inspection of the shaft base possible
- Offers opportunity of full-scale testing
- Achieving high-quality concrete possible
- Allows accurate placement of reinforcement
- Comparative ease of overcoming obstructions
- Construction depths in excess of 40*m* achievable
- Better control of connection details with floor slabs
- Reinforcement in concrete rings generally not required
- Can prove to be cost effective in constrained environment.

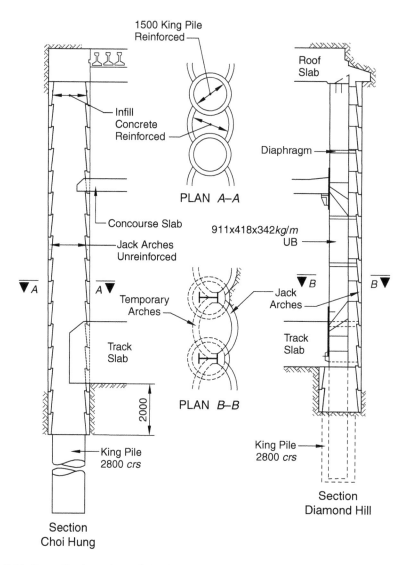

Figure 5.11 Cross-Sections through Main Station Walls (adapted from Benjamin *et al.*, 1978)

The principal disadvantages are:

*   Technique time intensive
*   Specialist labour force needed
*   Potential inflow of water and soil
*   Necessitating training new labour force
*   Potential problems of settlement outside the shaft
*   Potential problems of heave and blow-up inside the shaft.

From the above comparison it is clear that the drawbacks are very limited. Even so, it is possible to control the water inflow, the loss of ground and their potential consequences by grouting and the use of recharge wells. So, in spite of its manual approach, the *HDC* should not be dismissed out of hand as a less useful technique. In fact, under certain circumstances, it could easily present itself as the most appropriate solution.

Design considerations

In the use of hand-dug caissons, careful consideration should be given to the following aspects:

- Hand excavation in loose, fine-grained soils below water table can cause major inflows of soil and water into the caisson demanding careful monitoring. Problems associated with such inflows coupled with the base heave can be severe and include settlements of the surrounding ground and structures. Lowering of the water table can exacerbate matters further. Consideration must be given to grouting to stabilize the ground and to reducing the depth of the stages of excavation. To overcome problems of settlement likely to result from groundwater lowering, recharging the ground outside may need to be considered.
- Where poor or disturbed ground is encountered, increased ring thickness or use of enhanced strength, i.e. in excess of $20kN/m^2$, of concrete, or both, may be required. Larger caisson diameters also necessitate greater lining thicknesses.
- The design thickness of the liner is based on hoop compression and assumes absence of distortion in the ring. For this reason, no hoop reinforcement is generally stipulated. It is therefore imperative to ensure that continuous and uniform contact of the ring with the surrounding soil is maintained.
- Depth of the stage of excavation is generally a function of the stand-up time of the soil. Although a maximum depth of $1m$ is generally adopted, depths less than $1m$ may be necessary under poor ground conditions.
- Where ground conditions are poor, there may be a real chance of the ring dropping down due to the loss of the surrounding ground and the interface friction. To forestall such an occurrence, vertical reinforcement tying the successive rings together can be provided. To achieve this tie, reinforcement concreted into and projecting by the necessary lap length below the upper ring is forced into the ground below in readiness for being incorporated into the lower ring during its concreting. The tie reinforcement must be designed to carry the weight of the lower ring in direct tension. The assumption is that the weight of only one ring needs to be carried in this fashion at any given time. It is important to ensure that the ground reality is compliant, or made so, with this assumption.
- It is widely accepted that, for caissons, most of the ultimate skin friction is mobilized at vertical displacements of less than 1–1·5 per cent, and ultimate base resistance at settlements of 5–10 per cent of the shaft diameter (Fraser and Jenkins, 1990). In other words, very little movement is needed to mobilize full skin friction and unless the pile is bearing directly on rock, practically all the load is likely to be carried in shaft friction. To reflect this possibility, a smaller factor of safety, of the order of 1·2, on shaft friction and a much greater one, of the order of 3, on base resistance is generally recommended.
- Considerations of safety, from the point of view of both stability and nontoxic environmental control, must not be underestimated.

5.5 Ground-support systems: some useful data

Comparative study

McIntosh *et al.* (1980) and Hulme *et al.* (1989) have presented useful and informative comparisons of ground support systems adopted in the cut-and-cover construction of a number of stations on the initial Hong Kong *MRT* and the Singapore *MRT*, respectively. Abstracts of the comparisons are presented in Tables 5.1 and 5.2.

In the selection of the most appropriate ground support system under a given set of circumstances and site-specific constraints, awareness of the capabilities and limitations of the various plant and equipment available, their production rates, comparative costs, etc. can be extremely useful. For this reason, some relevant information is presented hereunder.

Production rates (assuming no obstructions)

Based on rope grab excavation and unit rates of $6m^2$/hour for excavation and $30m^3$/hour for concreting, a typical $57m^2$ diaphragm panel in stiff clay could be produced in a 12-hour shift. This assumes 9·5 hours for the dig, 1 hour for the insertion and positioning of the prefabricated reinforcement cage in the slurry-filled trench and 1·5 hours for the placement of tremie concrete. In stiff granular soil such as completely decomposed granite (SPT: 100–300), using Trenchcutters capable of achieving average dig rates of $10m^2$/hour, a comparable panel can be produced in about two-thirds the time, i.e. an 8-hour shift. With the drop in the stiffness of the soil, significantly higher average excavation rates can be achieved, i.e. $25m^2$/hour in soils with SPT: 50–100 and even $30m^2$/hour in soils with SPT less than 50.

For works in land reclaimed by hydraulically placed sand, following production rates may be adopted:

- Sheet Piling: $8m^2$/hour/rig $= 80m^2$ for a 10-hour shift per day
- Bored Piling: $10m^2$/hour/rig $= 100m^2$ for a 10-hour shift per day
- Diaphragm walling: $10m^2$/hour/rig $= 100m^2$ for a 10-hour shift per day.

It is worth noting that the speed of construction is seldom limited by the capacity of the equipment alone; other logistic factors, such as the ease or the difficulty of site access, fabrication and erection of the reinforcement cage, the method of concreting, site-delivery constraints, etc. can influence the rates and, at times, even significantly (Puller, 1996). So, the foregoing production rates and those quoted in Table 5.3 should be treated strictly as indicative only. Rates quoted in the table assume a 50-hour working week.

Minimum clearances

In view of the heavily built-up urban environment in which cut-and-cover metro station structures are often required to be constructed, the knowledge of how close the construction plant can get to the existing structures and operate can be of great importance. In fact, this can often dictate whether or not it is possible for a particular type of plant or a sequence of construction to be adopted on a given site. In this respect, the information contained in Table 5.4 can be particularly useful.

Table 5.1 Hong Kong Metro Stations: Ground Support Systems

Station	Exc. depth (m)	O/B depth (m)	Rock depth (m)	Nearby buildings	Ground support system	Sequence of const'n
Choi Hung	20	0–3	*	One end	HDC interlocked	Top-down
Diamond Hill	22	3	*	No	HDCs for steel piles, jack arches	
Wong Tai Sin	24 (max.)	3·5–6·5	*	MHHB	Diaphragm walls	
Argyle	25	3–5	*	HRCR	Secant pile wall (Benoto)	
Waterloo	28	2	0–27	HRCR	-do- to rock then *in situ*	
Prince Edward	28	2	16–30	HRCR	Secant pile wall (Ben.); HDC; *PIP* wall	
Chater/Pedder	28	3	33 ●	HRC, LRH	Strutted diaphragm walls	
Kowtoun Tong	18	2	*	No	Part diaphragm wall part *in situ*	Bottom-up
Tsim Sha Tsui	17–21	3·5–7·5	90·5–13·5	HRCR	*In situ* within strutted *PIP* walls	
Jordan	18–23	0–4.5	4–20	HRCR	-do-	
Shek Kip Mei	18–24	1·5–6·5	0–30	HRHB, Schools	*In situ*, Berlin wall part strutted part anchored	
Admiralty	25	0–3	20	No	Part diaphragm part *in situ* within anchored sheet pile wall	
Lok Fu	27	2	0–30	HRHB	*In situ* with sheet pile anchored Berlin wall	

* Not known; ● Rock level not proven in some sections, MHHB: Medium Height Housing Blocks, HRCR: High-Rise Commercial & Residential, HRC: High-Rise Commercial, LRH: Low-Rise Historic, HRHB: High-Rise Housing Blocks (adapted from McIntosh et al., 1980)

Table 5.2 Singapore Metro Stations: Ground Support Systems

Metro station	Maximum excavation depth (m)	Typical geological sequence	Ground support system
Tanjong Pagar	17·9	0·5F, S	Slopes, Anchors
Orchard	21·0	0·5F, G	Nailed Slopes
Toa Payoh	13·5	4F, 4K, G	Sheet Piles
Novena	14·7	0·5F, 14·2K, G	Sheet Piles
Dhoby Ghaut	16·1	1F, 10K, S	Sheet Piles
Outram	13·9	2F, 3K, S	8m deep S. Piles over K. Piles, T. Laggings
Tiong Bahru	14·1	1F, S	K. Piles, Sh. Lagging
City Hall	22·3	3F, 2K, 3S	K. Piles, Sh. Lagging
Marina Bay	16·4	12F, 24K, A	Comp. H pile/ S. Pile
Newton	14·3	3F, 13K, G	0·8 D. Walls
Braddel	14·9	1F, G4	0·6, 0·8 D. Walls
Lavender	16·5	3F, 20K, A	1·0 D. Walls
Bugis	18·3	1F, 34K, A	1·0/1·2m D. Walls
Somerset	16·2	2F, 8K, G	0·6 D. Walls/S. Piles

F: fill, K: Kallang, G: granite, S: boulder bed, S3: Jurong, A: old alluvium; example of geol. seq.: 2F, 3K, S: 2m fill overlying 3m Kallang deposit overlying boulder bed; Gnd. Sup. Sys. Legend: D: Diaphragm, S: Sheet, K: King, Sh: Shotcrete, T: Timber, Comp: Composite. (adapted from Hulme *et al.*, 1989)

Table 5.3 Production Rates (Assuming No Serious Obstructions)

Pile type	Diameter range (mm)	Length of piles/hour/rig (m)			
		Cohesive ground		Cohesionless ground	
		No water	With water	No water	With water
CFA	450–750	Not less than 20 all soils – f(concrete supply)			
Rotary	900–1200	10	8	2	to 4
Rotary	1350–1500	8	6	1·2	to 2

(adapted from Fernie, 1990)

However, it is important not to overlook the possibility that, with the ongoing improvements in the type, size and performance of the equipment and machinery used in the cut-and-cover metro construction works, their required minimum operational clearances from the nearest obstructions may also change. In view of this, the clearances listed in the table should be treated as indicative only and must be confirmed with the specialist suppliers.

Comparative costs

Given the range of the possible ground and groundwater complexities and constraints that can be encountered in the construction of cut-and-cover metro station structures, accurate prediction of the costs likely to be incurred in the installation of a particular ground support system is generally difficult. However, by comparing the cost-indices of the different systems as shown in Table 5.5, assuming that they are installed

Table 5.4 Minimum Clearance to Site Boundary/Nearest Obstruction

Type of ground support system	Type of installation plant	Minimum clearance (mm)*
Steel Sheet Piling	Crane & Piling Hammer	500
Berlin Walls: Soldier Piles & Horizontal Lagging	Rotary Piling Equipment 600 diameter bore	400
	Manual Excavation: Hydraulic Excavator & Trench Box	200 (up to 6m depth)
Contiguous Bored Pile Wall	Bored Pile: Tripod Equipment typical 600 pile	150
	Large-Diameter Rig:	
	Hughes CEZ 300 typ. 740 pile	385
	Hughes CEZ 450 typ. 750 pile	385
	Hughes KCA 100/130 typical 900 pile	450
Contiguous Bored Pile Wall & Hard-Soft Secant Wall	CFA Rig:	
	Soilmec CM 45 typ. 750 pile	350
	Soilmec CM 48E typ. 750 pile	350
	Rotary rig CFA:	
	Bauer BG 11 typical 500 pile	400
	Bauer BG 14 typical 600 pile	300
	Bauer BG 26 typical 600 pile	150
	Bauer BG 30 typical 750 pile	100
Hard-Hard Secant Wall	Bauer BG 7 FOW method 254–406mm diameter	Nil
Diaphragm Wall	Rope Suspended Grab	200
	City Cutter	150
	Hydrofraise	150

* Minimum clearances quoted are those at ground level; consideration must be given to the tolerances in the verticality of the support system. (adapted from Puller, 1996)

Table 5.5 Comparative Cost Indices

Wall type	Contiguous piled	Hard-soft secant	Hard-hard secant	Slurry wall
Cost Index	0·6–0·7	0·7–0·9	0·9–1·2	1·0

(adapted from Fernie, 1990)

under identical conditions and constraints, an estimate of the approximate relative costs can be obtained.

The comparison of the cost indices in the table is based on limited data; it should, therefore, be treated as indicative and used for guidance only. A more detailed break-down (Sherwood *et al.*, 1989) of the cost indices according to the type of ground and site conditions encountered, and the type of the equipment used is given in Table 5.6.

It is worth noting that the mobilization costs of the more sophisticated walling equipment such as the Hydrofraise or the Trenchcutter can be an order of magnitude higher than, say, a low-torque *CFA* rig, with the rope grab diaphragm rig and the high-torque cased secant machines falling somewhere in between. Therefore, from the point of view of economics, more sophisticated equipment is likely to be less attractive on relatively smaller-size jobs. Bearing this in mind, the minimum cost-effective job sizes are probably of the order of 1500–2000m^2 (Puller, 1996) for grab excavation and 5000m^2 and over for Hydrofraise or Trenchcutter work.

5.6 Wall support systems

It is advisable to read this section in continuation with the subsection headed 'Construction of Box Structure' under Section 4.3 in the Chapter 4.

For safe excavation and construction and recognizing that both the potential of loss of life and enormity of remedial costs in the event of failure can be too horrible to contemplate, it is imperative to ensure that the wall support system in place is effective and able to function as required. This not only requires that the bracing system above the formation is perfectly adequate but also demands that, if excessive deep-seated movements which could trigger progressive collapse are to be avoided, the mechanism of wall support below the formation is equally sound.

Wall support above formation

After the installation of the 'perimeter-cofferdam'/'integral-wall' as outlined in the previous chapter, the bulk excavation within the perimeter can begin. However, as the excavation advances downwards, progressively more and more of the wall stands exposed. With the wall thus stripped off its soil support over the depth of the excavation, it would tend to deflect inwards and, in the process, cause the surrounding ground also to move in sympathy. This would be accompanied, inevitably, by a commensurate vertically downward movement of the ground outside the perimeter as well thereby causing the settlement of the structures in the vicinity. Therefore, for general stability of the excavation and to limit the movements of the surrounding ground and the structures before the permanent structure is in place, some form of temporary lateral bracing of the wall above the formation would be necessary. The bracing would need to be installed progressively and in conformity with the design requirements as the depth of the excavation advanced downwards.

Lateral bracing of the walls above the formation can be achieved in four possible ways, i.e. with the use of:

- Horizontal struts inside the perimeter of the excavation;
- Subhorizontal ties anchored into the ground outside the perimeter;
- Hybrid system whereby the walls are partly strutted and partly anchored;
- Temporary/permanent bracing provided by the permanent floors or parts thereof.

Table 5.6 Relative Cost Indices for Various Walling Techniques

Ground conditions & Site constraints	Wall thickness (mm)	Secant wall				Diaphragm wall	
		CFA		Use of cased oscillator	Use of cased high-Torque rigs	Use of grab	Use of cutter
		Use of low-Torque rigs	Use of high-Torque rigs				
		Hard-soft	Hard-hard	Hard-hard	Hard-hard		
Sands and fine-grained fills; No obstructions; Open Site	<650	10	10·4	–	1·6	1·05	1·1
	650–800	1·0	1·2	–	1·35	1·0	1·1
	850–1000	–	1·25	1·4	1·15	1·0	–
	1050–1200	–	–	1·05	1·15	1·0	1·05
	1200–1500	–	–	–	–	1·0	1·1
Clays and fine-grained fills; No obstructions; Open Site	<650	1·0	1·3	–	1·5	1·0	–
	650–800	1·0	1·15	–	1·3	1·0	–
	850–1000	–	1·25	1·4	1·15	1·0	–
	1050–1200	–	–	1·1	1·15	1·0	–
	1200–1500	–	–	–	–	1·0	–
Sands and fills containing some wood, bricks etc. and brickwork; Open site	<650	–	1·0	–	1·2	1·05	1·1
	650–800	–	1·0	–	1·1	1·05	1·1
	850–1000	–	–	1·15	1·0	1·0	1·1
	1050–1200	–	–	1·0	1·0	1·0	1·05
	1200–1500	–	–	–	–	1·0	1·1
Clays and fills containing some wood, bricks etc. and brickwork; Open site	<650	–	1·0	–	1·25	1·05	–
	650–800	–	1·0	–	1·15	1·05	–
	850–1000	–	–	1·2	1·0	1·0	–
	1050–1200	–	–	1·0	1·0	1·0	–
	1200–1500	–	–	–	–	1·0	–

Continued

Table 5.6 Relative Cost Indices for Various Walling Techniques (continued)

Ground conditions & Site constraints	*Wall thickness (mm)*	*Secant wall*				*Diaphragm wall*	
		CFA					
		Use of low-Torque rigs	*Use of high-Torque rigs*	*Use of cased oscillator*	*Use of cased high-Torque rigs*	*Use of grab*	*Use of cutter*
		Hard–soft	*Hard–hard*	*Hard–hard*	*Hard–hard*		
Sands, clays and fine-grained fills; No obstructions; Small congested site	<650	1.0	1.4	–	1.6	1.25	–
	650–800	1.0	1.2	–	1.35	1.2	–
	850–1000	–	1.05	1.2	1.0	1.0	–
	1050–1200	–	–	1.0	1.0	1.05	–
	1200–1500	–	–	–	–	1.0	–
Sands, clays, fills containing some wood, bricks etc. and brickwork; Congested site	<650	–	1.0	–	1.2	1.25	–
	650–800	–	1.0	–	1.1	1.25	–
	850–1000	–	–	1.5	1.0	1.2	–
	1050–1200	–	–	1.0	1.0	1.2	–
	1200–1500	–	–	–	–	1.0	–
SCF + HO, MC, S; O, S/CS	All thicknesses	–	–	–	1.0	–	–
ASF + SRL; OS	–do–	–	–	–	1.25	–	1.0
ACF + SRL	–do–	–	–	–	1.0	–	–
AF + SRL; S/CS	–do–	–	–	–	1.0	–	–

SCF: *Sands, Clays, Fills; HO: Heavy Obstructions; MC: Mass Concrete; S: Steel; O: Open; S/CS: Small Congested Site; ASF: All Sandy Formations; ACF: All Clayey Formations; AF: All Formations (after Sherwood et al., 1989)* SRL: *Substantial Rock Layers; OS: Open Site;*

The type of bracing adopted under a given set of circumstances is a function of the following factors:

- Environment
- Sequence of construction
- Ease of excavation and construction
- Subsurface ground and groundwater conditions
- Magnitude of the lateral ground pressures on the walls
- Need to limit the wall movements to within specified levels.

From the point of view of construction, a site that is free of any obstructions and impediments to bulk excavation and movement of plant and equipment presents the most ideal situation as it can lend itself to speedy excavation and expeditious construction. This then points to the use of ground anchors as an ideal means of bracing.

However, where the ground conditions and the existing structural environment do not permit encroachment into the land beyond the site curtilage, some form of bracing within the perimeter is called for. Furthermore, if the ground pressures are not high and the movements not critical, a less expensive temporary ground support system may be acceptable. However, most temporary ground support systems are used in conjunction with the bottom-up sequence of construction requiring supports for the wall from the start of the excavation until the permanent structure is in place. The most common form of support is that utilizing conventional steel walings and struts.

Under certain circumstances, the possibility that a combination of strutting and anchoring may provide the ideal solution should not be ruled out.

In the case of poor ground and high water table subscribing to high lateral pressures coupled with the need to restrict movements of the surrounding ground and the structures, top-down construction and the use of vastly stiffer permanent floors may have to be relied upon to provide the wall support. However, in order to reach the final formation level as speedily as possible, it may not be necessary to construct, where the layout of the structure permits it, the entire floor slab at each level but only a perimeter band of a certain width may be all that is needed to provide the necessary temporary lateral support to the perimeter wall.

Horizontal strutting

Where steel struts are used in a conventional manner to maintain the opposite walls the minimum requisite distance apart, consideration must be given to the following aspects:

- In the event that heavy steel sections are needed as struts, in order to limit their deflection and the associated wall movements, they may require temporary intermediate support(s). In that case, it may be preferable to opt, instead, for large-diameter steel tubes which, on account of their superior stiffness/weight ratio, may obviate altogether or minimize the need for such additional supports.
- The effect of the diurnal temperature variations on the struts may require pressure cells to monitor the strut loads and the necessary jacking assembly to make load/movement adjustments in them as and when necessary.
- Where ground movements are critical and therefore need to be minimized, a proportion of the anticipated design loads may need to be jacked-into the struts.

However, both – jacking-in too little or too much load – can present problems of different kinds; achieving the right balance can often be challenging.

- Lining of the main box structure with a waterproof membrane especially around the struts can become problematic.
- Implications of the transfer of load from the temporary struts to the permanent structure need to be addressed. To avoid such effects, serious consideration should be given to so locating the struts as would make it possible to concrete them in and be incorporated as parts of the permanent floor slabs.

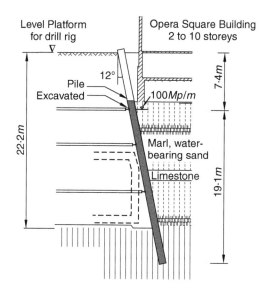

Figure 5.12 Use of Strutting – Frankfurt Metro (adapted from Puller, 1996)

It is important to recognize that the ability of the strutting system to mitigate ground movements decreases rapidly with poor workmanship and control. Coupled with this, the changes in the strut loadings upon their redistribution during the various stages of excavation cannot be discounted either. To have the flexibility of meeting all these changing requirements, the use of jacks in the construction of cut-and-cover metro structures, especially where control of ground movements is critical, has become virtually indispensable.

A typical bottom-up construction using strutted wall support system is illustrated in Figure 5.12.

Ground anchors

Ground anchors, subject to their feasibility, undoubtedly offer the most attractive solution, permitting open, unhindered excavation and, consequently, speedy construction. However, to decide whether ground anchors can be used in a given set of conditions and constraints, the following factors must be considered:

- Availability of wayleaves for the use of the surrounding ground to anchor into;
- Location, suitability and strength of the soil stratum for carrying the anchor loads and requiring prior *in situ* testing;

- Possible location of anchors taking into account the structural environment, i.e. type, quality and sensitivity of the existing structures and the magnitude of the load, the type and the depth of their foundations; existing utilities, caverns; potential future excavations and construction operations;
- Influence of anchor installations on the adjacent structures, such as settlement or heave of the ground, flow of grouting materials into cellars, and vibrations.

Typically, a ground anchor is prestressed to a significant proportion of its design load prior to excavation thus assisting in reducing ground movements. Trial and proof test, therefore, form an integral part of the ground anchor support system and stressing of the anchors individually can ensure that each is capable of supporting the intended design load adequately. However, redistribution of loads as the excavation progresses and, if the excavation remains open for an extended period, the potential for the loss of stress due to soil creep, cannot be disregarded and should be taken into account.

Furthermore, in spite of obtaining the appropriate wayleaves for the installation of the ground anchors, it is sometimes required that the anchors be removed or at least destressed when no longer necessary. While destressing is relatively simple and requires only heat release to reduce the tension in the strands, complete removal can be more difficult and can significantly add to the cost and programme constraints. Use of ground anchors is illustrated in Figure 5.13.

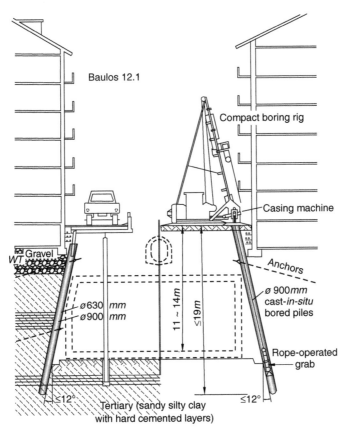

Figure 5.13 Use of Anchors – Munich Metro (adapted from Puller, 1990)

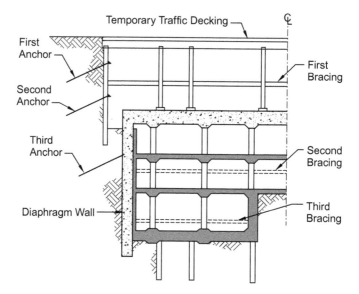

Figure 5.14 Use of Combined Strutting and Anchoring (adapted from Xanthakos, 1979)

Hybrid system

Use of horizontal struts and ground anchors is not necessarily mutually exclusive. Where ground conditions are not particularly good, it may not be possible to develop high anchor loads necessary with the use of anchor penetration lengths which could be deemed sensible. At the same time, it is conceivable that the need for more frequent stages of strutting could also give rise to significant problems in construction. Under such circumstances, a hybrid system combining both the techniques may provide the answer; the anchors would not be stressed beyond sensible limits but would be commensurate with the capability of the anchoring strata, and the frequency of the strutting too would not be stretched beyond the acceptable limits. This principle was used in the construction of a subway station in Akasaka, Tokyo and is illustrated in Figure 5.14.

Permanent floor slabs

In cut-and-cover metro construction, it is not uncommon to come across situations where the ground conditions are too poor to anchor into, water table is high indicating high lateral pressures necessitating unacceptably high frequency of bracing and, more importantly, sensitive structures are in the immediate vicinity necessitating the wall movements to be kept to the absolute minimum. Such demanding constraints would clearly preclude the use of the ground anchors as a means of wall support and bottom-up sequence as the preferred method of construction.

Under these circumstances, especially where containment of movements is critical to the success of the project, three requirements, at the very least, are vital. These are:

- A stiff ground support system
- A stiff wall support system
- Top-down construction.

The first requirement can be met simply by the use of diaphragm wall of the required thickness and installed in the shortest possible panel lengths to avoid instability of the trench during excavation. With regard to the second requirement, continuous and relatively thick permanent floor slabs can, by virtue of being inherently many orders of magnitude more stiff than discrete struts, substantially exclude wall movements. However, the use of the permanent floor slabs as struts in this fashion is possible only with the top-down construction which, therefore, represents its natural extension.

One major drawback with the top-down construction is the potential of relatively protracted excavation. This is chiefly due to two reasons – firstly, the delay in the commencement of excavation below the floor slab at a level until after the substantial completion of construction of the slab and the attainment of requisite strength at that level and, secondly, the slower rate of excavation under confined conditions. However, the process of excavation can be considerably speeded up if, at each level, instead of the entire floor, only discrete strips of the slab adjoining the perimeter wall are constructed to provide the lateral support initially, leaving the remainder of the area free and unconfined to be excavated *en masse*.

This is possible if such longitudinal and transverse, perimeter 'boundary strips', acting as horizontal deep girders, are capable of providing adequate lateral support to the walls at each level. To limit these longitudinal boundary strips to reasonable

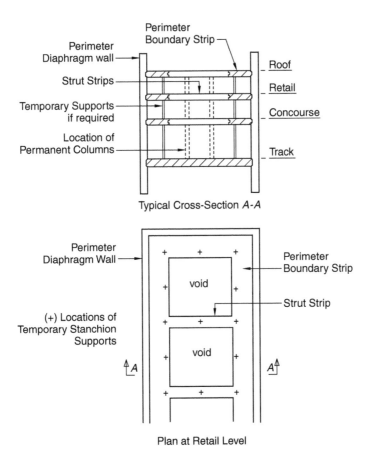

Figure 5.15 Use of Floor Strips as Lateral Wall Support

widths, it may be necessary to reduce the spans over which they are required to act as horizontal deep girders. This can be achieved by providing additional intermediate floor strips of discrete widths at designed intervals to run through transversely as struts across the void. The concept is illustrated in Figure 5.15.

In the event that these strips are unable to span wall-to-wall unaided, they may themselves need temporary intermediate supports vertically.

It is important to ensure that the boundary strips are also capable of functioning as transverse cantilevers carrying their self-weight with no, or very little, extra reinforcement; otherwise, the use of temporary hangers above, or conventional supports below, the cantilever ends may become necessary. However, such an approach will, inevitably, require substantial extent of falsework to complete the remainder of the floor slabs at each level above the base slab later on. In examining the feasibility of such an approach, all these factors, together with their associated costs, must be duly taken into account.

Wall support below formation

Wall bracing above the formation, even when more than adequate, cannot necessarily be relied upon to inhibit inward movement of the walls below the formation. In the case of soft ground and high water table, particularly, the movement can be excessive. Ignoring this aspect altogether, or not giving it the consideration it deserves, can lead to serious consequences. To be able to deal with it adequately, it is therefore important to understand clearly the mechanism of ground support, i.e. the nature of ground response and the performance of the wall, below the formation and appreciate its implications. This is dealt with in Chapter 21.

6   Groundwater Control

6.1 Introduction

Subsurface water can, potentially, subscribe to a significant proportion of the lateral pressures acting on a cut-and-cover structure; it depends upon how close the water table is to the surface. Besides, the rate of the movement of water *vis-à-vis* the drainage characteristics of the soil mass can influence the response of the ground during the various stages of excavation. This can, in turn, also have an impact on the existing structures in the vicinity.

The construction activity related to a cut-and-cover metro station structure involves the removal of large volume of soil. For such an activity to proceed in the dry, it also necessitates pumping out large volumes of groundwater. This dewatering in and around a metro site, if not properly controlled, can cause drawdown, the effects of which may extend far beyond the confines of the site and can, potentially, subscribe to a large proportion of settlements in the surrounding ground and structures. Large ground movements can, naturally, be of major concern to the engineer. Appropriate measures to control the movement of groundwater should therefore be planned in advance and undertaken with due consideration to the implications of safety and cost.

It is essential that adequate knowledge of the porewater pressure profile and drainage characteristics of the soil mass are available at the planning stage. Towards this end, the importance of carrying out pumping trials in developing strategic thinking for effective groundwater control cannot be overstated. However, even after planning groundwater control measures on the basis of all the available information, it is prudent to keep an open mind for any modifications that might become necessary or desirable as the excavation progresses.

Groundwater movement may be controlled, essentially, in one of two ways:

- Exclusion
- Removal.

6.2 Exclusion

In relation to the construction of cut-and-cover metro works, it can be safely assumed that the perimeter wall or cofferdam will be in place prior to the commencement of water table lowering over the site of works unless, of course, the construction is to be carried out in an open cut within stabilized slopes in which case the groundwater would have to be controlled through removal. If the cofferdam is constructed in such a way that it not only forms an impermeable barrier around the perimeter but is also adequately toed and sealed into a competent impermeable stratum below formation, entry of groundwater from the surrounding zone outside into the enclosed perimeter during the construction works can be prevented. Subsequent dewatering within the perimeter can then be carried out without the danger of any significant settlements of ground and structures outside the perimeter taking place. However, in that case, the perimeter walls would need to be designed to withstand the effects of full hydrostatic pressure differential. Since hydrostatic pressures very often constitute a major proportion of the design lateral loads, thicker, relatively expensive wall sections would be required.

In certain environments, such as a 'green-field' site or a dilapidated area, where the settlements of the surrounding ground and structures in the vicinity are of no major consequence or concern, or where the benefits of allowing the settlements to take place far outweigh the costs of putting right their consequential effects, it may be feasible to dewater outside the perimeter. This will considerably reduce the design pressures on the perimeter wall during construction thereby permitting the use of thinner, more economic wall sections. However, this may not always turn out to be the best solution where, for example, design of the permanent structure and consideration of flotation may dictate the use of heavier structure and therefore favour the use of thicker wall sections anyway. The significant advantage of lowering the water table outside the perimeter, where this may be feasible, is potentially the much-reduced lateral wall movements during the main excavation because of the considerably reduced lateral pressures.

In the bottom-up construction of Powell Street Station, *BART*, USA, exterior dewatering by deep wells was allowed, because the surrounding soil, consisting of compact fine sand with lenses of silty clay, was substantially incompressible (Gould, 1970). At that location, 750*mm* thick composite steel-concrete diaphragm walls, 24*m* deep were used. During the 22*m* deep excavation, 6 stages of cross strutting preloaded to 25 per cent of their estimated design loads were used. The third and the fifth struts were eventually incorporated into the permanent floors. Since the excavation was accompanied by dewatering outside the station perimeter, the very small wall displacements of about 10*mm*, representing a mere 0·045 per cent of the excavation depth, was largely attributed to the reduced lateral pressures.

In order to arrive at a solution that is most appropriate under a given set of circumstances, all the aspects must be carefully considered.

6.3 Removal

In practice, the perimeter cofferdam may not always act as a fully impermeable barrier; or, an impermeable stratum of requisite minimum thickness for sealing into may not be available at an economically viable depth. In such a case, control of groundwater is aimed at by the lowering of water table within the perimeter by sump pumping or vertical well-pointing. Sump pumping may be adequate for excavations within stabilized side slopes in coarse sand and clean gravels, provided that the fines are not being removed in any noticeable proportions since significant loss of fines can cause large settlements which could be unacceptable.

The chief drawback of the well-pointing system is the limit on its suction lift; around *5m* depth of lowering below the pump level is generally regarded as the practical limit. For this reason, for water table drawdown invariably much in excess of *5m* to be expected during the excavation for cut-and-cover metro structures, installation of two or more stages of well-points may generally become necessary. Accommodation of slopes commensurate with such a multi-stage system, including one or more mid-level terraces, would, inevitably, necessitate a larger width of excavation at the ground level.

Drainage characteristic of the soil mass plays an important part in the success or failure of a pumping scheme. For instance, adequate lowering of water can be difficult even in a highly stratified ground if the permeability of the soil mass is low. Likewise, for high permeability strata within a soil mass having significantly lower global permeability, adequate lowering can be equally difficult if the well points are not installed at the levels of the high permeability strata. It is therefore important to obtain the permeability characteristics from a properly conducted *in situ* test which is representative of the soil mass. It is important also to recognize that laboratory tests, which are unlikely to be based on representative samples, can, at times, be widely out.

Sometimes, bored filter wells are used to control groundwater. A typical bored filter well consists of a perforated inner tube called the screen, which allows ingress of water. The screen is surrounded by an annulus of graded filter medium which is not only related to the slot sizes in the screen but is also designed to disallow loss of fines from the soil. Where bored filter wells are used to control groundwater, the possibility of corrosion and encrustation should also be taken into account if the groundwater environment is aggressive.

In order to devise an efficient well-point system, it is important to know the following:

- Permeability of the aquifer
- Thickness of the aquifer to be drained
- Extent of porewater pressure reduction required.

As the dewatering advances downwards, equilibrium porewater pressures along the boundaries change resulting in the movements of the surrounding ground and the settlements of the structures in the vicinity. In order to be able to plan appropriate containment measures, it is essential to have an idea of the likely extent of such anticipated movements and settlements. To quantify the problem of movements due to dewatering, it is necessary to know the following:

- Grading of soil;
- Size of excavation;
- Stratification of soil;
- Compressibility of layers;
- Duration of the pumping period;
- Thicknesses of compressible layers likely to be affected;
- Length of time over which the excavation is likely to remain open;
- Potential draw-down *v/s* depth profile outside the perimeter due to steady-state seepage (For the definition of steady-state seepage, see Chapters 11 and 18).

- Porewater pressure and drainage characteristics of the ground, the rate of settlement being dependent upon the permeability of the compressible layers within the soil mass.

It is important to take into account the existing groundwater regime and the *in situ* drainage characteristics of the soil in establishing the most appropriate method of excavation and groundwater control. Due allowance also needs to be made for modifications to porewater pressure boundaries and drainage characteristics as the excavation and construction progress.

Reasonable estimates of the soil compressibility can be obtained from *in situ* penetration tests, laboratory tests and previous experience of foundation settlements in similar soils. However, establishing the draw-down *vs.* depth profile accurately is rather difficult. Much depends upon the local permeability and whether or not a satisfactory cut-off can be achieved. Where the existing structures within the zone of influence are sensitive to ground movements, parametric studies should be carried out using finite element seepage programs assuming various conditions and combinations of the permeability profile and cut-off.

6.4 Ground movements

Ground movements during dewatering can take place due to the following:

- Soil erosion accompanying uncontrolled seepage
- Soil consolidation.

Soil erosion

Sands and silts are particularly susceptible to seepage erosion and extra measures are often necessary to discourage abstraction of fines. Occurrence of troublesome settlements during excavations in $c' = 0, \phi'$ soils is almost invariably due to the failure to control groundwater flow adequately (Peck, 1969). Large erratic settlements due to the migration of fines into the excavation can also result in differential settlements.

Prentis and White (1950) describe a failure due to piping by heave. A metro excavation in New York, USA, was made in fine sand and coarse silt and close to buildings founded on short piles as shown in Figure 6.1. Seepage flow under the sheet pile wall led to the boiling at the base of the cut and caused the footings of the adjacent structures to settle by about 150*mm*.

Soil consolidation

Pumping from wells or sumps causes a reduction in porewater pressure and a corresponding increase in the effective stress of the surrounding soil. For normal dewatering operations in granular soils, the increase in effective stress is too small to cause any significant consolidation in all but the very loose sands. However, when compressible layers of soft clay, silts or organic materials are present, lowering the groundwater level or pumping from a confined aquifer may lead to large-scale settlements over a widespread area. As an extreme case, deep-well pumping from sand aquifers interbedded with clay in the Santa Clara Valley of California has been reported

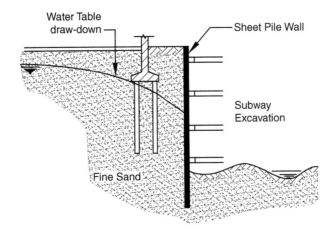

Figure 6.1 Piping by Heave at Formation (adapted from D'Appolonia, 1971)

(D'Appolonia, 1971) to have caused subsidence over a large area, with a maximum settlement of about 10*m*!

Settlements can also be caused by abstraction or drainage of water from an aquifer containing compressible soil lenses. The amount of settlement will depend upon the thickness and the compressibility of the compressible layers, the extent by which the phreatic surface is lowered and the length of the pumping period.

Figure 6.2 illustrates the effect of dewatering on the compressible clay strata for two typical soil cross-sections. The initial distribution of porewater pressure, assumed hydrostatic, is represented by the U_o-line. After the start of pumping, porewater pressure in the free-draining sands is reduced quickly. However, in the clay, owing to its low permeability, porewater pressures take some time to establish the conditions of steady-state seepage towards the sand layers. The dissipation of the excess porewater pressures towards their steady-state equilibrium condition (i.e. change from U_o to U_{ss} line) translates itself into an equal amount of increase in the effective stress (the total vertical stress remaining nearly constant) causing the clay layer to consolidate.

Many cases have been reported (Lumb, 1964; Lambe *et al.*, 1970; D'Appolonia, 1971; Morton *et al.*, 1981; Wong, 1981) of large-scale consolidation due to the settlements of clay layers causing, in some cases, damage to the nearby buildings as a result of dewatering. Two of these cases are briefly discussed hereunder.

The first case deals with the large settlement in the Mong Kok district of Hong Kong. The main seat of these settlements is the low strength, high compressibility and low permeability Marine Clay (Lumb, 1964). Although the overlying Marine Silty Sand is also of low strength ($N < 4$) and compressible, yet because of its high permeability and relatively free draining characteristics, its compression upon draw-down is almost instantaneous.

The underlying residual decomposed granite (RDG) is of high strength, fairly high permeability but low compressibility. Although its compression index is low, averaging about 0·025, it is not negligibly low, but the rate of compression is some 2–20 times greater than that of the clay. The upper more clayey zone of decomposed granite is rather more compressible than the lower, more silty zone.

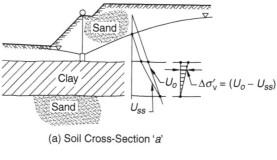

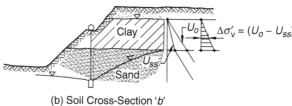

(a) Soil Cross-Section '*a*'

(b) Soil Cross-Section '*b*'

Figure 6.2 Changes in PWP and Effective Stress due to Dewatering
(adapted from D'Appolonia, 1971)

A draw-down of 6*m*, corresponding to an increase in the effective stress of about 60*kN*/*m²*, caused additional settlements. However, in the case of relatively free draining decomposed granite, these settlements were small and took place as the draw-down proceeded; whereas in the case of soft clay, for a similar increase in the effective stress, additional settlements of the order of 50–300*mm* were expected. Furthermore, because of the viscous retardation to free flow of water, consolidation of clay layer would take 3–20 years to complete, depending upon the thickness of the layer.

Estimation of the possible dewatering settlements of piled buildings can present problems. Settlement of the ground relative to the piles causes negative skin friction which can increase the load on the piles. This is particularly important for end-bearing piles when the penetration into the bearing stratum is small and, in those circumstances, it is possible for piled buildings to settle almost as much as the ground surface (Davies and Henkel, 1980).

The second case deals with a cut-and-cover subway excavation, a part of Boston's Haymarket Square to Charlestown extension of the Massachusetts Bay Transportation Authority (*MBTA*), USA. Wong (1981) has reported on the results of the instrumentation at the 'Test Section B'. Figure 6.3 shows the cross-section of the cut at the test section. The excavation is 11·3*m* wide and 17·7*m* deep. The ground was supported by sheet piles penetrating approximately 3*m* below formation; five levels of walers and struts at 3·7*m* centres horizontally and variable spacing vertically provided the lateral bracing system during construction. However, at the test section and over a stretch of over 15*m* extending northwards from its centre line, the piles hit refusal 600*mm* above formation and, as such, the intended penetration was not achieved.

The level of the ambient water table varied from approximately 2–2·6*m* below the surface. The head outside was required to be maintained at no lower than about

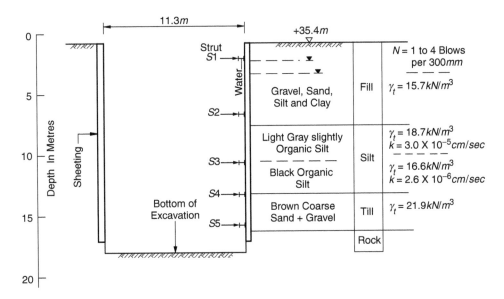

Figure 6.3 Cross-Section of Cut (adapted from Lambe, Wolfskill and Wong, 1970)

6·5*m* below the surface. The contractor carried out pumping from sumps within the excavation in order to achieve dry working conditions. However, substantial drops in the total head occurred outside the cut indicating porewater pressures far below their static values and in spite of a number of recharge wells having been installed in the till. In fact, a marked decrease in the total head was observed even as far away as over 21*m* from the cut.

At the test section, typically, the maximum lateral movements of the sheet piles were of the order of 130*mm*, although one slope indicator registered a significantly greater movement of the west sheet pile of 230*mm*. Large surface settlements of the ground outside the cut were also recorded with the maximum values reaching 130*mm* to the east and 250*mm* to the west of the cut. The settlements of the ground were caused by two factors: those due to the decrease in the porewater pressures associated with dewatering operations and those resulting from the upward movement (heave) of the formation due to the vertical stress relief, and inward movements of the faces of the cut due to the horizontal stress relief from the excavation. Most of the total ground settlements were those associated with dewatering. Substantial damage was sustained by the single-storey Empire Carpet building located 4–6*m* to the west of excavation.

6.5 Measures to limit settlements

Various measures which can be successfully employed to contain settlements due to groundwater draw-down within acceptable limits are as follows:

- Sealing
- Recharging

- Grout curtains
- Ground freezing.

Sealing

When space permits, an open excavation with stabilized side slopes is made and water level lowered in the aquifer with the aid of well points. While this approach is usually more economical, the effects of draw-down, however, can be expected to extend over large areas. On the other hand, sheet piles driven through granular soil and sealed into a relatively impermeable, underlying clay stratum can be reasonably effective in limiting the effects of draw-down and preventing an excessive depression of water level surrounding the excavation. Both these approaches were used for excavations on the *MIT* Campus, USA, (D'Appolonia, 1971) and their effect is compared in Figure 6.4.

The data for the severity and the zone of influence of the wellpoint dewatering for the open excavation and dewatering inside the sheeted perimeter were also gathered from four sites and presented in Figure 6.5. There is clear evidence to suggest that dewatering even for a shallow excavation in sand in an open cut can depress the water table over very large distances. It was also observed that while dewatering in the sand in the open cut caused settlement and cracking of many buildings on the campus, the effect of dewatering within the sheeted excavation was not significant.

Similar results are possible where the cofferdam can be toed into underlying rock provided that an effective seal is obtained and the rock is not so jointed or fractured as to allow unrestrained flow of water.

During the construction of Chater Station in Hong Kong, the dewatering settlements of the high-rise buildings to the south, i.e. Alexandra House and Princes Building, were markedly less than those recorded elsewhere on the site. This was attributed (Davies

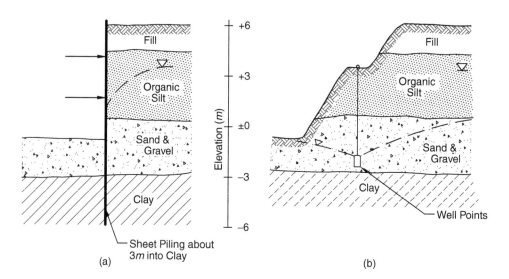

Figure 6.4 Comparison of Methods of Excavation and Dewatering (MIT campus) (adapted from D'Appolonia, 1971)

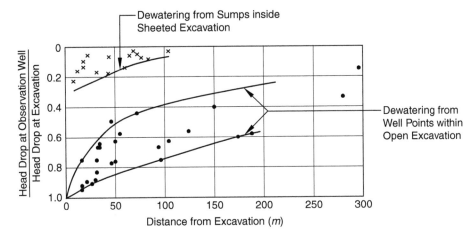

Figure 6.5 Extent of Drawdown from Dewatering (MIT Campus)
(adapted from D'Appolonia, 1971)

and Henkel, 1980) to the restricted draw-down effects as a result of obtaining a good cut-off between the diaphragm wall and the rock.

Recharging

For technical or commercial reasons, achieving an effective toe seal as explained before may not always be possible. Under such circumstances and where the extent of the anticipated settlements of the surrounding ground and structures is likely to give cause for concern, injection of water into the soil outside the perimeter wall to limit the radius of influence and to restore the porewater pressures as close to the ambient as possible, may be considered. However, recharge water should be clean and free of all suspended matter to avoid clogging the screen. It should also be recognized that recharging will prove successful only if the type of soil present is amenable to such a measure and that the porewater regime and flow characteristics of the soil are such that the recharging can overtake the draw-down outside the perimeter wall or, at any rate, keep it to within acceptable limits, as the dewatering within the perimeter proceeds.

Following the construction of the perimeter diaphragm wall at Chater Station in Hong Kong, main dewatering wells were installed at about 15*m* centres along the centre line of the station. Draw-down and settlement patterns for all the individual buildings were established on the basis of full-scale tests provided by the first stage lowering of the water level in all the wells by 9*m* for the construction of the roof slab. For the more critical buildings, i.e. Courts of Justice and the Hong Kong Club, the ground was then recharged through a number of wells installed between the buildings and the diaphragm wall. During the dewatering, the head in the wells was maintained above the ground level.

As the settlements of two points, one with and the other without recharging, are compared (Figure 6.6), effectiveness of this control measure is clearly in evidence.

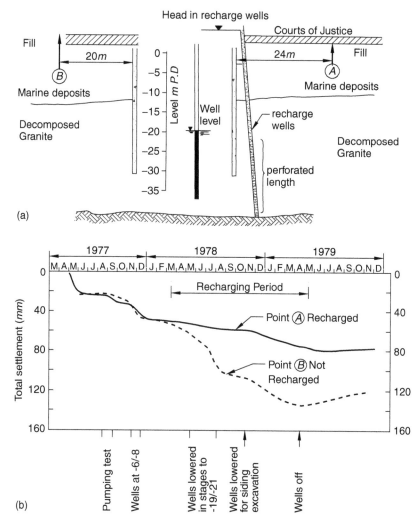

Figure 6.6 Effect of Recharge Wells in Controlling Settlement
(adapted from Davies and Henkel, 1980)

It is estimated (Davies and Henkel, 1980) that the recharge system reduced the final building settlements by about 60*mm*.

At Argyle Station, Hong Kong, also, recharge system installed before excavation virtually eliminated dewatering settlements. However, at Wong Tai Sin Station, it proved unsuccessful.

Similarly, recharging of the till even with the help of a large number of wells at Boston Subway in the USA, where the sheeting hit refusal about 3*m* above the intended toe level, also proved ineffective (Wong, 1981). This was due to the highly pervious gravel stratum in, and the erratic nature of, the till. Large amounts of water also flowed into the excavation through the joints in the rock and the ruptured joints and holes in the sheet piles.

In view of the foregoing, the assurance that recharge wells will always eliminate or reduce settlements of nearby buildings to acceptable proportions cannot be guaranteed. This lack of prior assurance of success, however, should not be used as an excuse to discourage the use of such a measure especially in situations where it could prove to be the most ideal solution. On the contrary, where appropriate, advance *in situ* tests should be carried out to confirm the potential success or otherwise of the measure if used.

Alternatively, a monitoring programme involving piezometers and settlement measuring devices should be set up to determine the effectiveness of the system used; if the measurements show that the desired results are not being achieved, appropriate alternative corrective measures can be taken.

Grout curtains

Settlements due to dewatering can also be limited by using nonstructural grout curtains as downward extensions below the toe of the structural perimeter wall. Such extensions may provide enough lengthening of the seepage path thereby reducing the groundwater flow into the excavation and enable dewatering to proceed without causing excessive draw-down and settlements outside the perimeter. Alternatively, and where appropriate, the curtains may be taken down and toed into the underlying, relatively impermeable, stratum.

Construction of hand-dug-caissons (*HDCs*) caused unacceptably large settlements of buildings adjacent to Argyle Street Station in Hong Kong. Settlements recorded varied from 8 to 113*mm*. Most of these settlements, and in some case all, were attributed to large-scale dewatering during the construction of the *HDCs* (Morton *et al.*, 1981). Grout curtain was taken down to the granite bedrock and the subsequent main box dewatering produced only small further settlements reflecting the efficiency of the grout curtain.

During the construction of Boston Subway, USA, on the other hand, pressure grouting of the till was attempted. But because of its erratic nature, it did not meet with much success and the total head continued to drop (D'Appolonia, 1971). A drop in the head of 6·4*m* (equivalent to about 64kN/m^2 of porewater pressure) was recorded at a piezometer even as far away as 21·4*m* from the wall!

To achieve success, the correct choice of grout for a given type of soil is important. In the case of the residual soils of Hong Kong, based on the available grading curves, Lumb (1964) suggests that cement grouting can be successfully employed only in the marine sand and the silty sand types of decomposed granite and that the marine clay and the clayey decomposed granite are largely unsuitable for any type of grouting.

Ground freezing

Freezing can be used in all types of saturated soils and rocks. But it can be very expensive. The condition that causes difficulty in forming an ice barrier is where there is groundwater movement. However, with the availability of significantly faster rates of freezing using liquid nitrogen instead of brine it is possible to overcome these problems.

6.6 Conclusions

To sum up, following important conclusions can be drawn:

- Decrease in boundary porewater pressures may cause consolidation of compressible clays and silts.
- Settlements caused by excavations in soils can be small provided that seepage pressures and groundwater flows are controlled adequately.
- Dewatering even for a shallow excavation in sand can depress the groundwater level over large distances and so unrestricted pumping without due regard to its implications should be avoided.
- Sheet piling driven through sand and sealed into a relatively impermeable clay stratum can be effective in limiting the extent of the draw-down and preventing an excessive depression of the water level outside the excavation.
- Dewatering of sands and silts having no cohesion are susceptible to erosion from uncontrolled seepage. Movements arising from progressive subsurface erosion may be large and extend to great distances. Piping by heave may cause large local movements. Under these conditions, control of seepage into excavation using wells or sheeting and an efficient filter for erosive springs may be required.
- Settlements can extend to large distances when interbedded granular soils are dewatered. Radial extent of draw-down can be limited by adequately sheeting the excavations or controlled by recharging.
- Where the ground is so amenable, the use of nonstructural grout curtains as downward extensions below the toe of the structural perimeter wall can extend the seepage path and thereby help contain draw-down outside the perimeter and minimize ground movements.
- The few available case histories in $c' = 0, \phi'$ to c', ϕ' soils referred to elsewhere in the book suggest that the lateral movements of the walls are likely to be small

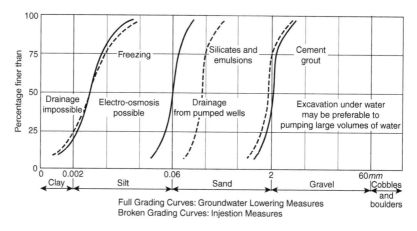

Figure 6.7 Approximate Range of Particle Sizes for Groundwater Control Measures (adapted from Glossop and Skempton, 1945)

(of the order of **25mm** or less) provided that the groundwater level is below the base of the cut or is otherwise brought under complete control.

• The methods of groundwater control will depend mainly on the conditions and the soil characteristics prevalent on site, especially the particle size distribution. Figure 6.7 shows the ranges of particle sizes which lend themselves to different groundwater control measures.

7 Interaction: I

7.1 Introduction

Excavation for the construction of a cut-and-cover structure inside an open cut with stabilized side slopes would require a large corridor of land around the perimeter of the structure; deeper the structure, wider the corridor needed. This type of construction therefore assumes that, other factors being equal, the existing structures in the vicinity, i.e. buildings, utilities, etc., fall outside the zone of influence of the construction works. In such a case, the effect of construction on the existing structures or the influence of the existing structures on the formulation of the criteria for the design and construction of the underground structure, i.e. the problems of interaction, are likely to be minimal.

However, the alignment of an underground metro route generally runs through heavily built-up, high density, urban areas and, of necessity, follows the alignment of the main thoroughfares on the surface. While such an alignment provides the most cost effective solution, the incidence of technical considerations of interaction due to the proximity of the existing structures bordering on the route are inevitable and need to be recognized and addressed as appropriate. The problems can be further exacerbated by any prevalent site specific constraints, ground and groundwater conditions and the method and sequence of construction required to be adopted.

The size and severity of the interaction problems likely to be encountered during the construction of a typical cut-and-cover metro structure depend, among other things, upon the following factors:

- Type, size, sequence and timing of the metro construction;
- Structural environment, i.e. the type of the existing utilities, foundations and age of the structures in the vicinity around the site;
- Soil environment, i.e. ground conditions, aggressivity, porewater pressure regime, drainage characteristics, threshold levels, etc.;
- Type, size and sequence of construction activity of potential future metro or non-metro works in the vicinity;
- Existing networks of roads on the surface and the nature and intensity of traffic in the area.

Problems of interaction likely to arise during and after the construction of a cut-and-cover metro structure may be placed under, or related to, the following broad categories:

- Effect of metro construction on existing structures;
- Influence of soil type on movements;
- Influence of wall stiffness and bracing on movements;
- Type of existing structures;
- Effect of existing structures on metro works;
- Potential effect of future construction on metro works;
- Any other considerations.

7.2 Effect on existing structures

During the construction of deep-underground metro stations in heavily built-up urban areas, lateral and vertical movements of the surrounding ground are inevitable. These will, in turn, have an impact on the existing structures in the vicinity. However, it is important to recognize that, in spite of following all the precautions and taking all the preventive measures, it may not be possible to eliminate the movements altogether; but then, total elimination may not always be necessary.

In his state-of-the-art report on deep excavations and tunnelling, Peck (1969) presented a comprehensive survey of ground movements around deep excavations constructed using conventional ground support methods. However, for well-supported excavations, Peck's movement envelopes have proved to be generally conservative. Peck's work was later updated by O'Rourke (1981) and Clough and O'Rourke (1990). In 1979, Burland *et al.* summarized the results of over ten years' research into the behaviour of ground around deep excavations in heavily overconsolidated London Clay. Karlsrud and Myrvoll (1976) presented excellent case histories of ground movements in excavations in soft clays in the Oslo area, whereas Wong (1987) summarized the results of observations around excavations in the soft clays of Singapore.

Most of the observations lead to the conclusion that the ground movements depend, primarily, upon the type of soil and the method of construction and broadly suggest that, as a ratio of the retained height, settlements of the surrounding ground are:

- Rarely in excess of 0·15 per cent in stiff clays;
- Generally around 0·5 per cent in loose sands or gravels;
- Of the order of 2–3 per cent, or even more, in soft clays.

Such order of movements can, in the event of serious damage to existing structures and utilities in the vicinity, potentially lead to substantial costs on account of disruption, delays and expensive remedial measures which might become necessary. Naturally, such movements and their effect on the existing structures can be of great concern to the engineer. It is therefore important to identify the potential causes of, and to quantify, these movements so that adequate allowances can be introduced into the design and appropriate measures incorporated into the construction procedures. Such measures should aim at avoiding, where possible, any distress to the surrounding structures or, at least, restricting the damage to acceptably low and manageable proportions.

Depending upon the size, type, age and the condition of the existing structures within the zone of influence of the metro works and with due deference to the costs and risks involved, it may sometimes be advisable to underpin the structure to avoid damage altogether; at other times, it might prove cost effective to simply let the damage occur and subsequently repair it. However, in the case of structures of religious, historic or national importance, damage may have to be prevented at all costs. In case that cannot be guaranteed, the design and construction of the works and, failing that, even the alignment of the metro system itself may have to be modified.

In order to make a rational evaluation of the available alternatives, the engineer must not only have a fair idea of the nature and the magnitude of the movements likely to be expected during the various stages of construction works, he must also be able to lay down performance criteria related to the degree of acceptability of movements for the adjacent structures. Besides, he must have a clear appreciation of the steps that can be taken at a short notice in order to prevent unacceptable movements and limit damage.

However, owing to their inter-relationship with factors including changing drainage boundaries and boundary conditions during the stages of construction, accurate prediction of ground movements based on theoretical analyses is extremely difficult. The problem is further complicated by the fact that the influence of factors such as the quality of workmanship, the sequence of construction and the relevance of conventional laboratory based soil testing in defining *in situ* soil parameters is difficult to assess and quantify accurately. Of necessity, therefore, predictions of movements due to various operations are based almost entirely on observational data and previous experience. Because of the diversity of factors affecting movements, it is also difficult to formulate generalized empirical rules. Thus, a basic understanding of the phenomenon of ground movement and the fundamental factors controlling it are essential. Observational data evaluated in terms of these fundamentals can help form a basis for a sound comprehension of the potential problems likely to be faced and also in devising appropriate solutions.

In geological conditions where no movement data based on previous experience are available and where, owing to the proximity and the sensitivity of the adjacent structures, control of ground movements is important, safe construction methods with positive and stringent controls similar to those successfully tried and tested elsewhere, must be employed. But even so, it must be recognized that it may not be possible to eliminate the movements completely.

Ground movements, albeit to different degrees, can be expected during each of the following stages of operations:

- Construction of Perimeter Wall or Cofferdam
- Groundwater Lowering within Cofferdam
- Main Excavation and Construction.

Construction of perimeter cofferdam

Construction of the perimeter wall or cofferdam invariably represents the first stage in the construction of a cut-and-cover metro structure. It defines the extent of immediate construction activity and attempts to limit its zone of influence. If the cofferdam is constructed in the form of a sheet pile wall (or similar), ground movements associated with its installation do not present a problem. However, the effect of vibrations on the existing structures in the vicinity during driving operations must not be overlooked.

If the perimeter wall is also intended to form an integral part of the permanent structure and a tighter control of ground movements is required, stiffer, stronger and deeper wall sections, such as contiguous, secant bored pile or diaphragm walls, may be required. The diameter of the bored piles and the thickness of the diaphragm walls commonly range from *500mm* to *1500mm*. The diaphragm walls are constructed with the use of slurry trench technique.

During the excavation for a diaphragm wall, the trench is supported by bentonite slurry. The unit weight of the slurry is maintained at a level at least marginally greater than that of groundwater. The resultant slurry pressure differential, i.e. the difference between the slurry pressure inside and the static water pressure outside the trench, is relied upon to support the lateral pressures due to the earth and the surcharge loads, if any, from the adjacent structures. As the trench is excavated, it is simultaneously filled with slurry to stabilize the sides. However, since the state of stress equilibrium existing hitherto is disturbed as a result of the excavation activity, ground relaxation in the form of some inward movement of the sides of the trench is inevitable. This can, potentially, lead to the movement of the existing structures in the vicinity. Such movements are a function of the *in situ* stresses in the soil and the level of the ambient water table; higher the lateral stress, potentially higher also is the likely movement. Also, where the water table is high, the slurry head differential against the water table can be a critical factor for stability and containment of ground movements. The stability of slurry trenches and the various means by which it can be enhanced are dealt with in Chapters 16 and 17.

Lateral movements of up to *90mm* during slurry trench wall construction have been observed in various parts of the world. During the construction of Boston Subway Extension, an inclinometer installed *900mm* behind the wall indicated lateral movements of *20–25mm* during the slurry trench excavation and prior to the placement of the tremie concrete (D'Appolonia, 1971). Considerably higher lateral movements and settlements have accompanied the construction of diaphragm walls in the residual soils (decomposed granite) of Hong Kong.

Ground movements during the construction of diaphragm walls for the Chater Station, Hong Kong, have been well documented (Davies and Henkel, 1980). The site is reclaimed land, the ground conditions poor with loose reclamation fill and marine deposit underlain by a mantle of residual soil (essentially silty sand) derived from the decomposition of granite. The water table is also high.

Figure 7.1 shows the location of Chater Station in relation to the adjacent buildings. The station structure was constructed in the top-down sequence with the perimeter diaphragm wall forming a permanent integral part of the structure. The diaphragm wall panels adjacent to the existing foundations were up to *37m* deep and constructed in lengths of *2·7* and *6·1m*. Horizontal movements of *27–60mm* during test panel excavation, and settlements of various buildings, ranging from *21–78mm*, during the diaphragm wall construction as shown in Figure 7.2 have been recorded. As a result, noticeable cracking was observed in the Courts of Justice building.

The effect of increasing the effective slurry pressure in reducing ground movements is clearly in evidence. However, the way in which these large settlements progressed is

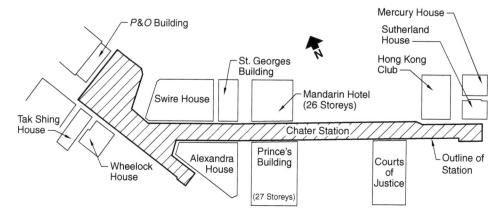

Figure 7.1 Location Plan – Chater Station and Adjacent Buildings
(adapted from Davies and Henkel, 1980)

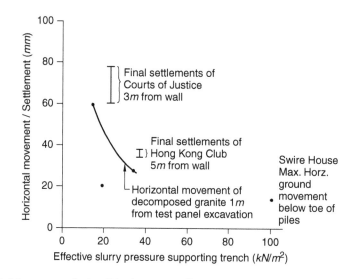

Figure 7.2 Movements during Diaphragm Wall Construction, Chater Station, Hong Kong
(adapted from Davies and Henkel, 1980)

an interesting feature. Most of the settlements were realized during the construction of a series of adjacent panels rather than that of a single panel. This can be seen by the steep gradients of the settlements which showed a marked similarity to the local horizontal movements in the decomposed granite suggesting that the final settlements were controlled by these horizontal movements. Excavation of an individual panel and the subsequent dissipation of the negative porewater pressures and the swelling of the decomposed granite created soft compressible zones around the panel.

Construction of the adjacent panels caused the arching around the compressible zone to break down resulting in recompression as the earth pressures built up.

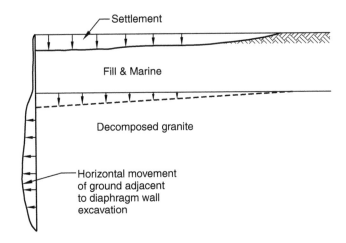

Figure 7.3 Ground Movements during Diaphragm Wall Construction
(adapted from Davies and Henkel, 1980)

Horizontal ground movements extended as far back as 50*m* from the wall (Figure 7.3) and the associated large settlements were attributed to this cause.

Swire House to the west is a 22-storey high building founded on small individual pile caps under each column with piles driven just to the top of the decomposed granite. It was considered doubtful that the combined superstructure and the foundation were stiff enough to safely redistribute the load during the excavation of the wall panels. To avoid having to underpin the entire building which would have been impractical, following measures were taken during the construction of the wall:

• Panels were restricted to 2·7*m* lengths.
• Slurry head was artificially raised above the ground level.
• Water level in the decomposed granite was lowered by 6*m* using well points.
• Unit weight of the slurry was increased, achieving an effective slurry pressure of about 100*kN/m²*.

The maximum horizontal movement recorded just below the toe of the piles was about 14*mm*; the maximum recorded settlement of the building was about 30*mm*, half of which was attributed to well point dewatering.

Groundwater lowering within cofferdam

This has already been covered separately under Chapter 6.

Main excavation and construction

Excavations for metro structures generally involve long, deep vertical cuts. After the dewatering has been carried out long enough to give it a head start of some 2–3*m* depth, excavation within the area enclosed by the perimeter walls is commenced. This enables the excavation and construction works to be carried out in a dry environment. As the excavation advances downwards, a system of supporting the perimeter wall panels is introduced in order to prevent the large-scale movement of the sides of the excavation. This may take the form of a temporary system of struts or ground anchors which can be later removed or destressed as the construction of the main structure progresses. Construction of the main structure may then be carried out by the conventional bottom-up method. Alternatively, the structure may be constructed in the top-down sequence and the permanent horizontal members, such as floor slabs and beams, may be relied upon to provide the wall support system during the stages of excavation as well.

Proven semigraphical design procedures (Terzaghi and Peck, 1968) are available to deal with the various requirements, generally. However, due to the proximity of high-rise structures and services generally associated with a heavily built-up urban environment especially those which are sensitive to ground movements, design requirements tend to be much more demanding in order to ensure that damage is avoided and the movements are kept to within acceptable limits.

Excavation removes a mass of soil and water thereby reducing the total stresses along the boundaries of the cut. This relief in total stress causes the sides to move inwards and the floor to rise upwards. The upward movement or heave of the floor of the cut is, in turn, accompanied by an inward movement of the soil below the excavation level as well. As these movements take place, the soil surrounding the excavation undergoes lateral displacement and vertical settlement. The settlement of the surrounding ground and structures is related to the inward movement, both above and below the formation level, along the sides of the cut.

Cracks in the Courts of Justice building first observed following the settlements due to the construction of the perimeter diaphragm wall panels for the Chater station as mentioned earlier became more serious (Davies and Henkel, 1980), subsequently, during the course of main excavation for the construction of the station structure. In fact, this led to the closure of the building during the later stages of the station construction.

Increase in the vertical effective stress upon dewatering can cause settlement and its decrease during the subsequent excavation, heave. It may be possible to so plan the two operations as to compensate the effects of each other so that the resultant vertical movement is not significant (Serota and Jennings, 1959). However, the magnitude and distribution of movement accompanying deep metro excavations are influenced by a large number of interrelated factors. These are:

- Type of soil
- Stiffness of wall
- Size of the excavation
- Method of groundwater control
- Stiffness of the wall support system
- Construction details and workmanship
- Time over which the excavation remains open
- Method and sequence of installing wall support system.

As the excavation reaches a horizontal support level, such as a floor slab, and before that support is installed and assumes its function, lateral support to the walls is provided by the soil below the formation. In mobilizing this support, lateral strain of the soil and the inward movement of the walls extending some distance below the formation level are to be expected. In order to appreciate the relative influence of the various factors that control overall behaviour, it is important to differentiate between lateral movements occurring above and below the formation level.

Once excavation has progressed below a given level, further inward movement above that level is controlled entirely by the bracing details (D'Appolonia, 1971). Unless careful attention is paid to these details, the lateral movement above the formation level may be a large proportion of the total lateral movement.

Magnitude of the movements taking place below the formation level is a function of three main factors:

- Type of soil below the cut;
- Stiffness of the wall and its penetration;
- Soil resistance to inward movement of embedded portion of wall.

In theory, if the wall penetrates deep enough below the bottom of the cut and is stiff enough to withstand the stress-resultants due to lateral ground pressures, then, regardless of the type of soil, lateral movements of the soil outside the excavation are unlikely to be significant. On the other hand, deformations of walls which cannot develop end fixity can be considerably (even 100 per cent) higher than those with the ends well set into, say, an underlying dense sand layer. Furthermore, practical difficulties and cost may preclude the use of thick wall sections prior to general excavation.

The properties of the soil are likely to have significant influence on ground movements. For instance, if the soil immediately below the base of the cut is required to undergo large strains to mobilize passive resistance, or if it cannot develop sufficient passive resistance, the inward movement of the soil towards the excavation and the consequent ground settlements are likely to be large.

7.3 Influence of soil type

In the case of deep excavations, particularly where relatively flexible system of supporting the sides, such as sheet piles or soldier piles with lagging, is used, the characteristics of the soil can have predominant influence on the movement behaviour. In evaluating the lateral movements and settlements associated with such excavations, Peck (1969) and D'Appolonia, (1971) classified observed data in relation to the following principal soil types: c', ϕ' soils, soft to medium clays, stiff clays and silts. To broaden the range, the data related to the residual soils such as those of Hong Kong have also been added.

Figure 7.4 summarizes the maximum movements as percentages of the excavation depths in the various soils using standard soldier piles or sheet piles braced with horizontal or raking struts or ground anchors.

c', ϕ' and $c' = 0, \phi'$ soils

The few case histories available in the range of c', ϕ' to $c' = 0, \phi'$ soils suggest that the lateral movements of the walls are likely to be small (of the order of $25mm$) provided that the groundwater level is below the base of the excavation or is otherwise brought

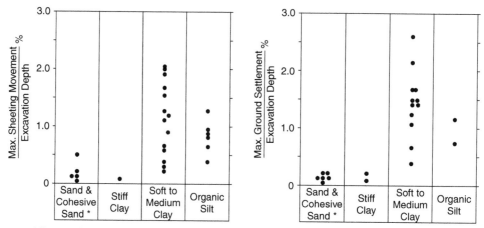

Figure 7.4 Movements against Flexible Ground Support (adapted from D'Appolonia, 1971)

under complete control. There is also evidence to suggest that the movements will be small regardless of the depth of excavation if adequate load is jacked into the struts or the ground anchors are stressed adequately. According to Terzaghi and Peck (1968), maximum settlements in loose sands and gravels may attain the order of 0·5 per cent of the excavation depth.

Soft to medium clays

Lateral movements associated with the plastic clays of very soft to medium consistency substantially exceed those in c', ϕ' or $c' = 0, \phi'$ soils (Peck, 1969). Maximum inward movements of the sheeting and maximum surface settlements of the order of 1–2 per cent of the excavation depth have been commonly observed. Both the magnitude of settlements and their distribution with distance from the edge of the cut are of practical importance. By plotting the settlements and distances from observed data of a number of cases, Peck (1969) defined three zones as shown in Figure 7.5. The plot enables rough estimates of settlements that might be expected under various conditions to be made. These reflect both the initial undrained settlements that accompany excavation and the consolidation settlements likely to occur within the construction period.

For soft clays, significant settlements (as much as 0·2 per cent of excavation depth) may be encountered even as far back from the excavation as 3–4 times its depth.

For excavations in clay, lateral wall movements are also associated with settlements of about equal volume. Measurements taken on well-designed and constructed sheeted excavations show that 60–80 per cent of the total lateral wall deflections take place below excavation level (D'Appolonia, 1971). These movements depend upon the passive resistance to inward movement that can be mobilized by the soil immediately below the base of the cut. When the excavation reaches a depth approaching bottom failure of the cut, both the lateral movement and the ground settlement increase sharply.

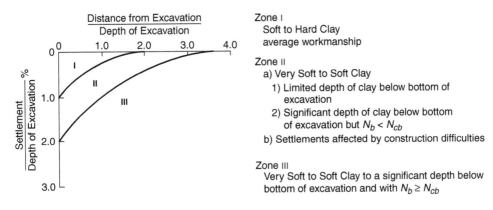

Figure 7.5 Settlements adjacent to Open Cuts in Clay (adapted from Peck, 1969)

Buildings supported on piles, even if extending down to stiffer soils but overlain by soft soils, are not necessarily immune from settlements as a result of the nearby excavation (Peck, 1969). The overlying soft clays, in their characteristic tendency to settle near the cut, can impose down-drag loads on the piles from negative skin friction thereby forcing an increase in their penetration.

Stiff clays

Movement in clay soils is a function of their stiffness. As the stiffness increases, the movement resulting from excavation decreases rapidly. However, only a small number of cases have been reported for stiff clays with undrained shear strength in excess of $95 kN/m^2$. For such clays, maximum movements are only a small fraction of one per cent of the excavation depth. Several large excavations over depths ranging from $12-18m$ have been made in Houston, Texas. The clays are stiff, being overconsolidated by desiccation, and have an undrained shear strength ranging between $48 kN/m^2$ and $285 kN/m^2$ with an average value of about $145 kN/m^2$. Movements of the adjacent ground surface were generally within the tolerance of ordinary engineering surveys. Mansur and Alizadeh (1970) have reported maximum ground settlements of $12 \cdot 5mm$ or less for a $12 \cdot 5m$ deep excavation using soldier piles and lagging in stiff clay with the shear strength in the range of $120-192 kN/m^2$. This represents a movement of $0 \cdot 1$ per cent or less of the excavation depth.

Compressible silts

Measurements of movements made at several test sections in silts along a subway excavation have been well documented by Lambe *et al.* (1970) and Lambe (1972). Typically, the sides of the $16 \cdot 7m$ deep excavation were supported by steel sheeting and cross bracing. The lateral movements recorded ranged from $0 \cdot 44$ per cent to $1 \cdot 34$ per cent of the depth of excavation. This confirms that the movements associated with excavations in compressible organic silts can be nearly as large as those in soft clays.

Residual soils

Residual soils are the product, *in situ*, of the disintegration and mechanical alteration of the lithological components of the parent rock as a result of weathering (Zeevaert, 1973). They are highly variable in composition ranging from large fragments to gravel, sand, silt, clay and colloids. Accordingly, the unit weight and cementation may also be variable. Organic matter may also be present. The deformability of residual soils can be high and, in some cases, very high with low shear strength. In the tropical and subtropical areas, weathering may extend, generally, several metres into the parent rock depending upon the climate and the physiographical environment of the region. In humid regions, deep soil profiles are encountered with medium to high deformability and low shear strength.

In the residual decomposed granites of Hong Kong, lateral wall movements ranging from 0·1 to 0·23 per cent of the depth of excavation have been recorded for various metro excavations. It is widely accepted that, in spite of the adverse soil, groundwater and structural environment, it has been possible to restrict the lateral wall movements to such modest limits, generally, by the use of stiff walls and adopting top-down sequence of construction. Movements have also been observed to have extended at least as far away horizontally as the depth of excavation.

The influence of wall stiffness and bracing on movements and other categories of interaction are dealt with in the following chapters.

8 Interaction: II

8.1 Introduction

This chapter deals with the influence of wall stiffness and bracing on movements and the parametric studies of wall movements carried out by some investigators.

In a multibraced wall, the effects of the wall and prop stiffnesses on movements and bending moments depend very much on the sequence of excavation and the sequence and timing of installing the bracing. Since it may neither be possible to define such parameters precisely nor to model the boundary conditions accurately, theoretical analyses and parametric studies often provide a valuable means of identifying, rationalizing and understanding the performance of an excavation at its various stages. This is dealt with later on in the chapter. However, the information obtained from instrumentation and back analyses of well-documented case histories can be invaluable in enabling predictions of ground movements to be made with a certain degree of confidence.

The walls supporting the sides of an excavation may be relatively flexible such as sheet pile walls or soldier piles with lagging; or, they may be stiff such as cast *in situ* concrete walls. Concrete walls can be up to two orders of magnitude stiffer and may take the form of contiguous bored pile, secant pile or diaphragm walls constructed by the slurry trench technique. In what follows, an attempt is made to highlight, all other aspects being equal, the influence of the inherent stiffness (or flexibility) of each of the ground support systems on the movements.

8.2 Excavations against flexible walls

For excavations in clays supported by sheet pile walls or soldier piles with lagging, the factors which have an important influence on the ground movements can be placed (D'Appolonia, 1971) under two categories – those affecting sheeting movements primarily above the excavation level and those affecting below it.

Lateral wall movement above the excavation level is a function of the soil and the wall support system and can be determined by procedures and details primarily associated with the installations of the bracing. Some of the more important factors controlling these movements are:

- *Horizontal and vertical spacing of braces.* Greater the spacing, greater will also be the load and the axial deformation of the braces. This will, in turn, allow greater movement of the walls and the surrounding ground.

- *Extent and depth of excavation along the length of wall opened up at any one level prior to the installation of the brace or support at that level.* If the excavation were to be advanced, initially, in discrete trenches just wide and deep enough to enable the installation of the brace to take place, relatively less lateral movements could be expected than if it were to be carried out *en masse* down to the underside of the brace level.
- *Time lag between excavation and the subsequent brace installation.* If longer time is allowed to elapse after excavation and before the installation of the brace, greater relaxation and relatively larger movements of the ground can be expected.
- *Details of prestressing of anchors, jacking-in of loads and wedging of struts.* What percentage of the anticipated theoretical earth pressure loads will need to be jacked in depends as much on the actual loads likely to be generated as the extent of the lateral movements that can be permitted.
- *Diurnal temperature effects on struts.* Where such effects are likely to be significant, they should either be allowed for in the design of the struts or monitored and accommodated in the adjustments of the loads in the struts.
- *Details of excavating and placing laggings between soldier piles.* The extent and manner of excavation, the size and stiffness of the lagging and the time and manner of its installation will influence the extent of the movement to be expected.

Lateral movement occurring below excavation level is controlled by the stiffness of the wall and the combined resistance of its embedment and the stiffness of the soil below the base of the cut. Experience has shown that soldier piles spaced even as close as 2–3*m* centres offer little resistance to overall movement below the excavation and that even the heaviest sections of steel sheeting are not usually stiff enough to inhibit lateral movement significantly.

If horizontal ground movements are monitored at frequent intervals during excavation, it is possible to identify, with reasonable degree of accuracy, the movements occurring above and below the excavation level as the depth of cut is advanced. Figure 8.1 shows sheeting deflections measured at various stages of excavation in medium clay in Boston, USA (D'Appolonia, 1971).

The broken lines represent sheeting deflections at various stages of excavation; the solid line is the final deflection at the end of excavation. The shaded area represents the amount of deflection that took place below the excavation level whereas the remainder of the area under the solid curve represents the deflection that took place above the excavation level. For example, just at 12*m* depth, 50*mm* inward deflection took place before the excavation level reached this depth while about 75*mm* of additional deflection ensued when the excavation was taken below this depth. The maximum excavation depth was about 14·33*m*, so all the deflection below this depth occurred below the excavation. In this case, greater proportion of the total wall movements occurred above the excavation level since not enough attention was given to the excavation and the bracing details. Where it was possible to install stages of braces at 2–3·5*m* centres vertically, probably representing excellent to good construction practice, it was possible to restrict the deflections above the excavation level to within 20–40 per cent of the total deflections.

These points are further emphasized in Figure 8.2. Sheeting movements at two nearby sections of the subway excavation in compressible organic silt in Boston, USA, are compared (D'Appolonia, 1971). Excavation at Section *B* was carried down deeper

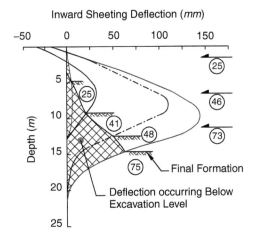

Figure 8.1 Sheeting Deflection Above and Below Formation in Medium Clay
(adapted from D'Appolonia, 1971)

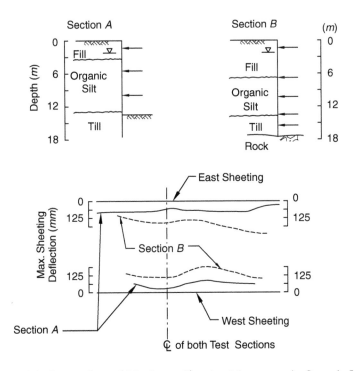

Figure 8.2 Comparison of Maximum Sheeting Movements in Organic Silt
(adapted from D'Appolonia, 1971)

than that at Section *A*, but the thickness of the compressible silt was greater at Section *A*. Furthermore, the excavation at Section *B* preceded that of Section *A* and, as a result, it became possible to incorporate several improvements in the excavation and strutting procedures employed at Section *A*. These improvements involved:

- Greater control over the depth and the length of excavation prior to installing struts at each level
- Greater control over time lag between excavation and installation of struts
- Improved prestressing and wedging procedures.

As a result of these refinements, the maximum sheeting deflections at Section *A* were reduced by about a factor of 2 as compared with those at Section *B*.

A qualitative assessment of the sheeting deflections and the associated settlements in the surrounding ground at any stage of the excavation can also be made by analysing the bottom stability of the open excavation in much the same way as conventional foundation stability. The extent to which a state of passive failure below the base of a cut is approached and, by implication, the movement of walls and the settlement of ground, can be judged by the value of the nondimensional stability number $N_b = \gamma \cdot H / s_{u(b)}$, where γ is the unit weight of soil, H the depth of cut and $s_{u(b)}$ is the undrained shear strength of the soil below the base of the cut. Bjerrum and Eide (1956) presented approximate theoretical solutions in terms of nondimensional bearing capacity numbers, N_{cb}. These are shown in Figure 8.3.

The definitions of normally consolidated and overconsolidated clays are given in Appendix C. For normally consolidated clay, first yield occurs when N_{cb}/N_b is about 1·5–2·0. However, dissipation of excess negative porewater pressure increases the tendency for yield below the base of the cut. In overconsolidated clays, first local yield occurs at higher values of N_{cb}/N_b.

Figure 8.4 shows the maximum deflections and settlements as percentages of excavation depth against N_{cb}/N_b ratios for a number of field cases from Oslo, Chicago, Mexico City and Boston. The undrained shear strength used in computing N_b was determined either by *in situ* tests or by unconfined compression tests and varied between 14 and $57kN/m^2$. Although the results showed a considerable scatter, there is no doubt that the settlements and sheeting deflections increased as the instability of the base of the cut was approached.

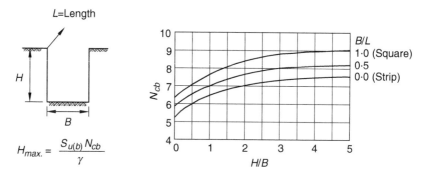

Figure 8.3 Stability Numbers (adapted from Bjerrum and Eide, 1956)

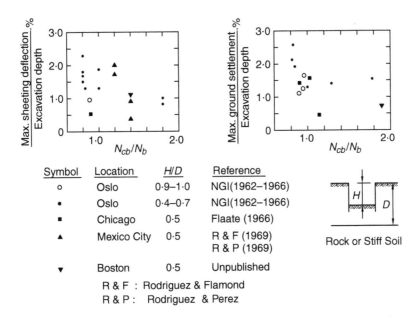

Symbol	Location	H/D	Reference
o	Oslo	0·9–1·0	NGI(1962–1966)
•	Oslo	0·4–0·7	NGI(1962–1966)
■	Chicago	0·5	Flaate (1966)
▲	Mexico City	0·5	R & F (1969) R & P (1969)
▼	Boston	0·5	Unpublished

R & F : Rodriguez & Flamond
R & P : Rodriguez & Perez

Figure 8.4 Maximum Movements related to Bottom Stability Number (Excavations in Soft and Medium Clay) (adapted from D'Appolonia, 1971)

- *Boston subway extension, MBTA: a case history*

Lambe *et al.* (1970) documented and interpreted the measured performance of an instrumented stretch of the subway excavation identified as 'Test Section *B*' of the Massachusetts Bay Transportation Authority (*MBTA*), USA. Typical cross-section of the excavation together with the soil profile, which varies significantly in the horizontal direction, is shown in Figure 8.5. The fill layer at the top is a loose mixture of gravel, sand, silt and clay. This overlies a layer of compressible organic silt containing shell fragments. Below this is an erratic till layer comprising stratified layers of sandy silt or clay and dense sand and gravel. This is underlain by the bedrock which is hard, dark, jointed volcanic intrusion.

The ambient hydrostatic head prior to the start of construction was approximately at 33·5m. Because of the existence of the compressible silt and the presence of important structures in the vicinity, the contractor was required to maintain the porewater pressure in the till at an equivalent total head of 29m. However, this could not be achieved.

The sheeting hit refusal roughly 1m above the floor of the excavation. Splitting of the sheet piles and the inability to drive them through the till allowed horizontal flow of water into the excavation through the pervious gravel stratum. In order to make up for the loss of the porewater pressure head outside the excavation, a large number of recharge wells were installed into the till. These proved largely ineffective because of the highly pervious gravel stratum in the till. Ruptured joints in the sheet piles and the joints in the rock also permitted large amounts of water to flow into the excavation. Furthermore, the nature of the till was so erratic that pressure grouting also proved

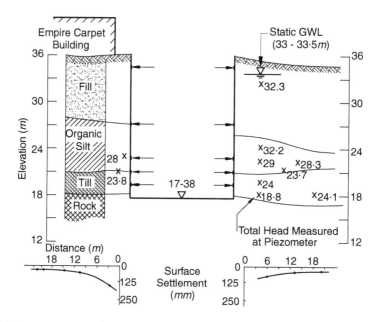

Figure 8.5 Cross-Section Subway Excavation, Boston (Changes in Total Head and Surface Settlements) (adapted from Lambe *et al.*, 1974)

ineffective and the total head continued to drop over a considerable distance beyond the wall.

The ground surface near the excavation experienced large settlements resulting in damage to the Empire Carpet Building. The maximum surface settlement of *175mm* occurred at the edge of excavation, but settlements as large as *75mm* were measured even as far back as *21m* from the excavation. About two-thirds of the settlement was attributed to consolidation due to the decrease in the porewater pressures (Lambe *et al.*, 1970). The soil also moved laterally away as much as *50mm* from the east face of the foundation wall of the building. The two walls and the bracing system moved towards the east at the top. Elsewhere the two walls moved inwards, the maximum inward movement being *225mm*, representing 1·3 per cent of the full depth of excavation. As the excavation advanced, the level of the maximum horizontal movement of the sheeting coincided with the excavation level indicating significant horizontal movements below the bottom of the excavation.

8.3 Excavations against stiff walls

Concrete walls constructed *in situ*, often using the slurry trench technique, are easily 1–2 orders of magnitude stiffer in flexure than steel sheeting. For this type of wall, the inherent rigidity of the wall and its penetration have an important influence on limiting movements below the excavation level. If it is possible for the wall to be toed into rock or a hard stratum, the situation can be improved even further. The manner in which the excavation is advanced and the manner and the timing of the installation of the stages of struts, on the other hand, have a marked influence on the extent of movement above the excavation. With their increased stiffness combined with considerable advances and innovations in the types of strutting techniques available, slurry trench walls, diaphragm walls in particular, are being used more and more in supporting deep metro excavations especially where containment of movements is paramount.

Ground movements are generally critical (Xanthakos, 1979) in soft or very loose soils (soft clays and loose sands), and seldom in stiff or dense ground (stiff clays and dense sands). However, with the use of relatively stiff diaphragm walls combined with a judicious adaptation of suitable strutting measures, it is possible to reduce ground movements significantly even in soft ground, even by an order of magnitude as compared with that of steel sheet piling. Excavation carried out *en masse* down to a particular stage, for instance, is likely to give rise to far greater movements than if it were carried out in discrete trenches wide and deep enough to allow installation of the corresponding stage of strutting to take place.

Excavation for an underground metro structure within the confines of an integral perimeter wall or a cofferdam can cause relaxation of the ground support structure as well as that of the ground outside the perimeter in sympathy. This results in the lateral movement of the perimeter walls inwards towards the excavation. Depending upon the type of soil encountered, ground movements can be analysed for supporting the walls in the following ways (Xanthakos, 1979).

- Bracing ahead of bulk excavation
- Excavation and bracing in tandem, using:

  - Temporary Struts
  - Permanent Floors
  - Tie-Backs or Anchors.

Bracing ahead of bulk excavation

It has been observed that the wall movements are generally critical at the top and the bottom of an excavation. Although for relatively deeper excavations, extra stages of intermediate strutting may be called for and installed as the excavation progresses, there is little doubt that the introduction of permanent struts at the top and bottom critical levels before the start of excavation can reduce wall movements significantly (Xanthakos, 1979).

However, wall movements can be virtually eliminated if an appropriate strutting system to brace the walls can be installed, preloaded and made operational before the main excavation starts. While it may be possible to install the struts before the commencement of bulk excavation, preloading these at depth at that stage may not always be practicable. As the natural soil support to the walls is gradually removed with the advancing excavation, the role of supporting the walls is progressively taken over by the bracing system. The extent of wall displacement is then governed by a combination of the extent of axial deformation of the bracing system as it takes up the load and its downward deflection under self weight. By the introduction of stiffer struts and the insertion of temporary intermediate vertical supports under them at more frequent intervals, it is possible to control the ground movements even further. In this way, the ground movements can, to a large extent, be rendered independent of the soil environment and become, essentially, a function of the stiffness of the bracing system as a whole.

It should be recognized that the lateral movements, particularly at formation level, can also be caused by other factors. For example, stress relief upon removal of a large mass of soil which can cause heave of the floor of excavation especially in the soft plastic clays, besides giving rise to the settlement of the ground outside the excavation, can also subscribe to the lateral inward displacement of the walls. Besides, any drawing down of the water table outside the perimeter can, likewise, cause consolidation of the surrounding ground and lead to further lateral inward movement of the walls. The bracing system will not, *per se*, control such ground movements at the bottom of the excavation unless the lower part of the wall is specially designed to do so. Desired control of such movements can be achieved by increasing one or more of the following:

- Stiffness of the wall
- Penetration depth of the wall
- Stiffness of the bottom stage brace, i.e. increased stiffness and reduced spacing of the lowermost level of struts.

Confirmation of the successful application of some of these measures can be seen in the following three case histories.

- *Cross-wall brace below formation, Oslo subway*

In a section of the subway extension in Oslo (Eide *et al.*, 1972) as shown in Figure 8.6, transverse struts in the shape of cross-walls below the formation level were installed to brace the main diaphragm walls. These were installed from the ground surface by the slurry-trench technique. The final excavation was 16*m* deep.

The soft Oslo clays are particularly susceptible to heave especially where excavations are in excess of 8–9*m* in depth. Against this background, the size and particularly the spacing of the cross-walls were designed to ensure the stability against base failure by heave. The initial excavation was carried out deep enough only to enable construction of the roof slab to take place. With the top and bottom braces, i.e. the roof slab and the cross-walls, thus in place, the main excavation under the roof was carried out. The intermediate floor slab was constructed following the top-down sequence and was relied upon to brace the walls at mid-height. The absence of an appropriate bearing area at the base of the walls coupled with very little friction at the soil–wall interface limited the vertical load-carrying capacity of the walls significantly. In preference to extending and keying the walls into the bedrock which would have been a costlier option, the entire structure was supported on steel piles, forming stilts and driven down to rock through special casings left in the walls at discrete centres.

The lateral movement of the walls at mid-height was only about 4*mm*! This rather small movement was attributed to the elastic deformation of the intermediate floor slab acting as the lateral strut between the walls.

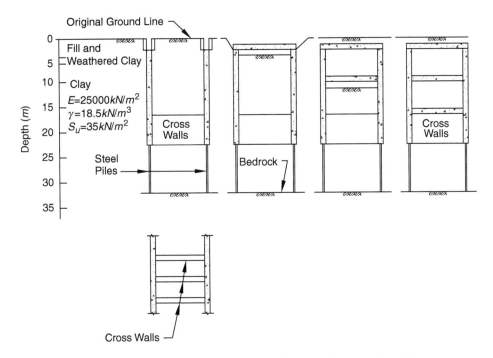

Figure 8.6 Bottom Cross-Wall Brace in Subway Construction, Oslo
(adapted from Eide *et al.*, 1972)

- *Embedment into rock, Oslo office block*

Where the construction of cross-wall brace below the formation is not economical and where bedrock exists at a shallow depth below the excavation, it may be economically viable to toe-in the main walls into the underlying bedrock in order to restrict the wall movements. Figure 8.7, shows the construction of a deep basement for an office block in Oslo (Dibiago and Roti, 1972). Although not strictly a subway structure, its construction does, nevertheless, illustrate the benefits of the principle of wall embedment into rock in containing the lateral movements.

The soil conditions (figure *a*), typical of central Oslo, consist of a thick deposit of normally consolidated marine clay of low sensitivity (defined in Appendix C) of about 5 and an undrained shear strength varying from 25 to $40kN/m^2$, overlying bedrock at 10–21*m* depth. The lower crust of clay layer is occasionally sandy and contains some gravel.

The perimeter wall comprised 1*m* thick diaphragm walls. The panels were constructed by the slurry-trench technique and were taken down 20*m* deep and keyed into the bedrock. The unit weight of $12kN/m^3$ for the bentonite slurry and an average unit weight of $23.5kN/m^3$ for concrete were used. The three permanent concrete floor slabs, each 400*mm* thick and constructed in the top-down sequence, braced the walls during excavation. During the excavation, a small drop in the porewater pressure outside the perimeter equivalent to a head of 1·5*m* was recorded. However, when

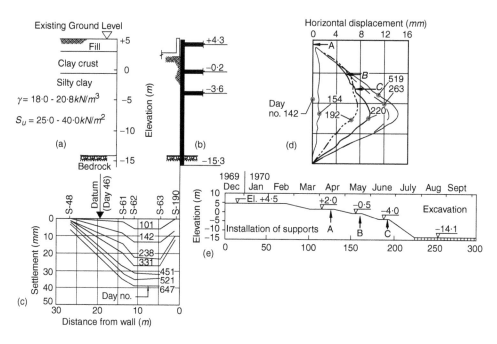

Figure 8.7 Deep Basement Construction against Diaphragm Wall, Oslo
(adapted from Dibiagio and Roti, 1972)

excavation reached the bedrock, flow through the rock caused a marked drop in the porewater pressures.

The last set of settlement observations before the commencement of excavation for the basement was taken on day 46. The ground settlements plotted in figure (*c*) are relative to those of this day and therefore represent settlements due to excavation. By day 238, when excavation reached the bedrock level, maximum settlement of 23*mm*, representing 0·12 per cent of the excavation depth, had occurred. The rate of settlement was relatively constant and since there was no significant change in the porewater pressure over this period, almost all of this settlement is attributed to lateral movement of the wall.

Beyond day 238, the excavation remained open and unsealed for a considerable period during the construction of the basement. The resulting drainage through the bedrock, until the basement was sealed into it, consolidated the ground by a further 10*mm*, increasing the total settlement to 0·17 per cent of the excavation depth.

The profiles also demonstrate that the maximum settlements did not occur directly behind the wall but some distance away from it. This could be attributed to the frictional resistance mobilized at the soil-wall interface which could have inhibited free settlement.

The lateral wall movements, assuming the movement on day 142 as the datum, have been plotted in figure (*d*). The data have been adjusted slightly to force the curves to pass through a point corresponding to zero movement at the top brace level. From day 142 to day 519, a period of over a year, maximum movement of less than 15*mm* is indicated. However, by day 142, excavation had been going on for 2·5 months and over *3m* of overburden had already been removed. Making some allowance for these, total wall movements since the beginning of excavation might conceivably have been close to *25mm*, representing about 0·13 per cent of the excavation depth. Figure (*e*) represents construction sequence as function of time.

- *Embedment into rock, Argyle station, Hong Kong*

Figure 8.8 shows the plan, the soil profile and the cross-section of the East and the West walls of Argyle Station, Hong Kong, with the west wall toed into the granite bedrock. Generally, the walls comprised 1·1*m* diameter Secant (Benoto) Piles. However, in areas with restricted headroom caused by overhanging balconies, 1·8–2·2*m* diameter hand-dug caissons were used. The station was constructed in the top-down sequence using the roof and the other floor slabs as the permanent wall braces. Intermediate struts were also installed to provide additional temporary bracing. The west wall was keyed into the granite bedrock by about 3*m*, whereas the east wall was cut off 3*m* short of the bedrock and a grout curtain extension was constructed down to the rock to prevent the flow of water and to minimize the ground settlements. The Secant Piles were constructed using casings so as to avoid settlements during their installation (Morton *et al.*, 1980).

The settlements measured on the adjacent buildings varied from 8*mm* to a maximum of 113*mm*. The largest part, and in some cases all, of the settlements were attributed

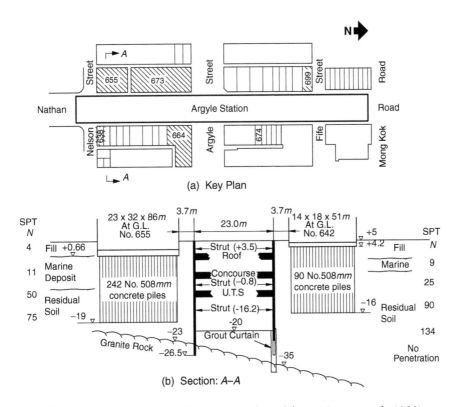

(a) Key Plan

(b) Section: *A–A*

Figure 8.8 Argyle Station, Hong Kong (adapted from Morton *et al.*, 1980)

to the dewatering that occurred during the construction of the hand-dug caissons. Not surprisingly, after the dewatering stopped, most of the buildings registered significant recoveries.

Table 8.1 Argyle Station – Building Settlements and Wall Movements

Building No.	Inclinometer No.	Settlement		Movement	
		(mm)	*(%)*	*(mm)*	*(%)*
642	AE 35	5	0·02	32	0·13
664	AE 103	4	0·02	43	0·17
655	AW 40	4	0·02	22	0·09
673	AW 76	0	0·00	33	0·13
699	AW 226	18	0·07	31	0·12
Average Ground Movement E			0·02		0·15
W			0·03		0·11

E: *East Wall*; W: *West Wall*
(adapted from Morton *et al.*, 1981)

The small ground settlements, generally ranging from 0 to 4*mm*, representing a maximum of only about 0·02 per cent of the excavation depth, were recorded during the subsequent 25*m* deep excavation and dewatering for the main station box. This amply confirmed the efficiency of the grout curtain at the east wall and the adequacy of the toe embedment at the west wall. The isolated case of the large settlement of 18*mm* (0·07 per cent) at No. 699 was attributed to the influence of the adjacent entrance excavation. The lateral movements for the west wall ranged from 22 to 33*mm* representing 0·09–0·13 per cent of the excavation depth. However, lateral movements of 32–43*mm*, representing 0·13–0·17 per cent of the excavation depth, were recorded for the east wall. Comparison of lateral movements between the east and the west walls highlighting the better performance of the west wall can be attributed to the beneficial influence of wall embedment into rock. Ground settlements and wall movements are presented in Table 8.1.

Excavation and bracing in tandem

In stiff cohesive and dense granular soils, ground movements can be relatively less critical. Coupled with this, if the location of the metro structure also happens to be either a 'green-field' site or one that does not lie within the zone of influence of any structure of major importance in the vicinity, then ground movements need not be of any major concern. In such a case, the extent of anticipated ground movements may, within reason, either be acceptable or, if there is some residual damage, it may be possible to put it right relatively cheaply. Under such circumstances, successive stages of excavation and bracing can follow each other in the conventional cyclic manner. In other words, a stage of excavation is taken down to a level that the inherent strength and stiffness of the wall will safely allow preparatory to the installation of the corresponding stage of strutting at that level, before advancing the excavation down to the next stage below, and so on. Depending upon the sequence of construction, whether bottom-up or top-down, the bracing may be temporary or permanent or incorporate a combination of the two.

Bracing by temporary struts

In the construction of underground metro structures by the bottom-up sequence, as the excavation progresses downwards, walls are generally braced by the use of temporary struts. With the subsequent construction of the structure progressing upwards, the struts are sequentially removed as the permanent floors are constructed, become operational and are able to take over the bracing function. If struts constitute the temporary bracing system, the lateral movement of the walls can be seen to be a function of the following factors:

(1) *Type of Ground*:   Largest lateral movements of the walls and consequently higher settlements and as such, potentially, serious problems are associated with excavations in soft and medium clays and compressible silts. For major metro excavations in such soils, maximum ground settlements, of the order of 1–2 per cent of the excavation depth can generally be expected. Furthermore, significant settlements can also occur even as far away as 2–3 times the excavation depth. In order to restrict the movements to within acceptable limits, more stages of strutting may become necessary. If that is not viable, modification in the sequence of construction may need to be looked at.

Incidentally, for excavations in clays, lateral wall movements are associated with settlements of about equal volume with 60–80 per cent of the total lateral wall deflection taking place below excavation level.

Excavations in c', ϕ' soils and stiff clays usually cause maximum movements of less than 0·2 per cent of excavation depth. Ground movements caused by excavations in $c' = 0, \phi'$ soils can also be small, provided that the seepage pressures and the groundwater flow are controlled adequately.

(2) *Stiffness of Bracing System*:   The size, length and support of the individual struts and their spacing, together, reflect the stiffness of the bracing system as a whole. If the struts are long, relatively small in size, placed on a widely spaced grid and are without any intermediate supports, they are likely to undergo a higher elastic deformation under the heavier loads that they are likely to be subjected to. Consequently, the inward movements of the walls towards the excavation will also be greater. However, if the bracing system is stiffer, wall movements are also likely to be smaller. It must be appreciated that, it is only with the relaxation of the ground that the load comes upon the bracing system, unless, of course, the bracing system is preloaded. However, once this has happened and the ground has moved in sympathy, it may neither be practical nor possible to push the walls back fully to restore the prerelaxation state.

(3) *Extent of Preloading in the Struts*:   As discussed in the foregoing section, the load coming onto the struts results in their elastic compression. This makes it possible, in turn, for the wall to move laterally. If a load equal in magnitude to the design strut load were to be jacked into the strut prior to it realizing that load then, in theory, it would

not suffer any further elastic compression and the potential of the wall to move laterally would have been eliminated. However, in practice, it may not be possible to know the actual strut load precisely at any given time, in which case if the load is overestimated, problems of a different nature, especially in soft compressible soils, might have to be faced. In any case, the strut loads are not unique and are unlikely to remain constant; they are very likely to change with their redistribution as the excavation is advanced downwards. Generally, therefore, only a proportion of the estimated design load is jacked-in to restrict movement. This can vary from as low as 10 per cent to as high as 50 per cent. However, provided that the wall is designed to withstand the consequential effects satisfactorily, in theory, loads even well in excess of the anticipated values can be jacked in.

   In further elaboration of the foregoing discussion, some informative case histories are considered next.

• *Embarcadero Station, BART, USA*

Figure 8.9 shows the soil profile and a typical cross-section of the excavation and the ground support system using diaphragm walls for the Embarcadero Station (Kuesel, 1969) in San Francisco.

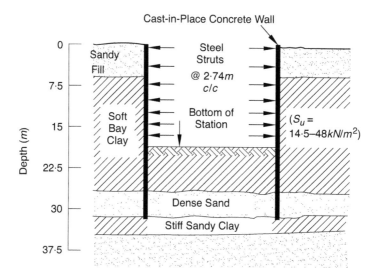

Figure 8.9 Braced Cut – Embarcadero Station, BART, San Francisco
(adapted from Keusel, 1969)

The cast-in-place structural wall is 1·35*m* thick and consists of I-beams and concrete panels. The excavation extending down into soft bay clay is 18*m* deep. The wall extends through the bay clay into an underlying deposit of dense sand and stiff sandy clay. At the end of excavation and with all the struts in place, the maximum recorded inward movements were only 25–37·5*m* representing 0·14–0·21 per cent of the excavation depth, even though the N_{cb}/N_b ratio ranged from 0·6 to 0·8. These movements are a factor of 10 less than the movements indicated for cases where the ratio is about 0·8–1·0. Such relatively modest movements in soft clay can be attributed to the following factors:

- High wall stiffness
- Toe anchorage into stiff stratum
- Close spacing of the stages of strutting
- Insertion of top strut prior to bulk excavation.

- *Boston Subway Extension, MBTA, USA*

Figure 8.10 shows the soil cross-section for a 15·25m deep excavation in medium Blue Clay in Section *D* of the subway extension of the Massachusetts Bay Transportation Authority (*MBTA*). The miscellaneous data compares the performance of the sheet pile wall section with that of the adjoining 915mm thick diaphragm wall section in the vicinity of the Don Bosco School. The school is a seven-storeyed building resting on shallow foundation.

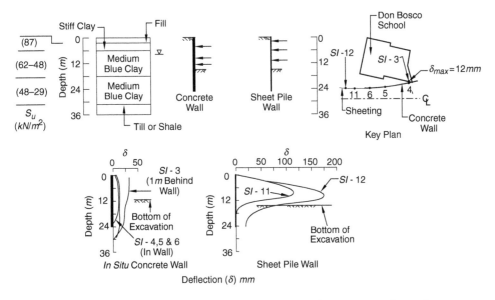

Figure 8.10 Effect of Wall Stiffness on Deflections for Cut in Medium Clay
(adapted from D'Appolonia, 1971)

For most of the excavation, steel sheet piling was used as a means of temporary support. However, in the proximity of the school building, diaphragm wall was adopted extending over a 61m-long stretch.

In the upper layer the medium blue clay is slightly overconsolidated. In the lower layer where it behaves as normally consolidated, the undrained shear strength can be seen to decrease with depth from about 48kN/m² down to about 29kN/m². The soil profile is uniform and the wall depth and the brace spacing are nearly identical for both the sections, the only significant difference being in the wall stiffness.

Whereas movements of 113–175*mm* were measured at the section using steel sheet piling, movements of about 12*mm* only were registered by the inclinometers placed in the concrete wall (D'Appolonia, 1971).

- *Utility Tunnel, Osaka, Japan*

Figure 8.11(*a*) shows a typical section for a 12*m* deep excavation with ground support provided by a 300*mm* diaphragm wall with a four-stage bracing sequence. The top 23*m* of soil consisting of very loose to loose sandy silt (*SPT* blow count < 5) overlies layers of hard clay and dense to very dense layers of sand and gravel (*SPT* blow count > 50). The seasonal fluctuations in the water table have led to an ongoing consolidation of the loose ground causing surface settlements at the rate of 10–20*mm* per year.

The excavation progressed to the final formation level with the help of temporary bracing supports to the wall. The structure was constructed in the conventional bottom-up sequence connecting the slabs rigidly to the walls. The base slab, acting as a raft, was expected to transmit all the gravity loads to the subgrade since there was practically no resistance at the base of the wall and the frictional resistance at the wall–soil interface was also negligible.

Figure 8.11(*b*) shows the predicted wall movements for each stage of excavation with a maximum value of about 15*mm* reached between the third and the fourth stages. It is assumed that, apart from some yielding, further movement of the braced portion of the walls above is restrained by the struts. The movement of the walls is therefore confined to the unbraced section as well as below the excavation level.

The actual wall deformations were remarkably close to the predicted values shown in the figure with the maximum observed lateral movement of 17*mm* representing about 0·14 per cent of the excavation depth. The maximum settlement of the ground surface was observed to be 40*mm* (Kitagushi, 1976) representing 0·33 per cent of the

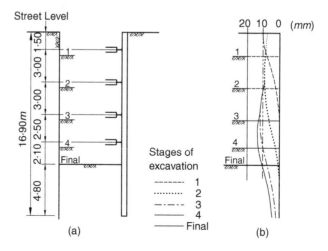

Figure 8.11 Braced Diaphragm Wall Utility Tunnel, Osaka (Predicted Lateral Wall Movements during Excavation) (adapted from Xanthakos, 1979)

excavation depth. However, the observed surface settlement also included the normal consolidation settlement.

Bracing by permanent floors

Alignment of metro systems generally follows busy urban areas and thoroughfares with high-rise structures often lining the route. Most of the time, therefore, deep metro excavations are required to be carried out very close to important structures which may be sensitive to, or intolerant of, ground movements. This can present major problems in planning and engineering particularly if it is necessary to maintain the traffic flow and the continued use of the existing services, and to ensure minimal disturbance to the surrounding areas generally. Estimating the movements accurately and quantifying the problem can be rather difficult. Nevertheless, ground movements around and as a result of the excavation assume major importance as the consequences of damage to adjacent buildings can be high and can incur substantial delays and costs.

By following a top-down sequence of construction, the permanent floor slabs adequately supported on temporary or permanent vertical intermediate supports, can be used as braces for the walls. Since the floor slabs act as continuous lateral supports and are, because of that fact, likely to be many orders of magnitude stiffer than even closely spaced individual struts, the ground movements can be significantly minimized.

The extent of the lateral wall movements will then depend upon the shrinkage and the elastic axial compression of the concrete floors acting as braces. However, it must be recognized that the floors cannot become fully effective as braces until they have achieved sufficient strength and maturity. Until then, therefore, the walls will be potentially free to move. Furthermore, the fact that the preloading is possible only if individual struts are used, and not when permanent floor slabs act as braces, precludes the possibility of partial compensation of movement. In spite of these unfavourable aspects, the movements tend to be relatively small as is in evidence in the following case records.

- *Chater Station, Hong Kong*

The construction of Chater Station, involving an excavation approximately $400m$ long and $27m$ deep and taking up almost the entire length and breadth of Chater Road, is well documented (Davies and Henkel, 1980). The ground conditions are as described in Chapter 7. The station is located in an area of reclaimed land in the heart of the commercial district of Hong Kong which is also the site of some important colonial buildings and prestigious high-rise structures, as indicated in the previous chapter. However, for easy reference, the location plan is reproduced here under Figure 8.12.

At the eastern end, the excavation extended to within a few metres of the older buildings such as the Courts of Justice and, to the west, even closer to the high-rise blocks such as the Mandarin Hotel, Princes Building, etc. as shown in Figures 8.13 and 8.14.

Ground movements around the station were to be expected due to the vertical and horizontal stress relief upon excavation. However, in order to restrict the ground movements during excavation to acceptable limits, top-down sequence of construction using four levels of the permanent structural slabs as braces, was adopted.

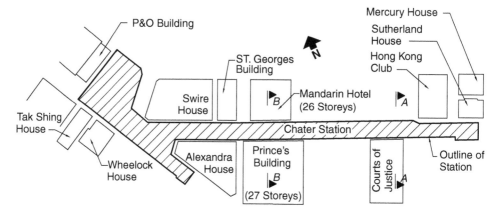

Figure 8.12 Location Plan – Chater Station and Adjacent Buildings
(adapted from Davies and Henkel, 1980)

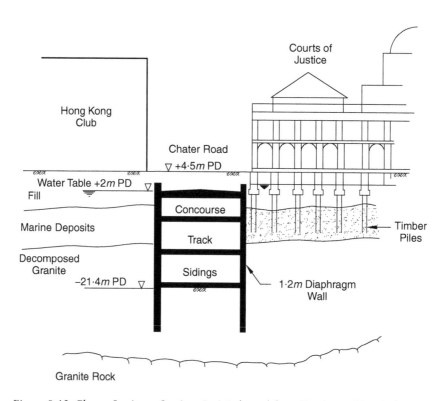

Figure 8.13 Chater Station – Section *A–A* (adapted from Davies and Henkel, 1980)

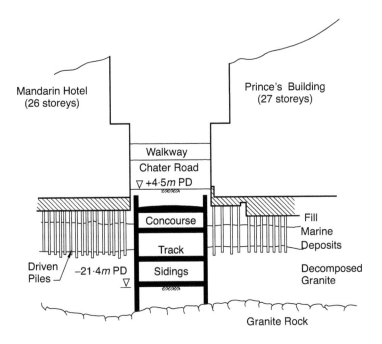

Figure 8.14 Chater Station – Section *B–B* (adapted from Davies and Henkel, 1980)

Maximum lateral movements of the order of 40–50*mm*, representing 0·15–0·20 per cent of the excavation depth, were predicted on the basis of a review of the available data for similar excavations around the world. After the construction of the roof slab, horizontal ground movements during excavation were monitored from an inclinometer installed between the diaphragm wall and the Courts of Justice building. The observations for each stage of excavation are given in Figure 8.15.

It is interesting to note that, as anticipated, horizontal movements reached a maximum of about 40*mm* by the time the excavation reached the sidings level. So, although considerable overall movements of the surrounding ground during the various stages of construction were observed, those due to excavation were restricted to within the predicted levels. This could be attributed to the use of stiff diaphragm walls and the use of floors as braces in the top-down sequence of construction.

- *New Palace Yard Car Park, London, UK*

The 16*m*-deep car park was constructed following 18·5*m*-deep excavation within a perimeter diaphragm wall cofferdam which also formed an integral part of the permanent structure. The foundation of the Big Ben Clock Tower is 16*m*, whereas that of the Westminster Hall only 3*m* away from the excavation.

In view of the sensitive nature of the load-bearing masonry construction of these structures, their immense historic value and their proximity to construction activity, it was imperative to come up with a structural profile and devise a construction procedure for the car park (Burland *et al.*, 1977) which minimized ground movements both during construction and the operational lifetime of the structures. Accordingly, the

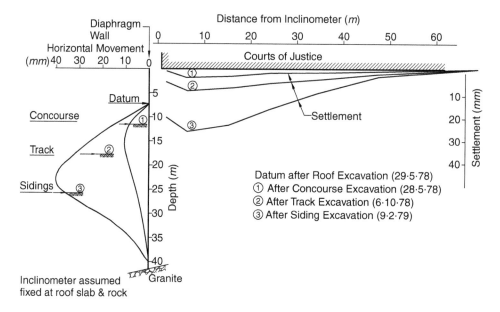

Figure 8.15 Ground Movements during Excavation (Courts of Justice, West Facade)
(adapted from Davies and Henkel, 1980)

diaphragm walls were toed into the *4m* thick layer of the very stiff intact clay so as to cut off seepage of water into the excavation thereby minimizing the draw-down-related settlements of the surrounding ground.

Furthermore, to restrict the inward movements of the diaphragm walls so as to minimize the excavation-related settlements also, it was decided to construct the permanent floor slabs in the top-down sequence making use of these as the temporary struts for the diaphragm walls during the stages of the main excavation for the car park. To retain a measure of control at all stages, these measures were supplemented by comprehensive instrumentation and a programme of monitoring of the ground movements and the response of the existing historic structures during all phases of excavation and construction.

Subsequent to the installation of the diaphragm walls and the construction of the foundations for the internal columns, steel stanchions were lowered into cased boreholes and grouted into position forming the intermediate supports for the floor slabs. Floor slabs were cast on plywood sheets laid on oversite concrete directly on the ground in a top-down sequence and were used as horizontal bracings for the walls.

The excavation was taken down through fill and water-bearing sandy-gravel into the underlying London Clay. London Clay is heavily overconsolidated and fissured and has, typically, average undrained shear strength of $100 kN/m^2$ and a high K_o-value varying between 2 and 3.

Figure 8.16 shows the observed vertical and horizontal surface movements at the end of excavation. Predicted movements have been superimposed for comparison. Horizontal movements can be seen to extend around four times and the vertical movements about three times the depth of excavation away from the wall. The effect of mobilization of the frictional resistance at the soil-wall interface in developing the

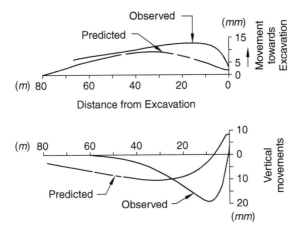

Figure 8.16 Observed and Predicted Ground Movements behind the South Wall
(adapted from Burland *et al.*, 1977)

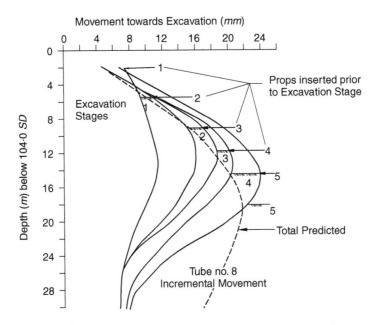

Figure 8.17 Observed Movements of the Diaphragm Wall (South Side Inclinometer Tube 8)
(adapted from Burland *et al.*, 1977)

maximum settlement some distance away from the wall, rather than directly behind
it, is clearly in evidence.

Profile of the lateral displacement of wall towards the excavation after the five floor
slabs were in place but just before the base slab was cast, is shown in Figure 8.17.
Maximum wall movement, including that due to the shrinkage of the concrete floors,

of only about *25mm*, representing 0·15 per cent of the excavation depth was observed. The effectiveness of the extent of wall embedment and its stiffness is clearly in evidence. However, the field measurements have confirmed that, no matter how well the excavation is supported, it may not be possible to eliminate the ground movements completely.

- *4-Level deep basement, Switzerland*

Figure 8.18 shows the cross-section of a *17m* deep excavation for the construction of a 4-level basement (Huder, 1969). The ground, with the water table just over a metre below the surface, comprises glacial till, varved decomposed moraine – mainly of clayey nature but somewhat sandy to gravelly with occasional lenses of silt, and lacustrine deposits. The undrained shear strength varied from $20kN/m^2$ in the lacustrine deposits to $140kN/m^2$ in the undisturbed moraine.

To guard against excessive draw-down, the perimeter walls were taken down to depths of *25–35m* below the surface. Dewatering inside the perimeter was carried out to a depth of over *15m*.

Construction followed the top-down sequence. The floor slabs, which provided the bracing, were cast on formation at each stage of excavation and were supported over intermediate steel piles.

Observations of lateral wall movements were taken at four locations marked I, II, III and IV on the key plan. Excavation down to level 1 did not cause significant wall movements at any of the four locations. However, wall displacements increased progressively as the excavation advanced to deeper levels. The profiles also show that, even after the placement of the floor slab at a level, wall movements at the floor above continued, clearly indicating the effect of shrinkage and elastic shortening of the slabs upon assuming increased loads.

The maximum wall displacement of about *36mm* at Tube IV represents only 0·20 per cent of the excavation depth. However, it is significantly higher than that

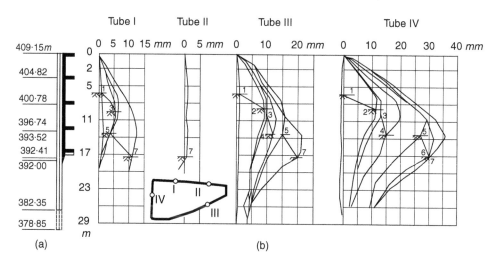

Figure 8.18 Observed Wall Movements during Construction of Four-Level Basement, Switzerland (adapted from Huder 1969)

indicated at Tube I for a wall of similar height. This is attributed to the following factors:

- Increased lateral load from loose, varved silt surrounding the wall at Tube IV;
- Greater movement below formation level resulting from very low soil stiffness;
- Relatively more shrinkage, creep and elastic shortening of the floor slabs because of their greater lengths.

The wall sections at Tubes II and III are adjacent to much stiffer moraine and are also of the same depth. However, the natural ground behind Tube III rises some 6*m* higher, thereby imposing higher lateral earth pressures on the wall. With the resulting net out-of-balance horizontal force, there is the tendency for the structure to move towards the wall containing Tube II. The adjusted deflection profile of Tube II therefore reflects the interaction between the two wall sections. The exact correlation of movement at various points along the wall was further complicated (Huder, 1969) by variations in wall stiffness due to construction imperfections in the 800*mm* wall thickness.

It must be clearly appreciated that the use of temporary and permanent bracing is not mutually exclusive. Where circumstances so warrant, it is perfectly feasible to rely for wall bracing, during the construction phase, on a combination of the two, i.e. discrete permanent floor slabs and discrete temporary stages of struts. It is also possible to convert certain discrete temporary struts into permanent bracing by so locating these that they can be concreted-in later on to form part of the permanent floor slabs. In fact, such a measure has the added advantage of locking-in the load permanently and avoiding further lateral movement which would otherwise occur during the load transfer from the temporary strut to the permanent floor. In any case, how significant such effects can be is addressed elsewhere in the book.

Tied back or anchored walls

During the construction of deep metro structures, bracing of perimeter walls with tie-backs or ground anchors offers the unique advantage of an obstruction-free site permitting unhindered excavation and construction of the structure at speed. It avoids many of the time consuming operational problems commonly associated with a conventionally braced excavation. Furthermore, with the vast improvement and ongoing refinement of the techniques of anchoring over the years, it has now become possible to achieve a significant increase in the load carrying capacity of ground anchors. The use of ground anchors, especially in deep metro excavations, has therefore become particularly attractive.

However, depending upon the type of the subsurface environment encountered, the use of ground anchors may not always be a practical proposition. Ostermayer (1976) has listed, amongst others, the following considerations in deciding whether or not the use of ground anchors is feasible:

- Location and suitability of soil strata for carrying anchor loads;
- Possible location of anchors considering adjacent properties (load, type and depth of foundations including quality and sensitivity of buildings), existing cables, trenches, caverns, wells and potential subsequent building operations and excavations on adjacent sites;

- Influence of anchor installation on adjacent properties (resettlement or heaving of ground, flow of grouting materials into cellars, vibration, etc.); and
- Permission to use the ground of adjacent sites. With relatively deeper cuts for metro structures, tie-backs or anchors may need to be carried under a number of properties. In such a case, it is important to ensure that no operations which are likely to impair, damage or otherwise interfere with the structural integrity of the anchor system are permitted to be carried out during the operational life of the anchor system.

With a multilevel anchored wall, since the anchors have to depend for their support on the supported ground itself, the general stability problems can be rather complex. While prestressed tie-backs are effective in reducing wall and surface movements, they do little to prevent the deep-seated movements which occur within the soil (Creed *et al.*, 1980). It is therefore important to aim for adequate factors of safety against tilting and deep-seated movement of the wall and the soil mass.

In the case of anchored walls, the top row of anchors has a significant influence on the efficiency and displacement of the wall. Its level is usually fixed by a judicious balance between the initial cantilever moment and its influence on the final bending moment envelope, and the limitations on the resultant inward movement of the wall. If the top row of anchors is installed, say, at *3m* or more below the ground surface, the portion of the wall above this level, acting as a cantilever, will help towards a relatively favourable distribution of the bending moments in the wall. However, in the process of taking the excavation down to this level, considerable movement of the wall may take place. These movements may, at times, reach as much as 50 per cent of the anticipated movements at the end of the excavation. While subsequent tensioning of the anchors may achieve a marginal recovery, restoring the wall to its initial profile, especially in stiff soils is highly unlikely. Thus if wall movements are to be kept to a minimum, the top row of anchors must be located as close to the surface as practicable. Littlejohn and McFarlane (1974) suggest limiting this depth to *1·5m*. However, in arriving at a suitable depth in a given soil environment, due consideration should also be given to the potential risk of local ground failure behind the wall during tensioning.

The main reason for the inward displacement of walls towards the excavation is the lateral compressibility of the soil between the walls below the formation (Ostermayer, 1974). Figure 8.19 shows, schematically, the stresses and the strains around an anchored cut. This is a simple, idealized model in which the effect of the moments has been ignored.

Displacement of the wall can occur either due to the displacement of each individual anchor or as a result of the active deformation of the whole of the retained soil block *ABEF* and the passive deformation of the trapped zone of soil between the walls directly opposite the underlying block *BCDE*. The effective ambient lateral stresses before excavation and the limit stresses after excavation are shown qualitatively in the figure. For the force equilibrium, i.e. $\sum H = 0$, the likely passive resistance of the soil below formation in front of *BC* must be equal to and commensurate with the driving pressure on *FD*. The driving pressure over the height *FD* is likely to lie somewhere between the earth-at-rest and the active pressures and the resistance mobilized below the formation somewhere between the earth-at-rest and the passive pressures.

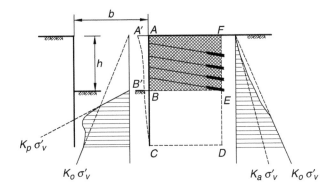

Figure 8.19 Idealized Stresses and Strains around Anchored Cut
(adapted from Ostermayer, 1974)

The extents of the pressures generated and mobilized are, of course, functions of the respective soil strains undergone in reaching equilibrium. Clearly, the intensity of the lateral pressures resisting is going to be far higher than that of the lateral pressures generating the strains. This, together with the removal of overburden above formation means that the soil below formation is going to experience a combination of reduction in the vertical stress and an increase in the horizontal stress beyond its at-rest value, causing its lateral compression and vertical expansion and resulting, inevitably, in the inward movement of the walls.

In the case of Frankfurt Main Railway Station, where 20*m* deep anchored excavation was carried out in unfavourable ground (stiff, highly plastic clay) conditions, wall displacements of 140*mm* (Breth and Rhomberg, 1972) representing 0·7 per cent of the excavation depth were measured. On the other hand, displacements of only 0·05–0·11 per cent of the retained height were observed (Nendza and Klein, 1973; James and Phillips, 1971; Egger, 1972) in very stiff to hard clays and silts and dense noncohesive soils where stiff ground support systems (diaphragm, bored pile walls, etc.) were adopted and the anchor loads were based on earth-at-rest pressures.

In the case of very soft compressible soils, unusually high level of anchor prestress may induce large movements of the wall into the soil. The possibility that the deflected profile thus obtained could remain permanently locked-in if the anchor is fixed into rock, must be taken into consideration.

For excavations in clay, there is also the potential for the time-dependent movement due to consolidation and creep of the highly stressed soil to take place. This can, potentially, be very significant, at times even doubling the end of excavation movements. The effect of these movements may also extend as far behind the wall as two to three times the depth of excavation. Therefore, the need for rapid construction of the main structure is obvious.

Invariably, the anchors are intended to be only temporary and may therefore be required to be destressed as soon as the loads can be safely transferred on to the permanent bracing floor slabs of the structure. Under such circumstances, due allowance for the shrinkage and axial compression of the slabs upon assumption of their respective loads must be made.

8.4 Parametric studies

A number of investigators have been able to arrive at important conclusions based on the parametric studies on wall movements carried out against different values of wall stiffness, bracing systems and soil types. Findings of these investigations and the confirmation available from a number of case histories briefly reproduced hereunder provide an extremely useful insight into, and a feel for, the potential effects of interaction under a variety of different ground conditions, structural environment and boundary conditions.

In sand

Egger (1972) compared the theoretical performance of a 'flexible' sheet pile and a 'stiff' cast-*in-situ* diaphragm walls. Finite element analysis was carried out for three stages of excavation and two tie-back prestress loads. Following important conclusions were reached:

- For the first stage excavation, when the exposed part of the wall acts as a vertical cantilever, horizontal displacement of the top of the flexible wall was three times as much as that for the stiff wall.
- By increasing the anchor prestress in flexible walls, movement at the top can be significantly reduced. In the case of stiff walls, the reduction can be even greater. For an increase in the anchor stress from 1000 to $6000 kN/m^2$, movement reduction of the order of 33 per cent in the flexible and 40 per cent in the stiff walls was obtained.
- As is only to be expected, 'stiff' walls spread the increase in the anchor prestress over a larger area as opposed to a smaller area locally in the case of 'flexible' walls.
- Stiff walls mobilize passive resistance over a noticeably greater depth. However, displacements diminish much less with depth.

In clay

Comparisons of field performance data of tied-back and braced walls carried out by various investigators in clay soils reveal that movements associated with tied-back system are usually less than those of the braced system. However, these results cannot be treated as conclusive since the systems compared, because of the variations in their wall stiffnesses and the soil properties, are not equivalent. To resolve this, Clough and Tsui (1974) carried out a series of finite element analyses as follows:

- *Analysis* 1: Tied-back and braced wall systems, assuming 'equivalence' in respect of construction sequence, prestress loads, soil conditions, wall stiffness and brace and tie-back stiffness, were considered. Temporary excavation platform was taken as being $600mm$ (*2ft.*) below the location of each tie-back or brace installation level.
- *Analysis* 2: Same as Analysis 1 except that: (*a*) braced wall was not prestressed, and (*b*) tie-back stiffness assumed was one-tenth of the brace stiffness.
- *Analysis* 3: Same as Analysis 2 except that the temporary excavation platform for the braced wall was 'over-dug' by $2·44m$ (*8ft.*) below the location of each brace level.

Excavation, 15*m* deep, assumed to be carried out in five stages was considered in a 28*m* deep homogeneous clay deposit with an undrained shear strength increasing with depth from 48*kN/m²* at the surface to 128*kN/m²* at the bottom of the deposit. Initial tangent modulus with a value of 200 times the undrained shear strength, varying with depth, and wall stiffness of 48,500*kN/m²/*m (i.e. value typical of a 300*mm* thick concrete wall) were assumed. Following results, as shown in Figure 8.20, were obtained:

- For 'equivalent' conditions, more movements and settlements occurred in the tie-back wall than in the braced wall.
- In analysis 2, although the struts for the braced wall were 10 times as stiff as the ties, the tied-back wall moved less than the braced wall. This can be attributed entirely to the effect of prestressing the tied-back wall, given the same sequence of construction. Even lesser movements could have been expected had the anchors been extended beyond the movement zone.
- Braced wall with 'over-dig' deflected twice as much as the braced wall without 'over-dig'.

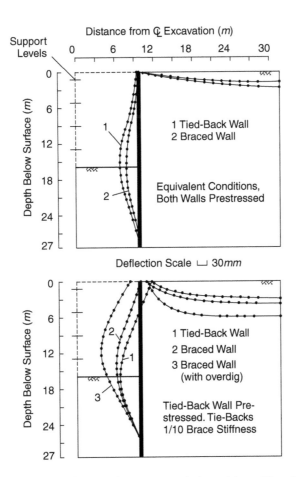

Figure 8.20 Predicted Wall and Ground Movements (adapted from Clough and Tsui, 1974)

From the foregoing observations, following conclusions were drawn:

- The assumption that tied-back walls yield smaller movements than the braced walls cannot be generalized. However, the effect of prestressing and the control of construction procedure can invariably result in a better performance for the tied-back wall as compared to that for a braced wall.
- Over-excavation can have a major adverse effect on wall movement.

Further studies

Further parametric studies in two excavations, again in clay, supported by four levels of tie-backs were also carried out by Clough and Tsui (1974). One cut was exactly 1·5 times the size of the other. Undrained shear strength was assumed to increase from $30kN/m^2$ at the surface to $88kN/m^2$ at $23m$ depth. Initial tangent moduli were taken as 400 times the undrained shear strength values.

Four different earth pressure diagrams as shown in Figure 8.21 were used to estimate the anchor loads. Three of these represent the Peck trapezoidal profile for braced cuts and the fourth the triangular at-rest profile. The models for testing assumed that:

- The anchors are set into rock to eliminate anchorage movement.
- The walls are fixed at base.

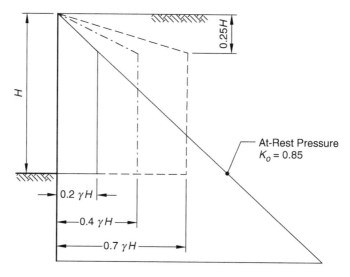

Figure 8.21 Design Pressure Diagrams used in Parametric Study
(adapted from Peck, 1969; Clough and Tsui, 1974)

Tie-back effects were simulated by the use of spring elements. The following interesting observations were made:

- Higher prestressing force resulted in lessening of wall and soil movements dramatically at the top and to a far less degree at the formation. However, it could neither restore the initial state of stress in the soil nor eliminate, completely, the movement which occurred at each stage of excavation (Figure 8.22). As the wall rigidity increased, the wall deformations and soil settlements reduced, although not in the same proportion (Figure 8.23).

- Stiffer tie-backs reduced movements but not in the same proportion as the increase in their stiffness (Figure 8.24).
- With a combination of high levels of prestress and high stiffness of walls and tie-backs, wall movements and ground settlements could be dramatically reduced (Figure 8.25).
- Movements increased with an increase in the depth of excavation under all circumstances.

To examine the combined influence of prestressing, wall rigidity and tie-back stiffness, two extreme cases were analysed. In one case, most flexible wall and tie-back system was considered with no prestress load; in the other, highest prestress load (based on $0.7\gamma H$ trapezoidal pressure) was applied to the stiffest wall and tie-back system. The result was dramatic; whereas for the stiff wall plus prestress case, the movements were very small, for the other case, the movements were considerable. However, movements increased with depth in all cases and circumstances.

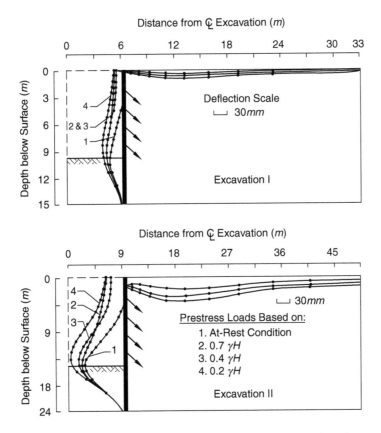

Figure 8.22 Effect of Prestress on Movements (adapted from Clough and Tsui, 1974)

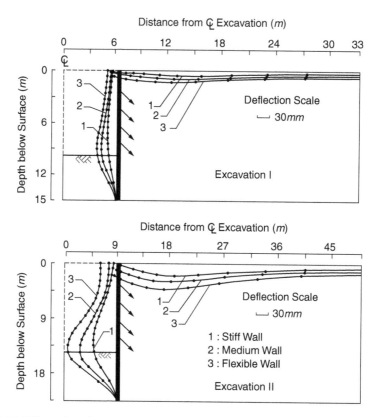

Figure 8.23 Effect of Wall Rigidity on Movements (adapted from Clough and Tsui, 1974)

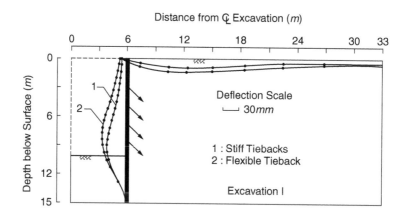

Figure 8.24 Effect of Tie-Back Stiffness on Movements (adapted from Clough and Tsui, 1974)

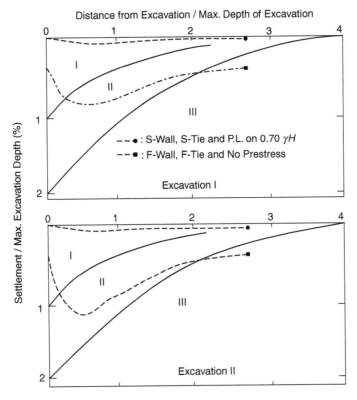

Figure 8.25 Effect of Combination on Settlements
(adapted from Peck, 1969; Clough and Tsui, 1974)

The parametric studies confirm that advancing the excavation down to a certain level causes, correspondingly, ground relaxation which extends well below that level. As a result, with every successive stage of excavation, this relaxation becomes cumulative. Since there is no means, after the event, of reversing this trend at depth which is always below formation, restoring the wall fully to its pre-excavation profile particularly at that depth, irrespective of the amount of anchor prestress, is virtually impossible. This is amply borne out by the studies. This also confirms that while prestressing the tie-backs reduces the wall and surface movements, it does very little to prevent deep-seated movements below formation. Increasing the stiffness of the wall, however, can alleviate the problem to some extent but cannot eliminate it altogether. The only way by which such displacements at depth can be contained, it would appear, would be by employing ground treatment or preinstalling stiff struts at that depth before the commencement of excavation, as discussed previously.

Figure 8.26 shows predicted net earth pressures for excavations I and II using four tie-backs on both the flexible and medium stiff walls. In the case of excavation II, the analysis was also carried out using only three tie-backs to highlight the effect of tie-back spacing. The reasonably close similarity in the distribution of lateral earth pressures suggests that, above the formation, the excavation depth does not have a significant effect on it. Below the formation also, both show a reduction of lateral pressures on the excavation side with the reduction being greater for the flexible wall.

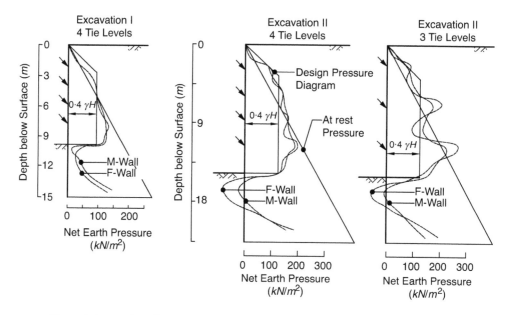

Figure 8.26 Predicted Net Earth Pressures (adapted from Clough and Tsui, 1974)

It is well established that the stiffness of wall, type of soil and the type of ground support system – all have a significant influence on the pattern and magnitude of ground movements. However, their influence also on the magnitude of the pressures to be resisted is no less significant.

Burland *et al.* (2004) carried out sensitivity analysis on the earth pressures and movements for a wall propped with single infinitely stiff prop at the top as illustrated

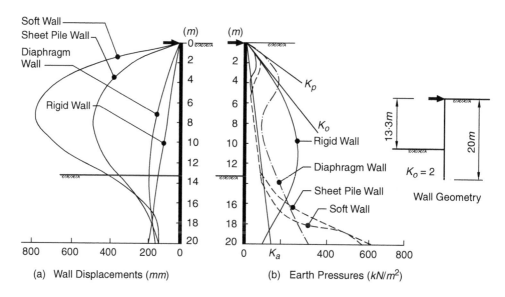

Figure 8.27 Propped Cantilever Wall (adapted from Burland *et al.*, 2004)

in Figure 8.27(*a* and *b*). Their results have shown the effects on these of changing the wall stiffness. As the wall stiffness reduces, there is a corresponding increase in the movements and a commensurate redistribution of earth pressures; the redistribution is such that there is a reduction in the earth pressure behind the central portion of the retained height of the wall and an increase at the top of the wall.

Table 8.2 Stiffness *vs.* Bending Moment

Type of Wall	Bending Stiffness, EI – (kNm²/m)	Maximum Bending Moment – (kNm/m)
Soft	$2{\cdot}3 \times 10^4$	700
Sheet Pile	$7{\cdot}8 \times 10^4$	1160
Diaphragm	$2{\cdot}3 \times 10^6$	4400
Rigid	$2{\cdot}3 \times 10^9$	8900

(adapted from Burland *et al.*, 2004)

Table 8.2 shows the variation in the maximum bending moments in the four types of walls. Significant increase in the bending moments (and the substantial reduction in the displacements) can be attributed to the increase in the flexural stiffness from that of the soft wall to that of the rigid wall.

In soft clay

Hashash and Whittle (1996) carried out a series of numerical experiments using nonlinear finite element analyses to investigate the effects on the undrained deformations around a braced diaphragm wall embedded in a very deep layer of soft clay. The modelling of the experiments and the various soil profiles, including the Composite profile, considered are discussed in Chapter 20. For easy reference, the model considered is reproduced here in Figure 8.28. The principal structural parameters

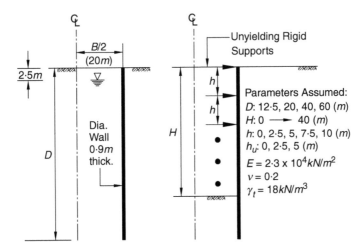

Figure 8.28 Model Geometry, Support Conditions (adapted from Hashash and Whittle, 1996)

considered in the study are the penetration depth of the diaphragm wall and the vertical spacing of the support bracing, while maintaining the excavation width and the wall thickness constant. The analyses incorporate the MIT-E3 effective stress soil model which has well-documented capabilities for describing the deformation and strength properties of Boston Blue Clay (*BBC*). For full details of the model input parameters etc., reference may be made to Hashash and Whittle (1992).

The effects of the penetration depth and the brace spacing of the diaphragm wall on the deformations around the cut in the case of (overconsolidation ratio) $OCR = 1$ clay profile are shown in Figures 8.29, 8.30, 8.31 and 8.32. For the definition of OCR, see Appendix C. Figure 8.33 summarizes the response to excavation in the $OCR = 2$ and 4 profiles whereas Figure 8.34 compares the wall deflections for the Composite and the constant OCR soil profiles.

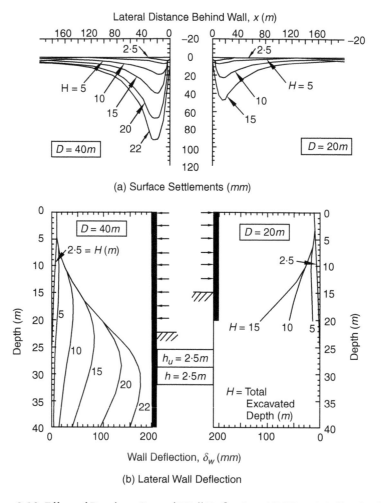

Figure 8.29 Effect of Depth on Lateral Wall Deflections ($OCR = 1{\cdot}0$ Clay Profile) (adapted from Hashash and Whittle, 1996)

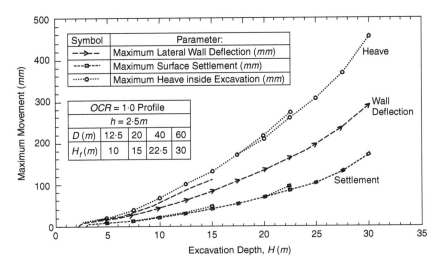

Figure 8.30 Effect of Wall Depth on Maximum Wall Deflections and Ground Movements
(adapted from Hashash and Whittle, 1996)

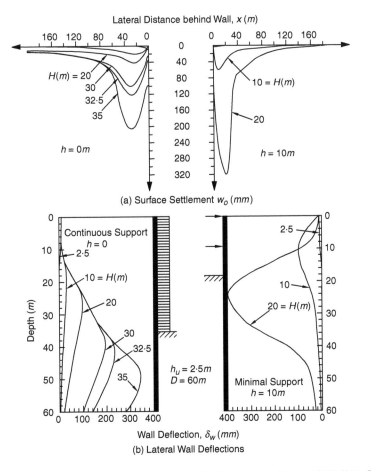

Figure 8.31 Effect of Brace Spacing on Surface Settlements and Lateral Wall Deflections
($OCR = 1\cdot0$ clay profile) (adapted from Hashash and Whittle, 1996)

The authors summarize the main results of the study as follows:

- Depth of wall penetration has a minimal effect on the prefailure deformations for excavations in deep layers of clay where there is no constraint on the toe movement, but does have a major influence on the location of the failure mechanisms within the soil.
- The predictions for excavations with continuous bracing (i.e. $h = 0$) show that the deep-seated soil movements occurring below the current formation represent the principal mechanism controlling wall deflections and surface settlements. Additional basal movements occur as the brace spacing (h) increases; however, the importance of this parameter is closely related to the stress history profile of the clay.
- Although the initial cantilever movements of the wall can represent a very large component of the ground movements for shallow cuts, they are eventually overwhelmed by basal movements in the underlying clay for excavations greater than 15–20m deep.
- For overconsolidated clay profiles with constant $OCRs = 2$ and 4, there is no tendency for basal instability, and the computed maximum ground movements are independent of wall penetration depth and are linear functions of the excavation depth.

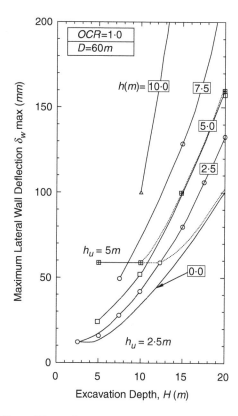

Figure 8.32 Effect of Brace Spacing on Maximum Lateral Wall Deflections
(adapted from Hashash and Whittle, 1996)

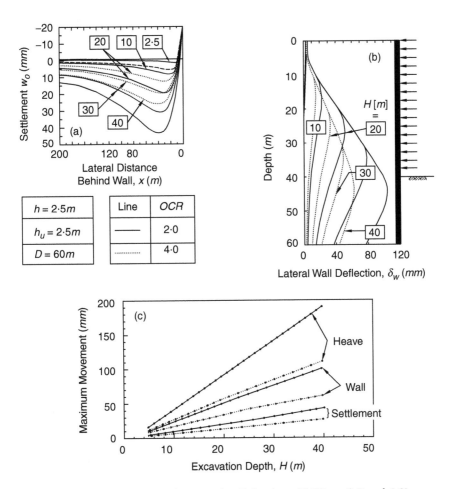

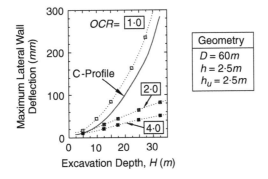

Figure 8.33 Summary of Excavation Behaviour ($OCRs = 2 \cdot 0$ and $4 \cdot 0$)
(adapted from Hashash and Whittle, 1996)

Figure 8.34 Comparison of Maximum Wall Deflections for Composite and Constant OCR
Profiles (adapted from Hashash and Whittle, 1996)

- For the composite stress history soil profile considered, the maximum wall deflections and ground movements appear to be strongly influenced by the properties of the soft normally consolidated clay which underlies the excavation at a depth below 25m.

For estimating the maximum lateral wall deflections δ^w_{max} (mm) and the maximum surface settlements δ^s_{max} (mm), the authors have come up, based on curve-fitting, with the following empirical relationship:

$$\Delta_{max} = H\left[ae^{bh} + (c+dh)H\right] \tag{8.1}$$

where $\Delta_{max} = \delta^w_{max}$ or $\delta^s_{max}(mm)$, and a, b, c and d are constant dimensional coefficients as listed in Table 8.3. Other parameters, H (m) and h (m), are as defined in Figure 8.29.

Table 8.3 Coefficients for Maximum Movements

Stress History Profile – OCR	Maximum Movement	Coefficients			
		a	b (m^{-1})	c (m^{-1})	d (m^{-2})
1·0	δ^w_{max}	1·7610	0·0184	0·1650	0·0323
	δ^s_{max}	0·2854	0·2216	0·1061	0·0175
2·0	δ^w_{max}	2·1972	0·0515	0·0	0·0
	δ^s_{max}	0·8768	0·0618	0·0	0·0
4·0	δ^w_{max}	1·3878	0·0548	0·0	0·0
	δ^s_{max}	0·5541	0·0638	0·0	0·0
Composite	δ^w_{max}	0·8557	-0·0984	0·1646	0·0152
	δ^s_{max}	0·0	0·0	0·0941	0·0070

(adapted from Hashash and Whittle, 1996)

The feedback available from the various case histories and the conclusions drawn from the parametric studies presented above provide large amount of valuable data which can be very useful in the process of selecting the method and sequence of construction most appropriate to deep metro structures under site-specific conditions and constraints.

9   Interaction: III

9.1 Introduction

Effect on the existing structures as a result of the various stages of excavation for and construction of deep metro structures, taking into consideration the influence of the soil type and the relative stiffness of the bracing system on the movements of walls, and the associated parametric studies, have been discussed in the previous two chapters. This chapter deals, briefly, with the aspects of potential interaction in respect of the following:

- Type of the existing structures
- Effect of existing structures
- Effect of future metro-related construction activity
- Effect of future non-metro construction activity
- Other considerations.

9.2 Type of existing structures

The effect of the various stages of metro construction upon the existing structures in the vicinity has already been discussed in some detail. However, the qualitative and the quantitative responses of the existing structures within the zone of influence also depend upon the type of their construction, age and their overall stiffness. The amount of settlement that a particular structure can safely tolerate is closely related to the type of its construction and its ability to distribute differential settlements. The age and condition of the structure are also important considerations for establishing or specifying the extent of the allowable movements.

The shape, size and type of construction of an existing structure will reflect the degree of its stiffness and the potential to redistribute the load. Low-rise residential structures, for instance, tend to be flexible and are therefore less able to redistribute the loads. Any distress experienced by such a structure during the construction of metro works, however, is unlikely to cause a major problem and one that is likely to be put right relatively simply and cheaply. High-rise structures of more robust construction, e.g. structures with shear walls, on the other hand, tend to be relatively stiffer. Their inherent rigidity enables such structures to redistribute the loads more readily and they are therefore less likely to invite structural problems. It is those high-rise structures, such as the ones using beam-column frame as the structural system, which may be unable to redistribute the loads safely that are susceptible to experiencing the most problems.

It is difficult to predict purely from a theoretical computation the distortional settlements which can cause danger to a building structure. This is because the actual behaviour of a structure is greatly influenced by a number of factors, such as the interaction between the principal structural and the secondary nonstructural elements, redistribution of loads, age and the degree of deterioration suffered, etc. These are difficult to quantify and model accurately and are therefore, generally, not taken into consideration in a theoretical calculation.

Foundations of a low-rise structure may be shallow, isolated pads or strip footings. In the case of high-rise structures, with or without basements, they may take the form of rafts or deep piles, depending upon the depth at which an adequate founding stratum is available. The effects that such foundations are likely to experience and the way they may respond to the various stages of construction of a metro structure will depend, among other factors, upon their respective founding levels relative to the depth to which the metro excavation has to be taken down, and their proximity to it. Shallow footings located close to the metro works, for instance, are likely to feel the impact more than the deeper foundations located some distance away. In the case of a piled foundation in close proximity to a metro structure, even when the piles are taken down deeper than the metro structure, interaction problems could still be expected if, for instance, the piles are fully or even partially frictional. The extent of the impact will depend upon what proportion of the load is carried in friction of the pile shaft and to what extent the effectiveness of such frictional support of the soil is likely to be compromised during the various stages of the metro construction.

To alleviate the impact of the metro construction works on the structures in the vicinity, it may be feasible, in certain circumstances, to underpin the existing foundations. This may entail strengthening of the foundations and the underlying soil by means of appropriate geotechnical processes, such as injection grouting, etc. If the ground treatment around the existing foundations is carried out at least as far down as the base of the metro structure, then the existing structure could virtually be rendered immune to the effects of the metro construction works. However, movements have been observed to have occurred up to a considerable distance beyond the excavation and, even in well-supported excavations, are known to have extended at least as far away as the depth of the excavation. Underpinning over such extents of areas can be prohibitively expensive and therefore other measures of controlling movements under such circumstances need to be looked at.

As a matter of prudence, a complete photographic record, as part of the condition or dilapidation survey, of the existing structures likely to be affected by the metro works should be assembled and agreed with the owners or their authorized agents before the commencement of the works. Such surveys generally form the threshold against which any damage sustained by the structures which is attributable to the construction or related activity can be identified and appropriate restitution made. In fact, this has come to be expected as a norm in connection with the construction of underground metro works these days. Where necessary, the vulnerable structures should be instrumented and closely monitored not only during construction but also for such a period beyond the construction activity until the effects on these structures of such activity can be seen to have tailed off. Observations from the monitoring should give timely notice and enable appropriate measures to be put into operation should the need arise.

During his measurements of settlements of buildings in the Mong Kok district of Kowloon, Hong Kong, Lumb (1964) reported that the effects of settlement due to the water table draw-down were most pronounced in the case of prewar buildings on spread footings. These buildings suffered the most severe structural and architectural damage. However, in the case of buildings with piled foundations, even including short timber piles, settlements gave rise to block tilting causing architectural damage affecting appearance rather than safety. It was also reported that average settlements of 50*mm* for buildings with shallow foundations and 15*mm* for piled structures were caused by the deep pumping carried out during the period 1964–1970.

In what follows, structural behaviour of some of the different types of buildings as a result of ground movements associated with the construction of the Argyle and Wong Tai Sin Stations, Hong Kong Mass Transit Railway (*HKMTR*) in the residual soils of Hong Kong (Morton *et al.*, 1981), is discussed. The areas surrounding the station are generally dotted with different types of structures, both old and modern, many of which would have experienced some movement as a result of deep pumping from private wells, prior to the commencement of the metro works.

Argyle Station, Hong Kong

Figure 9.1 shows the location of the station in relation to the existing structures in the vicinity. The properties of the soils and the cross-section through the metro works have been shown in Figure 8.8(*b*) of Chapter 8.

The buildings adjacent to the metro works are of reinforced concrete beam-column frame construction. Those of the buildings which were constructed on isolated pile caps under columns, thereby making the structures discontinuous at the pile cap level and in which the nonstructural brick walls had deteriorated over the years, lacked rigidity. Buildings of reinforced shear wall construction, on the other hand, could be classified as rigid structures. The rigidity of such structures is further enhanced by the then prevalent trend of using concrete, albeit nominally reinforced, even for the nonstructural walls.

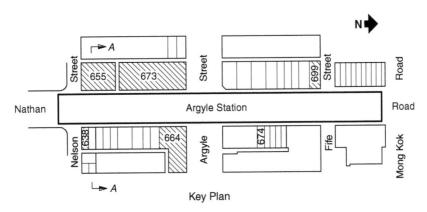

Figure 9.1 Argyle Station, Hong Kong (adapted from Morton *et al.*, 1980)

Table 9.1 Argyle Station – Building Settlements

Bldg. no.	Type of foundation	Settlements – 'front to rear' & (differential) in mm to		
		Aug 77	Dec 77	Feb 79
638E	19m deep precast piles driven 8m into residual soil	36–10 (26)	30–13 (17)	NA–12 (NA)
642E	Concrete piles to 23m depth driven 10m into residual soil	39–8 (31)	30–6 (24)	35–10 (25)
664E	18m deep Franki piles driven down to top of residual soil	56–7 (49)	40–41* (–1)	44–1 (43)
674E	No records available; spread footings on timber piles assumed	108–38 (70)	105–29 (76)	113–NA (NA)
655W	25m deep colcrete piles driven 15m into residual soil	8–3 (5)	0–2 (–2)	4–NA (NA)
673W	Caissons dug down to weathered rock	14–4 (10)	9–1 (8)	9–0 (9)
699W	15m Deep piles driven 9m into residual soil	46–2 (44)	50–0 (50)	68–NA (NA)

E: East Side; W: West Side
Asterisked figure appears to be suspect. Highest settlements were recorded on either side of the northern half of the station. In August 1977: Caisson dewatering stopped. In December 1977: GWL had 'fully' recovered. In February 1979: Station excavation stopped. (adapted from Morton *et al.*, 1981)

Tables 9.1 and 9.2 list the settlements at the front and rear and the apparent tilt of seven buildings in the vicinity of the station.

The differential settlements between the front and the rear of those buildings which appeared to tilt as units did not cause any perceptible angular distortion. However, some local cracking in basement/ground slab causing ingress of water was observed. There were also reports of localized architectural damage to windows, finishes, etc., but this was of minor nature and easily repaired.

Wong Tai Sin Station, Hong Kong

Figure 9.2 shows the location of settlement blocks around the station, a typical soil profile and section of metro works. The station was constructed inside a perimeter of 900*mm* thick diaphragm walls in the top-down sequence.

The settlement blocks on the south side are the seven-storeyed structures carried on piles which, at the time of metro construction, were about 20 years old. Block 9 is about 13·7*m* from the station wall. The buildings on the north side are more modern and Block 8 is as close as 3·7*m* from the station north wall.

All the buildings in the vicinity are of shear wall type of construction. Regular monitoring recorded maximum settlements of between 52 and 58*mm* for Blocks 8 and 9. Although the buildings experienced tilt due to the differential settlement between the front and the rear, no angular distortion within the structures was in evidence. This confirmed the belief that, due to their very rigid construction, the buildings tilted as units.

Table 9.2 Argyle Station – State of Adjacent Buildings and Damage Sustained

Bldg. no.	Age, type and condition of the adjacent buildings	Max. tilt	Damage sustained by the adjacent buildings
638E	Built 1959, 12 storeys; good state of repair; beam-column system, no shear walls.	1 in 730	Architectural damage at ground level.
642E	Built 1962, 15 storeys; rigid beam-column frame with cross-walls.	1 in 1631	Slight architectural damage.
664E	Built 1954, 9 storeys; beam-column frame; building split by expansion joint.	1 in 387	A few cracks in plaster, none in structural walls; expansion joint opened 100*mm* due to differential settlement.
674E	Built 1955, 5 storeys; flexible beam-column frame.	1 in 296	Archit'l damage, cracks in minor member, not serious; highest recorded settlement for station.
655W	Built 1965, 25 storeys with revolving roof-top restaurant; beam-col. + shear wall frame.	1 in 4500	No damage reported.
673W	Built 1969, 16 storeys; rigid beam-column & shear walls, 2-level basement.	1 in 2250	Slight water ingress in basement; no settlement due to station excavation nor due to sheet pile station entrance.
699W	Built 1958, 8 storeys; beam-column frame; some rigidity, no cross-walls.	1 in 312	Slight damage to finishes; some damage to outhouses but not serious.

Highest tilts (corresponding to highest settlements) were recorded on either side of the northern half of the station. (adapted from Morton *et al.*, 1981)

The conclusions of observations at the aforementioned two stations may be summarized as follows:

• The predominant cause of building settlement was groundwater lowering.
• Large settlements and tilts occurred even where metro construction activities were well controlled.
• Wide variety of buildings was able to tolerate movements with only minor damage.
• Contrary to expectations, even buildings of seemingly small stiffness, such as reinforced concrete beam-column frames on isolated foundations and without cross-walls, proved to be stiff enough to redistribute significant differential settlements.

9.3 Effect of existing structures

The effect on the metro works of existing structures in the vicinity is, essentially, that of generating the lateral earth pressures due to the surcharge effects through the foundations. The manner, in which these loads are transmitted on to the metro works at various stages of its construction, depends on the type of the foundation transmitting the loads, the type of the soil medium and the geometry of the problem. This is dealt with in Chapter 14.

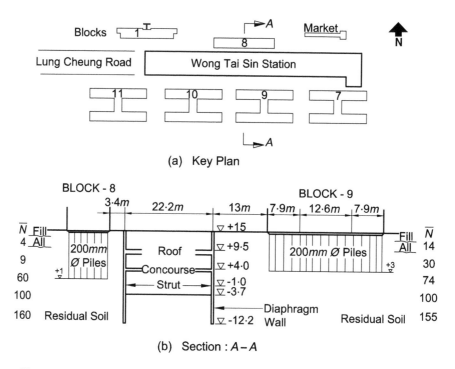

(a) Key Plan

(b)   Section : *A–A*

Figure 9.2 Wong Tai Sin Station Hong Kong (adapted from Morton *et al.*, 1980)

9.4 Future metro-related construction

Any construction activity required or likely to be carried out in the vicinity of a metro structure in the future may be metro or non-metro related. In the case of the metro-related construction activity, the expansion of the existing metro network may require the construction of an interchange connecting an existing station with a new one or, simply, the construction of new running line in close proximity to the existing one. Where this may be feasible, the anticipated future provisions should be built into the current metro structure in such a way as to facilitate the accommodation of the future integration with the new works. This is often easier said than achieved, since this calls for an attempt to be made in the current contract to second guess the requirements and the *modus operandi* of the future contract. In reality, it would amount to the future designer and the contractor being left with no choice but to follow the dictates laid down by the current thinking. Besides, any attempt at ensuring that the future development turns out to be cost-effective would entail a detailed examination and comparison of the various possible options in the current contract. This would, inevitably, load up the 'current' time, design and construction costs. However, even so, there would be no guarantee that, in the event of the future development going ahead, the provisions built in the current contract would, necessarily, continue to be appropriate and meaningful later on. In view of this, there may be a persuasive argument against this course of action. So, before arriving at a firm conclusion, the arguments for and against must be carefully weighed. Towards this end, some of the pros and cons are listed in the following section.

Arguments for

- If the scheme is sensible and based on sound principles, even if unlikely to have incorporated or allowed for all the eventual requirements, it can have the potential to save significant future costs. However, this is conditional upon the future construction definitely going ahead and in the manner envisaged currently.
- In the event of it going ahead as envisaged, the new structure can be designed relatively simply and constructed without the likelihood of major problems.
- Without the necessary provisions having been allowed for and incorporated in the current design and construction, the new construction could impose far too onerous stress conditions and, potentially, cause the distress of the existing structure avoidance of which could warrant recourse to major and expensive preventive measures.

Arguments against

- Building-in provisions for future metro works, which may or may not come about, can load up time, design and construction costs of current metro works.
- Introduction of such provisions for future requirements may be based on schemes which, given the demands and pressures of time and other prevalent constraints under current contract, are unlikely to have been fully thought through, to have foreseen all the possible design and construction constraints or to have appreciated all the potential implications and may therefore not prove to be quite adequate eventually.
- Such an approach could turn out to be intolerant of any change in thinking in the future. Consequently, any pressures for meeting the requirements of new thinking could render the existing provisions inappropriate and therefore wasteful.
- Defining the design and construction requirements for the future structure in the current contract is likely to impose constraints on the future planning and could severely restrict the freedom of choice in the approach to design and construction in the future.
- Imposition of such restriction on the freedom of choice can be expected, inevitably, to invite a cost penalty in the future design and construction.

Where the metro station location ends up being away from a thoroughfare either fortuitously or by design, and where the location so permits it, commercial considerations may dictate that an extension of the metro structure above ground is constructed for commercial use. This can be allowed for, if known at the time, by building-in specific advance provisions into the design and construction of the metro structure. Failing this, it is possible to incorporate some minimal blanket provisions which could allow for at least some future extension. However, lack of adequate provisions in time is likely to restrict the extent of possible future extension above ground. Notwithstanding this, the possibility of constructing a commercial structure directly above the metro even if no provisions have been incorporated previously cannot be ruled out. However, in that case, the extension may need to rest on huge transfer beams clear of the metro structure and be supported on independent foundations either side of it so as to avoid imposing extra load on the structure assuming always that such space is available. While such a possibility cannot be ruled out, it is likely to carry a significant costs penalty.

9.5 Future non-metro construction

As far as the future non-metro works are concerned, they may include structures which are likely to be located either directly on top of the metro structure or flanking either side of it. Examples of the former type could be a highways structure, e.g. an elevated road or an underpass, whereas that of the latter, buildings with deep basements or those founded on piles.

Where construction land is at a premium and economic considerations predominate, it is conceivable that, far from siting them a 'safe' distances away, the new construction works may be required to almost abut the station structure. In such a situation, there is often a great temptation to found the structure partly on the station box structure. But, if problems of interaction between the two structures, such as those resulting from differential settlement, are to be avoided, such temptations should, as far as possible, be resisted and the foundations for the structure maintained independent of the metro box structure. However, in the exceptional circumstances where this may not be possible, every attempt should be made to achieve compatibility, insofar as it may be possible, between the two structures thereby minimizing the potential of problems likely to arise out of interaction.

Elevated expressway above metro

In the case of an elevated expressway, whether it runs along the length of, or orthogonal to, the metro structure below, it is possible to avoid, relatively easily, their potential interaction by ensuring that the expressway neither relies on it for support in any way nor imposes any direct or indirect (surcharge) load on it. The pier supports to the expressway may be so located clear of the metro structure below as to achieve this without much difficulty.

Road or pedestrian underpass

In the event of construction of an underpass or pedestrian subway to be carried out across and on top of a metro structure, interaction problems can be expected to arise from two aspects which need to be addressed adequately. These are:

- Loss of gravity load upon removal of soil overburden
- Application of specific loads during the construction and after the completion of the underpass.

Removal of soil overburden in order to accommodate the underpass will cause a reduction in the factor of safety against flotation of the metro structure locally over the width of the underpass. This may not present a problem, *per se*, so long as the global factor of safety (i.e. over the entire length of the structure) continues to be adequate and provided that there is no danger of any distress to the structure locally. However, if the calculations reveal the global factor of safety to be inadequate, appropriate measures would need to be introduced to ensure continued existence of appropriate margins of safety both during the construction of the underpass as well as its operational life time.

Walls of the underpass could be designed as deep girders to carry, in the long term, whatever load they may be required to, across the metro structure. However, during construction, the weight from the wet concrete would have to be carried by the roof slab of the metro structure below in a manner that would not cause any distress to it. The path for transmitting the load has to be such that the structure is able to sustain these loads safely. However, this may not present a problem since, on balance, with the more weight removed than that replaced, there is the likelihood of a net reduction in the load imposed on the structure below. But then, because of that, the adequacy of long-term safety against flotation could be in question and would need to be investigated.

Deep basement construction

Construction of a deep basement adjacent to a metro structure can give rise to a number of interaction problems at various stages. In particular, a significant depth of soil excavated alongside could amount to a removal of a substantial lateral support to the metro structure. In view of this and the consequent relief of stress, movement of the ground and the structure can be expected. The magnitude of the movement and the severity of the potential problem will also depend, *inter alia*, upon the relative size of the excavation, i.e. its length and depth. If the basement is shallower than, and extends over a short length of, the metro structure, it may be possible, due to the inherent stiffness of the structure, to redistribute the loads in such a manner as to limit the potential movements to acceptable proportions without causing any structural distress. Otherwise, the manner in which the excavation and construction of basement is carried out and the lateral support to the structure replaced without impairing its integrity must be established and agreed upon in advance.

Furthermore, if the depth of the basement to be constructed is comparable with or deeper than that of the metro structure itself necessitating deeper excavation and dewatering, problems could also arise due to the differential settlement of the structure if the wall is not taken down deep enough to limit the effects of undercutting and draw-down under the structure. All these factors must be taken into account in deciding whether or not, and if yes, in what manner the basement construction in the vicinity of the metro structure can be permitted. For the successful design and construction of such structures, it is often stipulated as a mandatory requirement that the excavation and construction are carried out in a phased, piecemeal fashion to ensure that movements which could have significant impact, not only on the alignment and operation of the structure but also on design, are avoided at all costs.

It is not uncommon to construct a new underground structure in the vicinity of, or intersecting with an existing structure. The construction of the second stage of Prince Edward Station in Hong Kong is a prime example of such serious interaction. The design and construction of the station was carried out in two longitudinal halves of dissimilar widths in two stages at different times and under separate contracts.

Construction of the eastern half, i.e. Stage 1, as part of the Kwun Tong Line was carried out between the permanent eastern wall of the station and a parallel temporary boundary wall. The temporary wall was off-centre towards the east thereby making the western half, i.e. Stage 2, the wider of the two halves. By the time of the commencement of the Stage 2 and over the period of its construction, the tracks over the two levels of the first half of the station were already in place and operational.

In ensuring that the construction of Stage 2 and its integration with the already constructed and operational Stage 1 of the station structure were carried out safely, successfully and without disrupting the operations of the tracks in the existing half, it was imperative to take cognizance of the following requirements:

- Achieving harmony with the assumed boundary conditions and the behaviour on which the analysis and design of the complete station box structure had been envisaged in Stage 1 design;
- Achieving compatibility of the Stage 2 design with the existing rebar provisions in the Stage 1 construction;
- Ensuring control of movements of the existing Stage 1 of the structure to within acceptable limits during the construction of the Stage 2 of the structure;
- Satisfactorily addressing the potential effects of interaction with the existing services above the Stage 2 roof slab and the existing structures immediately to the west of the Stage 2;
- Appropriately addressing the potential effects of global flotation.

For a safe and successful completion of the structure, the following operations in the construction of Stage 2 were followed:

- Top-down sequence of construction was adopted under the temporary traffic deck.
- Depth of first stage excavation (*en masse*) was decided upon by the considerations of the extent of the acceptable heave.
- Second stage excavation to the underside of the Stage 2 roof slab to the west of the temporary boundary wall was carried out, the formation blinded immediately upon reaching it to prevent contact with outside water, and the construction of the slab carried out in step-by-step, piece-meal fashion.
- After substantial portion of the roof slab was constructed in this manner, excavation was advanced half-way down towards the concourse slab to allow installation of additional temporary struts midway between the floors. Thereafter, the excavation was progressed down to the underside of the concourse level and the slab at this level constructed, piecemeal, in similar fashion as that followed at the roof slab level.
- Same procedure was repeated at the upper and the lower track slab levels.
- Permanent columns were installed, bottom-upwards, from the lower track level up to the roof level and the temporary intermediate struts removed.
- After removal of the 'central' temporary support wall down to the underside of the roof slab, integration of the two halves, i.e. Stages 1 and 2, of the roof slab was completed in piecemeal fashion. The procedure, i.e. the cycle of removal of the wall and integration of the slab, was repeated sequentially at other floor levels.
- Monitoring was carried out throughout the construction period.

Piled construction

Pile driving operations can generate significant ground movements extending a considerable distance beyond its own boundaries. Such an operation, if undertaken in the vicinity of a metro structure, irrespective of whether it is already constructed or under construction, can therefore be of great concern to the engineer. The nature and the magnitude of potential problems faced can depend upon a number of factors. They are:

• Type and spacing of the piles
• Characteristics of soil penetrated
• Method and sequence of installation
• Relative timing of the piling operation.

In pile driving, the net available impact energy is expended partly in advancing the pile and partly in the elastic deformation of the surrounding soil mass. It is the latter component of energy that generates vibrations. In soft or loose soils which are easily penetrated, most of the available net energy is dissipated in advancing the pile. However, loose deposits of uniform fine and silty sands, especially those below groundwater table, when subjected to vibrations due to pile driving, are susceptible to large settlements. A number of authors (Swiger, 1948; Lynch, 1960; Feld, 1968; Terzaghi and Peck, 1968) have reported settlements ranging from 150 to 450*mm* due to pile driving in loose and medium-dense sands and gravels. Significant settlements are also known to have occurred even as far away as 15*m* from the pile driving. To limit the settlements in sands, therefore, the level of vibrations due to pile driving must be controlled. Low displacement piles do not appear to be particularly effective in achieving this. However, jetting ahead of the piles, if carefully controlled to prevent undermining the support, can be effective in reducing vibrations (D'Appolonia, 1971).

In fine-grained soils, the response to pile driving is both immediate and time-dependent, i.e. short-term, undrained, and long-term, drained. As the pile is driven in, it attempts to displace, radially outwards, a volume of the clay equal to its own driven volume. However, the surrounding soil, being unable to suffer a volume change in the short term, cannot accommodate the displaced soil which is therefore forced upwards. The extent of the resulting soil disturbance caused around a typical pile driven in clay can be divided into three zones (Zeevaert, 1950) as shown in Figure 9.3. The effects experienced in these zones and the surrounding ground are, generally:

• Bearing capacity failure under the pile tip enabling the pile to advance;
• Generation of excess positive porewater pressures. These pressures can reach values well in excess of the initial effective overburden pressures;
• Shear distortion and remoulding of the soil annulus in Zone I, which may be about half the pile diameter wide;
• Relatively smaller shear distortion but existence of high enough radial pressures causing shear failure in Zone II. This zone may extend to several pile diameters; and
• High radial pressures but not enough to cause a failure in Zone III.

The changes in the states of stress and strain experienced by an element of soil at the boundary between Zones II and III are traced (D'Appolonia, 1971), qualitatively,

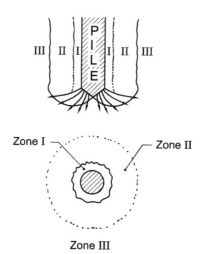

Figure 9.3 Zones of Soil Disturbance around Single Pile Driven in Clay
(adapted from Zeevaert, 1950)

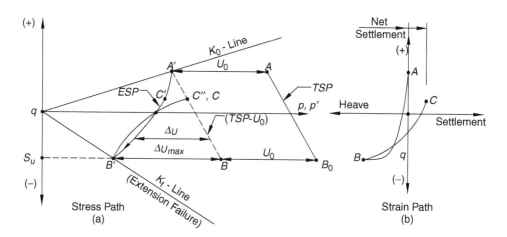

Figure 9.4 Stress Changes and Displacements accompaying Pile Driving in Clay
(adapted from Lambe, 1974)

in Figure 9.4. Although, in reality, the stress changes induced by pile driving are much more complex than that indicated by the figure, it does serve to illustrate that the change in porewater pressure is related to the stress changes and the movements that accompany pile driving.

Using the Stress Path Method (Lambe, 1967), the changes in the stresses have been plotted on a two-dimensional $q':p'$ and $q:p$ stress space. The stress invariants representing the axes are as follows:

$$q = \frac{1}{2}\left(\sigma_V - \sigma_H\right) = \frac{1}{2}\left(\sigma_V' - \sigma_H'\right) = q';$$

$$p = \frac{1}{2}(\sigma_V + \sigma_H); \quad p' = \frac{1}{2}(\sigma_V' + \sigma_H'); \quad p' = p - u$$

where, σ_V, σ_H : *vertical and horizontal total stresses*
             σ_V', σ_H' : *vertical and horizontal effective stresses*

During the pile driving operations, the vertical total stress σ_V remains effectively unchanged whereas the horizontal total stress σ_H and the excess porewater pressure u go on increasing. Increase in the porewater pressure thus causes a reduction in the effective stress and a corresponding loss in the shear strength of the soil.

Point A' represents the initial effective stress condition before pile driving and point A_o the corresponding total stress condition. Line $A_o B_o$ represents the total stress path (*TSP*) and $A'B$ the stress path for the total stress minus the initial porewater pressure (assumed static). Undrained effective stress path (*ESP*) resulting from pile driving is represented by $A'B'$. It meets the locus of the failure line, K_F, at B'. At this point, the element approaches a state of incipient failure as the shear stress equals the shear strength of the soil. The separation between B and B' represents the excess porewater pressure Δu (max) induced by pile driving and is given by Equation 9.1:

$$\frac{\Delta u \,(\text{max})}{\sigma_{Vo}'} = \left[\frac{2s_u}{\sigma_{Vo}'} + 1 - K_o \right] \cdot A_f \qquad (9.1)$$

where σ_{Vo}' : *Vertical effective overburden pressure*
          s_u : *Undrained shear strength*
          K_o : *Coefficient of earth-at-rest pressure*
          A_f : *Skempton's porewater pressure parameter at failure*

During the undrained loading, i.e. increase in the horizontal stress, the soil element also undergoes an extension in the vertical direction representing the elastic heave. This is represented by AB in the strain diagram on the right. Subsequent dissipation of the excess porewater pressures with time causes consolidation of the soil under its own weight and the settlement of the ground surface. This is represented by the curve BC on the strain diagram. During this phase, again, the vertical stress remains essentially unchanged as the horizontal effective stress increases. Point C represents the final effective stress condition when the ambient porewater pressure is restored. The end result is usually a net settlement.

Figure 9.5 serves to illustrate the foregoing close correspondence between the excess porewater pressure dissipation and the ground movement as obtained at the MIT campus.

The extent of the surrounding area over which the pile driving operation can have significant influence, is a function of the type and thickness of the clay layer being penetrated. Significant movements can be expected at least as far away as the thickness of the layer. However, if there is a weaker layer overlain by a stiffer surface layer, the effect could spread over a larger area. There is also some evidence (D'Appolonia, 1971) to suggest that the less plastic and more sensitive (defined in Appendix C) clays may exhibit smaller radii of influence than that by more plastic insensitive clays. Sensitive clays, when subjected to large shear distortions during pile driving, suffer a reduction in strength to their remoulded values. This softening of the soil annulus surrounding the

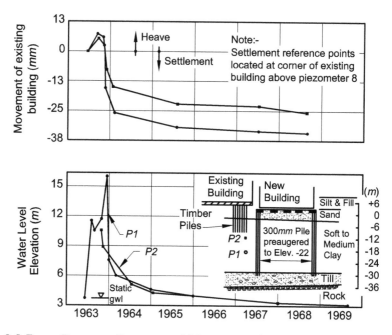

Figure 9.5 Excess Porewater Pressures and Movements of Nearby Building caused by Pile
Driving on the MIT campus (adapted from Lambe, 1967)

pile subscribes to an increase in its compressibility which tends to limit the propagation of movement away from the pile.

The type of piles used and the number of piles required per unit area (i.e. pile density) are also important factors influencing movement. Piles with large cross-sectional areas will displace a larger volume of soil and are likely to induce larger movements than those with small cross-sectional areas; also greater the density, greater, potentially, the movement of the surrounding ground. To limit the movements due to pile driving in soft to medium clays, the volume of soil displaced should be kept to a minimum by the use of low-displacement, high-capacity piles such as steel *H*-piles (D'Appolonia, 1971). Alternatively, the soil disturbance can be considerably reduced by installing piles into prebored holes.

Figure 9.6 illustrates the effectiveness of preboring prior to pile driving. The maximum excess porewater pressures induced at various times during pile driving have been taken from three construction sites at *MIT* and plotted as nondimensional ratios of the vertical effective overburden pressures against the depth below the ground level. These are compared with the theoretical maximum excess porewater pressures which correspond to the increase in the horizontal total stress.

Solid dots relate to cases where preboring was carried down to −25·9*m* level and the piezometers were located about 3*m* from the closest piles. Below the probed horizon, the measured values of the excess porewater pressures do not appear to be much different from the theoretical values (D'Appolonia, 1971). However, above this level, the maximum excess porewater pressures induced are only about 50 per cent of their maximum theoretical values. Its significance in limiting the ground movements is therefore obvious. The solitary circle at −29*m* level represents the case where piles

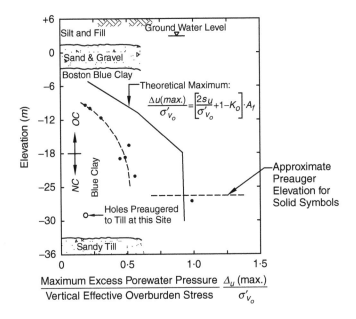

Figure 9.6 Effectiveness of Preboring Prior to Pile Driving on *PWPs*
(adapted from D'Appolonia and Lambe, 1971)

are driven very close to a tall smokestack supported on a shallow foundation, and the movements had to be kept to an absolute minimum. Full depth of the clay was therefore prebored before driving the piles. Although one pile was driven hardly a metre away from the piezometer, excess porewater pressures reached no more than 20 per cent of their maximum theoretical values.

The order in which the piles are driven can have an important effect on the development and distribution of porewater pressures in the surrounding ground. The presence of the previously driven piles tends to make the zone stiffer and therefore more resistant to deformation. The ground movements accompanying subsequent pile driving are therefore likely to be greater in a direction away from this zone. Field measurements (Hokugo, 1964; Hagerty, 1969; D'Appolonia and Lambe, 1971) confirm this. To minimize the effect of pile driving adjacent to metro works, it follows that the piles nearest to the works should be driven first and thereafter the driving should proceed away from the works.

Notwithstanding the foregoing, it is conceivable that a metro station structure, owing to its structural stiffness which is further enhanced by its confinement underground and the mobilization of frictional resistance at the soil–structure interface, may be able to sustain and distribute any movements and out-of-balance forces satisfactorily. However, in deciding whether or how close to a completed metro structure pile driving should be permitted, the following questions must be satisfactorily answered:

• Can the structure and the track(s) tolerate movements and increased loads? If not, is the degree of distress likely to be suffered, acceptable and manageable?

- Can measures minimizing the effects of pile driving be satisfactorily introduced and is the attendant cost escalation acceptable?
- Is there a chance of a serious problem developing? If so, can the potential implications be managed safely?

Regardless of the soil conditions encountered, large movements have been known to have taken place within $10m$ of pile driving operations. Large movements at even greater distances can also be expected if pile driving is carried out through deep deposits of soft clay. Pile driving close to a deep metro excavation can therefore be fraught, potentially, with major problems and should be avoided as far as possible. However, if other overriding considerations dictate that the pile driving in the vicinity cannot be avoided, it is important to ensure that adequate measures are taken to minimize its impact and that the metro structure is closely monitored.

9.6 Other considerations

Communicating aquifer

Where an underground metro structure is to be constructed close to a potential source of water, i.e. a river, reservoir or sea, it is essential to establish whether there exists an aquifer under the site of works which could be in communication with that source and what effect the fluctuation in the water level of the source might have on the flow characteristics of the subsurface water in general and metro works in particular. If such a communication does exist and the soil starts to wash out as a result of groundwater flow, there can be the danger of subsurface erosion forming a channel in the ground. With the consequent increase in the hydraulic gradient and erosion, large quantities of soil could be washed out leading, potentially, to serious settlement problems. To prevent such an occurrence, penetration of the perimeter wall deep enough to cut off the flow into the excavation may be necessary. However, if the groundwater flow is likely to pose problems during the construction of the perimeter wall itself, a secondary bentonite or grout curtain may need to be installed between the intended location of the perimeter wall of the metro and the source of water. In order to make a correct assessment of the problem and devise appropriate measures, the depth and the location of the aquifer in relation to the metro works and all the attendant implications during the various stages of construction should be carefully looked into.

Metro corridor

During the planning, design and construction of a metro, it is not always possible to anticipate and build-in allowances for all future non-metro construction works likely to be carried out in the vicinity of and affecting the metro route and structures. It must also be appreciated that, social, political and commercial pressures may not allow the future construction activity bordering on the route to be totally avoided. In order, therefore, to ensure that unacceptable interaction problems during future construction works are avoided, very often a tract of land of a certain minimum width, symmetrical about the centre line along the entire length of the route, is identified within which the new proposed works are subject to appropriate restrictions. This strip of land is commonly referred to as the Metro Corridor.

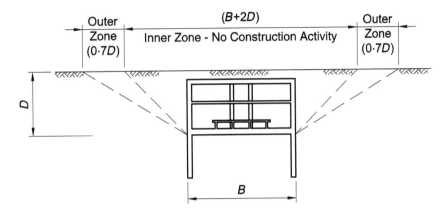

Figure 9.7 Typical Metro Corridor

Figure 9.7 defines the typical cross-section of such a corridor. It comprises an inner zone within which no future major non-metro works may be generally permitted, and the two outer zones within which such works may be permitted but only if their design and construction incorporate demonstrably adequate safeguards for the existing metro structure and its safe operation. Interaction problems due to construction activity outside the corridor are likely to be of little significance.

10 Conceptual Design

10.1 Introduction

Different stages in the design development process of cut-and-cover metro structures may be broadly classified as follows:

- Conceptual Design
- Defining Design Parameters and Criteria
- Assembling Loadings and Load Combinations
- Detailed Analyses and Design.

Conceptual design represents, arguably, the most important stage in the design development process. During this stage an attempt is made to identify the requirements of every discipline and to appreciate their mutual impact. In an attempt to reach satisfactory accommodation of the diverse and, at times, seemingly conflicting requirements, a great deal of interaction is required amongst the various disciplines. This calls for close liaison and understanding within the design team. As the design interaction progresses, thinking is continually influenced and progressively modified with a view to reducing any conflict of interests and in an attempt to establish the core requirements embracing all the disciplines. All the relevant requirements assembled thus during this stage may impose demands and limitations, i.e. constraints, on the final solution. These constraints have to be duly identified and assembled. Their interaction and the implications have to be examined so that the solutions that best satisfy the requirements can be identified.

A major part of the design development process at the conceptual stage involves important decision making. While some of the decisions may be based upon subjective reasoning, a large number of these are, inevitably, influenced by the qualitative and quantitative evidence available from past experience from similar types of projects.

The appreciation of the potential problems of construction and their implications on design, the extent of close inter-disciplinary liaison and the general quality of thought and care that go into the design process at this stage do, to a large extent, reflect in the quality of the final detailed design achieved and also, potentially, in the standard of construction that may be possible to achieve.

The four principal steps in the development of a conceptual design of a typical cut-and-cover metro structure are:

- Identifying prevailing constraints
- Resolving conflicts and defining order of priorities

- Identifying the preferred sequence and method of construction
- Establishing the threshold for the basic shape and size of the structure.

10.2 Identifying constraints

The first and foremost part of the conceptual design process is one of recognizing constraints. These may be internal or external. Internal constraints generally arise as a result of the requirements specific to the various disciplines within the design team. These are usually identified during the meetings of the design team. External constraints may be those that are likely to arise from the potential interaction of the metro structure to be constructed with the existing or proposed future structure(s) within its zone of influence. For proper identification of the constraints, a sound knowledge of the available construction methods and technology, materials and costs, an appreciation of the architectural, planning, environmental, sociopolitical and cultural issues and of course sound engineering skills are essential.

Some of the principal constraints are briefly discussed as follows:

Functional

The primary function of a metro system must, by definition, remain one of rapid mass transportation. However, there is a school of thought that advocates the possible use of underground metro structures to double up as potential wartime shelters as well. As a proper shelter, the structure may be specifically designed to provide 'full protection'. The manner of this protection, whether provided by an independent, external shield to the structure, or achieved by strengthening the structure itself, and the desired degree of protection against specified weaponry, i.e. conventional or nuclear, will call for different treatments of the problem which will impose their own specific constraints. Furthermore, in the case of sheltering the structure against conventional attack, whether the plant area is fully protected, or the plant and the plant area are unprotected but duplicated and spaced sensibly apart, enabling one lot to be treated as sacrificial to damage, need to be considered. All these constraints will, inevitably, have a marked influence not only on the size, design and detailing of the various structural elements but also on the overall size, layout and planning of the box structure as a whole and, of course, the cost. Alternatively, the metro structure may be designed primarily to conventional, nonshelter requirements but with the amount of reinforcement increased to such an extent and detailed in such a manner as to offer, additionally, at least some measure of protection against small-scale conventional weaponry as well.

The choice as to whether a metro system is to function exclusively as a mass transit system or as a partly protected, or a specifically designed, 'fully protected' underground shelter as well, is likely to be a political one and would, primarily, rest with the government of the day or the leadership of the country. However, be that as it may, the possibility that the engineer will be called upon to provide, at least, a comparative cost analysis to assist the authorities in reaching their decision should not be ruled out.

Design of the metro system additionally as a wartime shelter falls within a specialist field and is, as such, beyond the scope of this book. For information on this subject, reference should be made to pertinent specialist literature.

Architectural

Architectural criteria are generally associated with the planning standards and are perceived, primarily, to be concerned with the allocation and design of space, i.e. circulation areas, headroom, accommodation of services, etc. While recognizing that an efficient and safe movement of passengers in metro stations is the primary requirement, pleasing aesthetic appeal is an equally important aspect in the spatial design of the stations. Metro architecture is very often required to display indigenous cultural and historic bias in its aesthetic treatment of the stations. Besides, ensuring a modicum of individuality for every station may, insofar as it remains possible, require a change in the aesthetic treatment from one station to the other. All these factors can have a contributory influence on the overall size and layout of the structure, its internal planning, and location, size and spacing of any internal vertical support elements, such as columns, walls, etc. Both, the necessary planning requirements as well as the desired aesthetic aspirations, can highlight their own demands and therefore impose their own brands of constraints which need to be adequately identified and addressed.

Alignment related

Problems of horizontal and vertical interaction with existing structures and their foundations likely to be associated with a chosen alignment of the route can present their own peculiar constraints which need to be identified and satisfactorily addressed. For example, the extent of the backfill depth required for the accommodation of the services on top of the structure, whether the metro route has to cross under a river or an existing or a proposed future obstruction such as an underpass or another metro route, etc., are some of the factors which are likely to influence the required levels at which the rails are to be placed, the depth at which the roof of the structure can be located below the existing ground level and hence the vertical alignment of the route. Existence of major high-rise structures with their deep foundations, if in conflict with the most direct and preferred horizontal alignment, may also present major problems of design and construction. Rerouteing the metro, therefore, so as to minimize such conflicts may, however, involve extra land-take and, in turn, invite other constraints. All these aspects can have a significant effect on the design, method and sequence of construction and, inevitably, on the cost of the structure also.

Traffic related

Volume and intensity of the existing traffic on the surface, its likely disruption due to the construction activity and the availability, or otherwise, of a suitable, alternative route for its diversion will dictate whether or not a temporary traffic deck is required and if it is, what effect it might have on construction and design. However, in selecting an alternative route, proper assessment would need to be made of and careful consideration given to what impact it might have on the local environment. For instance, it would be important to ensure that the diverted route is able to cope with the increased volume and intensity of traffic in an acceptable manner and that it does not, through its inefficiency or proximity to the general area of primary disruption, actually subscribe to a widening of the zone of disturbance. In the case of the existing properties directly abutting the site of works, the problem can become further exacerbated by the need to maintain appropriate access points and routes for pedestrians and emergency services. All these factors need to be carefully assembled and studied for their potential impact on design and construction.

Utilities related

It is essential to carry out an extensive survey of the existing services along the metro route in order to establish what constraints they are likely to impose on the construction works. It is also important to establish whether they can be temporarily diverted for the duration of the works and subsequently reinstated, or permanently rerouted. Where the diversion of the existing services is not possible or feasible for reasons of practicality or cost, they may need to be supported on temporary structures during construction; in that case consideration of phasing-in of such works within the main construction activity and their implications on design need to be taken into account.

Environmental

Proximity, or otherwise, of the metro works to sensitive structures such as hospitals, important monuments of religious or historic interest, etc., and the availability or the lack of adequate construction corridor around the perimeter of the structure can influence the method and sequence of construction to be adopted and so too, design. For example, it may be perfectly feasible to construct a metro structure located in a 'green-field' site in an open cut with stabilized side slopes and in the conventional bottom-up sequence, without the settlements of the surrounding ground posing any serious concerns or constraints. However, in an urban environment especially in the proximity of high-rise structures, it may become necessary to adopt top-down sequence of construction in order to ensure that the constraints inherent in or implied by the acceptable levels of movements in the surrounding ground and structures are not violated. This will, in turn, also have a bearing on the design.

Geotechnical

Excavation for and construction of metro structures in high-density urban areas and within difficult ground conditions and high water table can present particularly severe engineering problems and impose constraints which need to be taken into consideration in evolving a commensurate design and an appropriate method and sequence of construction. The knowledge of the type of ground and its drainage characteristics, the *in situ* stresses and the ambient porewater pressure regime will, at least qualitatively, enable some idea of the likely ground response during the various stages of construction to be formed. This should forewarn the engineer and enable him to decide what constraints to expect, what measures and allowances to incorporate in his design and how to proceed with construction so as to avoid, or at the very least minimize, the potential problems which might otherwise be faced.

Geographic

The geographic location of a metro system, for instance in an area of seismic activity or exposure to the hazards of typhoons or floods, will dictate requirements and constraints specific to such locations and activities which could have an influence on the timing and the method of construction chosen and, of course, design. Such requirements are area or country-specific and are generally covered either by the national codes of practice, local regulations or appropriate bylaws and need to be incorporated in the design criteria to be followed for the design of the metro structures.

Economic

The focus of an efficient engineering solution must include cost-effectiveness, i.e. achieving maximum economy while retaining those features which are essential for safety, constructability, reliability, efficiency, comfort and maintainability. Costs of metro projects generally run into billions of pounds (sterling) and, in terms of their scale, are matched by few other engineering projects. Therefore, even small percentages of savings achieved by judicious engineering can represent significant sums of money. It is therefore important to identify whether there are any direct cost constraints likely to be imposed, for example, by the scarcity of certain construction materials and the need to import these, or, whether there are any indirect cost constraints by virtue of any programme implications inherent in a particular method or sequence of construction. These can be particularly important for making cost comparisons in weighing up the merits of the different schemes.

In addition to all the constraints explained in the foregoing section, statutory constraints by way of any specific rules, regulations or bylaws which may be in operation nationally or locally and may have a bearing on the construction and design of the structures also need to be identified.

10.3 Resolving conflicts

Having identified and assembled the various constraints as briefly discussed earlier, the next step in the conceptual design process is to identify areas of potential conflict and, where it may be possible to do so, resolve these through understanding and compromise. Failing this, priorities and the governing principles need to be agreed and established.

In their bid to jockey for preferred positions of advantage, it is not uncommon to see various disciplines, initially, scaling up the extent of their respective requirements which, on closer scrutiny, may turn out to be a lot more than might be actually necessary. That this is, at times, a deliberate ploy to retain enough comfort and spare capacity for favourable trade-offs at later stages is not all that unusual or uncommon and should be borne in mind. As the design process develops, any problems arising do generally get resolved by compromise within the design team on the basis of appropriate and sensible trade-offs amongst the various conflicting requirements, bearing in mind the considerations of cost, practicality of construction, engineering principles, aesthetics and any other overriding constraints.

Where there is a proposal for a new metro line or structure, with or without an interchange with the current metro works, to be constructed in the future, due consideration should be given to the potential problems of integration and interaction. This may call for certain specific provisions of design and construction to be taken on board in the current metro works. However, the areas of major conflict, generally, result from the potential interaction between the metro and the non-metro works, especially if both are scheduled for construction at roughly the same time. This can pose some interesting and challenging problems. For example, which construction should have precedence? If the metro is constructed first, should it incorporate provisions in its design and construction to allow for the effects of the subsequent construction? Or simply, should stipulations be made demanding of the later construction demonstrable assurances that it does not impose any onerous boundary or loading conditions on the structures existing at the time and, in consonance therewith, take on board the associated works and costs?

Although from a purely technical view-point, one could be forgiven to think that the dominant engineering constraint might form the basis for the preferred solution in a given set of circumstances, it must be appreciated that the priority of construction, and hence the governing solution, may not always be based on or dictated by the engineering considerations exclusively. In any case, all points of conflict, whether a metro structure interferes with an existing structure or is likely to be interfered with by a future development, would have to be resolved in a manner acceptable to all the parties concerned and any resultant interaction problems remaining would have to be duly taken on board in the design process.

10.4 Sequence of construction

In underground structures, unlike those above ground, the sequence of construction adopted can often have a significant influence on the design. In fact, it invariably turns out to be the dominant design criterion as is demonstrated by some solved examples in Chapter 26.

A top-down sequence of construction, for instance, is likely to allow relatively less movement in the surrounding ground and structures than would be the case with a conventional, bottom-up sequence. So, where a bottom-up sequence is expected to give rise to excessive and unacceptable movements, as may generally be the case with high-rise structures founded in adverse ground and groundwater conditions and very close to the site of the works, a top-down sequence may have to be adopted. It is therefore essential to establish the extent of both the anticipated as well as the acceptable movements in order to decide upon the preferred sequence of construction. However, ground movements are dependent upon a number of different variables which makes the process of quantifying these accurately rather difficult. Nevertheless, some guidance on these aspects can be found in Chapters 7, 8 and 9 on Interaction.

10.5 Threshold for basic size

Finally, when all the interdependent problems of the assembled constraints have been satisfactorily resolved, the extent of the engineering options within which alternative solutions have to be explored, can become clear. Simplified analyses and design procedures are then applied to the various options and a short list of the most likely solutions is obtained. The suitability of each solution is then assessed in terms of its structural efficiency, method of construction and cost. In certain circumstances, in view of the diversity of constraints and problems of interaction, it should be recognized that a compromise may be the only sensible optimum solution that can be hoped for.

The design process, in a strictly limited structural sense, starts with making certain simplifying assumptions with regard to the design parameters, relative structural stiffnesses, loadings and boundary conditions. A 'first-guess' structural form based on past experience and satisfying the basic, minimum planning standards and operational requirements is identified. An attempt is then made to break down the structural form from its indeterminate mode into discrete, determinate elements with one unique load path, which can then be easily analysed by the use of simple statics in order to arrive at a 'probable' structure, quickly.

However, it is useful to remember that approximations and simplifications do have their inherent limits of application which, at times, may neither be directly obvious nor readily identifiable. To guard against such seemingly acceptable simplifications leading to unsafe results, the engineer should identify the performance limits of the structure and its elements by calculating the maximum and minimum values of the stress-resultants, etc. True behaviour of the structure, however, is likely to lie somewhere between these limits which, in most cases, can be obtained from simple analyses or the application of standard formulae. The use of bounds should not only enable the engineer to develop a 'feel' for the behaviour of the structure but should also make it possible to obtain reasonable estimates of the dominant stress-resultants for use in the preliminary design. As a result, the feasibility of the proposed structural solution at the early conceptual stage can be sensibly assessed often without recourse to computer programmes or other sophisticated methods of analyses.

The structure thus defined sets down the threshold for the subsequent detailed design. How close the initial estimate turns out to be to the eventual solution will, of course, depend upon the past experience of the engineer in the design of similar structures and his appreciation of the site specific constraints and the implications of the proposed method and sequence of construction. If the proposed solution violates any of the assumptions made, falls foul of any of the constraints identified or compromises any of the essential principles or requirements, appropriate modifications need to be made progressively until the desired compatibility is achieved leading, eventually, to a more definitive design.

11 Engineering Properties of Soils

11.1 Introduction

In the case of a typical cut-and-cover structure, owing to its subsurface confinement, the surrounding ground represents the principal source of load on and the medium through which other secondary loads are applied to the structure. However, it also provides the most direct means of support and stability to the structure. This phenomenon represents one of the unique features in the design of cut-and-cover structures. In view of this, the nature and the extent of the movement of the surrounding ground and the potential changes it might experience in its strength during the various stages of excavation and construction and the extent of the resulting interaction with other structures in the vicinity can be of great importance to the engineer. The rate of ground movement, i.e. its deformability, is also, among other things, a function of the rate at which the movement of porewater within the soil mass takes place. In relation to cut-and-cover structures, deformability, strength and permeability, therefore, represent the three most important engineering characteristics of soil. For a competent design and safe and successful construction of a typical cut-and-cover structure, sound appreciation of these characteristics and an understanding of their interrelationship are essential.

11.2 Effect of construction

Construction of a typical cut-and-cover metro structure involves the removal of a large volume of soil through excavation at different stages. Furthermore, to enable the excavation and other construction works to proceed in the dry, it also entails the removal of a significant volume of water. The resulting weight loss through such removal of overburden can cause significant pressure relief at formation.

The net weight of the overburden likely to be removed to make way for the construction of the cut-and-cover structure is, given its cellular geometry, inevitably, going to be well in excess of the weight likely to be replaced by the structure when in place. In other words, construction of the structure and the replacement of the backfill on top can be expected to restore only part of the weight loss on the subgrade. During the operational lifetime of the structure, therefore, the subgrade is likely to remain subject to a net pressure relief.

However, under certain circumstances, other overriding interests may dictate the construction of, say, a commercial structure directly above ground to be integrated with the metro structure below. In that event, it is conceivable that, under discrete load paths, there may be a net increase in the effective vertical pressure on the subgrade locally.

However, be that as it may, whether there is global pressure relief leading to heave or localized pressure increase causing settlement, whatever the type and the extent of soil deformation and the rate at which it might take place, a clear understanding of the underlying concepts leading to such changes is essential. These include the principle of effective stress, the mechanism of change in the *in situ* stresses and the strength, and the ground response or the deformability of soil during the various stages of excavation and construction. Appreciation of these fundamental principles is necessary in order to establish what impact the various changes are likely to have on design and construction of the structure and how they can be appropriately dealt with.

11.3 Drained and undrained states

In dealing with geotechnical aspects, the term 'porewater pressure' in relation to the subsoil water, and the terms 'drained' and the 'undrained' in relation to the type of response of the soil come up with regularity. In our comprehension of the behaviour of the subsoil environment in general and its response during the cut-and-cover construction in particular, it would be helpful to understand the import of these terms clearly.

The pressure of water in the pore space within the soil structure is referred to as the porewater pressure. The level at which the porewater pressure is equal to the atmospheric pressure represents what is known as the water table.

A condition is said to be 'drained' when, in response to a pressure change imposed on an ambient stress regime, the porewater is able to flow freely restoring equilibrium quickly, such as would be the case in a granular medium. However, in the case of cohesive soils, owing to the inherently fine-grained structure of the soil medium, there is a viscous retardation to the free flow of porewater. In such a medium, there are essentially two stages under an imposed pressure change. The initial stage, over the duration of which there is, for all practical purposes, no movement of porewater, is referred to as the 'undrained' stage. During this stage, the imposed pressure change is taken up directly by the porewater and constitutes what is known as the excess porewater pressure. The excess porewater pressure is positive if the imposed pressure change is an increase and negative if it is a decrease. The subsequent stage over which, with time, slow but perceptible movement of porewater takes place is referred to as the drained stage. The porewater movement continues to take place until the excess porewater pressures are 'fully' dissipated and the equilibrium porewater pressure regime is restored. During this stage, the imposed pressures hitherto built-up as the excess pressures in the porewater are, through their dissipation, transferred on to the soil fabric as the commensurate changes in its effective stress. The concept of effective stress is explained later on.

11.4 Deformation–strength characteristics

In the design and construction of cut-and-cover structures, two aspects with regard to the behaviour of soils need to be clearly understood. These pertain to:

• Stress distribution and deformation
• Stability or limit equilibrium.

In classical mechanics, the problems with regard to stress distribution and deformation are generally dealt with on the basis of an idealized elastic soil where properties are defined by a single deformation modulus (Menzies, 1973), as represented by Figure 11.1(*a*). For the solution of problems related to stability or limit equilibrium, on the other hand, the soil is idealized as being rigid-plastic, and the properties are defined by a single value of strength for a given effective normal stress, as indicated in Figure 11.1(*b*).

However, real soils approximate more to the elastic-plastic idealization of Figure 11.1(*c*). In the figure, point *F* represents failure. Horizontal extension beyond this point suggests that the soil continues to yield at constant stress after failure; although,

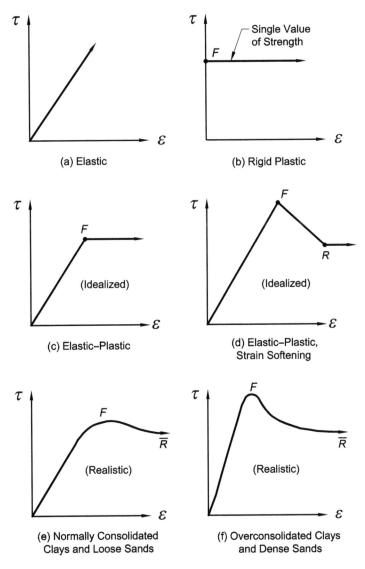

Figure 11.1 Idealized and Realistic Stress–Strain Relationships

with soils, this is not strictly so in reality. With a few exceptions, most naturally occurring soils are strain softening such that, beyond the point of failure, they continue to display a decrease in strength with increasing strain. Irrespective of whether the conditions are drained or undrained, they follow a stress-deformation relationship that is characterized by a peak followed by a drop in strength which continues until it stabilizes at an ultimate residual value as represented by point R in Figure 11.1(d).

Clays may be normally consolidated or overconsolidated. For their definitions, reference may be made to Appendix C. In normally consolidated clays and loose sands, the stress–strain behaviour is akin to the profile shown in Figure 11.1(e), indicating a very slight drop from the peak to the residual value – almost close to being flat. This characteristic may be approximated to the elastic-plastic idealization as represented in figure (c). However, in heavily overconsolidated fissured clays and dense sands, the drop from the peak to the residual value is much pronounced, as shown in Figure 11.1(f), which is also indicative of brittle behaviour. This strain-softening behaviour of the soil may be approximated to the elastic-plastic idealization shown in figure (d). Incidentally, in the case of overconsolidated intact clays, the drop from the peak to the residual value lies somewhere in between, i.e. more pronounced than that of the normally consolidated but less so than that of the heavily overconsolidated fissured clays.

11.5 Principle of effective stress

The difference between the total applied stress on a soil element in a given direction and the accompanying porewater pressure represents the effective stress on the element in that direction. Effective stress is a measure of the loading that the soil structure can transmit. However, it must not be confused with the intergranular stress although, for very small intergranular contact area as a proportion of the global area, the difference can be small. The principle of effective stress was postulated by Bishop (1959) in terms of two simple hypotheses:

1. The volume change and deformation in soils depend, not on the total stress applied, but on the difference between the total stress and the pressure set up in the fluid in the pore space (i.e. effective stress). This leads to the expression

$$\sigma' = \sigma - u \qquad (11.1)$$

   where, σ denotes the total normal stress, u denotes the porewater pressure, and σ' is termed the effective stress.
2. The shear strength depends not on the total normal stress on the plane considered, but on the effective stress. This may be expressed by the equation

$$\tau_f = c' + \sigma' \tan \phi' \qquad (11.2)$$

   where, τ_f denotes the shear strength, σ' the effective stress on the plane considered, c' the apparent cohesion, and ϕ' the angle of shearing resistance.

It is useful to remember that since water has no shear strength, shear stress τ will always be an effective stress, i.e. $\tau = \tau'$.

Simons and Menzies (1974) observed that while a change in the effective stress was always accompanied by a volumetric change, the reverse did not necessarily follow.

It was suggested that deformation was induced by a change in the state of effective stress whether or not there was an accompanied change in volume and it was therefore concluded that the sufficient and necessary condition for a change in the state of effective stress to occur is that the soil structure deforms – deformation occurring by volumetric or shear strain, or both.

In the relatively free-draining soils such as sands, in theory, there is no build-up of excess porewater pressure due to an imposed change in the total stress. This is because in such soils the dissipation of excess porewater pressure and, as a result and for all practical purposes, the restoration of the state of equilibrium are assumed to be instantaneous. Porewater pressure remaining the same, a change in the total stress therefore results simply in a change in the effective stress of equal magnitude.

In the highly fine-grained soils such as clays, on the other hand, movement of the porewater is time-dependent, and it can take up to many decades (Skempton, 1977) to restore equilibrium. In these soils, because of the inherent viscous retardation to the free flow of porewater, it is assumed that any change in the total stress, $\pm\Delta\sigma$, is reflected, in the short term (undrained stage), in a change in the porewater pressure, $\pm\Delta u$, of equal magnitude, i.e. $\Delta\sigma = \Delta u$, leaving the effective stress unchanged. This assumption is not unreasonable since any possible change in the effective stress generated by the dissipation of the excess porewater pressure through shear distortion is small and practically insignificant. The change in the porewater pressure sets up a hydraulic gradient in relation to the surrounding unaffected zones. Under this gradient, the excess porewater pressures start to dissipate causing, in the process, an exponential decay in the hydraulic gradient. With the dissipation of excess porewater pressure thus the soil structure experiences a change in the effective stress which is accompanied, naturally, by a commensurate change in the strength. When the dissipation is 'complete', i.e. $\Delta u = 0$, and the ambient porewater pressure has been restored, the change in the total stress will have been translated fully into the corresponding change in the effective stress. Stated mathematically,

$$\sigma_1' = \sigma_o' \pm \Delta\sigma = (\sigma_o \pm \Delta\sigma) - u_o$$

where, σ_o', σ_o and u_o refer to the initial (ambient) effective and total stress conditions and σ_1' refers to the final effective stress condition.

The foregoing conclusions are amply confirmed by comparing the findings of undrained and drained tests on a sample of soil. In the undrained test, increase in the total normal stress does not lead to any increase in the strength but results only in an increase in the porewater pressure of equal magnitude. In the drained test, on the other hand, since the excess porewater pressure is allowed to dissipate, the increase in the applied total stress is found to result in a corresponding increase in the shear strength. Experimental work by Rendulic (1937) also confirmed that, howsoever the values of the total stress, σ, and porewater pressure, u, were varied, it was always the stress difference '$\sigma - u$' (i.e. the effective stress) which controlled both the deformation and the failure (by implication therefore, also the strength).

It should be noted that, for very fine-grained soils at low levels of saturation, the simple effective stress relation does not hold. However, below the water table, it is not unreasonable to assume that, in most cases, the volume of entrapped air is likely to be so small that the saturation will be only marginally lower than 100 per cent. For all practical purposes, the assumption of full saturation is unlikely to be far wrong, and the simple effective stress relation is therefore assumed to be valid.

Effect of steady-state seepage

During the construction of an underground metro structure, to enable the construction activity to be carried out in the dry, lowering of water level within the perimeter cofferdam precedes excavation. Both these activities, i.e. dewatering and excavation, cause changes in the porewater pressure. However, if dewatering is continued until the porewater pressures no longer change with time, the rate of flow would become constant and a condition known as the steady-state seepage would be established.

During the steady-state seepage, since the porewater pressures remain constant, there is no deformation of the soil. In civil engineering, since strength and deformation are the two properties of most concern when dealing with soils, consideration of effective stress or, more specifically, change in effective stress, assumes great importance. In view of this, it will be particularly useful to examine the change in the effective stress from an ambient no-flow condition to a steady-state flow condition.

Consider a volume of soil of depth z and saturated unit weight γ_S. Assume the water table to be at a height d above the top of the soil. Figure 11.2(a) shows the total stress, the porewater pressure and the effective stress profiles for no-flow case assuming the conditions to be hydrostatic.

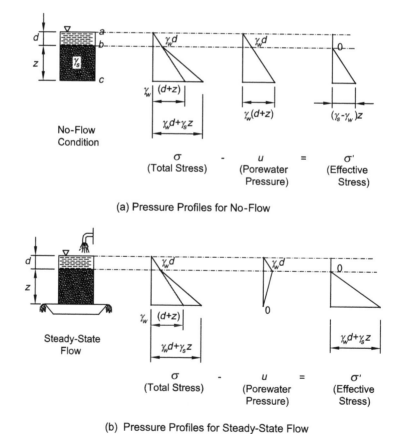

Figure 11.2 Pressure Profiles for No-Flow and Steady-State Flow (adapted from Harr, 1966)

Figure 11.2(*b*) shows the stress profiles for the steady-state flow conditions. Comparison of profiles (*a*) and (*b*) reveals that, for the same total stress, porewater pressure suffers a change in reaching the steady-state seepage condition. This difference in the porewater pressures is reflected in the change in the effective stresses from the no-flow to the steady-state flow conditions. Clearly, the change in the effective stress can be as significant as the extent of change in the porewater pressure. However, once the steady-state condition is established, there is then no further change in the effective stress.

11.6 Engineering behaviour of soils

A fine grained cohesive soil, such as clay, and a coarse grained granular soil, such as sand, may be subject to similar effective stress and display similar strengths, yet the permeability of clay can be several orders of magnitude lower than that of sand. This is because, in fine grained soils like clays, the sizes of the interconnecting pores are so small that the displacement of porewater is retarded by viscous forces. As a result of such pronounced differences in the magnitude of their drainage characteristics, the relative response to any imposed disturbance of the ambient stress regime in the two soil types can be vastly different. Porewater pressure movement, and so also the change in effective stress, can therefore be expected to vary from soil to soil. Although porewater pressure does not contribute directly to the strength of the soil structure, nevertheless the restraint or the freedom with which porewater is able to flow (i.e. the permeability of soil) can influence the rate at which strength and stiffness will change, as also the rate of response of the soil, in a significant way.

Besides permeability, deformability and strength are the two other important engineering characteristics in relation to soils. The deformability of a soil to loading or unloading is, essentially, its capacity to deform the voids, usually by the displacement of water. The resistance offered to such deformation is a measure of the strength of the soil. In other words, deformation of a soil structure under an imposed increase in stress or stress relief may be defined as the extent to which it will allow rearrangement of interstitial pore space by the displacement of water thereby forcing the solid particles either closer together into a denser state of packing or further apart into a looser fabric. In the former case, the strength of the soil will increase whereas in the latter, it will decrease. The movement of water may be caused either by the dissipation of the positive or the imbibition of the negative excess porewater pressures.

Porewater pressure 'u' at any point is direction invariant, i.e. it acts with equal intensity in all directions. If soil particles are subjected to such a uniform pressure all round their surfaces, clearly they will experience a small decrease in their volume but will suffer no distortion either individually or collectively as a soil skeleton. It follows therefore that the cause for the deformation of the soil skeleton must be that part of the total contact stress σ which is in excess of the (uniform) porewater pressure u, i.e. the component $\sigma - u$, which is a stress difference and not a principal stress. This confirms that two of the engineering characteristics, deformability and strength, in relation to soils are uniquely dependent on the effective stress. In other words, the amount by which a formation will deform as a result of an imposed change in the total stress will be governed by the extent that this change is translated into a corresponding change in the effective stress.

Consider a soil element subjected to an isotropic increase $\Delta\sigma$ in the total stress as shown on the left in Figure 11.3(*a*). Figure (*b*) shows the increase in the total stress by

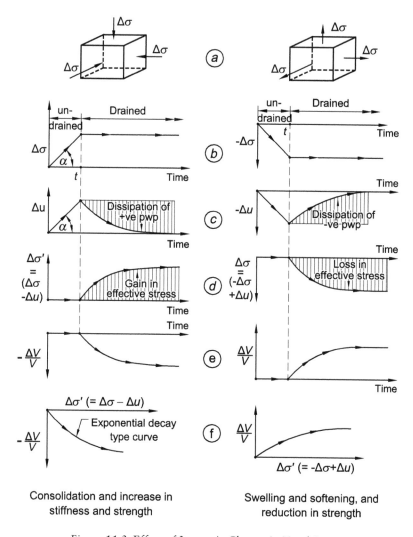

Consolidation and increase in   Swelling and softening, and
stiffness and strength   reduction in strength

Figure 11.3 Effect of Isotropic Change in Total Stress

this increment over the initial period 't' when there is no movement of the porewater, i.e. in the undrained condition. Beyond this stage, there is no further increase in the applied total stress. Figure (c) shows the variation of the porewater pressure. Initially, there is a build-up of the excess positive porewater pressure attaining its maximum value Δu equal to the corresponding total stress increment $\Delta\sigma$ at the end of the undrained stage. Thereafter, in the drained stage, the excess porewater pressure dissipates until the equilibrium is restored, i.e. Δu is reduced to zero. Figure (d) shows the variation in the effective stress with time. In the undrained stage, since there is no dissipation of the excess porewater pressure, there is no change in the effective stress. However, as the excess porewater pressure is allowed to dissipate, its loss is directly reflected in terms of an equal amount of gain in the effective stress. This is the reason why, in the drained stage, the '$\Delta\sigma - \Delta u$' curve (i.e. gain in the effective stress) is a mirror image of the 'Δu' (i.e. dissipation of the excess positive porewater pressure) curve.

Volumetric strain is a function of soil deformation which takes place only when excess porewater pressure starts to dissipate. So, there is no volumetric strain or deformation during the undrained mode as shown in figure (*e*). However, in the drained mode, as the effective stress increases, so will the volumetric strain reduce exponentially as shown in figure (*f*); the resulting consolidation leads, in turn, to increase in stiffness and strength.

On the right in Figure 11.3 is shown, similarly, a soil element subjected to an isotropic decrease in the total stress. The result, upon dissipation of the excess negative porewater pressure (i.e. sucking in water) in this case, is the swelling of the soil fabric causing it to be less densely packed and therefore becoming softer in structure. With the softening of the soil structure, there is, consequently, a reduction in its strength as well.

Notwithstanding the foregoing, it should be recognized that there is a component of deformation which is not accompanied by the dissipation of excess porewater pressures, i.e. in the undrained mode. This is due to change of shape at constant volume and is referred to as the undrained or the elastic deformation.

In the examples that follow, application of the principle of effective stress will be illustrated and the impact of variations in effective stress corresponding to the various operations of dewatering, excavation, etc. examined. Deformation of soils resulting from pressure relief due to excavation for cut-and-cover structures is dealt with in Chapters 19, 20 and 21.

Example 11.1

Figure 11.4(*a*) shows a typical cross-section of a layered soil. The different layers have different inherent drainage characteristics and this is reflected in the different piezometric levels displayed in the layers. For the water table at 4*m* below ground level, the given piezometric levels and the unit weights of the soils in different layers as shown, draw the initial porewater pressure and the vertical total and effective stress profiles representing the *in situ* state in the ground.

Solution:

- Porewater Pressures (*variation within the free-draining sand and silt layers is assumed hydrostatic*), $\left(u = \gamma_W \cdot h_p \right)$:

> At 4*m* below ground, i.e.
> At −4*m* level = 0
> At −15*m* level = $10 \times 6 = 60kN/m^2$
> At −20*m* level = $60 + 10 \times (20 - 15) = 110kN/m^2$
> At −30*m* level = $10 \times 18 = 180kN/m^2$
> At −35*m* level = $180 + 10 \times (35 - 30) = 230kN/m^2$
> At −45*m* level = $10 \times 20 = 200kN/m^2$.

The porewater pressure profile is shown in figure (*b*). For comparison, hydrostatic pressure profile is also shown alongside. The difference between the two profiles represents the reduction due to hydrodynamic effect. In the figure, *h* represents the

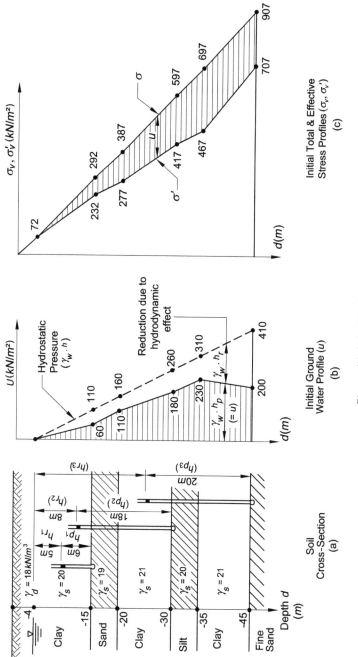

Figure 11.4 In Situ Stresses

hydrostatic head, h_p the porewater pressure or the piezometric head and h_r the reduction in the hydrostatic head due to the hydrodynamic effect.

- Vertical Total Stress: $(\sigma_V = \gamma \cdot d)$.

> At ground level $= 0$
> At $-4m$ level $= 18 \times 4 = 72kN/m^2$
> At $-15m$ level $= 72 + 20 \times (15 - 4) = 292kN/m^2$
> At $-20m$ level $= 292 + 19 \times (20 - 15) = 387kN/m^2$
> At $-30m$ level $= 387 + 21 \times (30 - 20) = 597kN/m^2$
> At $-35m$ level $= 597 + 20 \times (35 - 30) = 697kN/m^2$
> At $-45m$ level $= 697 + 21 \times (45 - 35) = 907kN/m^2$.

The vertical total stress profile is shown in Figure (c).

- Vertical Effective Stress: $\left(\sigma'_V = \sigma_V - u = \gamma \cdot d - \gamma_W \cdot h_p\right)$

The vertical effective stress at any level can be obtained simply by deducting from the vertical total stress at the level the coexistent porewater pressure at that level. The vertical effective stress profile is also shown in Figure (c).

Example 11.2

Figure 11.5(a) shows the geology and the initial piezometric levels at a site. The ground is layered and comprises clay alternating with water bearing layers of 'sand' or 'silt and sand'. The piezometric levels indicate that, with the exception of layers 1 and 2, the porewater pressures elsewhere are not hydrostatic. Preparatory to the commencement of excavation, piezometric levels are lowered as shown in Figure (b). Figure (c) shows the cross-sectional geometry of a proposed underground box structure measuring $10m \times 8m$ overall to be constructed.

Assuming the unit weight of soil, γ, to be constant (for illustrative purposes only) at $20kN/m^3$ throughout the depth and starting with the profile of initial *in situ* stresses,

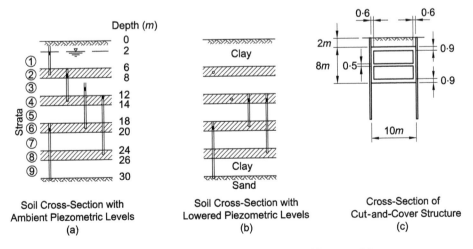

Soil Cross-Section with Ambient Piezometric Levels
(a)

Soil Cross-Section with Lowered Piezometric Levels
(b)

Cross-Section of Cut-and-Cover Structure
(c)

Figure 11.5 Ground Conditions and Geometry of Proposed Structure

examine the variations in the vertical effective stresses in the ground at the end of each of the following stages:

- Stage 1 – Lowering of piezometric levels
- Stage 2 – Excavation down to *5m* depth
- Stage 3 – Excavation down to 10*m* depth
- Stage 4 – Construction and reinstatement
- Stage 5 – Restoration of ambient porewater pressures.

Solution:
Rationale: To start off with, the ambient *in situ* stresses in the ground are established. Treating the stresses at each stage as the current stresses, the new stresses for each successive stage are worked out in turn. The effective stress profile for each new stage is then compared with that of the current profile of the previous stage to obtain the variation with each stage. At the end of Stage 5, the net final effect – decrease or increase – in the state of *in situ* ambient stresses in the ground can be established.

It is assumed that the porewater pressures at and above *7m* depth are fully hydrostatic. This fact is also made use of in working out the porewater pressures and effective stresses. In order to measure the variation in the stresses at the formation levels corresponding to Stages 2 and 3 later on, initial stresses at *5m* and 10*m* depths identifying these stages have also been included in the tables.

Initial in situ stresses

Initial, vertical total and effective stresses are computed from the expressions $\sigma_o = \gamma \cdot d$ and $\sigma_o' = \sigma_o - u_o = \gamma \cdot d - \gamma_W \cdot h_p$, respectively, where u_o is the ambient porewater pressure. Initial total and effective stresses are listed in Table 11.1. Vertical, total and effective stress profiles are shown in Figure 11.6(*a*).

Stresses at end of Stage 1

By reducing the piezometric levels at the centre of layer 4 and above to zero, vertical effective stresses at and above this level will increase and become equal to their

Table 11.1 Initial *In Situ* stresses

Depth (m)	Total stress σ_o (kN/m²)	Porewater pressure u_o (kN/m²)	Effective stress σ_o' (kN/m²)
0	0	0	0
2	$20 \times 2 = 40$	0	40
5	$20 \times 5 = 100$	$10 \times 3 = 30$	$100 - 30 = 70$
7	$20 \times 7 = 140$	$10 \times 5 = 50$	$140 - 50 = 90$
10	$20 \times 10 = 200$	55 (*by interpolation*)	$200 - 55 = 145$
13	$20 \times 13 = 260$	$10 \times 6 = 60$	$260 - 60 = 200$
19	$20 \times 19 = 380$	$10 \times 9 = 90$	$380 - 90 = 290$
25	$20 \times 25 = 500$	$10 \times 13 = 130$	$500 - 130 = 370$
30	$20 \times 30 = 600$	$10 \times 11 = 110$	$600 - 110 = 490$

respective total stresses. Since the piezometers in layers 6 and 8 also indicate lowering of the levels, albeit partially, there will also be commensurate changes in the effective stresses at these levels. The stresses are listed in Table 11.2.

Vertical, total and effective stress profiles at the end of Stage 1 are shown in figure (*b*). The resulting increase in the effective stress profile is highlighted in Figure (*c*).

Stresses at end of Stage 2

Removal of the overburden by carrying out excavation down to 5*m* below ground level will cause a commensurate relief in the vertical stresses. The stresses corresponding to

Table 11.2 Stresses at End of Stage 1

Depth (m)	Total stress σ_1 (kN/m^2)	Porewater pressure u_1 (kN/m^2)	Effective stress σ_1' (kN/m^2)
0	0	0	0
2	$20 \times 2 = 40$	0	40
5	$20 \times 5 = 100$	0	100
7	$20 \times 7 = 140$	0	140
10	$20 \times 10 = 200$	0	200
13	$20 \times 13 = 260$	0	260
19	$20 \times 19 = 380$	$10 \times 6 = 60$	$380 - 60 = 320$
25	$20 \times 25 = 500$	$10 \times 12 = 120$	$500 - 120 = 380$
30	$20 \times 30 = 600$	$10 \times 11 = 110$	$600 - 110 = 490$

Table 11.3 Stresses at End of Stage 2

Depth (m)	Total stress σ_2 (kN/m^2)	Porewater pressure $u_2 = u_1$ (kN/m^2)	Effective stress σ_2' (kN/m^2)
5	0	0	0
7	$20 \times 2 = 40$	0	40
10	$20 \times 5 = 100$	0	100
13	$20 \times 8 = 160$	0	160
19	$20 \times 14 = 280$	$10 \times 6 = 60$	$280 - 60 = 220$
25	$20 \times 20 = 400$	$10 \times 12 = 120$	$400 - 120 = 280$
30	$20 \times 25 = 500$	$10 \times 11 = 110$	$500 - 110 = 390$

Table 11.4 Stresses at End of Stage 3

Depth (m)	Total stress σ_3 (kN/m^2)	Porewater pressure $u_3 = u_1$ (kN/m^2)	Effective stress σ_3' (kN/m^2)
10	0	0	0
13	$20 \times 3 = 60$	0	60
19	$20 \times 9 = 180$	$10 \times 6 = 60$	$180 - 60 = 120$
25	$20 \times 15 = 300$	$10 \times 12 = 120$	$300 - 120 = 180$
30	$20 \times 20 = 400$	$10 \times 11 = 110$	$400 - 110 = 290$

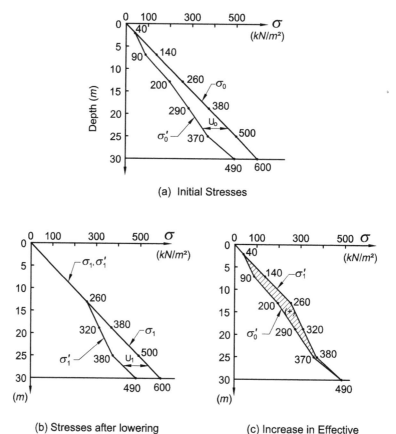

Figure 11.6 Variations in Stresses

this stage are listed in Table 11.3. Note that the porewater pressures remain unchanged from those of Stage 1.

Vertical, total and effective stress profiles at the end of Stage 2 are shown in Figure (*d*) and the resulting decrease in the effective stress profile is highlighted in Figure (*e*).

Stresses at end of Stage 3

Advancing the excavation downwards by a further 5*m*, i.e. to the final formation level of 10*m* below ground, will cause a further relief in the vertical stresses. The stresses at the end of this stage are listed in Table 11.4. Once again, at this stage also, the porewater pressures remain unchanged from those of Stage 1.

Vertical, total and effective stress profiles corresponding to this stage are shown in Figure (*f*). Excavation activity in this stage will also cause a drop in the effective stress profile beyond that of Stage 2 as highlighted in Figure (*g*).

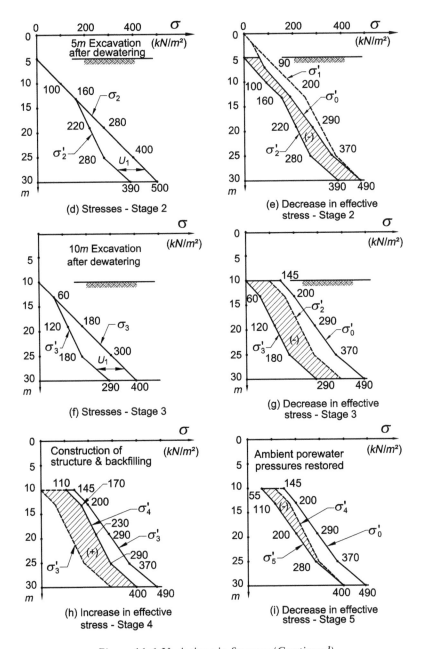

Figure 11.6 Variations in Stresses (*Continued*)

Stresses at end of Stage 4

Construction of the structure will, by virtue of its dead weight, restore at least part of the overburden pressure removed through excavations in Stages 2 and 3. Increase in the pressure on the formation beyond that of Stage 3 resulting directly from the weight

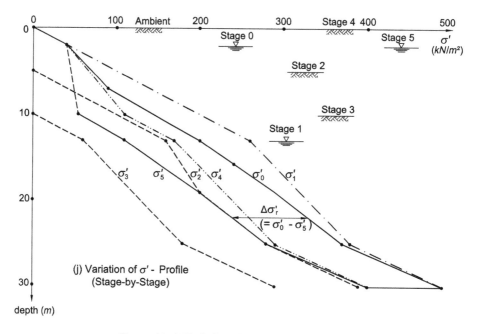

Figure 11.6 Variations in Stresses (*Continued*)

of the structure, i.e. the extent by which the weight of overburden lost is made good, works out, approximately, to $70kN/m^2$.

Reinstatement of the $2m$ deep backfill on top of the roof after the structure is in place will further reduce the shortfall occasioned by the $10m$ deep excavation. The increase in the effective stress due to the replacement of the backfill alone is worth $40kN/m^2$.

The aggregate effect of Stage 4, i.e. due to construction and replacement of backfill, is thus worth $110kN/m^2$ of increase in the vertical total stress. With the porewater pressures remaining unchanged, the soil mass will experience a consistent increase in the effective stresses of this amount from formation downwards as shown in figure (*h*). Notwithstanding this, the net effect continues to remain one of decrease in the effective stress as is evident in figure (*i*).

Stresses at end of Stage 5

As the piezometric head was lowered in Stage 1, there was a commensurate increase in effective stress as represented in figure (*c*). In course of time, when the ambient porewater pressures are 'fully' restored, there will naturally be a reversal of this increase causing, this time round, a corresponding decrease in the effective stress profile as highlighted in figure (*i*).

To appreciate the effect of the various stages, variation in the σ'-profile, stage-by-stage, is shown in figure (*j*). It is clear from the comparative study of the effective stress profiles that the residual effect at the end of Stage 5 continues to remain, not surprisingly, one of stress relief. This can, potentially, subscribe to time-dependent, upward deformation or heave of soil the implications of which would need to be duly addressed in design. Various aspects of heave are dealt with in Chapters 19, 20 and 21.

Example 11.3
With the porewater pressures computed in Example 18.1(A) and listed in Tables 18.1 and 18.2 in Chapter 18, compare the lateral total and effective pressures on the outside face of the cofferdam using a stipulated value of the coefficient of earth pressure, K, of 0·3 Assume both the dry unit weight of soil above and the saturated unit weight below the water table (for illustrative purpose only) to be $20kN/m^3$.

Solution:
Appropriate stresses at any depth z below the water table can be obtained from the following expressions:

Total vertical stress, $\qquad \sigma_V = \gamma \cdot d_1 + \gamma_S \cdot z$

Porewater pressure, $\qquad u :$ As available directly from Tables 18.1

and 18.2, Example 18.1, Chapter 18.

Effective vertical stress, $\qquad \sigma'_V = \sigma_V - u$

Effective horizontal stress, $\qquad \sigma'_H = K \cdot \sigma'_V = K\left(\sigma_V - u\right)$

Total horizontal stress, $\qquad \sigma_H = \sigma'_H + u$

By way of illustrating the use of the foregoing expressions, the stresses at level 'c' have been worked out in detail as follows:

Total vertical stress, $\qquad \sigma_V = 20 \times 3 + 20 \times 12 = 300kN/m^2$

Porewater pressure, $\qquad u = 90kN/m^2 (flow\ net)$

$\qquad = 72\cdot67kN/m^2 (Pavlovsky)$

Effective vertical stress, $\qquad \sigma'_V = 300 - 90 = 210kN/m^2 (flow\ net)$

$\qquad = 300 - 72\cdot67 = 227\cdot33kN/m^2 (Pav.)$

Effective horizontal stress, $\qquad \sigma'_H = 0\cdot3 \times 210 = 63kN/m^2 (flow\ net)$

$\qquad = 0\cdot3 \times 227\cdot33 = 68\cdot2kN/m^2 (Pav.)$

Total horizontal stress, $\qquad \sigma_H = 63 + 90 = 153kN/m^2 (flow\ net)$

$\qquad = 68\cdot2 + 72\cdot67 = 140\cdot9kN/m^2 (Pav.)$

Stresses on the outside face of the cofferdam using porewater pressures from flow net are presented in Table 11.5.

Likewise, stresses on the outside face of cofferdam using porewater pressures from Pavlovsky's method of fragments are presented in Table 11.6.

Table 11.5 Pressures on Cofferdam using Flow Net

Level	Depth z (m)	σ_V $= \gamma \cdot d_1 + \gamma_S \cdot z$	u (F. Net)	σ'_V $= \sigma_V - u$	σ'_H $= K \cdot \sigma'_V$	σ_H $= \sigma'_H + u$
		Pressure (kN/m²)				
a	0·0	60·0	0·0	60·0	18·0	18·0
b	6·2	184·0	47·0	137·0	41·1	88·1
c	12·0	300·0	90·0	210·0	63·0	153·0
d	15·8	376·0	115·0	261·0	78·3	193·3
e	18·0	420·0	120·0	300·0	90·0	210·0

Table 11.6 Pressures on Cofferdam using Pavlovsky's Method

Level	Depth z (m)	σ_V $= \gamma \cdot d_1 + \gamma_S \cdot z$	u (Pav.)	σ'_V $= \sigma_V - u$	σ'_H $= K \cdot \sigma'_V$	σ_H $= \sigma'_H + u$
		Pressure (kN/m²)				
a	0·0	60·0	0·00	60·00	18·0	18·0
b	6·2	184·0	37·54	146·46	43·9	81·4
c	12·0	300·0	72·67	227·33	68·2	140·9
d	15·8	376·0	95·68	280·32	84·1	179·8
e	18·0	420·0	109·0	311·00	93·3	202·3

In the context of the preceding example, comparison of the stresses in the two tables reveals the following:

- Porewater pressures given by the Pavlovsky's method are 10–25 per cent lower than those obtained using the flow net; the difference is greater higher up and reduces with depth.
- Horizontal total stresses given by the Pavlovsky's method are also lower than those from the flow net method. However, the decrease is consistently around 8–9 per cent except at the bottom of cofferdam where it is around 4 per cent.
- Horizontal effective stresses obtained with the use of the Pavlovsky's method are generally higher than those obtained from the flow net results by 7–8 per cent except at the bottom of cofferdam where the increase is about 4 per cent.

12 Identification of Loads

12.1 Introduction

Identification and assessment of appropriate loads expected to be imposed during the various stages of construction and operational lifetime constitutes an important step in the design of a cut-and-cover structure. In the case of structures above ground, by virtue of their exposure to view, the nature, the magnitude of the anticipated loads and their points of application are, generally, clearly established and well defined. As such, they can be calculated and their impact on design analysed reasonably accurately. However, cut-and-cover structures exist in confined environments and the pressures likely to be exerted by the surrounding ground constitute a large proportion of the applied loads. The complexity of ground conditions and soil–structure interaction often make the accurate assessment of the various aspects of these loads rather difficult.

Furthermore, the data on the imposed load from the surrounding ground to be used in design are largely derived from relatively small and often unrepresentative samples of the strata in which the works are to be executed. In order to render the design process more tractable, of necessity, empiricism based on approximations and assumptions which can be reasonably justified are employed. Also, because of the difficulty in assessing the values of the design parameters, particularly those for the soil, accurately, greater reliance has often to be placed on past experience. Feedback from the back analyses of past case histories in similar ground conditions has been known to provide the most helpful and reliable data which have, over the years, proved invaluable aids in design.

The loads to which a typical underground metro structure is likely to be subjected during the various stages of its construction and operational lifetime can be broadly classified as follows:

- Gravity loads
- Lateral loads
- Upward loads
- Accidental loads
- Other loads.

Figure 12.1 shows typical loads generally acting on a completed cut-and-cover metro station structure.

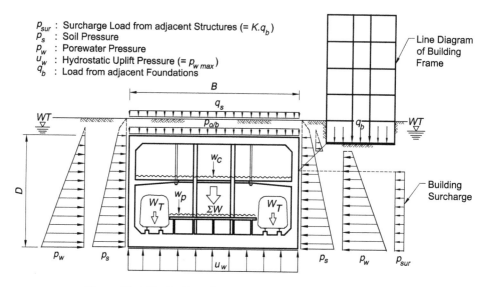

Figure 12.1 Typical Loads on Cut-and-Cover Metro Structure

12.2 Gravity loads

The gravity loads constitute all those downward acting loads which not only act on the structure itself but also those, including the self-weight, which are transmitted through it down to the subgrade below. Essentially, gravity loads fall into the following three categories:

- Permanent loads
- Semi-permanent loads
- Transient loads.

Permanent loads

These loads comprise the self-weight, W, of the basic structure itself, the weight of the backfill, i.e. corresponding to the overburden pressure $p_{O/B}$, on top of the roof of the structure and, where appropriate, the self-weight of the structure above.

Self-weight: basic structure

Estimation of the self-weight at any stage of progress in the construction of the basic structure comprising those elements which are unlikely to be replaced during their lifetime is fairly straightforward. However, in computing this load for a particular stage, it is important to bear in mind the purpose for which it is required. Whether the assessment of the load from self-weight is required to examine stability against flotation or to be used for the strength design of the structure, care must be exercised to ensure that, in the case of the former, it is neither overestimated so as to give a false sense of safety nor, in the latter, underestimated to compromise strength.

To guard against any such adverse estimate, potential variations in the unit weight of the material are generally allowed for in such a way as to minimize the load during the investigation for stability and to maximize it for the design of strength, by a sensible margin. Accordingly, the variation from the minimum to the maximum unit weights of reinforced concrete of $24kN/m^3$ to $26kN/m^3$ is recognized as an acceptable range for use. Admittedly, the higher value if used globally is unlikely to yield adverse values for the stress-resultants at every critical location; however, the error is likely to be very small and, therefore, need not be of any serious concern.

Backfill on roof

The vertical alignment of tracks, vertical planning clearances required and the extent of accommodation necessary for the utilities, etc. together, define the required depth at which the roof of a typical underground metro structure must be placed below the ground level. However, adoption of a minimum depth of soil cover of $2m$ is not uncommon. Out of this, the thickness of the road surfacing that may be removed and replaced from time to time is generally disregarded in the stability check against flotation, and the weight of the remainder of the backfill is treated as a permanent load.

To make the potential provision for future excavations in the backfill in order to accommodate new utilities, some authorities around the world are known to stipulate the complete exclusion of up to $1·5m$ or so of backfill from the stability calculations. Disregarding the contribution from the backfill of such a depth over the entire extent of the structure is difficult to justify and would appear to be wasteful. It can, in certain circumstances, be unrealistically severe. However, if in the weight of the backfill, the relief that can be expected during the course of excavation for trench(es) to accommodate new service(s) in the future is to be allowed, this can be achieved either by establishing the equivalent depth of the soil backfill or downgrading its unit weight appropriately allowing for the anticipated trench geometry as outlined in the solution to Example 22.1, in Chapter 22.

Effect of water table

The level of the water table undoubtedly plays a prominent role in the stability calculations. For a fully submerged structure, the force of buoyancy is invariant with the level of the water table. However, the level of the water table and the way it affects the extent of the depth of the backfill over which the submerged unit weight, i.e. $(\gamma_S - \gamma_W)$ is operative, do affect the effective weight of the backfill counteracting flotation. So, for stability calculations, the worst credible highest level of the water table must be established for design. For strength calculations, on the other hand, the highest and the lowest levels should be used, selectively but not coexistently, in appropriate combinations with other loads, to generate the most severe stress-resultants in different elements and locations as appropriate. For example, to maximize the corner moments at the 'roof (or base) slab-perimeter wall' junction, it is appropriate to consider the highest credible design water level, whereas to maximize the span moments in the roof slab, the lowest credible level should be considered.

Effect of arching

In the context of cut-and-cover metro structures, the designer may need to establish how real the prospect of arching in respect of the soil cover over the roof of the structure is. If the geometry of the problem is such that the arching cannot be ignored, then whether the effective weight of the backfill is likely to be greater or less than that implied by its actual depth, must be established so that the corresponding impact on the structure can also be appropriately taken into account.

Arching may be positive or negative depending upon whether the weight of the backfill directly on top of the structure experiences a decrease or an increase, respectively, in its notional load. The depth of the overburden on top and the geometry of the buried structure may be such that a part of the weight of the overburden is shed on to the sides either side of the structure through arch action, and only the remainder of the total weight of the backfill may be experienced by the structure itself. In such a case, the load experienced by the structure would be less than that corresponding to the actual depth of the backfill on top of the structure. Such relief in load would be due to positive arch action.

Positive arch action is possible when the combined vertical compression of the structure and that of the soil backfill directly on top of it is greater than the compression of the full height columns of soil above the subgrade on either side of the structure. However, where opposite may be the case, i.e. the columns of soil either side of the structure experiencing greater vertical compression than the combined compression of the soil backfill directly on top of the structure and the vertical compression of the structure itself, there is then load shedding from the sides on to the backfill on top of the structure. This can cause the structure to experience a load which can be in excess of that implied by the actual depth of the soil backfill and which, in certain situations, can be very significant. The mechanism that causes such an increase in the load is referred to as the negative arch action.

Whether and to what extent arching can be relevant to the design of cut-and-cover metro structures need to be addressed. Marston Theory of Loads on Underground Conduits (1930) as verified by Spangler (1933) is generally used as the basis of arch action on buried structures. Figure 12.2 shows a typical cross-section of a metro structure. Left-hand half is shown to be constructed within a cofferdam, whereas the right half, in an open cut within stabilized side slopes. Assuming the subgrade to be unyielding, the three important considerations of this theory in its application to cut-and-cover metro structures may be listed as follows:

- The aspect ratio, i.e. the ratio of depth to width (d/B) of the fill directly on top of the structure;
- The method of backfilling;
- The extent and the direction of the differential vertical movement along the potential vertical shear planes (*1–1* and *2–2*) defined by the upward extensions to the surface of the earth-side faces of the outer longitudinal side walls of the structure.

If the width of the structure (and therefore also that of the backfill directly above it) is relatively small as compared to the depth of the backfill on top of the structure, i.e. the aspect ratio is high (>0.5), the effect of load shedding, particularly from a downward

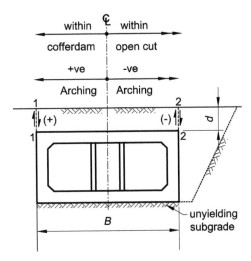

Figure 12.2 Positive and Negative Arching

differential movement of the soil columns outside the shear planes, is likely to extend over the full width of the structure. If not, the effect is likely to be concentrated locally close to the sides. In the case of cut-and-cover metro structures, the aspect ratio is unlikely to exceed 0·2. As a result, the effect of load shedding is not only going to be concentrated close to the sides of the structure, it is also likely, if at all it does act, to be very small.

The type of the effect felt by the structure, i.e. whether the load on it is likely to experience an increase or decrease, will depend on the width of excavation and the method of backfilling. If the structure is constructed within stabilized side slopes, the depth of the backfill on either side of the structure is going to be greater than that directly over the structure. Furthermore, if the settlement of the deeper backfill on the sides turns out to be more than the combined vertical compression of the structure and that of the backfill directly above it, then the differential down-drag along the shear planes due to negative arching will cause extra load to be shed on to the backfill directly above the structure thereby increasing the load on it beyond that represented by the notional depth of the backfill.

If, on the other hand, the structure is constructed within an outer temporary cofferdam that extends all the way up to the surface so that the ground outside the cofferdam remains untouched and unaffected, then the settlement of the backfill directly above the structure will cause at least part of the backfill load close to the shear planes to be shed on to the cofferdam thereby reducing, correspondingly, the load on the roof of the structure. Notwithstanding this, it is important to recognize that, in the course of the replacement and compaction of the backfill on top in discrete layers, the possibility that the structure will experience the full weight of the overburden cannot be ruled out. Full weight of the overburden will also be experienced by the roof slab under the compaction of the backfill if the perimeter wall extends up to the surface and forms an integral part of the structure.

In the case of construction carried out in an open cut within stabilized side slopes, in theory, negative arch action could be expected as long as the fill outside the longitudinal perimeter walls settles more than the combined total compression experienced by the structure and the settlement of the fill above. However, once again, owing to the very small magnitude of the aspect ratio generally encountered in cut-and-cover metro structures, the effect of load enhancement on the structure is likely to be insignificant. Therefore, for most metro station structures encountered in practice, it would be safe to disregard the potential load enhancing effect of negative arching.

Semipermanent loads

Semipermanent loads include all 'permanent' loads such as those arising from the road surfacing, finishes within the structure, plant loads and other superimposed dead loads including partitions, etc. which may be subject to removal and replacement from time to time as part of the regular maintenance.

Transient loads

Temporary loads arising out of construction activity, surcharge loads from the traffic, q_S, on the surface and all the live loads within the structure itself and, where appropriate, from the structure above, fall under this category.

Construction live load

Live loads from construction activity are invariably job and site specific loads. Since their relevance is limited to the period of construction activity only, it is prudent to restrict such loads to modest values $(3-5kN/m^2)$ so that they do not impose major and expensive permanent modifications on design. Alternatively, actual track loads from the construction rigs, as appropriate, should be considered. Where such discrete loads are likely to be severe, to alleviate their intensity, consideration should be given to the use of temporary grillages to distribute the loads over larger areas more evenly and tolerably.

Traffic surcharge

In the case of an underground metro structure where the top of the roof slab is flush with and constitutes also the road surface, the slab can be treated as a bridge deck. Being directly exposed to the traffic, the intensity of the load on the structure is obtained simply by dealing with the *HA* and *HB* traffic loading in the conventional way.

Where the metro structure is depressed below the road surface, the wheel loads will undergo dispersal with depth and, in the process, experience a reduction in their intensity before reaching the roof of the structure. However, in the event of the depression being 600*mm* or less, the diminution in the load intensity due to dispersal is, as a safe simplification, disregarded and the traffic loading is assumed as if directly applied to the roof slab. For depressions greater than 600*mm*, on the other hand, the extent of dispersal and therefore the intensity of the surcharge loading likely to act on the roof slab will vary commensurate with the extent of the depression.

Before dealing with the dispersal of loads, it is important to recognize, in their relevance to buried structures, the distinction between the *HA* wheel loads and the equivalent uniformly distributed load together with the knife-edge load commonly

used in lieu of the actual wheel loads in the design of bridge decks. The live loads transmitted by the wheels of the *HA* vehicles are the actual loads experienced by the deck. On the other hand, the combination of the equivalent uniformly distributed load (*udl*) and the knife-edge load (*KEL*) are artificially contrived loads used purely as a convenient expedient which offers, in a simple manner and, in most cases, a safe value for the potential effects of the actual loads.

However, for estimating the effect and dispersal of the traffic surcharge on a buried structure, consideration of loads other than the actual wheel loads is clearly meaningless. To establish the live load intensity for design on the roof slab, actual wheel loads from *HA* vehicles should be considered. By way of illustration, the effect of traffic surcharge on a buried structure from the wheels of a '44 *tonne* articulated' and a '44 *tonne* drawbar combination' vehicles (Lorry Weights, The Department of Transport) representing the most onerous loads has been considered. The axle loadings under these vehicles are shown in Figure 12.3.

In arriving at the surcharge intensity on the structure, the following assumptions have been made:

* Buried structure is directly under a thoroughfare in such a manner that the direction of travel on the road above is the same as the length of the structure, i.e. the span of the roof of the structure is at right angles to the direction of travel.
* The span of the roof of the structure is 23*m* which carries dual carriageway comprising three lanes each.
* Two of the six lanes carry full *HA* vehicular loads, whereas each of the remaining four lanes carries 60 per cent of the load. This is the worst possible combination identified in Table 14, Clause 6.4 of *BS 5400*, Part 2, 1987.
* Minimum depth of the compacted backfill on the roof of the structure is 1·0*m*.
* Dispersion of wheel loads within the fill follows the guidance given in Clause 3.2.1 of *BD 31/01*.

In the case of both the vehicles, the worst-loaded axle combinations with axles at 1·5*m* centres and track at 1·8*m* have been considered.

In the case of the '44 *tonne* articulated' vehicle, the three rear axles carry a combined total load of 24 *tonnes*. With the wheel contact area of 215*mm* diameter and using the dispersion as referred to before, the wheel loads are likely to hit the roof slab, virtually, as a *udl* over a band width of 4·215*m* as shown in Figure 12.4 along the entire 23*m* span. Given the inherent longitudinal stiffness of the reinforced concrete roof slab which can easily be more than 1*m* thick, it would be perfectly reasonable to assume

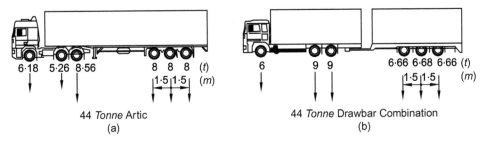

Figure 12.3 Axle Loadings from 44 Tonne Articulated and Drawbar Combination Vehicles

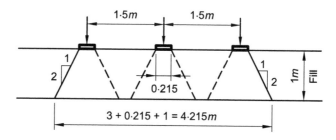

Figure 12.4 Load Dispersal

that at least an extra 1*m* of the slab either side of the load band would also be mobilized in carrying the load. However, for this exercise, let us assume, conservatively, an extra width of only 500*mm* on either side giving an effective band width of the order of 5·215*m*. On the basis of the *HA* lane factors as outlined earlier, the equivalent *udl* on the roof of the structure at 1*m* depth then works out to 8·8*kN*/*m²*.

In the case of the '44 *tonne* drawbar combination' vehicle, the middle two are the worst loaded axles carrying a combined total load of 18 *tonnes*. With the exception of this and the wheel contact area, which in this case works out to 228*mm* diameter, using the same reasoning as before, the effective band width carrying the load works out to 3·728*m* and the corresponding equivalent *udl* as 9·24*kN*/*m²*.

Both – the use of 44-*tonner* rather than a 40-*tonner* vehicle and the extents of the effective widths as assumed – introduce degrees of conservatism in the foregoing approach. Besides, there is a subtle difference in the application of wheel loads directly on to the structure and indirectly via the soil medium. Being transient loads, the wheel loads, unlike where these are applied directly to the structure, will also undergo a certain amount of dissipation while dispersing through the relatively compressible fill material before hitting the deck. So, in addition to the reduction of load intensity due to dispersion, there is also the loss due to the effect of dissipation. However, the latter is generally disregarded and so a blanket figure of an equivalent *udl* of 10*kN*/*m²* on the structure can be regarded as being conservative. It should also be noted that at depths greater than 1*m*, because of greater dispersal and dissipation and consequently greater reduction in the intensity of loading, the use of 10*kN*/*m²* *udl* as a surcharge load on the structure would prove to be even more conservative. However, it is assumed that the effect of any downward variation in the surcharge loads at overburden depths greater than 1*m* is not likely to be significant.

Live load within the metro structure

Live loads within a completed metro structure comprise the following:

- Carriageway wheel loads on the tracks;
- Loads imposed directly by plant and machinery;
- Pedestrian loads on the platforms from metro usership traffic;
- Pedestrian loads on the paid and the unpaid areas of the concourse;
- Loads associated with the occasional movements of plant and machinery.

Typical values for some of the aforementioned loads are given in Appendix A. However, these should be treated as indicative and for general guidance only. Furthermore, the possibility that the loads related particularly to plant and machinery may vary from plant to plant, country to country, time to time and from one metro system to the other, should not be ruled out.

12.3 Lateral loads

An underground structure is, by definition, in intimate contact with the soil on all sides, as such, it is free to deform as much or as little as the stiffnesses of the surrounding ground mass and the structure itself will allow it. Soil–structure interaction, therefore, assumes great importance in the analysis and design of such structures.

Being in intimate contact with the structure, the surrounding ground not only generates pressures that often constitute the predominant loads acting on the structure, but also provides all-round support and stability to it. The extent of the soil support available and the magnitude of soil pressures generated or mobilized at any given time are therefore closely interrelated. However, during the excavation for and the construction of the structure, as the geometry and the boundary conditions of the structure change from those of one stage of construction to the next, so also will the extent of the soil support available. It follows therefore that the magnitude of the different loads to be expected at any given time and the manner in, and the timing at, which these become operative will also change from one stage to the other depending upon the extent of excavation and the state of completion of the structure reached at any given stage.

Lateral loads on an underground metro structure, which also include groundwater pressures, accrue directly from the surrounding ground. However, the surrounding ground also represents the medium through which the effects of the loads from nearby structures and foundations are transmitted as lateral surcharge loads on to the metro structure. So, the lateral loads on the structure include the following:

- Horizontal earth pressures;
- Pressures from the groundwater regime;
- Surcharge loads from the foundations of nearby structures.

Assessment of these pressures and those from the surcharge loads as applicable to cut-and-cover metro structures are dealt with separately in Chapters 13 and 14. Implications of groundwater flow and the assessment of the porewater pressures are dealt with in Chapter 18.

Out of the lateral earth and hydrostatic pressures, generally, the latter turn out to be the greater and therefore the more dominant in design. This is particularly so where the ground is hard (i.e. ϕ' value is high) as a result of which the coefficient of earth pressure is generally significantly less than 1, and where the water table is also high. In fact, where the phreatic surface is close to the ground level, three-quarters or more of the lateral pressures can be those due to the groundwater (Golder *et al.*, 1970). For such a situation, naturally, a reliable prediction of the distribution of porewater pressures assumes particular importance.

However, where the ground is inherently soft and the water table is very low, earth pressures can be significant. In any case, where a parametric study indicates the analysis of the structure to be sensitive to potential variations in the values of the earth pressure coefficients, it is advisable to work out the loadings from the earth pressures for a range of ϕ' values for use in the analysis to ensure that no realistically onerous case is overlooked.

12.4 Upward loads

The upward loads acting on an underground metro structure can be those due to the gross contact pressure (representing the reactive force), heave or the hydrostatic pressure.

Gross contact pressure

All the gravity loads, including that from the self-weight of the structure transmitted down to the subgrade, subscribe to the gross contact pressure, p_C, reacting on the underside of the base slab of the structure. As the self-weight of the structure, owing to its cellular construction, is bound to be much less in comparison to that of the weight of the backfill removed to accommodate it, the reactive gross contact pressure mobilized will, likewise, be much less than the overburden pressure experienced by the subgrade previously. However, how uniform the gross contact pressure is likely to be will depend upon the structural stiffness of the base slab and how effective it is in spreading uniformly any inequalities in loads through redistribution. By reference to Figure 12.1, the minimum and the maximum gross contact pressures per unit run of the structure are given by the following expressions:

$$Min. p_C = p_{O/B} + \left(\sum W \div B \right) \tag{12.1}$$

$$Max. p_C = q_S + p_{O/B} + \left(\sum W + W_C + W_P + \sum W_T \right) \div B \tag{12.2}$$

where $\sum W$ is the self-weight of structure, W_C is the total live load on the concourse, W_P is the total live load on platform, and $\sum W_T$ is the total load on the tracks.

Uplift pressure from elastic heave

To quantify the effect of undrained heave on the structure, elastic theory can, once again, come to our rescue. The global effect of heave on metro structures can be estimated, approximately, by treating it as an equivalent upward load on the structure. In order to estimate this load, the depth of the soil below formation over which heave is effective (i.e. the seat of heave) needs to be established. Seat of heave (D) is defined as the depth of the soil between the formation where the heave is the maximum and the level below where it tails off to zero. After estimating the value for the seat of heave (as outlined in Chapter 21), the following procedure may be adopted:

• Treat the seat of heave (depth D) as the length of the elastic medium.
• Since heave is unlikely to be uniform all over the base of the structure, estimate the total value using an average of the values over the entire base area.

- By reference to the variation of heave with depth (D), estimate the average design value of heave over this depth.
- Then, with due deference to the units, the equivalent upward load on the structure can be obtained from the expression:

$$P_{eq} = A E_{u(e)} \delta_{e(ave)}/D \qquad (12.3)$$

Alternatively, the equivalent average uplift pressure from heave may be obtained as:

$$p_U = P_{eq}/A = E_{u(e)} \cdot \delta_{e(ave)}/D \qquad (12.4)$$

where A is the area of the base of the structure, $E_{u(e)}$ is the average undrained elastic modulus for unloading over depth D, $\delta_{e(ave)}$ is the average heave over depth D, and D is the seat of heave.

Hydrostatic uplift pressure

During the various stages of excavation and construction of a cut-and-cover metro structure, groundwater lowering is carried out, generally, in such a manner as to enable construction activities to be carried out in the dry. As long as pumping out of water is maintained and the steady-state of seepage (for definition, see Chapter 11) continues to prevail, there will be no porewater pressure uplift at the subgrade. However, once the pumping stops, the porewater pressures will begin to build-up and continue to rise until the ambient hydrostatic conditions are restored. (In reality, owing to the threshold effects, cent per cent restoration may not be achieved.) The hydrostatic uplift force will then act on the base of the structure. The uplift force on the structure at any time is a function of the 'wet depth', D_W, of the structure at that time and is given by the expression:

$$U = \gamma_W \cdot D_W \qquad (12.5)$$

Wet depth, as defined in Chapter 22, is the depth over which the structure is in contact with groundwater.

12.5 Accidental loads

In the context of cut-and-cover metro station structures, accidental loads are those loads which may be imposed on the structure as a result of impact from train in the event of its derailment. In the case of surface trains, generally, the main causes of derailment, other than those due to collision, are the obstructions across, and the poor maintenance of, the tracks. However, in the case of underground metros where the tracks exist in confined and closely controlled environment because of which they are less likely to be exposed to the occurrence of obstructions across them, and where the operations of the trains are also inherently better controlled, derailments are very rare. Nevertheless, in view of the enormity of its consequences, potential of a derailment should not be ignored and, for that reason, controlling the effects of derailment assumes major importance.

General philosophy

The principle adopted in dealing with accidental loads, particularly in the design of cut-and-cover metro station structures, is to ensure protection of such elements impact upon which could otherwise trigger progressive collapse of the entire structure, while permitting controlled damage, which is not so threatening, of other elements which are either sacrificial or can be safely repaired or replaced.

In the case of an out-of-control train charging at speed along a poor track alignment, derailment cannot be ruled out. However, realistically, derailment is likely to occur in the event of the front wheels of the leading carriageway of the train coming off the rails. Incidentally, this will also represent the scenario for the worst conceivable impact load.

Given the aforesaid premise, it can be argued that if the train, even if out of control and running at speed, enters the station on track, it is unlikely to have a reason to derail at least within the station as long as the tracks within and for a similar distance out of the station follow a straight alignment. That being so, the most likely and vulnerable situation for derailment of the train and possible damage to the station could, potentially, be upon the entry of the train following a somewhat nonlinear alignment into the station.

Possible scenarios

As an errant train enters a metro station, depending upon the structural layout of the station, i.e. whether side or island-platform type, two possible scenarios could present themselves:

- The derailed train could approach a primary vertical support member, such as a column, located close to the railway structure gauge, and hit it directly. Such a scenario is conceivable in the case of a side-platform station with tracks close to and on either side of a central row of columns.
- Alternatively, the derailed train could hit and scrape along the nearest horizontal structural member, such as the edge of the platform slab, causing its damage. It may be unlikely to reach a primary vertical support member directly on account of it being located some distance away from the path of the errant train. Such a scenario is possible in the case of an island-platform station.

In the first scenario, upon impact, some energy is bound to be dissipated not only in the physical deformations likely to be experienced by the train as well as the structure itself, but also in the erratic movement of the train off the tracks and the heat, the noise, the vibrations, etc. generated. Being a function of a number of variables, including the angle of impact, the respective strength and stiffness of the elements of the train and the structure, the time span between the derailment and the point of impact, etc. quantifying such losses of energy precisely is not easy. However, even after making some allowance for these losses and even assuming some drop in the speed upon derailment, the longitudinal component of the residual impact force would, more than likely, continue to be enormous and unmanageable. Designing the column to safely withstand such a force head on would not only be impractical but well nigh impossible. Therefore, derailment under such circumstances if the impact is not avoided could, potentially, lead to a catastrophic situation even triggering a progressive collapse of the structure and must, therefore, be prevented at all costs.

In view of the enormity of such consequences, it is important to ensure that, in the event of derailment, the train does not hit the most vulnerable primary support elements, such as the columns. This could be achieved in two stages. The first line of defence would be provided by the use of strategically located deflectors at the entry to the station which could go some way in guiding the errant train away from impacting on such elements. The deflectors, while shielding the vulnerable elements and restraining the train to some extent would, nevertheless, be destroyed themselves causing partial loss of the kinetic energy of the train in the process.

The second line of defence would be presented by an appropriate protective barrier or a fender beam, designed and constructed within the station to restrain the train within predefined boundaries in such a manner as to contain the effect of the derailment without endangering the most vulnerable elements of the structure. In the case of a side-platform station, it would be prudent to connect the protective barriers located on either side of the central row of columns by cross beams at discrete centres so as to make it possible to bring both the barriers into play in dissipating energy in the event of impact from an errant train.

In the second scenario, the impact may cause damage to or even destruction of the platform slab, but at least it is unlikely to threaten collapse of the station structure as a whole. However, even so, the potential for injury cannot be ignored and, in such a case, the safety of the general public becomes a major consideration.

The platform slab, by virtue of its inherent stiffness as a horizontal plate, can bring its entire length, together with its support structure, into play to absorb the impact energy. This can cause a significant drop in the residual impact load. Notwithstanding the fact that the columns may not be directly threatened, the platform slab itself could, owing to its self-same stiffness, transmit the residual impact load into them causing substantial damage. However, it is possible to neutralize the potential of such impact on the columns in a two-stage process:

- First, it is possible to reduce the residual load. This can be achieved by reducing the inherent stiffness of the horizontal plate while still continuing to mobilize the resistance of the entire slab. The platform slab can be shaped transversely into a folded plate profile incorporating virtual hinge details. These details can be achieved by ensuring discontinuity of the reinforcement at the junction between the platform slab and its support wall, and at the ridge of the main span, as shown in Figure 12.5. Such a geometry is likely to make the slab amenable, upon impact, to relatively greater upward deflections by virtue of the virtual hinge configurations. This would, in turn, lead to an increase in work done and a commensurate increase in the destruction of the impact energy thereby causing a reduction in the residual load. Incidentally, such a proposal would impose a variable thickness of the finish on the slab.

- Second, it can be ensured that even the reduced residual load does not reach the columns. This can be achieved by maintaining a discrete gap of a sufficient width in the platform slab all round the columns. The gap, which would need to be filled with an appropriate compressible filler material, has to be sufficient enough to accommodate, at the very least, the anticipated displacement of the slab upon impact without touching the columns. It is equally important also to consider

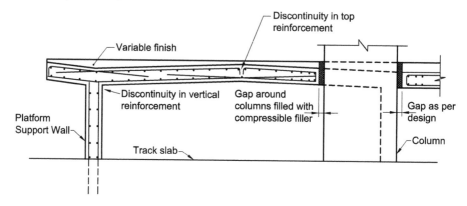

Figure 12.5 Proposed Platform Slab Profile

whether the connections with other vertical primary support elements, such as lift shaft and shear walls, as well need to be treated similarly.

Energy calculation

Kinetic energy, E, of an errant train upon its impending impact on the structure can be obtained from the well-known expression:

$$E = \frac{1}{2}M \cdot v^2 \qquad (12.6)$$

In the above expression, M is the mass of the carriage and v the speed at which it is likely to hit the structure.

According to Newton's second law of motion, force $F = M \times a$, where 'a' is the acceleration. Incidentally, the validity of this expression for objects travelling at speeds approaching the speed of light being questionable should, for obvious reasons, cause us no concern!

A force that gives a mass of $1kg$ an acceleration of $1m/s^2$ is 1 Newton. In other words:

$$1N = 1\left(kg\right) \times 1\left(m/s^2\right) = 1\left(kgm/s^2\right) \qquad (12.7)$$

Assuming the units of mass as kg and velocity as m/s, equation (12.6) can be rewritten, incorporating the respective units, as:

$$E = \frac{1}{2}M\left(kg\right) \cdot v^2\left(m/s\right)^2 = \frac{1}{2}M \cdot v^2\left(kgm^2/s^2\right)$$

$$= \frac{1}{2}M \cdot v^2\left(mkgm/s^2\right) = \frac{1}{2}M \cdot v^2\left(mN\right) = 5 \times 10^{-4}Mv^2\left(mkN\right)$$

To obtain a feel for the magnitude of this energy, assume (for illustrative purposes only) the laden mass of a carriage to be $60,000kg$ (i.e. 60 *tonnes*) and the speed of

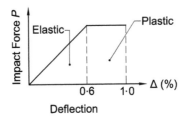

Figure 12.6 'P — Δ' Graph

the errant train as $22 \cdot 22 m/s$ (i.e. $80 km/hour$). Substituting these values in the above equation, the total kinetic energy of the carriage before undergoing any losses would work out to be:

$$5 \times 10^{-4} \times 60,000 \times 22 \cdot 22^2 = 14,812 mkN.$$

After the derailment of the train off its tracks and before coming into contact with the protective structure, a significant amount of loss of energy is bound to occur. But, being a function of a number of variables, the proportion of such a loss is difficult to quantify precisely. However, conservatively, a figure of 50 per cent is commonly adopted. This is based on its usage in relation to the impact of ships against jetties in the maritime industry. Assuming the same rationale, the protective structures then remain to be designed to absorb, through their deformation, the remaining 50 per cent of the energy. In the case of the present illustration, this would amount to $7,406 mkN$ of energy.

Force calculation

If the energy thus obtained were to be neutralized in a head-on impact, a commensurate amount of work would need to be done by the impact force during the deformation of the protective structure. The deformation would include both the elastic and the plastic components somewhat in accordance with the 'P — Δ' graph as shown in Figure 12.6.

If Δ is assumed to be the total deflection that the protective structure undergoes without making contact with any vulnerable member, then P represents the force for which the protective structure must be designed. Assuming, again for illustrative purposes only, a Δ-value, of say $100 mm$, made up of 60 per cent elastic and 40 per cent plastic deformation, the design force would be obtained by equating the work done (represented by the area under the 'P — Δ' graph) with the design energy, i.e.

$$P(0 \cdot 5 \times 0 \cdot 6 + 0 \cdot 4)\Delta = E \quad \text{or} \quad P \times 0 \cdot 7 \times 100 \times 10^{-3} = 7,406$$

whence, Design Force, $P = 105,800 kN$.

Clearly, this force is enormous. Admittedly, it would not at all be practical to design the structure to contain such a force presented by a head-on impact. In view of this, the design of the protective barriers may be based on the following rationale.

Upon derailment, the leading carriage is expected to deviate from the alignment of the track and its intended line of travel before hitting the protective barrier.

As the impact can only occur at an angle, the impact energy can be resolved into two components: longitudinal component parallel to the barrier and the transverse component at right angles to it. The magnitude of the angle of impact is a function of the size (length and breadth) of the typical carriage and the clear width between the elements of the structure or the protective elements on either side of the track which will directly experience the impact. This is generally small – around 3 to 4°.

For a small angle of impact, the longitudinal component of the energy is likely to be very high and not far removed from that of the head-on impact value. As such, the corresponding impact force is unlikely to be manageable. Therefore, no attempt is made to design for this component. However, by the same token, the transverse component is likely to be small and of a magnitude that can be handled. It is argued that, as long as the protective barriers can contain the effect of this transverse component within predefined limits, they will be able to deflect the errant train and guide it along in its forward motion past, and without endangering, the vulnerable elements until all the energy is dissipated and before it eventually comes to rest. To achieve this, it is also important to ensure that the finished face of the barrier is rendered so smooth as to minimize friction along its face. Deflection of the barrier, however, should also discourage development of longitudinal force along the barrier face.

Therefore, all that is needed is to design the barriers for the much reduced impact force corresponding to the transverse component of the energy which will be capable of confining the errant train within the acceptable boundary conditions. The design impact force is obtained by equating this component of energy with the work required to be done in the same direction to neutralize it.

For illustrative purposes, assume the carriage in the previous example to be $3 \cdot 65m \times 20m$ in plan and the clear width between the 'fender/station-wall' on one side of the track and the 'protective barrier' on the other, of $4 \cdot 5m$ as shown in Figure 12.7.

By reference to the figure, $B = 3 \cdot 65, L = 20, C = 4 \cdot 5$

$$\text{Diagonal, } D = \sqrt{(L^2 + B^2)} = \sqrt{(20^2 + 3 \cdot 65^2)} = 20 \cdot 33m$$

$$\text{Maximum angle of impact, } \theta = \theta_1 - \theta_2 = \sin^{-1}\left(\frac{C}{D}\right) - \sin^{-1}\left(\frac{B}{D}\right)$$

$$= \sin^{-1}\left(\frac{4 \cdot 5}{20 \cdot 33}\right) - \sin^{-1}\left(\frac{3 \cdot 65}{20 \cdot 33}\right) = 2 \cdot 45°$$

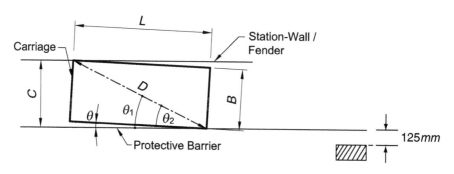

Figure 12.7 Plan Layout (not to scale)

Transverse component of the kinetic energy would be:

$$7,406 \sin^2 \theta = 7,406 \sin^2 2 \cdot 45 = 13 \cdot 5 mkN$$

Assume that the protective barrier is 125*mm* in front of the structural element, i.e. a column, it is meant to shield as shown in Figure 12.7. In other words, the barrier could deflect by this available amount before making contact with the column. However, it would be sensible to build-in a margin of safety by limiting the deflection of the barrier to less than the available amount. Let us assume this to be 100*mm* in the illustration.

By equating the work done with the energy, the design impact force can be obtained, i.e.

$$P_D \times 0 \cdot 7 \times 100 \times 10^{-3} = 13 \cdot 5$$

whence, Design Force, $P_D \simeq 193kN$.

This is quite manageable. A solved example outlining the design of a typical derailment barrier is presented in Appendix B.

12.6 Other loads

Other loads that a typical cut-and-cover metro structure could be potentially exposed to during the various stages of its construction or operational lifetime are those due to:

- Typhoon flooding
- Compaction of backfill
- Seismic activity
- Specified weaponry
- Negative skin friction (*NSF*).

In places of coastal areas like Hong Kong, Singapore, etc. which are known to experience flooding from typhoons, the impact of such flooding if not taken into account and allowed for has the potential of causing serious problems particularly during construction. Measures may be devised to exclude the flooding altogether. This can be achieved by constructing appropriate temporary barriers to protect the construction works from being flooded. Alternatively, the likely effects of flooding may be appropriately allowed for in the design of the structure. However, in that case, the impact of flooding on the subsequent construction activity should be properly considered.

In the case of construction carried out in an open cut within stabilized side slopes, compaction of backfill in discrete layers around the perimeter of the structure can cause modification not only in the intensity but also in the profile of the lateral earth pressures. Seismicity can also cause modification in the lateral earth pressures or distortion to which an underground metro structure may be subjected. The possible ways of handling the effects of compaction and seismicity in the design of underground metro structures are covered in Chapters 13 and 15, respectively.

Where cut-and-cover metro structures are also intended to act as wartime air-raid shelters, the structures need to be specifically designed as such or shielded against the impact of loads from specified weaponry. This topic is considered outside the scope of this book, and for further information on it, reference should be made to the appropriate specialist literature available.

In the construction of metro structures in soft soils, such as Norwegian soft clays or marine clays of Hong Kong and Singapore, the ongoing compaction of the soft sediments surrounding the structure and below the formation and the possibility of causing down-drag forces on the structure due to the negative skin friction cannot be ruled out. Negative skin friction (*NSF*) can occur at any soil–structure interface provided that the downward movement of the soil exceeds the combined effect of the settlement and the elastic deformation (vertical compression) of the structure itself. In such a case, the structure not only loses the frictional support of the surrounding soil to help carry some of the permanent gravity loads but the surrounding soil, instead, imposes additional down-drag forces on the structure. As a result, the structure, including its foundations, could experience further settlements. If, at this stage, total downward movement of the structure relative to that of the surrounding soil becomes greater, then a positive skin friction scenario could emerge. This would, in turn, impose extra loads on the surrounding soil and cause it to settle some more initiating another round of *NSF* and down-drag loads. This process would continue until soil–structure equilibrium was reached. However, if during the design life of the structure the consolidation of the surrounding soil was to be ongoing, then it would be reasonable to assume the down-drag load from the *NSF* as a longer-term load on the structure until such time as 'full' consolidation was achieved.

This certainly would be the case surrounding the perimeter wall of the structure. However, situation within the perimeter wall and directly below the base of the structure would be somewhat different.

In the construction of a typical cut-and-cover station structure, the net weight of the overburden removed during excavation is likely to substantially outweigh the weight of the permanent structure replacing it. In such a case, it is argued that there will be net unloading on the formation and thus the cause for any future consolidation of the soft clay directly below the structure would no longer exist. While this may be substantially true, nevertheless, if, for some reason, the latter exceeded the former, then consolidation of the soft clay below the formation would, in principle, continue. However, it would happen to a somewhat lesser degree. This is mainly due to the fact that the water table draw-down within the perimeter to enable excavation and construction to take place in the dry would have taken care of some of the settlements anyway. The issues become somewhat more complicated if the structure is carried on piles taken down in end-bearing into a more competent stratum.

To sum up, in estimating the loading due to the *NSF* to be expected on the structure, it is important to establish the nature of the relative movement at the soil–structure interface and the mechanism of load transfer correctly.

13   Assessment of Earth Pressures

13.1 Introduction

In the analysis related to cut-and-cover structures, the pressures imposed on the structure by the surrounding ground and the groundwater environment constitute the dominant lateral loadings for the design of the perimeter wall structure. These can also have a significant influence on the design of other elements as well. The magnitude of the pressures developed depends, among other things, upon the geometry of the problem and varies with the following:

- Sequence of excavation
- Type and drainage characteristics of soil
- Type and the stiffness of the ground support system
- Extent of soil strain permitted by the movement of the ground support system.

Because of the number of variables involved and the difficulty commonly experienced in assessing their values precisely, 'accurate' prediction of earth pressures commensurate with the amount of strain undergone often presents a problem. However, educated estimates based on classical approach and past experience generally provide the starting point. Where appropriate, modifications are then made in the light of the feedback from observation and monitoring.

13.2 Horizontal earth pressures

Before proceeding to make an assessment of earth pressures for use in the design process, it is important to:

- Identify the prevailing *in situ* stresses before any construction activity begins;
- Understand where the lower and the upper ultimate limits of the lateral pressures lie;
- Appreciate if, and the mechanism by which, these limits could possibly be reached during the course of the construction activity.

A sound understanding and appreciation of all these aspects is necessary for the designer to ensure that the earth pressures are neither grossly over- nor under-estimated and be able to proceed towards a safe and sensible design.

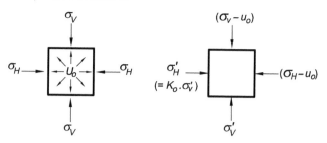

Figure 13.1 In Situ Soil Stresses

In situ stresses

Stresses within a soil mass when the ground surface is horizontal and the variation in the nature of soil in the horizontal directions minimal are known as geostatic stresses. Since there are no accompanying shear stresses upon the vertical and the horizontal planes, they represent the principal planes and the stresses on these planes, accordingly, the principal stresses.

Consider an element of soil at a depth z in a state of elastic equilibrium (i.e. state of rest or zero lateral strain) under the action of the principal total vertical and horizontal stresses σ_V and σ_H, and the ambient porewater pressure (*pwp*), u_0. The corresponding *in situ* vertical and horizontal effective stresses, respectively, on the element are then given by the expressions:

$$\sigma'_V = \sigma_V - u_o \qquad (13.1)$$

$$\sigma'_H = \sigma_H - u_o \qquad (13.2)$$

where $\sigma_V = \gamma \cdot z$, γ = unit weight of soil, $\sigma_H = K_o\left(\gamma \cdot z - u_o\right) + u_o$, and K_o = coefficient of earth-at-rest pressure.

These are the pressures present in the soil when it has not undergone any strain or movement. They are, because of that fact, referred to as the 'earth-at-rest' pressures. Collectively, they represent the *in situ* state of stress in the soil before the commencement of any construction activity or the imposition of any imbalance in the ambient stress regime. This state of stress is shown in Figure 13.1. If it were possible to wish an underground structure into place, it is likely to be subjected to these lateral pressures around its perimeter.

The ratio of the effective horizontal stress, σ'_H, to the corresponding effective vertical stress, σ'_V, at zero lateral strain, represents what is known as the coefficient of earth pressure at rest, K_o. These effective stresses also represent the principal stresses since they are not accompanied by any shear stresses on their respective planes.

If, by altering the magnitude of one of the principal effective stresses, the soil element were to undergo enough strain to cause failure on any plane (other than the principal planes), then the stresses acting upon such a failure plane would be a combination of the following:

- The shear stress, τ_f, parallel to the shear plane representing the structural shear strength of the soil;
- The effective stress, σ'_n, normal to the shear plane.

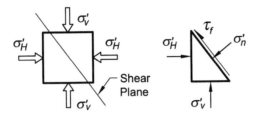

Figure 13.2 Stresses at Failure

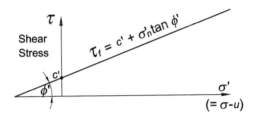

Figure 13.3 Failure Envelope

These are shown in Figure 13.2.

In other words, σ'_n and τ_f represent the normal effective stress and the structural shear strength, i.e. shear stress at failure, on the shear plane, and these are related by the idealized expression:

$$\tau_f = c' + \sigma'_n \cdot \tan\phi' \tag{13.3}$$

In the expression, c' and $\tan\phi'$ are the experimentally determined shear strength parameters with respect to the effective stress for the given soil. It also implies that the shear strength of soil is controlled by its effective stress. Clearly, this relationship is linear. This expression was first proposed by Coulomb (1776) and later modified by Terzaghi (1943). When represented graphically in its idealized form, as shown in Figure 13.3, it is referred to as the failure envelope.

If the stresses σ'_V and σ'_H are plotted on the same coordinate axes, σ' and τ, as for the failure envelope, a circle with the stress difference $(\sigma'_V - \sigma'_H)$, also known as the deviator stress, as the diameter can be drawn as shown in Figure 13.4. This represents the Mohr circle of stress for the ambient condition, i.e. earth pressure at rest. Note that the circle does not touch the failure envelope; this is only to be expected since the 'at-rest' condition represents a state of 'elastic' equilibrium and not a failure condition.

For a given soil, if a number of samples were made and tested at different consolidation pressures, it would be seen that, at failure, the envelope would very closely approximate a straight line and would always be tangential to the Mohr circle of stress. In other words, the shear stresses at failure, i.e. the shear strength at different

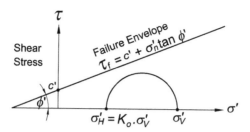

Figure 13.4 Mohr Circle for Earth-at-Rest Pressure

consolidation pressures, would always lie on this idealized line and not cross it. It is for this reason that this line is commonly referred to as the failure 'envelope'.

Mohr-Coulomb failure criterion

Where the Mohr circle touches the failure envelope, the tangent point is common to both. As the tangent point lies on the failure envelope, it has the coordinates τ_f and σ_f' representing failure. However, since it also lies on the Mohr circle, it therefore represents a state of failure stress within the soil as well. This is referred to as the Mohr-Coulomb Failure Criterion (Menzies, 1973).

Active failure

If a trench is excavated horizontally adjacent to a soil element, the vertical principal effective stress, σ_V', on the element remains unaltered. However, the horizontal principal effective stress, σ_H', is reduced because of the removal of the lateral restraint of the soil thereby allowing the soil structure to expand horizontally. With this freedom, as the soil structure goes on expanding horizontally, the Mohr circle describing the consequent changing state of stress will go on growing until it touches the failure envelope as shown in Figure 13.5.

Since the Mohr circle cannot cross the failure envelope, the minimum value of the horizontal effective stress, below which it may not fall, is therefore reached when the expansive circle touches the failure envelope. This ultimate lower limit of the lateral principal effective stress is referred to as the 'active earth pressure', p_a, and is equal to $(K_a \cdot \sigma_V')$; K_a is known as the coefficient of active earth pressure and its value is given by the ratio of p_a and σ_V'.

If deformation prior to failure is ignored and if it is assumed that, at failure, the whole soil mass is brought into a state of limiting active 'plastic' equilibrium, that is to say that yield occurs simultaneously throughout the entire mass at constant yield stress, it can be demonstrated using the Mohr circle geometry, or analytically, that the coefficient of active earth pressure,

$$K_a = \frac{p_a}{\sigma_V'} = \frac{1 - \sin\phi'}{1 + \sin\phi'} = \tan^2\left(45° - \frac{\phi'}{2}\right) \tag{13.4}$$

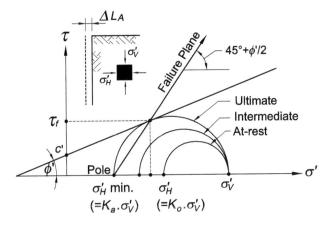

Figure 13.5 Active Failure

Passive failure

Pile driving operations in the vicinity of metro works, or if the earth retaining element were to be pushed into the retained soil through preloading to limit ground movement, could both cause compression of the soil laterally while maintaining the vertical principal effective stress unchanged. If the horizontal principal effective stress were to be increased thus, the Mohr circle of stress describing the consequent changing state of stress in the soil would diminish to a point whereat the horizontal and the vertical effective stresses would be equal and would, thereafter, grow as the horizontal effective stress exceeded the vertical effective stress (Menzies, 1973), as shown in Figure 13.6.

A limiting condition would once again be reached as the compressive Mohr circle of stress touched the failure envelope. This ultimate upper limit of the lateral major principal effective stress beyond which it may not increase is referred to as the 'passive earth pressure', p_p, and is equal to $(K_p \cdot \sigma'_V)$; K_p is known as the coefficient of passive earth pressure and its value is given by the ratio of p_p and σ'_V.

If it is assumed that, at failure, the whole soil mass is brought into a state of limiting passive 'plastic' equilibrium, it can be demonstrated, once again, using the Mohr circle geometry, or analytically, that the coefficient of passive earth pressure,

$$K_p = \frac{p_p}{\sigma'_V} = \frac{1 + \sin\phi'}{1 - \sin\phi'} = \tan^2\left(45° + \frac{\phi'}{2}\right) \tag{13.5}$$

It should be noted that, unlike the earth-at-rest condition, both these extreme states of earth pressure, i.e. the active and the passive, represent the limiting values and obtain only when the soil is in a state of failure. In that state, it is assumed that the shear strength of the soil structure would have been fully mobilized along the corresponding shear planes. This requires the soil element to strain sufficiently in expansion for σ'_H to drop down to its minimum value $(K_a \cdot \sigma'_V)$ or, in compression for it to rise to its maximum possible value $(K_p \cdot \sigma'_V)$.

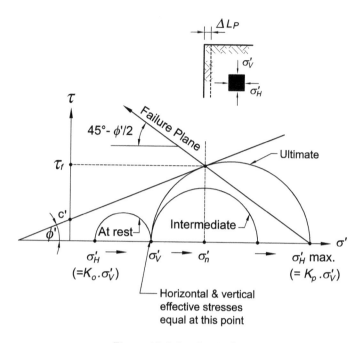

Figure 13.6 Passive Failure

Total stresses include porewater pressure components in them. However, porewater pressure at any point is constant and invariant with direction; as such, unlike earth pressures, it is not subject to differential factorization. It is therefore worth remembering that the coefficients of earth pressure, i.e. K_a, K_o and K_p, are never applied to total stresses. It is for this very reason also that, for estimating the design loadings, the soil and the porewater pressures are computed discretely. This is generally so unless, of course, the coefficient of earth pressure assumed happens to be 1 as adopted in the total stress '$(\phi = 0)$' analysis, in which case if the saturated unit weight of soil below the water table is not very different from the dry unit weight of soil above the water table, it may not be necessary to deal with the pressures separately.

Soil strains

Clearly, if the full strains necessary for the lateral pressures to attain their ultimate limiting values are not possible or permitted, they may only reach some intermediate values lying somewhere between the ambient earth-at-rest and the ultimate limiting active or passive values.

During the excavation for and construction of a cut-and-cover metro structure, movement in the surrounding ground of some magnitude, as a result of the disturbance in its native state of rest, is inevitable. This implies that the lateral pressures to be used in the design of the structure for the construction stage will be less than the earth-at-rest pressures. Whether these pressures are likely to drop as far down as their active values, or how close they are likely to get to them, depends upon the extent of movement permitted by the stiffness of the ground support system. This is also, in turn, a function of the method and sequence of construction.

If the structure were to be constructed within a sheet pile perimeter cofferdam in the conventional bottom-up sequence, it could, arguably, be assumed that the relative flexibility of such an earth support system might permit the surrounding soil to undergo enough movement to suggest that the lateral pressures might be close to their limiting active values. However, if the construction were to be carried out within an integral, relatively stiff, perimeter diaphragm wall cofferdam in the top-down sequence, it is conceivable that the extent of potential movement of the wall and the surrounding ground could then be restricted only to a value which would be permitted by the deformation of the floor slabs acting as horizontal struts consequent upon their shrinkage, axial compression, creep, etc. It is then likely that, owing to the relatively higher stiffness provided by the floor slabs, the resulting strain in the soil would not be sufficient enough to generate full active pressures and would place the lateral earth pressures somewhere between their at-rest and active values.

13.3 Earth pressure coefficients

For detailed treatise on the various earth pressure theories in general and coefficients in particular, the reader is referred to any standard text book on geotechnical engineering. However, in order to maintain the continuity of the text, a passing reference is made to the coefficients in this chapter. Coefficients of active and passive earth pressures are functions of the ϕ' value and have been dealt with in the previous section. Assessment of the coefficient of earth-at-rest pressure, K_o, is addressed next.

Coefficient of earth-at-rest pressure (K_o)

Value of K_o may be obtained by any one of the following means:

- Laboratory testing
- *In situ* testing
- Use of empirical formulae.

Laboratory testing

The coefficient of earth-at-rest pressure is generally obtained in the laboratory by performing the K_o consolidation test. However, the accuracy of the results obtained through this test cannot be relied upon since it can involve a number of potential sources of errors. Also, as explained in Chapter 23, the value of K_o for a given soil is not unique but a function of its stress history. With the stress history likely to get destroyed during sampling and unlikely to be fully restored thereafter, the accuracy of the value of K_o so obtained cannot be guaranteed. In fact, based on his work on the *in situ* measurement of initial stresses and deformation characteristics, Wroth (1975) was able to conclude that, for natural soils, laboratory testing was unlikely to yield satisfactory K_o values. In the case of sand, the problem is further compounded due to the inability to obtain a truly representative, undisturbed sample.

In situ testing

Testing of soil *in situ* to obtain the coefficient of earth-at-rest pressure, K_o, is carried out, traditionally, with the use of Menard pressuremeter. Pressuremeter was initially developed by Menard (1965) for the *in situ* measurement of stress–strain modulus. However, the borehole pressuremeter test results are commonly used to obtain the value of K_o also. Pressuremeter tests are very sensitive to the condition of the borehole before the test and so care needs to be exercised in the use of the results obtained from these tests. According to Tavenas (1975), K_o values obtained from any *in situ* testing technique are unlikely to be any closer to the truth than those obtained empirically. Alternatively, value of K_o may be directly obtained *in situ* with the use of an appropriate measuring device such as a Camkometer (Cambridge K_o meter).

Evaluation of K_o, the coefficient of earth-at-rest pressure, empirically, by various researchers has been dealt with in Chapter 23.

13.4 Choice of coefficients for design

During the excavation for and construction of a cut-and-cover structure, inevitably, the stress history of the surrounding soil is disturbed and so, thereafter, the *in situ* stresses are unlikely to be those 'at-rest' at least over the duration of construction. If the ground support structure (i.e. the retaining element) were to be extremely flexible with, virtually, negligible stiffness, the soil would be free to strain as if in a free-field environment and the load imposition on the structure would be minimal. However, since, from the considerations of practicality, the retaining element has to be of at least a certain minimum stiffness, it is bound to attract to itself, as it deforms, a commensurate amount of load. Achieving optimization of the movement of the ground support system in harmony with the stiffness of the ground *vis-à-vis* the load in the support structure is the hallmark of a good design.

Depending upon the extent of soil strain permitted by the stiffness of the ground support system, the stresses in the surrounding soil above the formation could lie anywhere between their 'at-rest' and the 'active' values. Build-up of stresses back to their 'at-rest' values thereafter could be a matter of some considerable period of time. Notwithstanding this, since it is difficult to quantify precisely the rate of this build-up and with a view to ensuring that the design pressures are not underestimated, the use of 'at-rest' pressures also as one of the load cases for the long-term design has come to be accepted as the norm.

When a retaining structure is pushed into the ground, the passive resistance mobilized can be well in excess of the earth-at-rest pressure. Such a situation could arise under the action of an asymmetric loading such as an out-of-balance lateral load resulting, for instance, from surcharge load applied to one side of the structure only. In the context of cut-and-cover metro structures, generally, the effect of such an imbalance in the lateral loads generated locally is unlikely to be significant.

However, even if it were significant, it would be a near certainty that, under the load imbalance, the extent of the inherent lateral stiffness of the structure and the soil–structure interface friction generally available would bring into play the entire length of the structure, thereby minimizing the effect to such an extent as to render the impact insignificant globally. In view of this, it would be reasonable to assume that the pressures mobilized in the surrounding soil above the formation level are unlikely to exceed, significantly, the at-rest values during the lifetime of the structure and would imply these values to represent the 'likely maximum' load case.

Likewise, given the understanding that the earth pressures simply cannot drop below their active values, there is the perception that designing for these pressures during the stages of construction would cover the 'minimum possible' load case. Nevertheless, it would be a mistake to infer that designing for both the 'minimum possible' and the 'maximum likely' load cases would also automatically cover the effect of every other load case in between thereby ensuring a safe design. If the structure could be wished into place and the structure and the ground were free to undergo unrestricted strain, then designing for these two cases could, possibly, lead to a safe design.

However, in view of the complexities of staged construction and the changes in the associated boundary conditions, the design concept as outlined above could represent an oversimplification of what might happen in reality. There are a number of reasons for this:

- In the case of cut-and-cover metro structure, the geometry of the fully completed structure, and therefore the structural model for the long-term operational case, may be much different from those of the partially completed models (cross-sections) during the various (temporary) stages of construction.
- The boundary conditions for the in-service stage can be vastly different from those of the different stages of construction.
- If the earth pressures, in reality, did not drop down to their active values, the stress-resultants based on the 'minimum possible' load case could easily 'underestimate' the locked-in effects.

In view of the foregoing, the design based on the two load cases only, i.e. the active and the at-rest pressures, may not necessarily cover all the onerous loading possibilities at all the critical locations.

While there can be no doubt that, with the relaxation in the surrounding soil mass due to excavation, the lateral earth pressures above the formation level will no longer be those at rest, it is equally possible that they may not drop down far enough to attain their minimum values either. This is because the temporary internal strutting or the permanent bracing provided by the floor slabs above the formation level are unlikely to permit unrestrained inward movement of the opposite walls but are likely to restrict it to the extent permitted by the deformation of the wall support (i.e. bracing) system. In that case, it is conceivable that, depending upon the extent of soil strain undergone, the pressures could be higher than the active and lie somewhere between the active and the at-rest values. Besides, it is not unusual for the temporary span of an element (i.e. during its construction) under consideration to be, sometimes, longer than its span when functioning as part of the permanent structure. The possibility, then, that the combination of the increase in the span and the loading could prove to be more onerous, cannot be ruled out.

To illustrate this point, consider a typical cross-section with a perimeter diaphragm wall with idealized boundary conditions at a certain stage of excavation and construction as shown in Figure 13.7.

In this exercise, it is assumed for simplicity that $\gamma = 20kN/m^3 = \gamma_S$ for the soil and that the water table is at $2 \cdot 5m$ below ground level. Soil pressures have been worked

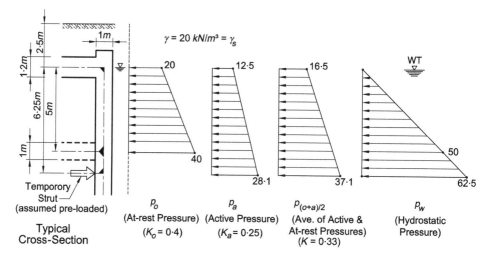

Figure 13.7 Typical Cross-Section and Pressures (during a Stage of Construction)

out using coefficients of at-rest and active pressures and also the average of the two. Porewater pressure is assumed to be hydrostatic. For simplicity, no other loads are considered. It is assumed that the perimeter wall extends a significant depth below the temporary strut level and that the strut is preloaded. The boundary condition at the strut level is therefore idealized as that provided by a rigid, unyielding support.

Bending moments for the three load cases have been plotted in Figure 13.8.

In the figure, M_{K_a} represents the moment diagram due to the active earth pressure on an idealized wall element during construction and M_{K_o} that due to the earth-at-rest

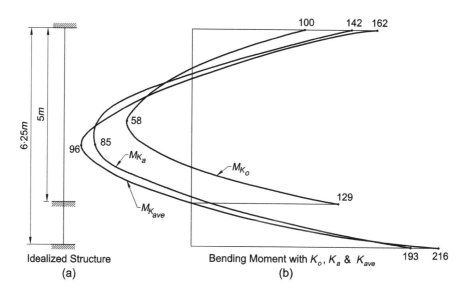

Figure 13.8 Bending Moment Profiles

pressure on the idealized wall element of the permanent structure. However, $M_{K_{ave}}$ represents the moment diagram that corresponds to the average of the active and the at-rest pressures on the wall element during a stage of construction with the temporary strut in place.

Comparison of the moment diagrams reveals that the moments given by the use of average K-value $(< K_o)$ corresponding to the commensurate soil strain during the stage of construction are significantly higher than those given by that of K_o on the permanent structure. This confirms that the two 'extreme' load cases referred to previously may not necessarily cover all the onerous cases which can be reasonably expected to occur.

In the foregoing example, the difference between the active and the average pressure coefficients is of the order of 32 per cent. However, the difference in the corresponding moments is only of the order of 12–14 per cent at the two ends and 13 per cent in the span. This is essentially because, out of the hydrostatic pressure that remains unchanged and the earth pressure, in the two cases, the former is the dominant contributor to the moments. It should therefore be appreciated that, bigger the difference between the active and the average pressure coefficients, bigger also can be the difference expected in the end moments. Furthermore, deeper the water table, proportionally more significant can the contribution from the earth pressure become.

In the case of the perimeter walls particularly, once the movement under lateral pressures has taken place and the surrounding soil mass has moved with the deflected profile of the wall in sympathy, it is well nigh impossible to fully recover the predeflection profile and the moments are likely, thereafter, to remain permanently locked-in. It is important not to overlook the implications on the overall design of the magnitude as well as the location of such locked-in effects in the walls above the formation level.

Below the formation level, however, restraint to the inward movement of the walls is provided by the passive resistance mobilized by the soil under the formation and enclosed between the walls.

As explained in Chapter 23, it is conceivable that, for a given soil, the value of K_o can vary in conformance with the respective stress histories at different locations. However, one thing is certain – the value must lie somewhere between that of K_a and K_p. Furthermore, the position of K_o in relation to these limits will govern the amount of movement required to mobilize either of these limits as indicated in Figure 13.9.

The figure shows, schematically, the wall tilt necessary to achieve the ultimate limit pressures. It is important to remember that considerably larger strains need to be undergone by the soil in order to mobilize pressures corresponding to a passive failure than those necessary to reach an active failure. This is the reason why, if ground movements are to be controlled, full passive pressures that would require unacceptably large movements are not mobilized but need to be restricted to half or a third of their values.

Typical values for maximum wall tilt are listed in Table 13.1.

In the table, ΔL_a and ΔL_p are the extents of the active tilt away from and the passive tilt into the wall, respectively; H is the height of the retained ground. It would appear that, for granular soils, the tilt into the wall necessary to mobilize full passive pressures can be up to an order of magnitude greater than the tilt away from the wall required to generate full active pressures. This is also confirmed by finite element analysis carried out by Clough and Duncan (1971). In the case of cohesive soils, however, the passive tilt required appears to be twice as much as the active tilt necessary.

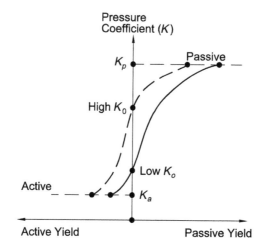

Figure 13.9 Wall Yield and Pressure Coefficient (K) (adapted from Burland *et al.*, 2004)

Table 13.1 Typical Values of Maximum Wall Tilt for Rankine State

Soil type	$\Delta L_a/H$	$\Delta L_p/H$	$\Delta L_p/\Delta L_a$
Loose sand	$(1-2)10^{-3}$	10×10^{-3}	$10-5$
Dense sand	$(0{\cdot}5-1)10^{-3}$	5×10^{-3}	$10-5$
Soft clay	20×10^{-3}	40×10^{-3}	2
Stiff clay	10×10^{-3}	20×10^{-3}	2

(adapted from Das, 1990)

In the case of heavily overconsolidated soils with very large K_o-values that are only slightly less than the K_p-values, on the other hand, the fact that relatively smaller strains would be needed to reach a passive limit state and relatively larger strain to achieve an active limit state should not be overlooked.

Mobilized earth pressure coefficient

As discussed in the foregoing, the magnitude of earth pressure generated is intrinsically related to the extent of soil strain undergone. If it were possible to quantify different ϕ' values mobilized at different stages of construction corresponding to the extents of soil strains undergone at those stages, computations of earth pressures would be facilitated considerably by using the mobilized value, i.e. ϕ'_{mob}, directly. By reference to Table 13.2, it can be seen (Bolton, 1986) that, in the case of normally consolidated soils, mobilized coefficients of active pressure based on the mobilized angles given by $\phi'_o = \phi' - 11{\cdot}5°$ are almost identical with the coefficents of earth pressure at rest, K_o, as those given by the widely accepted empirical version of the Jaky relationship

Table 13.2 K_o-Values for Normally Consolidated Soils

ϕ' (°)	$\phi'_o = \phi' - 11.5°$	$1 - \sin\phi'$ (*Jaky*)	$\tan^2\left(45° - \phi'_o/2\right)$
20	8.5	0.66	0.74
25	13.5	0.58	0.62
30	18.5	0.50	0.52
35	23.5	0.43	0.43
40	28.5	0.36	0.35
42	30.5	0.33	0.33
45	33.5	0.29	0.29
50	38.5	0.23	0.23

(Equation 13.6). In other words,

$$K_o^N = 1 - \sin\phi' \simeq \frac{1 - \sin\phi'_o}{1 + \sin\phi'_o} \simeq \tan^2\left(45° - \frac{\phi'_o}{2}\right) \tag{13.6}$$

Results based on this expression for different values of ϕ' in soils are listed in Table 13.2.

From the table it is obvious that, in the range of ϕ' values commonly encountered in practice, i.e. 30° and above, there is hardly any difference in the K_o-values based on the Jaky expression and those using the mobilized active ϕ' value. For this range, therefore, expression for the mobilized earth pressure coefficient may be written as:

$$K_m^N = \tan^2\left(45° - \phi'_m/2\right) \tag{13.7}$$

where the subscript m represents the percentage strain invoked. For example, for 1 per cent axial strain in a triaxial test, $\phi'_{m=1} = \phi' - 8°$ would represent a conservative estimate for a moderately stressed granular soil. In view of the foregoing, it is possible to choose ϕ'_m at a desired degree of lateral strain and then use the usual K-factors, for example those given by Caquot and Kerisel (1948) presented in Figure A6 in Appendix A.

13.5 Effect of stratification

A soil is said to be 'homogeneous' if it displays identical properties at all points. It is said to be 'isotropic' if it displays identical properties in all directions and 'anisotropic' if it does not do so. It is not uncommon for a soil to be homogeneous and anisotropic. The main reason for anisotropy in soils is the stratification or layering due to nonuniform deposition. Preferred particle packing as a result of the inherent characteristic of soil particles to settle with long axes horizontally is also a contributory factor, but this is only a very minor one. Anisotropy in soils is reflected in the change in the values of the parameters c' and ϕ' (or s_u) in the successive strata or layers. Lateral pressures are a function of these parameters.

For constant values of the parameters c' and ϕ' (or s_u) within each layer, the theoretical lateral limit (i.e. active and passive) pressures increase linearly with depth. At the level of the interface common to two touching layers, the vertical (overburden) pressure, being simply a function of the depth at that level, is invariant. However, this

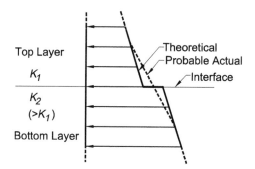

Figure 13.10 Typical Lateral Earth Pressures in a Layered Soil

is not so with regard to the lateral pressure. The lateral pressures at the interface, i.e. at the bottom of the top layer and the top of the bottom layer, will vary reflecting the difference in the theoretical values of the respective earth pressure coefficients above and below it. This is the reason for the apparent jump in the value of the theoretical lateral pressure at the interface as indicated in Figure 13.10.

The foregoing rationale notwithstanding, it is difficult to entertain the notion and justify the physical possibility, in reality, of such a jump in the lateral pressure at the interface. The reality is more likely to be close to that represented by the dotted line. Nevertheless, for ease of computation and without much loss of accuracy, the theoretical pressure profile with the jump in the pressure at the interface is commonly adopted and accepted in design.

13.6 Choice of method of analysis

Water has no inherent shear strength and as such subsurface water does not contribute to the shear strength of the soil. However, the rate of groundwater movement does affect the effective stress in the soil and, to that extent therefore, does influence its strength. In relatively free draining ($c' = 0$) soils such as sands and gravels, by virtue of their immediate dissipation, excess porewater pressures do not build up so that, effectively, the long-term, drained condition is reached during construction itself. Therefore, it is appropriate that the earth pressures in such soils for both the construction stages in the short-term as well as the long-term design life of the structure are evaluated by the 'effective stress' (c', ϕ') analysis.

However, in cohesive soils, in the short term during construction, due to viscous retardation to free flow of water, excess porewater pressures will not have had enough time to dissipate to their equilibrium values. So, the effective stress, as also the undrained shear strength s_u (also symbolized by c_u), will remain unaltered. Any change in the total stress is simply reflected by a commensurate change in the porewater pressure. Therefore, provided that the soil is saturated and behaves in the undrained mode, i.e. the volumetric strain remains zero, it is possible to characterize the strength of a soil element in its *in situ* effective stress state by its undrained shear strength. In that case, the 'Total Stress' analysis based on the '$\phi = 0$' concept may be used to estimate the lateral earth pressures. Nevertheless, if, even in the short term, the dissipation of the excess porewater pressures takes place which could cause a corresponding change in the effective stress and the strength of the soil, total stress approach is invalid and the effective stress analysis must be used.

In the long term, on the other hand, when the excess porewater pressures have had time to dissipate, the effective stress and hence the strength will change reflecting the amount of the excess porewater pressures dissipated. So, for the long term, Effective Stress (c', ϕ') analysis will apply.

In reality, however, it cannot be guaranteed that the type of soil encountered will be exclusively granular or totally cohesive; it may be some combination of the two (i.e. c', ϕ' type, such as silt). In such a case, it may be necessary to carry out both the total stress as well as the Effective Stress analyses in order to examine the possible range of soil behaviour and its effect on the stress-resultants of the structural elements.

Total stress analysis

The basis of the total stress analysis is the '$\phi = 0$' concept. According to it, if a soil undergoing undrained loading is saturated so that, with regard to the total stresses, it behaves as though it were purely a cohesive material, then the computation is greatly simplified by using the so-called '$\phi = 0$' concept. The undrained shear strength, s_u, is taken as half the deviator stress $(\sigma_1 - \sigma_3)/2$. This assumes that the undrained shear strength can be expressed independently of the effective stress at failure. In the analysis, therefore, wherever undrained shear strength is used along the failure surface, porewater pressures are taken as zero. This should not be taken to imply that the porewater pressures necessarily are zero but that these are treated as such only to be consistent with the assumption.

The '$\phi = 0$' concept

If three identical soil samples were to be tested under quick undrained conditions at three different cell pressures, σ_a, σ_b and σ_c, the strengths obtained in the three cases would be identical. This is because, as long as drainage is prevented, any change effected in the loading is simply reflected in a commensurate change in the porewater pressures with the effective stress remaining unchanged. As a result, the deviator stress at failure, $(\sigma_1 - \sigma_3)_f$, which represents the compressive strength of the soil structure, remains unaltered. This is represented in Figure 13.11.

Clearly, a tangent to the circles under these circumstances describes a horizontal line, which implies a ϕ-value equal to zero in terms of the total stress in the coulomb equation:

$$\tau_f = c + \sigma \cdot \tan \phi_u \qquad (13.8)$$

This is due to the fact that the strength, which does not change with total stress, is being plotted against the total stress. It is clear therefore that '$\phi = 0$' apart from stating the obvious is not a very helpful concept. However, the fact remains that the strength is controlled by the effective stress, so that

$$\tau_f = s_u = \frac{1}{2} (\sigma_1' - \sigma_3')_f \qquad (13.9)$$

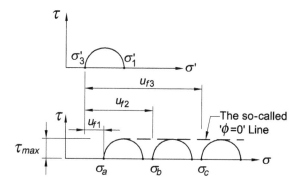

Figure 13.11 '$\phi = 0$' Concept

The application

In the total stress approach, the general expressions for the total lateral active and passive design pressures, respectively, are those given by the following expressions:

$$p_a = K_a \cdot \sigma_V - K_{ac} \cdot s_u \qquad (13.10)$$

$$p_p = K_p \cdot \sigma_V + K_{pc} \cdot s_u \qquad (13.11)$$

where K_a and K_p are the active and the passive earth pressure coefficients as defined previously and s_u is the undrained shear strength (also symbolized as c_u). Also,

$$\sigma_V = Vertical\ total\ stress = \gamma z$$

$$K_{ac} = 2\sqrt{K_a\left(1 + c_w/s_u\right)} \quad \text{and} \quad K_{pc} = 2\sqrt{K_p\left(1 + c_w/s_u\right)}$$

where, γ = bulk unit weight of saturated soil, z = depth below ground, and c_w = wall adhesion.

In the analysis using the total stress approach, the two parameters to operate with are: $\phi' = 0$ and s_u. With $\phi' = 0$, $K_a = 1 = K_p$. In view of this, it may not be necessary to separate the contributions to lateral pressures from the soil and the porewater. Furthermore, if the wall adhesion is ignored, i.e. $c_w = 0$, then the lateral earth pressure expressions reduce to:

$$p_a = \gamma z - 2s_u \qquad (13.12)$$

$$p_p = \gamma z + 2s_u \qquad (13.13)$$

Total stress approach should be applied with care. It is at best applicable to temporary works design but only if the conditions are undrained; otherwise, effective stress analysis should be used.

Effective stress analysis

In the effective stress approach, the general expressions for the effective lateral active and passive pressures, respectively, are given by the following expressions:

$$p'_a = K_a \cdot \sigma'_V - K_{ac} \cdot c' \qquad (13.14)$$

$$p'_p = K_p \cdot \sigma'_V + K_{pc} \cdot c' \qquad (13.15)$$

where: σ'_V = vertical effective stress $\sigma_V - u$; u = porewater pressure;

c' = effective stress, shear strength parameter

$$K_{ac} = 2\sqrt{K_a \left(1 + c_w/c'\right)}; \quad K_{pc} = 2\sqrt{K_p \left(1 + c_w/c'\right)};$$

By adding the contribution of the porewater pressures to the effective pressures, the total lateral design pressures can be obtained as follows:

$$p_a = p'_a + u \qquad (13.16)$$

$$p_p = p'_p + u \qquad (13.17)$$

For the various procedures available to evaluate the porewater pressure (u), reference may be made to Chapter 18.

13.7 Earth pressures in soft clays

As explained in Chapter 20, in an open cut if, at any stage, the vertical stress relief at the base is so large that the shear strength of the subsoil is approached, plastic zone formation is initiated around the lower corners of the cut. Then, regardless of the integrity of the bracing system, the deformations increase rapidly causing a progressive extension of the plastic zone and eventually leading to instability and base failure.

Taking H as the depth of a cut, and $s_{u(b)}$ the undrained shear strength of clay below the formation, an index of excavation approaching base failure is given by the dimensionless stability number

$$N = \gamma \cdot H / s_{u(b)} \qquad (13.18)$$

As N approaches the value $(\pi + 2)$, significant plastic deformations develop below the base of the cut. Experience based on the soft Norwegian and Mexican clays has shown that, under such a situation, the earth pressures that have to be carried by the wall support (bracing) system become greater than those computed by normal procedures using Bell's equation, i.e.

$$p_a = \gamma \cdot H - 2s_u \qquad (13.19)$$

According to the theoretical treatment developed by Henkel (1971) based on the development of plastic Prandtl zones beneath the cut, the coefficient of earth pressure reflecting the increased pressures is expressed as:

$$K_a = \left[1 - \frac{4s_u}{\gamma H}\right] + 2\sqrt{2} \cdot \left[\frac{d}{H}\right]\left[1 - \frac{(2+\pi)}{N}\right] \qquad (13.20)$$

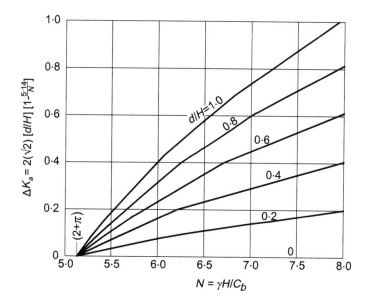

Figure 13.12 Variation of ΔK_a with N and (d/H) (adapted from Henkel, 1971)

The parameter s_u is the average undrained shear strength of clay above the cut and d the depth of the Prandtl zone. The depth d, irrespective of the extent of penetration of the perimeter wall below the formation, is taken as the depth to the bedrock horizon below the base of the cut or $B/\sqrt{2}$, whichever is the lesser, B being the width of the cut.

The first term on the right hand side of the equation can be recognized straight away as representing the conventional value for the earth pressure coefficient, K_a, from Bell's equation. As the stability number N exceeds the value $(2+\pi)$, the second term on the right hand side of the equation will become positive. It will then represent an enhancement ΔK_a in the conventional coefficient and, correspondingly, a commensurate increase in the earth pressure.

Figure 13.12 shows the variation of ΔK_a with the stability number N for different values of d/H. It can be seen that, in the N-range of $(2+\pi)$ to 8, for higher values of d/H and N, the value of ΔK_a and therefore also of the corresponding additional earth pressure can be significant.

13.8 Compaction loads

In the case of construction of a typical cut-and-cover metro structure within stabilized side slopes as shown in Figure 13.13(*a*), replacement of backfill around the structure would need to be carried out in discrete layers. These layers would also need to be appropriately compacted. Ingold (1979) has demonstrated that, due to compaction against rigid walls, the lateral earth pressures experience an increase beyond their well-known classical values especially against the upper reaches of the wall. It has been found that this enhancement closely follows the profile as idealized in Figure 13.13(*b*).

Compaction of layers above the roof of the structure is unlikely to cause any additional enhancement in the compaction pressures below the roof. The effect of

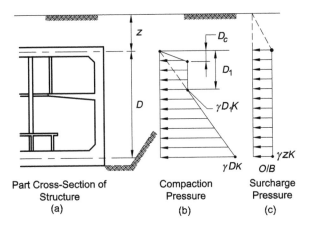

Figure 13.13 Compaction Pressures

the backfill above the roof level on the lateral earth pressures acting on the perimeter wall of the structure, i.e. below the roof, should therefore be treated as a surcharge load on the structure and treated in the conventional manner, as shown in Figure (c).

Assume an equivalent line load of P (kN/m) from compaction roller, unit weight $\gamma\,(kN/m^3)$ of fill material at specified moisture content, coefficient of earth pressure of K as appropriate, and equivalent *udl* q (kN/m^2) from traffic surcharge.

$$\text{Depth } D_C = K\sqrt{\frac{2P}{\pi\gamma}} \tag{13.21}$$

Depth whereat pressure due to compaction is equal to the effective lateral earth pressure is given by:

$$D_1 = \frac{1}{K}\sqrt{\frac{2P}{\pi\gamma}} = \frac{D_C}{K^2} \tag{13.22}$$

Effective lateral earth pressure due to compaction at depth D_C:

$$\sigma'_C = \sqrt{\frac{2P\gamma}{\pi}} \tag{13.23}$$

This pressure is assumed to be constant down to depth D_1, where it meets the conventional earth pressure diagram. Therefore, at this point:

$$\sigma'_C = \sigma'_{D_1} = \gamma D_1 K = \gamma\frac{D_C}{K} \tag{13.24}$$

Effective lateral pressure at the base of the structure is $\sigma'_D = \gamma DK$, and the lateral surcharge load on the structure due to backfill above the roof level, $\sigma'_S = \gamma ZK$. In the event of the structure being a vertical cantilever, characteristic moment at the base, including that due to compaction loads, is given by the expression:

$$M_B = \frac{D^2}{2}\sqrt{\frac{2P\gamma}{\pi}} + \frac{K\gamma}{6}(D - D_1)^3 - \frac{PK}{\pi}\left(D - \frac{D_C}{3}\right) + \frac{1}{2}K \cdot q \cdot D^2 \qquad (13.25)$$

The foregoing treatment also assumes the complete absence of porewater pressures behind the wall in the fill material. While this may well be the case during construction, in the long-term operational stage, porewater pressures can be expected to have been restored. For the in-service case, accordingly, modified design pressures should also be used as one of the loading cases.

13.9 Multi-level braced walls

In relatively soft ground where it may be feasible to do so, construction of cut-and-cover metro structure within a temporary sheet pile cofferdam can often present a cost-effective solution. In view of the nature of such ground, however, the lateral pressures on the cofferdam would be high enough to necessitate multiple stages of bracing especially where large ground movements are to be avoided. The need to contain ground movements may also dictate preloading of the bracing or, in the case of anchors, their prestressing by an appropriate amount.

For many years, assessment of the bracing or anchor loads at different levels has been based on the classical Coulomb's earth pressure distribution, which presupposes a certain amount of wall movement, rather than on the earth-at-rest pressures. If these preloads are based on the theoretical coulomb pressure distribution but the actual pressures realized are less than those anticipated, then the preloads would inhibit the movements necessary to allow the design loads in the anchors to be generated. To avoid occurrence of such an anomaly and, potentially, the adverse effects of the use of excessive preload, and to allow for the redistribution of loads due to relaxation at different stages of excavation, only a proportion of the theoretical design loads is generally jacked in.

The methods available for the assessment of loads (Puller, 1996) in multilevel braced walls are:

- Soil–structure-interaction-related computer methods using numerical approach adopting finite element or finite difference techniques. Full analysis of soil and structure stiffnesses is made and their interaction relied upon using realistic soil constitutive models and 'accurate' boundary conditions.
- Other computer programs based on Winkler spring theory, wherein the initial at-rest earth pressures on either side of the wall are allowed to reach equilibrium, as the excavation progresses, in a series of iterations using numerical methods.
- Empirical methods.

The first two approaches listed above are likely to be of most use in the computations related to relatively flexible ground support systems requiring complex solutions. However, given the lack of precise control over the various stages of excavation and construction, the inexactitude of parameters intrinsically associated with such systems and the extent of accuracy warranted in computations, approaches of such accuracy,

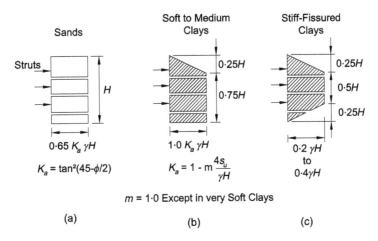

Figure 13.14 Apparent Pressure Diagrams for Computing Strut Loads in Braced Cuts (adapted from Terzaghi and Peck, 1967)

except perhaps in the very extreme cases, may not be warranted. Accordingly, these are considered beyond the scope of this book and are therefore not addressed.

Flexible ground-support system

Unlike the case of the relatively rigid walls, which are likely to fail as integral units, the more flexible walls such as a sheet pile cofferdam can, conceivably, fail progressively as one or more struts or anchors fail. The classical wedge theory is unable to provide a mechanism by which the redistribution of bracing loads under such a situation can be assessed. Accordingly, empirical approach is generally adopted.

Peck (1969) proposed an empirical lateral earth pressure diagram against sheetings, as shown in Figure 13.14, for cuts in sands, soft to medium clays and stiff clays, essentially, for the design of the bracing systems. It is important to recognize that these pressure diagrams are not intended to represent actual earth pressures or their distributions with depth; they are simply meant to represent the envelopes for computing strut loads. Besides, they do not allow for the effects of any penetration of the sheeting below the formation.

In the context of excavations for deep cut-and-cover metro stations, penetration of the sheeting below the formation is most desirable both from the point of view of reducing the load in the lowest brace, which would otherwise be the most heavily loaded, as also lengthening the path of the steady-state seepage to ensure adequate control of groundwater movement. To allow the effect of penetration to be taken into account in assessing the bracing loads, two empirical methods are available.

According to the first method, the active pressure generated below the formation and a portion of the load from the strut load envelope between the lowest strut level and the formation are assumed to be resisted by the passive resistance mobilized below the formation. If, under prevailing circumstances, the movement necessary to mobilize full passive pressures is not permissible, passive pressures for computations should be appropriately factored. This method is applied to strutted excavations in uniform soil

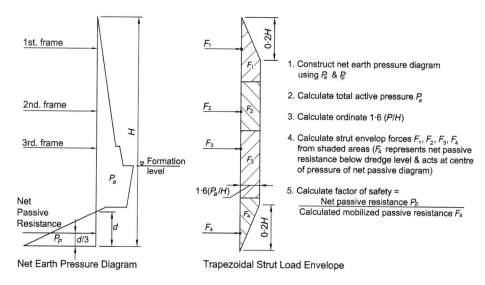

1. Construct net earth pressure diagram using P_a & P_p

2. Calculate total active pressure P_a

3. Calculate ordinate $1{\cdot}6$ (P/H)

4. Calculate strut envelop forces F_1, F_2, F_3, F_4 from shaded areas (F_4 represents net passive resistance below dredge level & acts at centre of pressure of net passive diagram)

5. Calculate factor of safety =
$$\frac{\text{Net passive resistance } P_p}{\text{Calculated mobilized passive resistance } F_4}$$

Figure 13.15 Construction of Strut Load Envelope for Braced Cuts

conditions of reasonable strength such as medium-dense to dense granular and stiff cohesive soils. In the case of less competent soils, such as loose sands and gravels and soft clays, the full effectiveness of the passive resistance against the sheeting below the formation may be doubtful. In such a case, Goldberg *et al.* (1976) advise that the sheeting below the lowest strut level should be designed as a cantilever.

In the second method from an unknown source, an additional 'hypothetical strut' representing the passive resistance mobilized below the formation is assumed to act against the strut load envelope. The level of this 'strut' is determined from the net pressure diagram as shown in Figure 13.15. The 'strut' load calculated from the load envelope is then compared with the passive resistance available using the limiting pressure from the net pressure diagram for the given penetration depth. The bending moment in the sheeting is calculated from the net pressure diagram.

The most onerous of the strut loads calculated for each successive stage of excavation are adopted for the design of the struts and the walings. For the design of the sheeting, however, the worst stress-resultants based on the Coulomb pressure distribution with hydrostatic pressure for each stage are calculated and the most critical values used.

In his review, Potts (1992) points out that the simple empirical methods of analyses for earth retaining structures are flawed. The reasons advanced are that they:

- Assume the entire soil mass involved reaching failure simultaneously;
- Are unable to distinguish between excavation and backfilling;
- Are unable to recognize or reflect the *in situ* stress regime;
- Do not incorporate the surcharge load effects accurately;
- Do not shed any light on wall or ground deformations;
- Do not represent working load conditions accurately;
- Are unable to model redistribution of strut loads.

The foregoing limitations notwithstanding, Potts concedes that such methods continue to remain the mainstay of most design work and, in the circumstances where the empiricism employed can be verified by field observations, their application may not be inappropriate.

Rigid ground-support system

For an excavation supported by a rigid (diaphragm) wall, the use of conventional empirical approach is likely to significantly underestimate the lateral loads in the upper reaches of the wall and the braces. In the case of a cut in a deep layer of soft clay rigidly supported by multilevel cross-braces, as shown in Figure 13.16, Hashash and Whittle (2002) have presented, based on their numerical experiments, a detailed interpretation of the evolution of stresses around the cut. As the excavation progresses downwards, compressive arching develops due to a differential lateral stiffness between the bracing and the underlying soil. As a result, loads are transferred on to the embedded section of the wall below. At every stage of excavation, as the bracing system above is rigid and prevents further inward movement of the wall above the current formation, lateral pressures in excess of the initial ambient level are developed in the higher reaches of the retained soil. Also, deep-seated soil movements occur due to the flexing of the wall below formation.

Similar arching mechanisms have been described qualitatively by Lambe and Whitman (1969). In their assessment of lateral thrusts and earth pressures acting

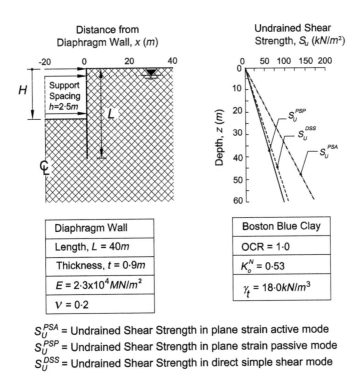

S_U^{PSA} = Undrained Shear Strength in plane strain active mode
S_U^{PSP} = Undrained Shear Strength in plane strain passive mode
S_U^{DSS} = Undrained Shear Strength in direct simple shear mode

Figure 13.16 Geometry and Soil Strength Profile (adapted from Hashash and Whittle, 2002)

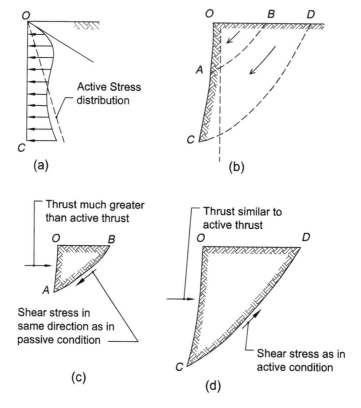

Figure 13.17 Active Arching Mechanism against Braced Excavations
(adapted from Lambe and Whitman, 1969)

on a braced wall, they also indicate higher than the active (i.e. passive type) lateral earth pressure distribution in the retained soil at the top of the wall as sketched in Figure 13.17. This provides the basis for the empirical apparent earth pressure diagram (Peck, 1969) widely used in design.

Hashash and Whittle (2002) compared the computed distributions of 'total lateral stress minus the initial porewater pressure' normalized by the vertical effective stress, i.e. $(\sigma_h - u_o)/\sigma'_{v_o}$, acting on both sides of the wall at excavation depths $H = 10$ and $20m$ with those corresponding to the initial $K_o^N [= (\sigma_{h_o} - u_o)/\sigma'_{v_o} = 0.53]$ conditions. These are shown in Figure 13.18. Water table is considered at $2.5m$ below surface and the initial porewater pressures are treated as being hydrostatic.

For reference, values of the Rankine limit stress states are also indicated in the figure; these are defined as follows:

$$\text{Active:} \quad \frac{\sigma_h - u_o}{\sigma'_{v_o}} = 1 - \frac{2s_u^{(PSA)}}{\sigma'_{v_o}} = 0.32 \qquad (13.26)$$

$$\text{Passive:} \quad \frac{\sigma_h - u_o}{\sigma'_{v_o}} = \frac{\sigma_v^* - u_o}{\sigma'_{v_o}} + \frac{2s_u^{(PSP)}}{\sigma'_{v_o}} \qquad (13.27)$$

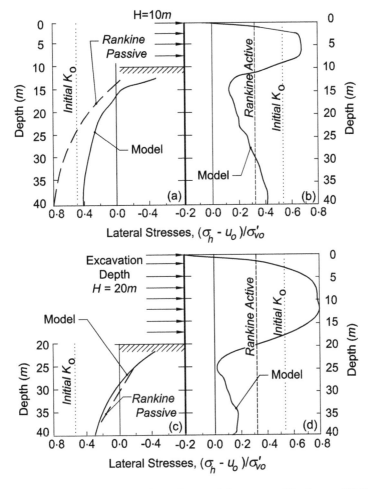

Figure 13.18 Predictions of Total Lateral Stresses on Diaphragm Wall
(adapted from Hashash and Whittle, 2002)

where, σ_h is the horizontal total stress, u_o is the initial porewater pressure, σ_v^* is the vertical stress below current formation level.

$$= \gamma_t (z - H)$$

γ_t is the unit weight of soil, z is the depth below surface, H is the depth of current formation level, s_u^{PSA} is the undrained shear strength (Plane Strain Active mode), s_u^{PSP} is the undrained shear strength (Plane Strain Passive mode), σ_{h_o} is the initial total horizontal stress, and σ_{v_o}' is the initial vertical effective stress.

$$s_u^{(PSA)} / \sigma_{v_o}' = 0.34; \quad s_u^{(PSP)} / \sigma_{v_o}' = 0.17$$

Findings of the authors reveal that, inside the excavation, below the formation, the computed lateral earth pressures are less than those given by the Rankine passive condition at $H = 10m$, but move towards convergence at $H = 20m$ as the system approaches failure. However, in the retained soil above formation, the lateral earth pressures not only exceed Rankine active pressures as anticipated in Figure 13.13, they even exceed the initial at-rest pressures over a large proportion of the braced height, thereby giving rise to apparent earth pressures higher than those anticipated from empirical design methods.

The aforementioned findings are based on Boston Blue Clay (*BBC*) profile with *OCR* = 1. However, similar distributions of lateral stresses have been computed (Whittle and Hashash, 1994) using the Modified Cam Clay (*MCC*, Roscoe and Burland, 1968) and a composite clay, with *OCRs* of 4, 2 and 1 for successive deposits, in Boston (*BBC*, Hashash and Whittle 2002). In view of this, the use of conventional empirical approach in the case of rigidly supported deep cuts in soft clays is likely to underestimate the lateral pressures in the upper reaches of the rigid wall and the brace loads and, where these are likely to be critical, more accurate computational methods, such as nonlinear finite element techniques, should be adopted.

14 Assessment of Surcharge Loads

14.1 Introduction

Surcharge loads transmitted through the retained ground around the perimeter of a cut-and-cover structure influence both the magnitude as well as the distribution of the lateral earth pressures on the structure. These loads can be generated from a number of different sources, e.g. foundations of the adjacent structures, regular traffic, construction activity, provision, in the form of blanket loads, for future construction, etc. In the case of the nearby structures, the foundations may be shallow, i.e. isolated pads, continuous strips, rafts, etc. or they may be deep, such as piles; accordingly, they are likely to generate different types of surcharge loads with their own specific profiles and intensities.

In dealing with the impact of the various types of the aforementioned structural environments on a typical cut-and-cover metro structure in the vicinity, the surcharge loads causing additional lateral earth pressures on the structure can be placed, in principle, under six main categories, i.e. those due to:

- Uniform load
- Point load
- Line load
- Strip load
- Orthogonal line load
- Pile load.

The transverse effect on the retaining elements of uniform surcharge load is reasonably well represented by established procedures using plastic theory. However, this is not so in the case of the point, the line and the strip loads placed on top of the soil behind the retaining element. One reason for this may be that when these loads are applied, the soil behind the wall may not be in a state of incipient failure. So, further research is still needed to establish a truly representative treatment into the effects of such loads on the intensity and distribution of lateral pressures. Meanwhile, it can be assumed, arguably, that if the soil is not plastic, it is likely to be in some sort of elastic state. In view of this, Boussinesq elastic theory, with some empirical modifications, is relied upon to provide solutions, howsoever approximate, for such surcharge loads.

It must be clearly recognized that, for the theory of elasticity to be legitimately applicable, the medium has to be homogeneous, perfectly elastic and isotropic. However, the real soil deposits are anything but these; they are generally layered with strength and stiffness parameters all too often changing with location and depth. The use of the elastic theory thus disregards these inherent variable characteristics of the ground and it is, even more fundamentally, indifferent to the type and profile of the soils encountered in the ground and their ambient stress regimes. As a result, some disparity in the values of lateral pressures between the theoretical estimates and the actual field values is only to be expected. Based on a limited number of field observations available in the literature, this could be of the order of 25–30 per cent either way (Das, 1990). In view of this, one must not lose sight of the limitations of this theory when applied to a soil medium.

14.2 Uniform surcharge

Loads from raft or mat foundations are generally idealized as uniform surcharge loads. Unless specifically known otherwise, provision for future construction activity is also generally made in terms of a blanket uniform surcharge load. However, apart from the magnitude, the only other differences between the two are in the levels at which they are applied and their disposition; the blanket load may be applied at the surface and assumed to extend right up to the structure, whereas the load from a raft would be applied at its appropriate depth below the surface and a known location away from the structure.

Uniform surcharge can be likened to the effect of an equivalent (fictitious) height q/γ of soil above the surface, where q is the intensity of the surcharge load and γ the unit weight of the soil, as shown in Figure 14.1(a). Product of the unit weight and the height, i.e. $(q/\gamma) \times \gamma$, represents the uniform vertical effective stress q in the retained soil mass that the fictitious height of the soil could be expected to generate below the surface. This, in turn, activates the corresponding lateral effective pressure $q \times K$ over the entire depth of the structure, where K is the coefficient of earth pressure as appropriate. In the case of a layered soil, K-values appropriate to each layer should be

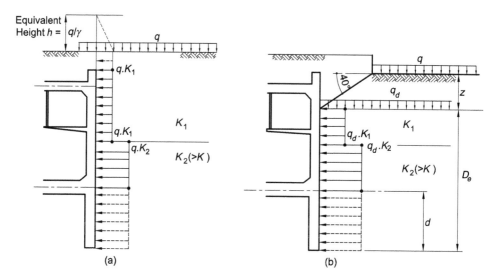

Figure 14.1 Lateral Pressure from Uniform Surcharge

used in arriving at the corresponding effective lateral pressures in the respective layers as shown in the figure.

If the uniform surcharge does not extend as far as the retaining structure but stops some distance short of it, it would then be necessary to identify the level at which the load dispersal would hit the structure and below which the structure would begin to feel the impact of the surcharge. This is generally established by striking a line of dispersal at the boundary of the load towards the structure at, usually, $40°$ to the horizontal as shown in Figure 14.1(*b*). It is assumed that the level at which this line hits the structure defines the upper boundary of the resulting lateral pressure effective on the structure. Let us assume the reduced depth of the retaining structure below this level to be D_e. By the time the primary surcharge load reaches the structure, it would, owing to its dispersal, experience a reduction in its intensity. Assuming this level to be z below that at which the primary surcharge load is applied, the reduced intensity of the surcharge load for design may be approximated as:

$$q_d = \frac{D_e}{(Z + D_e)} \cdot q \tag{14.1}$$

Whether the reduced depth of the retaining structure is taken as D_e or $(D_e - d)$, where d is the effective depth of wall embedment, the difference in the intensity of the surcharge load is unlikely to be more than 10 per cent.

The profile of the lateral pressure on the retaining structure can then be based on this surcharge load as shown in Figure 14.1(*b*).

Cor. uniform rectangular surcharge

A set of graphs of influence coefficients, as shown in Figure 14.2, for lateral pressures against an unyielding wall due to a uniform load q spread over a rectangular area $L \times B$ on the surface is presented in Figure 14.2.

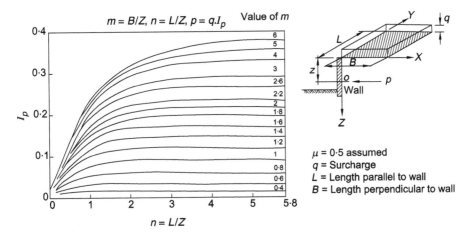

Figure 14.2 Lateral Pressure on an Unyielding Wall due to Uniform Rectangular Load (adapted from Sandhu, 1974)

In the event of the wall yielding, the lateral pressures will tend to drop, and so care must be exercised in the use of the influence coefficients from the graphs.

14.3 Point load surcharge

Loads from isolated column footings may be treated as point loads. The surcharge effect from a construction vehicle can also be obtained by the summation of the effect from each wheel idealized as a discrete point load.

Solutions based on elastic theory inherently disregard plasticity. However, since part of the soil mass is likely to be in a plastic state, to take account of the imperfect assumption of elasticity, empirical modification (Gerber, 1929; Spangler, 1938), based on tests and experimental findings, has been made to the elastic solution. This has resulted in the nondimensional plot of n against $(\sigma_H D^2)/Q_P$ for various values of m as shown in Figure 14.3.

Effect of the point load can be felt both in parallel and at right angles to the retaining structure. Accordingly, figures (*a*) and (*b*) represent the vertical and the horizontal pressure profiles, respectively, on the structure of depth D under a point load Q_P located a distance x away from the structure. Dimension h locates the height of the resultant lateral pressure from the base of the excavation.

The variation of the lateral earth pressure vertically (σ_H) and the equivalent resultant horizontal load (P_H) are given as follows:

$$\text{For } m \leq 0.4, \ \sigma_H = 0.28 \frac{Q_P}{D^2} \cdot \frac{n^2}{\left(0.16 + n^2\right)^3} \quad \text{and} \quad P_H = 0.78 \frac{Q_P}{D} \qquad (14.2)$$

$$\text{For } m > 0.4, \ \sigma_H = 1.77 \frac{Q_P}{D^2} \cdot \frac{m^2 \cdot n^2}{\left(m^2 + n^2\right)^3} \quad \text{and} \quad P_H = 0.45 \frac{Q_P}{D} \qquad (14.3)$$

The variation of the lateral earth pressure horizontally is given by the expression:

$$\sigma'_H = \sigma_H \cos^2 (1.1\theta) \qquad (14.4)$$

Location of the resultant horizontal load from the base is as tabulated in Figure 14.4(*c*).

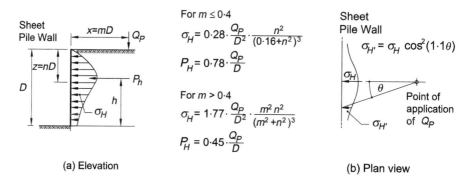

(a) Elevation

(b) Plan view

Figure 14.3 Lateral Pressure on Wall from Point Load (adapted from Terzaghi, 1948)

In the case of cut-and-cover metro structures, generally, the roof, i.e. the top of the earth retaining element, is likely to be located *2m* or more below the surface. In using Figure 14.3 in arriving at the lateral pressures on the structure, it is therefore important to establish, *a priori*, the level at which the point load is applied *vis-à-vis* the top of the structure. If the two levels are approximately the same, or can be assumed so without much loss of accuracy, then the graph as it stands can be used directly with *D* as the depth of the structure; otherwise some adjustments may be necessary.

If the level at which the point load is applied is below the roof of the structure, then the depth of the structure or the retaining element effective in experiencing the surcharge effect from the point load must be the reduced depth below this level. Accordingly, the vertical pressure profile should then be based on this reduced depth. This is not strictly correct as it disregards any interaction between the soil and the portion of the retaining element above the level at which the point load is applied. However, the effect of this approximation is unlikely to be significant.

If the level of the load application is above the structure, such as that at the surface, it would not be appropriate to assume a full height pressure profile against a fictitiously extended retaining element up to the surface and then considering a truncated version of it corresponding to the actual height of the element. This is because, in the overburden above the roof level, the point load applied at the surface would be free to disperse all around unobstructed. Under the circumstances, one way of dealing with the situation could be as follows:

- Assume an all-round dispersal, say, at $40°$ of the point load as far down as the top level of the retaining structure or element.
- Divide the area of dispersal at this level into four equal squares.
- Treat the centroid of each square as the point of application of one-quarter of the point load at the surface.
- Treating these subloads as discrete point loads, summate the effects from these appropriately to arrive at the required result.

In the event that the point load itself is not or, after its dispersal, the four quarter subpoint loads are not, significant, or that they are not significantly apart to register a marked effect on the pressure distribution, the point load may be considered as if applied at the same level as the top of the retaining element. It is always advisable to consider the sensitivity of the computations in terms of weighing up the approximations of assumptions and the sophistication of analysis against the accuracy of the results that may be warranted.

14.4 Line load surcharge

Load from a continuous footing may be treated as a strip load unless its width is too narrow, in which case it may be idealized as a line load. This can be regarded as a plane strain problem with a variable pressure profile vertically as shown in Figure 14.4.

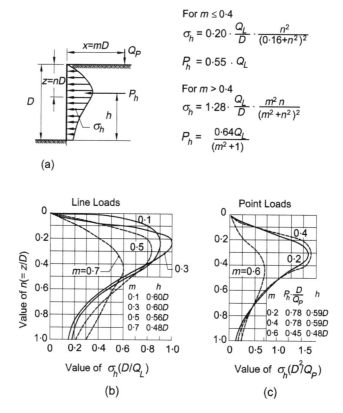

For $m \leq 0.4$

$$\sigma_h = 0.20 \cdot \frac{Q_L}{D} \cdot \frac{n^2}{(0.16+n^2)^2}$$

$$P_h = 0.55 \cdot Q_L$$

For $m > 0.4$

$$\sigma_h = 1.28 \cdot \frac{Q_L}{D} \cdot \frac{m^2 n}{(m^2+n^2)^2}$$

$$P_h = \frac{0.64 Q_L}{(m^2+1)}$$

(a)

Figure 14.4 Lateral Pressure on Walls from Point and Line Loads
(adapted from Terzaghi, 1954)

For line load intensity q_L, the variation of the lateral earth pressure (σ_H) vertically, and its summation representing the equivalent resultant load (P_H) acting on the structure horizontally, are given as follows:

$$\text{For } m \leq 0.4, \ \sigma_H = 0.20\frac{q_L}{D^2} \cdot \frac{n}{\left(0.16+n^2\right)^2} \quad \text{and} \quad P_H = 0.55 q_L \qquad (14.5)$$

$$\text{For } m > 0.4, \ \sigma_H = 1.27\frac{q_L}{D} \cdot \frac{m^2 \cdot n}{\left(m^2+n^2\right)^2} \quad \text{and} \quad P_H = 0.64\frac{q_L}{\left(m^2+1\right)} \qquad (14.6)$$

Location of the resultant horizontal load from the base is as tabulated in figure (*b*).

As for the point load surcharge, here too, the level at which the line load is applied can be important. If it is applied at a level below the top of the retaining element, then reduced depth may be assumed for computation in a manner similar to that explained in the case of the point load.

However, if the load is applied at the surface, its dispersal towards the retained element may be considered, again, at $40°$ causing a reduction in the load intensity. Then, treating the reduced load intensity on the dispersal area as a strip load, the procedure as outlined in the following section for a strip load surcharge may be followed.

14.5 Strip load surcharge

Loads from reasonably wide strip footings to continuous walls, vehicular traffic from roads or surface railways, adjoining a cut-and-cover structure, are typical examples capable of generating strip load surcharge on or against a metro structure. As for the line load surcharge already discussed, strip load also presents a plane strain case with a variable pressure profile vertically as shown in Figure 14.5.

The variation of the lateral earth pressure vertically corresponding to the strip load intensity q_S is given (Jarquio, 1981) by the semiempirical expression:

$$\sigma_H = \frac{2q_S}{\pi}[\beta - \sin\beta \cdot \cos 2\alpha] \tag{14.7}$$

where, $\beta = (\theta_a - \theta_b)$ and $\alpha = \frac{1}{2}(\theta_a + \theta_b)$

The corresponding force per unit run of the structure can be obtained by integrating σ_H with respect to z within the limits from zero to D. The force can be expressed as:

$$P_H = \frac{q_S}{90}[D(\theta_2 - \theta_1)] \tag{14.8}$$

where, θ_1 and θ_2 are θ_b and θ_a, respectively, as measured at the subgrade level, and are given by: $\theta_1 \, (\deg) = \tan^{-1}(m_1/D)$, $\theta_2 \, (\deg) = \tan^{-1}(\overline{m}/D)$, $\overline{m} = (m_1 + m_2)$; D is the retained height, and m_1, m_2 and $\overline{m}$ relate to the geometry of the loading.

The location of the total resultant horizontal surcharge load measured downwards from the top of the pressure diagram can be obtained from the expression:

$$\bar{z} = \frac{1}{2}\left[D + \frac{(R - Q) - 57 \cdot 3m_2 \cdot D}{D(\theta_2 - \theta_1)}\right] \tag{14.9}$$

where, $R = \overline{m}^2(90 - \theta_2)$, $Q = m_1^2(90 - \theta_1)$.

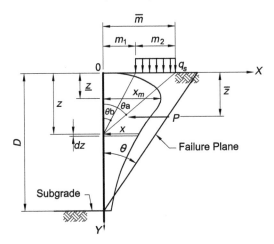

Figure 14.5 Lateral Pressure from Strip Load and Notations (adapted from Jarquio, 1981)

However, location $\underline{z}$ of the maximum lateral pressure $\sigma_{H(max)}$, again, measured downwards from the top of the pressure diagram, can be obtained by equating the first derivative of the expression in Equation (14.7) to zero and solving for $\underline{z}$. This yields:

$$\underline{z} = \left[\frac{\left(N^2 + 4M \cdot O\right)^{1/2} - N}{2M} \right]^{1/2} \tag{14.10}$$

where, $M = 3\overline{m} \cdot m_1 + m_2^2$, $N = 2\overline{m}^2 \cdot m_1^2$ and $O = \overline{m}^3 \cdot m_1^3$

Using this value in the geometry of Figure 14.5, the values for the angles α and β corresponding to the maximum lateral pressure location can be worked out. Upon substituting these values in Equation (14.7), the expression for the maximum lateral pressure can be obtained as:

$$\sigma_{H(max)} = \frac{2q_S}{\pi} \left[\frac{\pi}{180} \cdot \beta - \underline{z} \left(\frac{\overline{m}}{D_2^2} - \frac{m_1}{D_1^2} \right) \right] \tag{14.11}$$

where, D_1 and D_2 are the diagonal distances from the location of the maximum stress point on the wall to the two extremities of the applied strip load such that $D_1^2 = m_1^2 + \underline{z}^2$, $D_2^2 = \overline{m}^2 + \underline{z}^2$ and $\beta = \left[\tan^{-1}\left(\overline{m}/\underline{z}\right) - \tan^{-1}\left(m_1/\underline{z}\right) \right]$.

In the case of the load applied at a level below the top of the retaining element, the corresponding reduced depth of the structure should be considered in the computations. If it is applied above, e.g. at the surface, dispersal of the load (say, at 40°) will result in a load of reduced intensity spread on a wider strip width at the level of the top of the retaining element. This enlarged size of the strip and the associated reduced intensity of the load should then be considered in the design.

14.6 Orthogonal line load surcharge

Consider a line load of intensity q and finite length extending from point $B(x_1, 0, 0)$ to $C(x_2, 0, 0)$ acting at right angles to a wall as shown in Figure 14.6.

The intensity of lateral pressure on the element at a point P located at a depth z below the level of the line load and a distance y parallel to it may be obtained (Spangler and Handy, 1960) from the following expression:

$$\sigma_H = q \cdot z \int_{x_1}^{x_2} \frac{x^2}{R^5} \cdot dx \tag{14.12}$$

Solution of the integral yields:

$$\sigma_H = \frac{q \cdot z}{\pi R_0^2} \left\{ \sin^3 \left[\tan^{-1}\left(x_2/R_0\right) \right] - \sin^3 \left[\tan^{-1}\left(x_1/R_0\right) \right] \right\} \tag{14.13}$$

where, R in the diagonal distance of elemental load from point P, x the distance of the elemental load from the wall, x_1 the distance of near end of load from the wall,

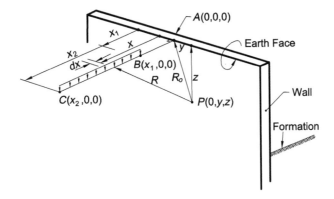

Figure 14.6 Orthogonal Line Load of Finite Length

x_2 the distance of far end of load from the wall, and dx is the elemental length of the line load.

$$R_0 = \sqrt{(y^2 + z^2)}$$

Cor. orthogonal line load of infinite length

When the line load extends from the wall at least an order of magnitude as far away as the depth z to the point P under consideration which, for all practical purposes, can be deemed as infinity, then by substituting $x_1 = 0$ and $x_2 = \infty$ in Equation (14.13), the corresponding expression for the lateral pressure at point P on the wall reduces to:

$$\sigma_H = \frac{q \cdot z}{\pi R_0^2} \tag{14.14}$$

Clearly, the profile of the maximum intensity of lateral pressure will occur on the section of the wall directly aligned with the line load where R_0 is the least being equal to z. On this section, the intensity of the lateral load is therefore given by the expression: $\sigma_H = q/\pi z$. At other sections along the length of the structure, R_0 becomes greater than z and the lateral pressures will, accordingly, decrease. However, this expression suggests that, irrespective of the intensity q of the line load, the lateral pressure at the top, i.e. with $z = 0$, will end up being infinite. Clearly, this does not make sense since, according to the elastic theory, the lateral pressures at the top cannot be anything other than zero. So, the expression for the maximum lateral load intensity should not be used to obtain the lateral pressure at the top.

14.7 Pile load surcharge

In the case of deep foundations, such as piles, the nature and extent of the surcharge loads depend upon the disposition of the piles, i.e. whether isolated single piles or a group, and the type of the piles used, i.e. whether end-bearing, frictional or both.

Piles supporting a nearby high-rise structure if founded in granular soils are likely to be totally end-bearing. If such piles are founded at the same level as or below the base of the cut-and-cover metro structure then, irrespective of their disposition, in theory, there would be no surcharge effect from the piles on to the metro structure to be considered. However, if the piles are not deep enough but stop above the level of the base of the metro structure, the effect of the dispersal of the end-bearing load from a typical pile can be treated as for a point load at the appropriate depth. In the case of a pile group, the load may be approximated to a uniform surcharge and dealt with accordingly as outlined previously.

In the case of piles in cohesive soils, except where they are under-reamed, most of the load is generally carried in shaft friction. The extent of dispersal of the shaft load from an isolated pile on to the retaining structure may be taken at $40°$ vertically and $30°$ horizontally, as illustrated in Figure 14.7. Since the load is likely to be carried by the progressively increasing resistance down the pile shaft, the dispersal of the shaft friction should, likewise, reflect a commensurate increase in lateral pressure linearly with depth against the metro structure as shown in elevation (*a*).

Pressure due to the surcharge effect of a row of piles may present itself, vertically, as a variable load similar to that shown in figure (*a*). However, horizontally, owing to the effect of overlap, the surcharge effect on the wall will be in the form of a uniform lateral load as shown in figure (*c*).

In the case of piled foundations, where the load is carried partly in end-bearing and partly in shaft friction, the total pile loads need to be split up into the two components before evaluating their respective effects on the structure as explained earlier. Where the pertinent information on the basis of which the piles were designed is available, the required split up of the load can be obtained easily.

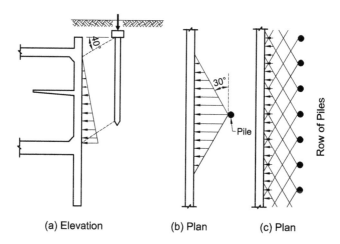

(a) Elevation   (b) Plan   (c) Plan

Figure 14.7 Lateral Pressure from Pile Loads

In the case of older structures, all the pertinent design information may not always be available. This can present a problem. In such a case, the sensible thing to do would be to work out afresh from the available details, floor-by-floor and taking into account the use of the structure, the loadings for which the piles would most likely have been designed. However, if the structural details are not available, this may not prove to be easy. Under such circumstances, it may be expedient to estimate the split up of the load by back-analysis on the basis of the current carrying capacity of the piles using the given size and depth of the piles and the current soil parameters and factors of safety. This approach demands caution since it is likely to yield exaggerated surcharge loads. This is chiefly because of the potentially improved current strength of the soil since the construction of the existing structure, and the use of factors of safety much leaner than those likely to have been used in the past at the time of original design, leading to enhanced carrying capacities for the piles. Then, depending on how sensitive the design of the metro structure is to the variation in the components (frictional and end-bearing) of the pile loadings, it may be advisable to assume for design complementary ranges for the component loadings to ensure that no onerous case is overlooked.

The preceding discussion assumes that, if not the entire floor-by-floor details of the structure above, at least the details of the foundation below, i.e. the size, the depth and the grid spacing of the piles, are available or known. However, that also may not always be the case. In the absence of such details, it is necessary to establish these by some exploratory means on site. While establishing the size and the spacing may be relatively easy, once again, it may not be so to find the depth of the piles. Under these circumstances, it would be helpful to examine the following aspects closely:

- The existing geology of the ground;
- The age of the building and the nature of its use;
- The type of construction, i.e. beam-column frames, shear walls, basement etc.;
- Any telltale damage or distortion suffered by the structure over the years;
- The type of foundation in common use in those days in the given ground conditions; and
- Comparison with the foundations of structures of similar age in the vicinity or similar ground conditions.

On the basis of the indirect evidence gathered on the above lines thus, and using engineering judgment, it may be possible to piece together the most probable picture of the foundation. However, even so, in following such a route, utmost caution must be exercised.

Example 14.1

For the idealized part cross-section of a typical cut-and-cover metro structure as shown in Figure 14.8, estimate the lateral force per unit run of the wall due to the following surcharge load cases:

(a) A uniform load of $50kN/m^2$ applied over a raft width of $15m$, founded $1m$ below ground level and $3m$ away and parallel to the wall.

(b) A vertical point load of $20kN$ located at a distance of $2m$ from the wall and founded $2\cdot5m$ below ground level.

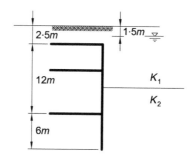

Figure 14.8 Idealized Part Cross-Section of Metro Structure

(c) A line load of 150kN/m applied 2m below surface, parallel to and 4m away from the wall.
(d) A line load of 150kN/m applied on surface, parallel to and 4m away from the wall.

Solution:
(a) *Uniform surcharge of 50kN/m²*

Assume a 40° dispersal of load from the nearest boundary of the raft towards the retaining structure. This line will hit the structure at $3\tan 40° = 2.50m$ below the ground level. In the absence of any information as to whether the raft load is free to disperse at 40° on the far side as well, it is assumed that it is free to do so only on the near side. Recalling Equation (14.1), average dispersed raft load at the level it hits the structure can be approximated as:

$$q_d = \frac{D_e}{(Z + D_e)} \cdot q = \frac{18}{(2.5 + 18)} \times 50 = 44kN/m^2$$

Intensity of the lateral pressure on the retaining element in the upper and the lower soil layers will be: $44K_1$ and $44K_2 kN/m^2$, respectively, where K_1 and K_2 are the respective coefficients of earth pressure in the two layers corresponding to the anticipated soil strain. The corresponding lateral forces on the retaining element can then be obtained by summating these pressures over their respective heights.

(b) *Vertical point load of 20kN*

The load location parameter: $m = \dfrac{2}{12} = 0.17 < 0.4$

Recalling Equation (14.2), variation of the lateral pressure intensity vertically at any depth $z\,(= nH)$ of the retaining element is given by:

$$\sigma_h = 0.28\frac{Q_P}{D^2} \cdot \frac{n^2}{\left(0.16 + n^2\right)^3} = 0.28\frac{20}{12^2} \cdot \frac{n^2}{\left(0.16 + n^2\right)^3} = 0.039\frac{n^2}{\left(0.16 + n^2\right)^3}$$

Also, recalling Equation (14.4), the variation of the lateral earth pressure horizontally is given by the expression:

$$\sigma_h' = \sigma_h \cos^2(1 \cdot 1\theta) = 0 \cdot 039 \frac{n^2}{\left(0 \cdot 16 + n^2\right)^3} \cos^2(1 \cdot 1\theta)$$

The total lateral pressure is given by:

$$P_h = 0 \cdot 78 \frac{Q_P}{D} = 0 \cdot 78 \frac{20}{12} = 1 \cdot 3 kN/m$$

By reference to Figure 14.4(c), the height of the resultant from the base works out as $0 \cdot 59D = 0 \cdot 59 \times 12 = 7m$.

(c) Line load of 150 kN/m 2m below the surface

Load location parameter, $m = \dfrac{4}{12} = 0 \cdot 33 < 0 \cdot 4$

Recalling Equation (14.5), the intensity of lateral pressure on the structure is given by:

$$\sigma_h = 0 \cdot 20 \frac{q_L}{D^2} \cdot \frac{n}{\left(0 \cdot 16 + n^2\right)^2} = 0 \cdot 20 \times \frac{150}{12^2} \cdot \frac{n}{\left(0 \cdot 16 + n^2\right)^2} = \frac{0 \cdot 21n}{\left(0 \cdot 16 + n^2\right)^2} \qquad (14.15)$$

The corresponding total lateral pressure, $P_h = 0 \cdot 55 q_L = 0 \cdot 55 \times 150 = 82 \cdot 5 kN/m$ and by reference to the table of Figure 14.4(b), the height of the resultant lateral pressure for $m = 0 \cdot 33$ is approximately: $0 \cdot 59D = 0 \cdot 59 \times 12 = 7m$ above the base of the structure.

(d) Line load of 150kN/m on the surface

The line load is applied at the surface whereas the top of the retaining element is $2m$ below the surface. Assume a $40°$ dispersal of the load as far down as the roof level. Then the corresponding spread of the load at this level is $4 \cot 40° = 4 \cdot 77m$, located at: $4 - 2 \cot 40° = 4 - 2 \cdot 38 = 1 \cdot 62m$ from the face of the structure. The load may thus be treated as a strip load of intensity $150 \div 4 \cdot 77 = 31 \cdot 5 kN/m^2$ at the same level as the top of the structure in lieu of the given intensity of the line load at the surface.

Then, for the strip load, by reference to Figure 14.5, the location parameters are: $m_1 = 1 \cdot 62, m_2 = 4 \cdot 77m$ and $\bar{m} = m_1 + m_2 = 1 \cdot 62 + 4 \cdot 77 = 6 \cdot 39$. Recalling Equation (14.7), the lateral earth pressure vertically at any depth z of the structure corresponding to the strip load is given by:

$$\sigma_h = \frac{2q_S}{\pi} [\beta - \sin \beta \cdot \cos 2\alpha]$$

where, α and β define the aspect of the strip load as related to the depth z considered.

Then, recalling Equation (14.8),

$$\text{for}\quad \theta_1 = \tan^{-1}\left(\frac{m_1}{D}\right) = \tan^{-1}\left(\frac{1\cdot62}{12}\right) = 7\cdot7^\circ$$

$$\text{and}\quad \theta_2 = \tan^{-1}\left(\frac{\overline{m}}{D}\right) = \tan^{-1}\left(\frac{6\cdot39}{12}\right) = 28^\circ,$$

the total lateral force on the structure is given by:

$$P_h = \frac{q_s}{90}\left[D\left(\theta_2 - \theta_1\right)\right] = \frac{31\cdot5}{90}\left[12\left(28 - 7\cdot7\right)\right]$$

$$= 85\cdot3kN/m \;(cf.\; 82\cdot5kN/m \;for\; the\; line\; load)$$

Recalling Equation (14.9), the location of the total lateral load is given by:

$$\bar{z} = \frac{1}{2}\left[D + \frac{(R - Q) - 57\cdot3m_2\cdot D}{D\left(\theta_2 - \theta_1\right)}\right]$$

where, $R = \overline{m}^2\left(90 - \theta_2\right) = 6\cdot39^2\left(90 - 28\right) = 2531\cdot59$

$$Q = m_1^2\left(90 - \theta_1\right) = 1\cdot62^2\left(90 - 7\cdot7\right) = 215\cdot99$$

$$\therefore \bar{z} = \frac{1}{2}\left[12 + \frac{(2531\cdot59 - 215\cdot99) - 57\cdot3\times4\cdot76\times12}{12\left(28 - 7\cdot7\right)}\right]$$

$$= 4\cdot03m \text{ measured from top of pressure diagram, or}$$

$$= 12 - 4\cdot03 = 7\cdot97m \text{ above base (cf. } 7m \text{ for line load)}$$

Note: Examination of the total lateral loads obtained in cases (c) and (d) shows these to be within 3·4 per cent of each other. This close similarity suggests a near equivalence of the line and the strip loads. This can also be confirmed to be so irrespective of the location of the line load, i.e. whether $m \leq or > 0\cdot4$. The point of application of the resultant in the case of the strip load, on the other hand, turns out to be about 14 per cent above that of the line load. However, given the scale of this disparity, especially in the context of the empiricism involved in the computations for the design of cut-and-cover metro structures, it can be concluded that, by adopting the elastic theory with empirical modifications as outlined in the text, the use of an equivalent strip load in lieu of a line load, or vice versa, are likely to yield similar results.

Example 14.2

For the load case (d) of Example 14.1, estimate the location and the intensity of the maximum lateral pressure.

Solution:
Recalling Equation (14.10), location of the maximum pressure

$$\underline{z} = \left[\frac{\sqrt{(N^2 + 4MO)} - N}{2M} \right]^{1/2}$$

where, $M = 3\bar{m} \cdot m_1 + m_2^2 = 3 \times 6\cdot39 \times 1\cdot62 + 4\cdot77^2 = 53\cdot81$, $N = 2\bar{m}^2 \cdot m_1^2 = 2 \times 6\cdot39^2 \times 1\cdot62^2 = 214\cdot32$ and $O = \bar{m}^3 \cdot m_1^3 = 6\cdot39^3 \times 1\cdot62^3 = 1109\cdot3$.

$$\therefore \underline{z} = \left[\frac{\sqrt{(214\cdot32^2 + 4 \times 53\cdot81 \times 1109\cdot3)} - 214\cdot32}{2 \times 53\cdot81} \right]^{1/2}$$

$$= 1\cdot72m \text{ from the top of the pressure diagram.}$$

Also, recalling Equation (14.11), the maximum intensity of the lateral pressure

$$\sigma_{H(\max)} = \frac{2q_S}{\pi} \left[\frac{\pi}{180} \cdot \beta - \underline{z} \left(\frac{\bar{m}}{D_2^2} - \frac{m_1}{D_1^2} \right) \right]$$

where, $D_1^2 = m_1^2 + \underline{z}^2 = 1\cdot62^2 + 1\cdot72^2 = 5\cdot58$, $D_2^2 = \bar{m}^2 + \underline{z}^2 = 6\cdot39^2 + 1\cdot72^2 = 43\cdot79$, and $\beta = [\tan^{-1}(\bar{m}/\underline{z}) - \tan^{-1}(m_1/\underline{z})] = [\tan^{-1}(6\cdot39/1\cdot72) - \tan^{-1}(1\cdot62/1\cdot72)] = 31\cdot65°$

$$\therefore \sigma_{H(\max)} = \frac{2 \times 31\cdot5}{\pi} \left[\frac{\pi}{180} \times 31\cdot65 - 1\cdot72 \left(\frac{6\cdot39}{43\cdot79} - \frac{1\cdot62}{5\cdot58} \right) \right] = 16\cdot06kN/m^2$$

Example 14.3
For the idealized structure as shown in Figure 14.8, obtain the variation in the values of the maximum lateral pressures anywhere on the structure with depth due to a line load of 150kN/m run acting at right angles to the wall. Assume that: (*a*) the line load is acting at the same level as the roof top of the structure and is of a finite length with its near end 3*m* and the far end 15*m* away from the wall. Compare the pressures so obtained with those assuming: (*b*) the far end of the line load to extend to infinity, and (*c*) the near end also to extend right up to the wall.

Solution:
Profile for the maximum pressure will occur on the section directly aligned with the line of application of the load such that $R_0 = z$. Then, making the substitution in Equation (14.13) and putting $x_1/z = m_1$ and $x_2/z = m_2$, lateral pressure at any depth z for case (*a*) will be given by the expression:

$$\sigma_{H(a)} = \frac{q}{\pi z} \left\{ \sin^3 \left[\tan^{-1} m_2 \right] - \sin^3 \left[\tan^{-1} m_1 \right] \right\} kN/m^2$$

where, q is in *kN/m* and z in *m*. Then, for $x_1 = 3m$ and $x_2 = 15m$, the maximum lateral pressures down the structure computed on the basis of the above expression are listed in Table 14.1.

Table 14.1 Maximum Lateral Pressures

z	m_1	m_2	$\sigma_{H(a)}$	$\sigma_{H(b)}$	$\sigma_{H(c)}$
1	3·000	15·00	6·66	6·98	47·75
2	1·500	7·500	9·50	10·12	23·87
2.3	1·300	6·52	9·70	10·42	20·76
3	1·000	5·000	9·38	10·29	15·92
4	0·750	3·750	8·19	9·36	11·94
5	0·600	3·000	6·85	8·25	9·55
6	0·500	2·500	5·66	7·25	7·96
8	0·375	1·875	3·84	5·71	5·97
10	0·300	1·500	2·64	4·66	4·77
12	0·25	1·25	1·84	3·92	3·98

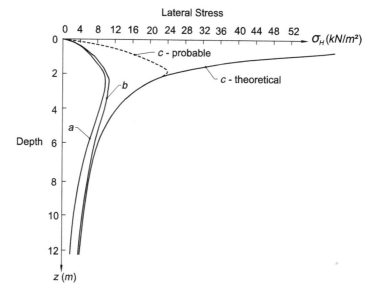

Figure 14.9 Vertical Profiles for Maximum Lateral Pressure

Lateral pressures for cases (b) and (c) can be obtained, respectively, from the expressions:

$$\sigma_{H(b)} = \frac{q}{\pi z}\left\{1 - \sin^3\left[\tan^{-1} m_1\right]\right\} \quad \text{and} \quad \sigma_{H(c)} = \frac{q}{\pi z}$$

The values for the lateral pressures for the three cases are listed in Table 14.1 and the corresponding profiles plotted in Figure 14.9.

Conclusions
- Comparison of the results in the table indicates that, with the far end of the finite line load extending to infinity, the lateral pressures experience increase in their

values. However, since the finite line load is fairly long in its own right, the increase is only marginal. Furthermore, as is to be expected with the extension of the far end of the load, the increase in the lateral pressures is the least (< 0·5 per cent) at the top increasing progressively with depth and attaining the maximum value (> 100 per cent) at the formation level.

- For the finite load and the load case with its far end extended to infinity, the lateral pressure variation profiles are similar and the respective maximum values are reached at approximately the same depth. However, not surprisingly, the resultant lateral load in the case of the latter will be greater and its point of application lower than those of the former.

- With the infinite load also extending right up to the wall, once again the lateral pressures experience increase in their values. It is the upper reaches of the structure this time which are likely to experience significant increases in the lateral pressures making the profile top-heavy and therefore pushing the line of action of the resultant force upwards.

- The reversal in the trend in the value of the lateral pressure above the level of its maximum value at the depth of 2·3m in the penultimate column of the table does signify the tendency for the pressures to drop towards zero at the top. However, there is no such trend apparent in the last column where the lateral pressures is indicated as increasing progressively to infinity towards the top. Clearly, this is not correct and should therefore be disregarded. The more likely profile towards the top is as shown dotted in Figure 14.9.

- It is also clear that if the extension of the near end of an infinite load taking it right up to the wall is of the order of one quarter of the retained height of the structure, then the increase in the lateral pressure of such extension of load tails off by the time it reaches the bottom of the retained height.

Example 14.4
For the idealized soil profile and part cross-section of a typical cut-and-cover metro structure as shown in Figure 14.10, draw the lateral pressure diagram per metre run of the structure likely to result from a building surcharge load of 102 *tonnes* borne by a pile group over a plan area of 20m×40m. Assume a load corresponding to an average of 50kN/m^2 to be carried in friction around the pile group perimeter.

Solution:
The effect of the frictional load on the pile group may be represented by an equivalent enhancement in the unit weight of the soil given by:

$$\frac{50 \times 12 \times 2\,(20+40)}{25 \times 45 \times 12} = 5.33 kN/m^3$$

Accordingly, the equivalent modified effective unit weight, γ', is:

$$8 + 5.33 = 13.33 kN/m^3 \text{ in layer (a) and}$$

$$9 + 5.33 = 14.33 kN/m^3 \text{ in layer (b)}$$

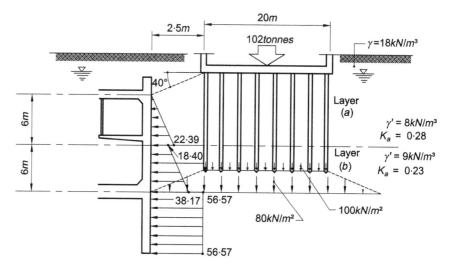

Figure 14.10 Lateral Pressures from Pile Group

∴ Lateral pressures against the structure are:

At the bottom of top layer (*a*): $13 \cdot 33 \times 6 \times 0 \cdot 28 = 22 \cdot 39 kN/m^2$

At the top of bottom layer (*b*): $22 \cdot 39 \times \dfrac{0 \cdot 23}{0 \cdot 28} = 18 \cdot 40 kN/m^2$

At the base of the box structure: $18 \cdot 40 + 14 \cdot 33 \times 6 \times 0 \cdot 23 = 38 \cdot 17 kN/m^2$

The total lateral force resulting from the frictional component of the pile group can then be simply obtained by summating the lateral triangular and trapezoidal pressures over their respective depths.

Given the 40° dispersal of the frictional load below the pile raft which, in this case, hits the structure at roughly its base, the lateral pressures below the base of the box structure will reflect the combined effect from the entire pile raft. In other words, it will include not only the frictional component as a surcharge load but the component from the base pressure of the pile raft also, i.e.

$$38 \cdot 17 + 100 \times \frac{20}{25} \times 0 \cdot 23 = 56 \cdot 57 kN/m^2$$

Lateral pressure diagram against the structure is shown in Figure 14.10.

15 Assessment of Seismic Loads

15.1 Introduction

Most parts of the world are prone to the occurrence of earthquakes, albeit to varying degrees. World history is replete with cases which bear testimony to the scale and intensity of devastation to life and property that seismic activity is capable of causing. However, historically, underground structures have experienced a relatively lower rate of damage than aboveground structures (Dowding and Rozen, 1978; Rowe, 1992). This suggests that, of the two types, the former are less vulnerable to damage by earthquakes than the latter. This could be due to the fact that, unlike aboveground structures, the underground structures are unlikely to distort to any significant extent independently of the displacement of the soil medium or be subjected to vibration amplification; although there is evidence (Hashash, 2001) to suggest, as will become clear later on, that under certain circumstances, the ground displacement can be amplified even due to the presence of the underground structure itself.

The occurrences of damage to underground structures are not very common. However, the 1995 Hyogoken-Nambu earthquake in Japan did cause severe damage, resulting in a major collapse of the Daikai Subway station in Kobe (Iida *et al.*, 1996). The collapse experienced by the central row of columns was accompanied by the collapse of the roof slab and the settlement of the overlying soil cover by more than $2.5m$. The station was designed in 1962, and what is particularly significant is that it did not incorporate any specific provisions to withstand the effects of seismicity. A failure of such a magnitude, thus, serves as a reminder of the grave consequences if the impact of seismic activity is underestimated or ignored and brings into focus the need to take it seriously.

In view of the foregoing, it is imperative that, in earthquake-prone areas, the potential impact of seismic activities on the structures is duly investigated and appropriately allowed for in their design. Nevertheless, it may not be necessary to design a cut-and-cover structure that is 100 per cent earthquake-resistant; it may be enough if the structure is able to withstand a given level of seismic motion with a degree of damage that is neither life-threatening nor exceeding a predefined acceptable level and one that can be put right relatively easily.

15.2 Cut-and-cover *vs.* bored tunnels

It is interesting to note that, of the underground structures, cut-and-cover tunnels are more vulnerable to seismic damage than are bored tunnels (Wang, 1993; Hashash *et al.*, 2001). This is, potentially, due to three main factors:

- First, in comparison to bored tunnels, cut-and-cover tunnels are generally constructed at shallower depths in soils where the stiffness of the soil mass tends to be relatively low. As a result, most of the seismic forces have to be borne by the structure. Furthermore, the shaking intensity and the resultant ground deformations are also greater at shallower depths as compared to those at deeper locations.
- Second, cut-and-cover box structures, owing to their geometry, do not transmit lateral loads as efficiently as can be expected of their bored tunnel counterparts. Cut-and-cover box structures invariably turn out to be stiffer transversely, attract greater loads and are, accordingly, less tolerant to distortion.
- Third, the placement and the compaction of the backfill on top of the structure, however well controlled, are unlikely to fully restore the ambient structure and the history of the virgin ground. As a result, the surrounding ground mass is susceptible to a differential response during a seismic event.

Bored or mined tunnels, on the other hand, in addition to their structurally efficient geometry, are generally constructed at larger depths and without significantly affecting the overlying soil cover. As such, they avoid many of the surface and subsurface disturbances generally associated with cut-and-cover construction. Accordingly, comparatively better performance of such structures under seismic activity can be attributed to these aspects.

15.3 Seismic activity

Seismic hazard analysis is outside the scope of this book. However, ground response to seismic activity can be evaluated, broadly, in terms of either ground shaking and deformation or ground failure (Hashash *et al.*, 2001). Ground shaking refers to the vibration of the ground produced by seismic waves propagating through the earth's crust. The ground response due to the various types of waves, along with their two subtypes, as shown in Figure 15.1, comprises:

- Body waves that travel within the earth's material. They may be either longitudinal, 'Primary' or 'Push' (*P*) waves or transverse, 'Secondary' or 'Shear' (*S*) waves and may travel in any direction in the ground.
- Surface waves that travel along the earth's surface. They may be either Rayleigh (*R*) waves or Love waves.

In general, during a seismic activity, there is an upward propagation of shear waves from the underlying rock or 'rocklike' layers, and ground motion near the surface of the soil deposit. As the ground is deformed by the travelling waves, a cut-and-cover structure will also suffer deformation to the extent permitted by the relative stiffness of the ground *vis-à-vis* that of the structure.

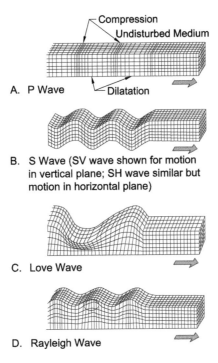

A. P Wave

B. S Wave (SV wave shown for motion in vertical plane; SH wave similar but motion in horizontal plane)

C. Love Wave

D. Rayleigh Wave

Figure 15.1 Ground Response to Seismic Waves (adapted from Bolt, 1978)

Ground failure, as a result of seismic shaking, can manifest itself (Hashash *et al.*, 2001) in terms of slope instability, liquefaction where ground is so susceptible, or the displacement of a fault which can be particularly important in the case of a structure straddling across the fault. As alignments of typical cut-and-cover metro structures generally run in heavily built-up urban environments and follow the main thoroughfares on the surface, the situations presenting incidences of landslides due to slope instability are likely to be very remote.

Liquefaction is a phenomenon associated with an increase in the porewater pressure in soils to such an extent as to neutralize the effective stress fully so that the soil medium loses its strength completely and thereafter behaves, virtually, as liquid. In other words, the soil medium is no longer able to provide support to the structure; in fact, the structure could, potentially, float or sink in the liquefied mass. This phenomenon is typical of saturated cohesionless soils and is more prevalent in relatively loose sands and artificial fill deposits. Once again, the incidence of liquefaction in relatively large cut-and-cover metro structures, as opposed to the smaller bored tunnels for instance, is rare.

To avoid major expense, construction of a major cut-and-cover structure, such as a subway station, across an active fault must be avoided. If it has to be close to a fault, it should be located to one or the other side of it. For a running tunnel, however, it may not always be possible to avoid straddling the fault. Under such circumstances, the joint would need to be aligned with the fault and designed to tolerate the anticipated fault displacements allowing only minor damage.

The shaking of the soil mass also gives rise to hydrodynamic effects in the ambient porewater pressure regime. To ensure that a cut-and-cover metro structure is able to safely withstand the effects of seismic activity, the structure needs to be adequately designed also for all loads resulting from such an activity.

For the assessment of design load on underground structures under seismic activity, generally, two approaches are in use. These are:

- Modified classical load approach
- Soil–structure interactive approach.

15.4 Modified classical load approach

By suitably modifying the classical Coulomb's active earth pressure equation based on the fully plastic solution, it is possible to take into account, after a fashion, the effect of the vertical and the horizontal coefficients of acceleration induced by an earthquake. This approach is based on the 'Mononobe–Okabe' analysis (Okabe, 1926; Mononobe, 1929). It is also referred to as the 'inertial force' or the 'dynamic earth pressure' method. It is highly idealized and is based on the following assumptions:

- The movement of the structure is sufficient enough to reduce the lateral pressures to their ultimate lower limit, i.e. the active values.
- At failure, the shear strength along the entire failure plane is fully mobilized.
- The retained soil behaves as a rigid body.
- The failure in the retained soil is planar.
- Shear strength of dry granular soil is given by the expression $s = \sigma' \tan \phi'$, where σ' is the effective stress.

Use of the inertial force approach was initially made for the seismic design of aboveground structures and later on also extended, achieving reasonable results, to the case of U-shaped structures with free-standing cantilever retaining walls. As such, it may be used for the design of cut-and-cover tunnel sections over the extent where they transition from being fully confined underground to emerging fully open on to the surface. In areas of modest seismic activity also, the analysis, presented by the conventional force method duly enhanced using seismic soil pressures, may not be sensitive to the variations in the design loadings. It is also likely to yield reasonable results for tunnels buried at shallow depths (Wang, 1993).

To demonstrate the use of this approach, two vertical profiles of the structure – one extending right up to the surface and the other stopping some distance below it – have been considered.

Retaining element extending up to surface

Lateral load per unit run (P_a) of the retaining element based on Coulomb's equation for active earth pressure resulting from retained dry granular soil is given by:

$$P_a = \frac{1}{2} \gamma H^2 \cdot K_a \qquad (15.1)$$

where, γ is the dry unit weight and H the height of the retained soil.

For horizontally retained soil (i.e. with the slope of the backfill = 0) against the vertical face of the retaining element (i.e. with the slope of the back of the wall = 0) and assuming δ as the angle of wall friction, the coefficient of active earth pressure K_a is given by the expression:

$$K_a = \frac{\text{Cos}^2\phi'}{\text{Cos}\,\delta\left\{1+\left[\dfrac{\text{Sin}\,(\delta+\phi')\,\text{Sin}\,\phi'}{\text{Cos}\,\delta}\right]^{1/2}\right\}^2} \qquad (15.2)$$

Values of K_a for different values of ϕ' and δ based on this expression are listed in Table 15.1.

To extend the analysis to include earthquake forces, consider a trial failure wedge ABC as shown in Figure 15.2(a). The forces acting on the failure wedge per unit run of the retaining element, assuming it to extend right up to the ground level, are as follows:

- Weight of the soil wedge, W;
- Vertical inertial force, $k_V W$;
- Horizontal inertial force, $k_H W$;
- Resultant, R, of the normal and the shear forces (N & S) on the failure plane BC;
- Earthquake modified active force, P_{ae}, on the retaining element.

where, $$k_V = \frac{vertical\ component\ of\ earthquake\ acceleration}{g}$$

$$k_H = \frac{horizontal\ component\ of\ earthquake\ acceleration}{g}$$

and $g = Acceleration\ due\ to\ gravity.$

Table 15.1 Values of Active Earth Pressure Coefficient K_a

$\phi' \backslash \delta$	0°	5°	10°	15°	20°	25°	30°	35°
12°	0·66	0·62	0·60	0·57	0·57	0·56	0·56	0·57
16°	0·57	0·54	0·51	0·50	0·49	0·49	0·49	0·49
20°	0·49	0·46	0·45	0·43	0·43	0·42	0·42	0·43
24°	0·42	0·40	0·39	0·38	0·37	0·37	0·37	0·37
28°	0·36	0·34	0·33	0·33	0·32	0·32	0·32	0·32
30°	0·33	0·32	0·31	0·30	0·30	0·30	0·30	0·30
32°	0·31	0·30	0·29	0·28	0·28	0·27	0·28	0·28
34°	0·28	0·27	0·26	0·26	0·25	0·25	0·26	0·26
36°	0·26	0·25	0·24	0·24	0·24	0·24	0·24	0·24
38°	0·24	0·23	0·22	0·22	0·22	0·22	0·22	0·22
40°	0·22	0·21	0·20	0·20	0·20	0·20	0·20	0·20
42°	0·20	0·19	0·19	0·18	0·18	0·18	0·18	0·19

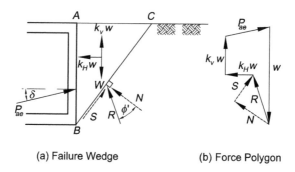

(a) Failure Wedge   (b) Force Polygon

Figure 15.2 Active Earthquake Force on Structure

The polygon corresponding to the aforementioned forces is shown in figure (*b*). From the geometry of the force polygon, expression for the modified active force can be obtained as:

$$P_{ae} = \frac{1}{2}\gamma H^2 \left(1 - k_V\right) K_{ae} \tag{15.3}$$

$$\text{where,} \quad K_{ae} = \frac{\text{Cos}^2 \left(\phi' - \beta'\right)}{\text{Cos }\beta' \text{Cos }\left(\delta + \beta'\right)\left\{1 + \left[\dfrac{\text{Sin }\left(\delta + \phi'\right)\text{Sin }\left(\phi' - \beta'\right)}{\text{Cos }\left(\delta + \beta'\right)}\right]^{1/2}\right\}^2} \tag{15.4}$$

$$\text{and} \quad \beta' = \tan^{-1}\left(\frac{k_H}{1 - k_V}\right) \tag{15.5}$$

Note that, in the absence of the inertial forces, β' would be zero. If this is substituted in Equation (15.4), expression for the active earth pressure coefficient without the effect of earthquake as given in Equation (15.2) can be recovered.

   Close comparison of the Equations (15.2) and (15.4) indicates that, by a judicious use of equivalence, it is possible to establish a correspondence between the expressions for K_a and K_{ae}. In other words, if in Equation (15.4), the expressions $\left(\phi' - \beta'\right)$ and $\left(\delta + \beta'\right)$ are replaced, respectively, by their hypothetical equivalent values ϕ'_{eq} and δ_{eq} such that $\left(\delta_{eq} + \phi'_{eq}\right) = \left(\delta + \phi'\right)$, then upon making these substitutions, Equation (15.4) takes the form:

$$K_{ae} = \frac{1}{\text{Cos }\beta'} \cdot \frac{\text{Cos}^2 \phi'_{eq}}{\text{Cos }\delta_{eq}\left\{1 + \left[\dfrac{\text{Sin }\left(\delta_{eq} + \phi'_{eq}\right)\text{Sin }\phi'_{eq}}{\text{Cos }\delta_{eq}}\right]^{1/2}\right\}^2}$$

which is of the form:

$$K_{ae} = \frac{1}{Cos\ \beta'}\left(Equiv.K_a\right) \tag{15.6}$$

The value for the *Equiv.K$_a$* can be obtained from Table 15.1 by simply treating the ϕ'_{eq} and δ_{eq} as the respective ϕ' and δ values in the table. Alternatively, it may be obtained directly from the Equation (15.2) itself. Thereafter, K_{ae} can be obtained from Equation (15.6).

The procedure for evaluating the active force due to earthquake for a retaining element extending up to ground level may thus be summarized as follows:

- Obtain β' from Equation (15.5).
- Evaluate the equivalent parameters from $\phi'_{eq} = (\phi' - \beta')$ and $\delta_{eq} = (\delta + \beta')$.
- Obtain the *Equiv.K$_a$* from Table 15.1 using the equivalent parameters, or directly from Equation (15.2).
- Compute the modified active force due to earthquake from Equation (15.3).

Examination of the contents of Table 15.1, especially over the range of the ϕ' values commonly encountered in practice, reveals that K_a (and by implication, also K_{ae}) is not very sensitive to the variation in the value of wall friction, δ.

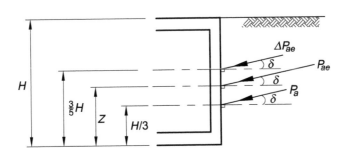

Resultant Active Force: P_{ae}
Location above base: z

Figure 15.3 Location of the Resultant Active Force (roof of structure at surface)

Based on the principles adopted by Seed and Whitman (1970), the following procedure for the determination of the location of the line of action of the modified lateral active force, P_{ae} is proposed:

- Evaluate P_a for no earthquake case;
- Evaluate P_{ae}, the modified value due to earthquake;
- Obtain the incremental force ΔP_{ae} due to earthquake, i.e. $(P_{ae} - P_a)$;
- Assuming that the force P_a acts at a third and ΔP_{ae} at three-fifths of the height from the base as shown in Figure 15.3, height z of the line of action of the resultant force.

P_{ae} from the base can be obtained from the expression:

$$z = \frac{\left(0{\cdot}33P_a + 0{\cdot}6\Delta P_{ae}\right)}{P_{ae}}\cdot H$$

Retaining element not extending up to surface

In the case of cut-and-cover metro structures, the retaining elements very rarely extend up to the ground level. They generally stop some distance below the surface. Accordingly, the procedures for evaluating the modified active force and the location of its line of action as outlined above need to be modified.

Consider an idealized geometry of a typical cut-and-cover metro structure not extending up to the surface. The resultant active earthquake force can be obtained as follows:

- Assume, initially, that the structure extends the full height up to the surface identified as Height (1).
- Obtain active earthquake force $\left(P_{ae1}\right)$ on this height in the manner as outlined previously.
- Evaluate, similarly, the active earthquake force $\left(P_{ae2}\right)$ on the top fictitious extension alone identified as Height (2).
- Deduct the latter from the former to arrive at the resultant active earthquake force $\left(P_{ae}\right)$ over the retaining element of depth D of the structure.

The location of the resultant is qualitatively shown in Figure 15.4.

The value of K_{ae} can be obtained from Equation (15.6) and that of the *Equiv.*K_a from Table 15.1. Then, the active earthquake forces can be evaluated from the expressions in Equations (15.7) and (15.8)as follows:

$$P_{ae1} = \frac{1}{2}\gamma\,(D+Z)^2\left(1-k_V\right)K_{ae} \tag{15.7}$$

$$P_{ae2} = \frac{1}{2}\gamma{\cdot}z^2\left(1-k_V\right)K_{ae} \tag{15.8}$$

The resultant force, $P_{ae} = \left(P_{ae1} - P_{ae2}\right)$ \hfill (15.9)

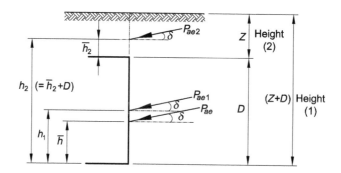

Figure 15.4 Location of the Resultant Active Force (roof of structure below surface)

The height of the line of action of the resultant force can be found by considering the equilibrium of moments of forces $(\sum M = 0)$ about the base of the structure. This yields:

$$\text{The line of action,} \quad \bar{h} = \frac{(P_{ae1} \times h_1 - P_{ae2} \times h_2)}{P_{ae}} \tag{15.10}$$

$$\text{where,} \quad h_1 = \frac{(0\cdot33P_{a1} + 0\cdot6\Delta P_{ae1})}{P_{ae1}} \cdot (z + D)$$

$$\text{Also} \quad h_2 = \bar{h_2} + D \tag{15.11}$$

$$\text{where,} \quad \bar{h_2} = \frac{(0\cdot33P_{a2} + 0\cdot6\Delta P_{ae2})}{P_{ae2}} \cdot z$$

By substituting the various values in Equation (15.10), rearranging and simplifying, we obtain:

$$\bar{h} = (z + D)(0\cdot33a + 0\cdot6b) - D\cdot c \tag{15.12}$$

$$\text{where,} \quad a = \frac{(P_{a1} - P_{a2})}{P_{ae}}; \quad b = \frac{(\Delta P_{ae1} - \Delta P_{ae2})}{P_{ae}}; \quad c = \frac{(0\cdot67P_{a2} + 0\cdot4\Delta P_{ae2})}{P_{ae}},$$

$$P_{a1} = \frac{1}{2}\gamma(z+D)^2 \cdot K_a; \quad P_{a2} = \frac{1}{2}\gamma \cdot z^2 \cdot K_a$$

Value of K_a for no earthquake case is obtained from Table 15.1. Also note that the values for z and D are not absolute but case-specific, as will become clear from the solved example at the end of the chapter.

It is important to remember that, in general, deeper the tunnel embedment, the less reliable are the seismic lateral earth pressures as estimated above because it becomes increasingly important to account for variations in seismic ground motions with depth.

15.5 Soil–structure interactive approach

There is no doubt that, under a seismic activity, dynamic earth pressures will be induced around a box structure and the loads resulting from such pressures are likely to be in the form of a complex distribution of shear and normal stresses along the entire periphery of the structure. Accurate quantification of these loads requires rigorous dynamic soil–structure analysis; however, there is no denying the fact that the overall effect of such a loading is to cause the racking of the structure. Recognizing this, the designers find it more realistic to approach the problem by specifying the loading in terms of the racking deformation; accordingly, its assessment forms the main focus of attention in the soil–structure interactive approach.

It is important to recognize that, in respect of their seismic behaviour, the notable features that distinguish underground metro structures from the aboveground structures are:

- Complete confinement under ground
- Soil–structure interaction
- Significant length.

In view of these, in the case of closed box structures confined within a shaking soil mass, it must be recognized that the conventional approach disregards two essential aspects – the inherent stiffness of the structure and the soil–structure interaction. Accordingly, the use of conventional force method with seismic soil pressure enhancement cannot be relied upon to yield realistic results, and it is more appropriate to adopt a seismic soil–structure interactive (*SSI*) approach.

Assuming that the soil does not liquefy or that the ground does not otherwise fail and lose its integrity during a seismic event, the effects on the structure can then be limited to those resulting from the shaking of the ground.

Ground shaking refers to the deformation of the ground produced by seismic waves propagating through the earth's crust. The major factors influencing shaking damage of an underground structure (Dowding and Rozen, 1978; St. John and Zara, 1987) include: (1) the shape, i.e. size and depth of the structure; (2) the properties of the surrounding soil medium; (3) the properties of the structure; and (4) the severity of the ground shaking.

The treatment hereunder is confined to the impact of ground shaking. In such a case, the interactive response of underground metro structures to seismic activity is generally expected to be reflected in three types of deformations (Owen and Scholl, 1981) – those due to:

- Axial extension and compression
- Longitudinal bending
- Racking.

For this reason, the approach to design based on these deformations is also referred to as the 'Seismic Deformation Method' (*SDM*).

Axial deformations causing alternating tension and compression along the length of the structure as shown in Figure 15.5(*a*) are generated by the components of seismic waves that produce motions parallel to the axis of the structure. Seismic waves producing particle motions normal to the longitudinal axis, on the other hand, give rise to bending deformations of the structure as shown in Figure 15.5(*b*). The longitudinal design of the structure must account for the effects of both these deformations; these are presented in Appendix F.

It is also important to remember that, for long underground structures, different parts of the structure may experience different ground motions. Recorded ground motions have shown (Kramer, 1996) that spatial coherency decreases with increasing distance and frequency. This spatial incoherence could have a significant impact on the response of the structure and in the computation of differential strains and the force build-up along the length of a tunnel. Hashash *et al.* (1998) show how the use of time histories with spatial incoherence affects the estimation of

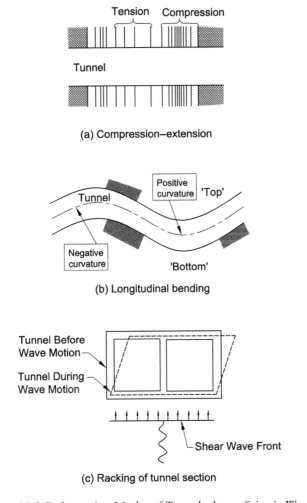

Figure 15.5 Deformation Modes of Tunnels due to Seismic Waves
(adapted from Owen and Scholl, 1981)

the development of axial force in a tunnel that can lead to significant longitudinal push-pull and other effects. In view of this, it is advisable for the designer to work closely with an engineering seismologist to identify the relevant factors contributing to ground motion incoherence at a given site and to generate ground motion time histories.

When shear waves propagate normally to the longitudinal axis, transverse racking deformations develop in a box structure in a manner resulting in the distortion of the cross-section of the structure as shown in Figure 15.5(c). Transverse shear waves transmit the greatest proportion of the seismic energy and, as such, the effect of the resulting distortion can be significant on the transverse design of the box structure. Naturally, such distortions are treated as a form of imposed loading on the structure and are of most interest to the structural engineer.

From the foregoing, it follows that the ground deformation and the interaction between the structure and the surrounding soil control the nature and intensity of the seismic loads on an underground structure. In recognition of this fact, it is more appropriate to consider the racking displacement of the structure, taking into account the effects of seismic soil–structure interaction (*SSI*), as the principal component of the seismic loading for the transverse design of the structure. It is also important to recognize that, under certain circumstances, the structural racking displacements can exceed the free-field racking displacements implying, thereby, that the ground displacements can get amplified due to the presence of the structure itself (Hashash, 2001).

The term 'free-field deformations' refers to strains in the virgin ground (i.e. before it has experienced any disturbance) caused by seismic waves. These deformations ignore any interaction between the underground structure and the surrounding ground. Such an approach can, at best, provide a first order estimate of the anticipated deformation of the structure. The designer may choose to impose these very deformations on the structure. However, the possibility that this approach may overestimate or underestimate deformations according as the stiffness of the structure relative to that of the ground is higher or lower cannot be ruled out. In soft soils particularly, where the structure may be stiffer than the surrounding soil, the deformations of the structure are likely to be overestimated.

A number of factors can be listed that are likely to influence the interaction of the surrounding soil medium with the underground structure; these include the stiffness of the soil relative to that of the structure, geometry of the structure, input earthquake motions and the embedment depth of the tunnel structure. Of these, the most important factor is the stiffness, in simple shear, of the soil relative to that of the structure replacing it, i.e. the flexibility ratio (Wang, 1993).

Flexibility ratio: (F)

Consider a rectangular soil element $(W \times H)$ in a column of soil under simple shear conditions as shown in Figure 15.6. If shear stresses τ_s are applied to the element

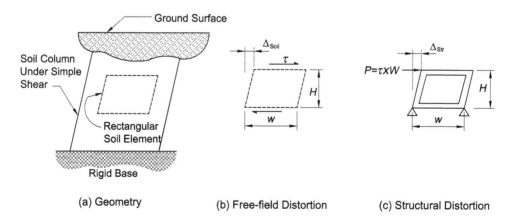

Figure 15.6 Relative Stiffness between Soil and Structure (adapted from Wang, 1993)

causing shear displacement $\Delta \, (= \Delta_{f-f})$, there will be free-field shear strain or angular distortion (Wang, 1993; Hashash *et al.*, 2001) as given by:

$$\gamma_s (= \gamma_{f-f}) = \frac{\Delta}{H} = \frac{\tau_s}{G_S} \tag{15.13}$$

where, G_S is the shear modulus of soil. The stiffness of the soil element can be presented, by rearranging the terms, as the ratio of the shear stress and the corresponding free-field angular distortion, as:

$$\frac{\tau_s}{\gamma_s} = \frac{\tau_s}{\Delta/H} = G_S \tag{15.14}$$

Shear stiffness of soil represented by its shear modulus, G_S, may be obtained from the expression:

$$G_S = \rho_s \cdot (C_s)^2 / g \tag{15.15}$$

where, ρ_s: unit weight of soil, C_s: apparent velocity of shear wave propagation in soil, and g: acceleration due to gravity.

Summation of any applied shear stress τ_{str} across the structure over its interface with the soil, i.e. over the width W of the structure, can be represented by an equivalent point load $P(= \tau_{str} \cdot W)$ applied at the node of the structure as shown in figure (c). The lateral racking displacement of the structure under the action of this point load can be expressed as:

$$\Delta_{str} = \frac{P}{K_X} = \frac{\tau_{str} \cdot W}{K_X} \tag{15.16}$$

In the expression, K_X is the racking stiffness, defined as the force required to cause unit racking displacement of the structure, i.e. P/Δ_{str}. It can be obtained by performing a simple frame analysis using a conventional plane-frame program. Clearly, if the value of P is set to unity, the value for K_X will, simply, be the inverse of the lateral racking displacement, i.e. $1/\Delta_{str}$. The associated angular distortion of the structure will be given by the expression:

$$\gamma_{str} = \frac{\Delta_{str}}{H} = \frac{\tau_{str} \cdot W}{K_X \cdot H} \tag{15.17}$$

Use of the angular distortion ensures that the rigid body rotation of the structure as a whole is not included in the racking results.

The stiffness of the structure also, like that of the soil element in Equation (15.14), can, upon rearrangement of the terms in Equation (15.17), be expressed as:

$$\frac{\tau_{str}}{\gamma_{str}} = \frac{K_X \cdot H}{W} \tag{15.18}$$

The ratio of the stiffnesses of the soil element (given by Equation (15.14)) and that of the structure (given by Equation (15.18)) which, by definition, represents the flexibility ratio of the structure, can then be computed from the expression:

$$F = \left(\frac{G_S}{K_X}\right) \cdot \left(\frac{W}{H}\right) \tag{15.19}$$

For single-cell box structures, the flexibility ratios can be calculated, without recourse to computer analysis, directly from the following formulae (Wang, 1993):

Case (a): Box structure with identical moments of inertia (I_R) for the roof and the base slabs, and (I_W) for the side walls:

$$F = \frac{G_S}{24} \cdot \left(\frac{H^2 W}{EI_W} + \frac{HW^2}{EI_R}\right) \tag{15.20}$$

where, E is the plane strain elastic modulus for the material of the box structure.

Case (b): Box structure with the moments of inertia (I_R) for the roof and (I_B) for the base slabs, and (I_W) for the side walls:

$$F = \frac{G_S}{12} \cdot \left(\frac{HW^2}{EI_R} \cdot \psi\right) \tag{15.21}$$

where, $\quad \psi = \dfrac{\left(1+a_2\right)\left(a_1+3a_2\right)+\left(a_1+a_2\right)\left(1+3a_2\right)^2}{\left(1+a_1+6a_2\right)^2}$

$$a_1 = \left(\frac{I_R}{I_B}\right) \quad \text{and} \quad a_2 = \left(\frac{I_R}{I_W}\right) \cdot \left(\frac{H}{W}\right)$$

Interpretation of the flexibility ratio

A structure can be classified into five possible categories (Wang, 1993; Penzien, 2000, Hashash et al., 2001) in accordance with its flexibility ratio. These are:

$F \to 0 \cdot 0$: The structure is infinitely stiff; it will not rack regardless of the extent of ground distortion and must carry the entire seismic load by itself.

$F < 1 \cdot 0$: The structure is stiff relative to the soft soil medium and will therefore deform less; as such, some of the load will be carried by the medium depending upon the soil–structure interaction.

$F = 1 \cdot 0$: This implies that the structure and the soil medium are of comparable stiffness, so that the structure can be expected to undergo approximately free-field distortions. Examination of Figure 15.8 indicates that, for flexibility ratio of 1, racking ratio is also close to 1. This confirms that the racking of the structure is nearly of the same magnitude as that of the free-field soil medium.

$F > 1 \cdot 0$: The structure is less stiff relative to the medium. In analytical terms, this would amount to the structure being perceived as an inherent weakness such as that resulting from the existence of a cavity in the free-field. Accordingly, it would provide lower shear stiffness than would be expected in a nonperforated medium causing, thereby, an amplification of distortion.

$F \rightarrow \infty$: In this case, the embedded structure is deemed to have virtually no stiffness at all; as such, it can undergo distortion commensurate with that of the perforated medium. For such structures, the racking ratio can reach values between two and three (Penzien, 2000). This point is of particular importance since, in the past, it was believed that the racking of the structure could not exceed the free-field racking of the soil medium.

Racking coefficient: (R)

For box structures, racking coefficient R can be defined as the normalized racking distortion of the structure with respect to the free-field ground distortion. It can be expressed (Wang, 1993) as:

$$R = \frac{\Delta_{str}}{\Delta_{f-f}} = \frac{\left(\Delta_{str}/H\right)}{\left(\Delta_{f-f}/H\right)} = \frac{\gamma_{str}}{\gamma_{f-f}} \qquad (15.22)$$

According to Penzien (2000), if the structure has no stiffness (i.e. $F \rightarrow \infty$), R is approximately equal to $4\left(1 - \nu_S\right)$, where ν_S (varying from $0 \cdot 3$ to $0 \cdot 45$) is the Poisson's ratio of the soil medium.

Steps for obtaining the racking displacement

A simplified frame analysis can provide an adequate and reasonable approach to the assessment of seismic loading imposed on a cut-and-cover box structure. A step-by-step procedure for such an approach is as follows:

- Obtain lateral stiffness K_X of the structure using a structural analysis program assuming simple boundary conditions as shown in Figure 15.7.
- Obtain the average strain-compatible shear modulus G_S of the soil layer in the depth range of the structure in the ground. This may be obtained from the initial and the strain-compatible shear wave velocity graphs based on computer program SHAKE (Schnabel *et al.*, 1972). Alternatively, it may be estimated from Equation (15.15).
- Adopting normalized form for the soil and the structure stiffnesses, compute the flexibility ratio F from Equation (15.19).
- From Figure 15.8, obtain the corresponding racking parameter, coefficient R.
- With the help of Tables A2 and A3, Appendix A, obtain the Peak Particle Velocity (V_S) which when divided by the apparent velocity of shear wave propogation (C_S) gives the maximum free-field shear strain, γ_S.
- Multiply γ_s by the depth of the structure to obtain the maximum relative free-field displacement Δ_{f-f} between the roof and the base of the structure.
- Using Equation (15.22), compute the design racking displacement Δ_{str} for the structure.

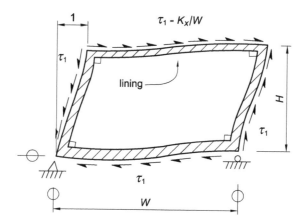

Figure 15.7 Idealized Structure and Boundary Conditions
(adapted from Ostaden and Penzien, 2001)

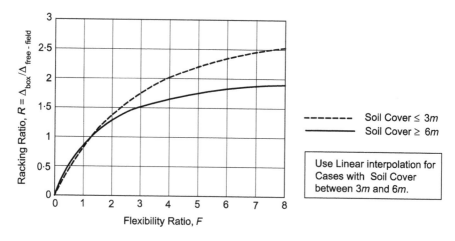

Figure 15.8 Racking Curves for Single-Cell Structures in Alluvial Soil
(adapted from Ostaden and Penzien, 2001)

Similar procedure can be followed for the double- and triple-cell structures also to obtain seismic loads. The procedure can be extended to multilevel stations as well; however, for all these, the corresponding racking ratios need to be obtained for use in design. For sections of the structure at the ends of stations, it is recognized that the presence of cross-walls stiffens the sections considerably (Ostadan and Penzien, 2001). For such sections, *SSI* analysis can be used in design to obtain soil pressure distribution.

In addition to racking displacement, the design of cut-and-cover box structures should also take into consideration (Hashash *et al.*, 2001) loads likely to be generated by vertical accelerations and longitudinal strains resulting from frictional soil drag along the soil–structure interface. The vertical seismic forces imposed on the roof slab of a cut-and-cover box structure may be estimated by multiplying the backfill mass by the estimated peak vertical ground acceleration.

The recourse to the use of Seismic Deformation Method notwithstanding, Power *et al.*, 2006, recommend that for a tunnel that is stiff relative to the surrounding ground (i.e. where $F \leq 1\cdot0$), the walls of the structure should additionally be checked for their capacity to withstand the dynamic earth pressures in combination with the static (earth-at-rest) pressures acting on them. The dynamic pressure, representing the inertia-induced increment, must be added to the lateral static (earth-at-rest) pressure on one side of the structure and simultaneously deducted from that on the opposite side. A check should also be made to ensure that the tunnel racking induced by the dynamic pressures does not unreasonably exceed the shear deformations of the surrounding ground mass.

For cut-and-cover box structures where the depth (z) of the soil overburden over the roof of the structure is less than half the depth (D) of the structure (i.e. $z/D < 0\cdot5$), the seismically induced dynamic lateral earth pressure Δp_E distributed uniformly (Seed and Whitman, 1970; Wood, 1973) over the side wall may be estimated by the expression:

$$\Delta p_E = C_V k_H \gamma (z + D)$$

where, C_V: seismic earth pressure coefficient, k_H: horizontal seismic coefficient, and γ: unit weight of soil.

The seismic earth pressure coefficient, C_V, may be assumed equal to $0\cdot4$ for tunnels founded in soil and $1\cdot0$ for those founded in rock.

15.6 Hydrodynamic effect

Westergaard theory (1933) has been used to obtain the hydrodynamic effect on concrete dams. Based on this theory, the intensity of the hydrodynamic pressure $p_{1(w)}$ due to an earthquake, at a depth z below the water level on the seaward face may be expressed (Das, 1982) as:

$$p_{1(w)} = \frac{7}{8} \cdot k_H \cdot \gamma_w \cdot \sqrt{hz} \tag{15.23}$$

where, h is the height of water and k_H the inertia parameter as defined previously. By integration, total hydrodynamic force $P_{1(w)}$ on the seaward side can be obtained as:

$$P_{1(w)} = \int_0^h p_{1(w)} dy = \frac{7}{12} \cdot k_H \cdot \gamma_w \cdot h^2 \tag{15.24}$$

The depth of the location of the resultant hydrodynamic pressure, z_r, below the water level is given by:

$$z_r = \frac{1}{P_{1(w)}} \int_0^h \left(p_{1(w)} dy \right) y = 0\cdot6h \tag{15.25}$$

According to Matsuo and O'Hara (1960), the increase in the porewater pressure due to hydrodynamic effect on the landward side is approximately 70 per cent of that on

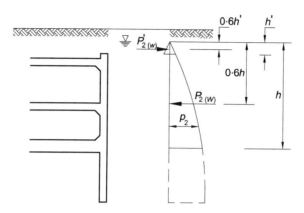

Figure 15.9 Hydrodynamic Effect on Structure

the seaward side. Based on this, the dynamic porewater pressure $p_{2(w)}$ on the landward side may be taken as:

$$p_{2(w)} = 0 \cdot 7 \left(\frac{7}{8} \cdot k_H \cdot \gamma_w \cdot \sqrt{hz} \right) = 0 \cdot 61 k_H \cdot \gamma_w \cdot \sqrt{hz} \qquad (15.26)$$

Correspondingly, the total dynamic increase in the porewater force over that of the static force per unit run of the structure is given by:

$$P_{2(w)} = 0 \cdot 7 \left(\frac{7}{12} \cdot k_H \cdot \gamma_w \cdot h^2 \right) = 0 \cdot 41 k_H \cdot \gamma_w \cdot h^2 \qquad (15.27)$$

Using this analogy, the lateral force due to the hydrodynamic effect of seismic activity on a cut-and-cover metro structure as shown in Figure 15.9 may also be obtained with the help of Equation (15.27). This must be treated as an increase in the lateral force over that due to the ambient static effect.

In the event of the water table not being flush with the top of the cut-and-cover structure, appropriate adjustments should be made both in the hydrodynamic pressure as well as its point of application in order to neutralize the effects due to the fictitious extension of the pressure diagram extending from the top of the structure to the water table.

Example 15.1
For the soil profile and part cross-section of the metro structure shown in Example 14.1 of Chapter 14, assume the depth below the surface to the water table as $1m$, to the top of the structure as $2m$, to the underside of the base slab as $14m$ and to the interface between the soil layers as $10m$. Estimate the effect of earthquake on the lateral forces for the following parameters: $\gamma_1' = 8, \gamma_2' = 9, \phi_1' = 30°, \phi_2' = 35°, \delta = 15°, k_V = 0 \cdot 1$ and $k_H = 0 \cdot 2$.

Subscripts 1 and 2 refer to the upper and the lower soil layers, respectively.

Solution
Active earthquake force

For the layered soil, consider the depth from the surface down to the interface between the layers as depth (a), to the top of the structure as depth (b) and to the underside of the base slab of the structure as depth (c). Then, depth $(a) = 10m$, $(b) = 2m$ and $(c) = 14m$. Depths (a) and (b) relate to the upper soil layer and depth (c) to the pertinent part of the lower layer. Also note that subscripts u and l refer to the upper and the lower layers, respectively.

$$\beta' = \tan^{-1}\left(\frac{k_H}{1-k_V}\right) = \tan^{-1}\left(\frac{0.2}{1-0.1}\right) = 12.53°$$

For the upper layer, $\phi'_{eq1} = (\phi'_1 - \beta') = 30 - 12.53 = 17.47°$

For the lower layer, $\phi'_{eq2} = (\phi'_2 - \beta') = 35 - 12.53 = 22.47°$

Also, $\delta_{eq1} = \delta_{eq2} = (\delta + \beta') = 15 + 12.53 = 27.53°$

For the upper layer, by reference to Table 15.1, for $\phi' = 17.47°$ and $\delta = 27.53°$, the value for $Equiv.K_{au}$, obtained by interpolation, works out to 0.46. Likewise, for the lower layer, for $\phi' = 22.47°$ and $\delta = 27.53°$, the value for $Equiv.K_{al}$ is obtained as 0.39. Dividing the equivalent values by $Cos\beta'$, in accordance with the Equation (15.6), the corresponding K_{ae} values for the two layers can be obtained as: $K_{aeu} = 0.4712$ and $K_{ael} = 0.3995$.

Recalling Equation (15.3), components of the active earthquake forces can be obtained as follows:

Over depth (a), $P_{ae1} = \frac{1}{2}\gamma_1 H_a^2 (1-k_V) K_{aeu}$
$$= \frac{1}{2} \times 8 (2+8)^2 (1-0.1)(0.4712) = 169.64kN$$

Over depth (b), $P_{ae2} = \frac{1}{2}\gamma_1 H_b^2 (1-k_V) K_{aeu}$
$$= \frac{1}{2} \times 8 (2)^2 (1-0.1)(0.4712) = 6.79kN$$

Over depth (c), $P_{ae3} = \frac{1}{2}\gamma_2 H_c^2 (1-k_V) K_{ael}$
$$= \frac{1}{2} \times 9 (2+12)^2 (1-0.1)(0.3995) = 317.14kN$$

Over depth (a), $P_{ae4} = \frac{1}{2}\gamma_2 H_a^2 (1-k_V) K_{ael}$
$$= \frac{1}{2} \times 9 (2+8)^2 (1-0.1)(0.3995) = 161.80kN$$

Forces P_{ae3} and P_{ae4} assume lower layer properties over their full respective depths below the surface.

Resultant active lateral earthquake force on the structure:

$$P_{aeu} = (P_{ae1} - P_{ae2}) = (169.64 - 6.79) = 162.85kN$$

$$P_{ael} = (P_{ae3} - P_{ae4}) = (317.14 - 161.80) = 155.34kN$$

$$\sum P_{ae} = (P_{aeu} + P_{ael}) = (162.85 + 155.34) = 318.19kN$$

Location of the resultant lateral earthquake force

By recalling Equations (15.9 and 15.12), applying their principles to the upper and the lower soil layers in turn and remembering that the parameters z and D are case-specific, the lateral forces and their points of application in the two layers can be obtained. Thereafter, by taking moments of the respective forces about the base, location of the resultant lateral earthquake force can be evaluated.

- For the upper soil layer: $z = 2m$ and $D = 8m$;

Also, for $\phi' = 30°$ and $\delta = 15°$, $K_{au} = 0.30$ (for no earthquake case)

$$P_{a1} = \frac{1}{2}\gamma_1 (z+D)^2 K_{au} = \frac{1}{2} \times 8 (2+8)^2 \times 0.30 = 120kN$$

$$P_{a2} = \frac{1}{2}\gamma_1 z^2 K_{au} = \frac{1}{2} \times 8 \times 2^2 \times 0.30 = 4.8kN$$

$$P_{aeu} = 162.85kN \text{ as before}$$

$$\Delta P_{ae1} = P_{ae1} - P_{a1} = 169.64 - 120 = 49.64kN$$

$$\Delta P_{ae2} = P_{ae2} - P_{a2} = 6.79 - 4.8 = 2kN$$

$$a_u = \frac{(P_{a1} - P_{a2})}{P_{aeu}} = \frac{(120 - 4.8)}{162.85} = 0.71$$

$$b_u = \frac{(\Delta P_{ae1} - \Delta P_{ae2})}{P_{aeu}} = \frac{(49.64 - 2)}{162.85} = 0.29$$

$$c_u = \frac{(0.67P_{a2} + 0.4\Delta P_{ae2})}{P_{aeu}} = \frac{(0.67 \times 4.8 + 0.4 \times 2)}{162.85} = 0.025$$

$$\text{So, } \bar{h}_u = (z+D)(0.33a_u + 0.6b_u) - D \cdot c_u$$
$$= (2+8)(0.33 \times 0.71 + 0.6 \times 0.29) - 8 \times 0.025 = 3.88m$$

- For the lower soil layer: $z = 10m$ and $D = 4m$;

Also, for $\phi' = 35°$ and $\delta = 15°$, $K_{al} = 0.25$ (for no earthquake case)

$$P_{a3} = \frac{1}{2}\gamma_2 (z+D)^2 K_{al} = \frac{1}{2} \times 9 (10+4)^2 \times 0.25 = 220.5kN$$

$$P_{a4} = \frac{1}{2}\gamma_2 z^2 K_{al} = \frac{1}{2} \times 9 \times (10)^2 \times 0.25 = 112.5kN$$

$$P_{ael} = 155.34kN \text{ as before}$$

$$\Delta P_{ae3} = P_{ae3} - P_{a3} = 317.14 - 220.5 = 96.64kN$$

$$\Delta P_{ae4} = P_{ae4} - P_{a4} = 161.80 - 112.50 = 49.3kN$$

$$a_l = \frac{(P_{a3} - P_{a4})}{P_{ael}} = \frac{(220 \cdot 5 - 112 \cdot 5)}{155 \cdot 34} = 0 \cdot 70$$

$$b_l = \frac{(\Delta P_{ae3} - \Delta P_{ae4})}{P_{ael}} = \frac{(96 \cdot 64 - 49 \cdot 3)}{155 \cdot 34} = 0 \cdot 30$$

$$c_l = \frac{(0 \cdot 67 P_{a4} + 0 \cdot 4 \Delta P_{ae4})}{P_{ael}} = \frac{(0 \cdot 67 \times 112 \cdot 5 + 0 \cdot 4 \times 49 \cdot 3)}{155 \cdot 34} = 0 \cdot 61$$

So, $\overline{b}_l = (z + D)(0 \cdot 33 a_l + 0 \cdot 6 b_l) - D.c_l$

$$= (10 + 4)(0 \cdot 33 \times 0 \cdot 7 + 0 \cdot 6 \times 0 \cdot 3) - 4 \times 0 \cdot 61 = 3 \cdot 31 m$$

- Location of the resultant force:

$$\overline{b} = \frac{(P_{aeu} \times \overline{b}_u + P_{ael} \times \overline{b}_l)}{\sum P_{ae}} = \frac{(162 \cdot 85 \times 3 \cdot 88 + 155 \cdot 34 \times 3 \cdot 31)}{318 \cdot 19} = 3 \cdot 60 m$$

Hydrodynamic force

$$P_{2(w)} = 0 \cdot 41 k_H \cdot \gamma_w \cdot b^2 = 0 \cdot 41 \times 0 \cdot 2 \times 10 \times (14 - 1)^2 = 138 \cdot 58 kN$$

- Location of the hydrodynamic force:

Location of the resultant of the hydrodynamic force on the structure below the water table may be obtained by considering the algebraic summation of the moments about the water table of the full depth of water to the base of the structure and the partial depth to the top of the structure.

So, depth of the resultant hydrodynamic force below the water table, taking $z + D = H$, is given by:

$$z_r = \frac{0 \cdot 41 k_H \cdot \gamma_w H^2 \cdot [0 \cdot 6 H] - 0 \cdot 41 k_H \cdot \gamma_w (z)^2 \cdot [0 \cdot 6 z]}{0 \cdot 41 k_H \cdot \gamma_w H^2 - 0 \cdot 41 k_H \cdot \gamma_w (z)^2}$$

$$= \frac{0 \cdot 6 (H^2 + H \cdot z + z^2)}{(H + z)} = \frac{0 \cdot 6 (14^2 + 14 \times 2 + 2^2)}{(14 + 2)}$$

$$= 8 \cdot 55 m, \text{i.e. } 5 \cdot 45 m \text{ above the base of the structure.}$$

Example 15.2
Using the simplified procedure, estimate the design racking deformation of a single-cell rectangular cut-and-cover tunnel structure, 10m wide and 5m deep, under a soil cover

of 5*m*. Assume the soil to be stiff with a unit weight of $19kN/m^3$, and the following earthquake parameters:

- Moment magnitude, $M_w = 7.5$
- Source-to-site distance $= 15km$
- Peak ground particle acceleration at surface $= 0.5g$
- Apparent velocity of S-wave propagation in soil, $C_s = 200m/s$.

List any other assumptions made.

Solution
- Lateral Stiffness K_X of the structure:

Lateral racking stiffness can be defined as the force required to cause a unit racking displacement (say, 1*m*) for a unit length (say, 1*m*) of the rectangular frame structure. Assume, for the purpose of this exercise, that the structural analysis, if carried out on the frame, yields a K_X value of, say, $380,000kN/m^2$.

- Average Shear Modulus G_S of soil over the relevant layer depth:

$$G_S = \rho_s \cdot (C_s)^2 / g = \left(\frac{19}{9.81}\right) \cdot (200)^2 = 77,470kN/m^2$$

- Flexibility Ratio F, and Racking Coefficient R:

$$F = (G_S/K_X) \cdot (W/H) = (77,470/380,000) \cdot (10/5) = 0.4$$

From Figure 15.8, for $F = 0.4$, Racking Coefficient, $R = 0.5$

- Maximum lateral free-field shear displacement $\Delta_{free-field}$:

Estimate of ground motion at tunnel depth from Table A3, Appendix A,

$$a_d = (1.0)(0.5g) = 0.5g$$

From Table A2, Appendix A, ratio of peak ground 'velocity to acceleration' at surface $= 130$

$$\text{Peak Particle Velocity, } V_S = 130 \times a_d = (130 \ cms/s/g)(0.5g)$$
$$= 65 \ cm/s = 0.65 \ m/s$$

Maximum free-field shear strain of the soil medium,

$$\gamma_s = V_S/C_s = 0.65/200 = 0.00325$$
$$\text{Then } \Delta_{free-field} = \gamma_s (H) = 0.00325 \times 5 = 0.01625m$$

- Maximum Racking Displacement of the structure Δ_{str} for design:

$$\Delta_{str} = R \cdot \Delta_{free-field} = 0 \cdot 5 \times 0 \cdot 01625 = 0 \cdot 00812m$$

Finally, in order to obtain the stress resultants for design, a relative lateral displacement of 8·12*mm* (to be treated as being reversible) should be imposed on the relevant node of the structure in combination with other appropriate coexistent static loads.

16 Slurry Trench Stability: I

16.1 Introduction

The first stage in the construction of a typical cut-and-cover metro station structure, invariably, is the installation of the perimeter wall. It either forms an integral part of the permanent structure itself or is treated as an outer temporary cofferdam within which the main structure is constructed. In view of the significant depths to which such a structure is generally required to penetrate below ground, construction of the perimeter wall in the conventional manner in an open cut, would require vast corridors of land outside the perimeter to accommodate the formation of stabilized side slopes. However, in heavily built-up urban environments generally encountered in metro alignments where existing structures can be expected to be closely abutting the works, such a luxury is rarely available. Under such circumstances, the perimeter walls are installed vertically with the least amount of intrusion beyond the perimeter and often using the slurry trench techniques.

Activities related to the construction of cut-and-cover structures, inevitably, give rise to interaction with the existing structures in the vicinity which, in certain cases, can be significant. The type of response of the surrounding ground and the structures in the vicinity is dependent, among other factors, also upon the integrity of the excavation for the installation of the perimeter wall. During the excavation, the slurry-filled trench is subjected to lateral earth and groundwater pressures as also the effects from surcharge loads if present. If the lateral pressures are excessive, inward movement can ensue and, if not adequately contained, even collapse of the trench faces cannot be ruled out. Under such circumstances, the stability of a high-rise structure in the vicinity can also be threatened. In view of this, the stability of the slurry-filled trenches during the construction of the wall can be of paramount importance in the design and construction of such walls. This is particularly so where water table is close to the surface and soil pressures are high and, as is also very often the case, structures generating high surcharge loads are in close proximity. Other factors of equal concern are the nature of the soil, potential for loss of fluid through highly pervious layers and effects of construction procedures. This chapter deals with the trench excavation under the protection of slurry and the measures necessary to maintain the required stability.

As the trench is excavated, although it is simultaneously filled with slurry, the *in situ* state of stress equilibrium at the slurry–soil interfaces is disturbed. Consequently, a certain amount of inward movement of the trench faces is to be expected. This can, in turn, lead to varying degrees of movements in the surrounding ground and structures.

However, it is important to ensure that, by judicious design and the use of well thought out construction procedures, the excessive movements which could potentially lead to catastrophic trench collapses are prevented. Sensible design must incorporate an adequate margin of safety against caving-in and collapse of the trench faces.

In order to make a realistic assessment of the factor of safety available under a given set of boundary conditions and loads, it is essential to understand the nature and behaviour of the particular soil medium and its porewater pressure regime. This understanding should go a long way in enabling the designer to select an appropriate method of analysis and appropriate design parameters to go with it.

16.2 Methods of analyses

Consider an element of soil at a depth H below ground subjected to vertical and horizontal principal effective stresses σ'_V and σ'_H. If an excavation for a trench to one side of the soil element is taken down to this depth, the vertical effective stress, σ'_V on the element will continue to remain unaltered. However, the horizontal effective stress, σ'_H, on it will experience a drop in its value owing to the removal of the lateral restraint on the soil element thereby allowing its structure to expand horizontally. The resulting movement of the face of the trench will continue until further drop in σ'_H is prevented by the effective trench fluid. An appropriate margin of safety is incorporated to ensure that the extent of the anticipated movement of the face is acceptable. It is this state of stress in the soil that is the focus of attention when examining the stability of a slurry trench.

Examining the stability of a trench mainly involves the identification of the potential failure surface and the assessment of shear strength mobilized along that surface. Since the shear strength of soil is controlled by its effective stress, effective stress analysis would appear to be the most appropriate method of analysis. The soil strength mobilized along the failure plane, as discussed in Chapter 13, is given by the expression:

$$\tau_f = c' + \sigma'_n \cdot \tan \phi'$$

where, c' and $\tan \phi'$ are the experimentally determined shear strength parameters with respect to effective stress, and σ'_n is the effective stress normal to the plane of failure.

Effective stress approach presupposes for its application, a prior knowledge of the porewater pressure regime. So, broadly speaking, if in a given situation, the porewater pressures are known or can be estimated, the stability analysis should be carried out in terms of effective stress. For various reasons, such information may not always be available at the time.

However, provided that the soil is saturated and behaves in the undrained mode, i.e. the volumetric strain remains zero, it is possible to characterize the strength of a soil element in its *in situ* stress state by its undrained shear strength, s_u, in which case the total stress analysis based on the '$\phi = 0$' concept, as discussed in Chapter 13, may be used.

Total stress approach assumes that there is no change in the effective stress implying thereby a constant value of s_u. With the exception of free draining soils like sands and gravels, it may be reasonable to assume undrained conditions in view of the relatively short-term duration over which a slurry trench is likely to remain open.

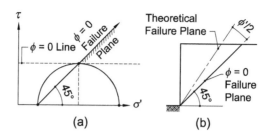

Figure 16.1 '$\phi = 0$' and Theoretical Failure Planes

Nevertheless, if the dissipation of the excess porewater pressures takes place which could cause a corresponding change in the effective stress and the strength of the soil, Total stress approach is invalid and the effective stress analysis must be used.

Inherent also in the '$\phi = 0$' concept is the implication that the failure plane rises at 45° to the horizontal irrespective of the actual value of ϕ' as shown in Figure 16.1(*a*). However, there is evidence available to suggest that most clays exhibit angles of internal friction in excess of zero and so, this concept does not help locate the failure plane correctly.

For c', ϕ' soils also, the failure plane suggested by the '$\phi = 0$' analysis (i.e. at 45°) would therefore not coincide with, but lie further back from, the theoretical failure plane, i.e. at $45° + \phi'/2$, as shown in the Figure (*b*).

Furthermore, in heavily overconsolidated stiff fissured clays, the negative porewater pressures in the undrained test will tend to overestimate s_u, whereas in the normally consolidated relatively softer clays, positive porewater pressures will underestimate its value. The implication is that the '$\phi = 0$' analysis is likely to indicate far higher factors of safety than might really be the case for the heavily overconsolidated clays and, likewise, far lower values for the soft, normally consolidated clays.

16.3 Pole and failure plane

Mohr circle permits an excellent visualization of not only the magnitude of stresses in the soil but also the orientations of the planes on which these stresses act.

The notion of Pole is especially useful in this context. Pole is a unique point on the Mohr circle with the following characteristics:

- A line passing through the Pole P and any other point A on the Mohr circle is parallel to the plane on which the stresses represented by the point A act. This is shown in Figure 16.2(*a*).
- All planes pass through the Pole. Because of this property, it is also referred to as the origin of planes.

The line joining the Pole P to the tangent point T as shown in Figure (*b*) represents, by definition, the orientation of the failure plane.

In Figure (*c*), point M represents the point of maximum shear stress. By comparison with Figure (*b*), it can be seen that although PM represents the plane on which the maximum shear stress acts, it does not rise at the same angle as PT and is therefore

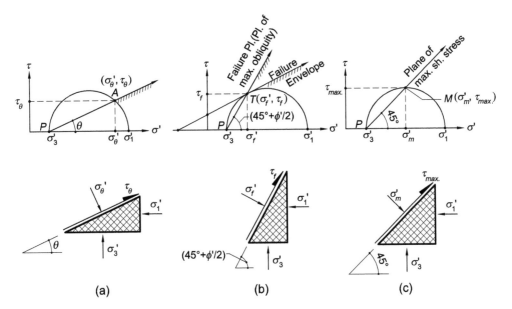

Figure 16.2 Stresses and Orientation of Planes

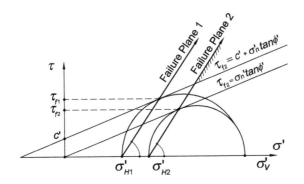

Figure 16.3 c', Shear Strength and Failure Planes

not the failure plane. This is because the criterion of failure is maximum obliquity and not maximum shear stress.

It is also interesting to observe that while the parameter c' affects the magnitude of shear strength, it does not influence the orientation of the failure plane. This is illustrated in Figure 16.3.

Clearly, $\tau_{f_1} \neq \tau_{f_2}$, nevertheless the failure planes 1 and 2 are parallel, i.e. their orientation is identical and unchanged.

16.4 Limit equilibrium

The principles explained in the foregoing will assist in understanding the potential failure mechanism of soil adjacent to a slurry trench and enable a rational assessment of the margin of safety against such a failure to be made. However, because of the many variables involved, exact three-dimensional theoretical solutions tend to become mathematically complex. The stability of vertical narrow cuts in soils is generally analysed by the use of limit equilibrium theory assuming simple two-dimensional wedge failure and without any reference to displacements. Simultaneous with the excavation, as the trench is filled with a stabilizing fluid, it will apply a hydrostatic force against the face of the trench. Critical height over which the trench is likely to remain stable is obtained on the basis of the equilibrium of forces including the surcharge load as appropriate. Experience and empirical evidence suggest that such an approach can often give conservative results.

Owing to its inherent simplicity, the use of limit theory has become a convenient and often used analytical tool with the designers. Where the knowledge of the lateral pressures generated and the resistances mobilized constitutes the predominant aim of the analyses, the concept of limit equilibrium can be relied upon to provide acceptable solutions. However, it must be recognized that, in its application, little or no consideration is given to the displacements associated with the generation of loads or the mobilization of reactions within the soil medium. In situations associated with deep excavations where limiting the movements in the surrounding ground and structures is paramount, recourse to sophisticated methods of analyses, such as a finite element method, should not be excluded if meaningful prediction and control of displacements are to be achieved. However, while not being able to quantify the associated displacements, the use of a factor of safety in the application of the limit equilibrium theory does introduce, it can be argued, a certain measure of control in the extent of the anticipated displacements.

16.5 General assumptions

Principal simplifying assumptions generally used in the investigation of the stability of slurry-filled trenches are as follows:

- The failure surface is planar. This is an idealization which simplifies the computation without any significant loss of accuracy.
- The failure is a two-dimensional, i.e. plane strain, problem. A unit length is considered in the computation of stability. It is assumed that the state of affairs existing in the x–z plane holds equally true for all the other planes parallel to it.
- The effects of arching are ignored. This is conservative as will be demonstrated in the solved examples in the next chapter.
- Friction at the soil–slurry interface is ignored. The extent of friction likely to be mobilized at the interface, besides being dubious is also difficult to quantify accurately. However, the effect of ignoring it for the active soil wedge is on the safe side generally and, in the context of slurry trench stability, insignificant.
- Force equilibrium alone is considered, i.e. $\sum H = 0 = \sum V$. Static equilibrium is unlikely to be satisfied since all the forces acting on the wedge, when extended may not meet at a point, in which case $\sum M \neq 0$.
- Full hydrostatic pressure of the slurry acts against the sides of the trench. This presupposes the formation of a fine, impermeable film of bentonite on the excavated trench face.

Stability of the trench requires that the fluid pressure in the trench is higher than the porewater pressure in the soil. This is achieved in two ways – first, by the use of slurry which is heavier than water and second, by maintaining a positive differential head of the slurry above the piezometric level. The latter assumes particular importance if artesian pressures are likely to be encountered in the ground. The pressure differential is responsible for the filtration of the slurry water into the soil leaving behind, in the process, a deposition at the interface of a thin film of bentonite known as the filter cake.

In soils of high permeability, such as coarse sands and gravels, appreciable penetration of the slurry fluid into the soil takes place. The bentonite thus lost impregnates the soil but leaves behind a poor filter cake. However, in such soils, the loss of bentonite can be considerably reduced and the sealing mechanism improved by the addition to the slurry of about 1 per cent fine sand, or similar. In such soils as clays, viscous retardation of the free filtration of slurry water may inhibit the formation of a significant filter cake.

In what follows, stability of slurry trenches in different types of soils is examined.

16.6 Stability of trench

(A) c', ϕ' soils

Consider the geometry of a typical slurry trench in a c', ϕ' soil as idealized in Figure 16.4(*a*).

Let γ_w, γ and γ_s represent the unit weights of water, dry soil above water table and saturated soil below water table, respectively. Parameters c' and ϕ' are as defined previously. It is assumed that the soil adjacent to the trench is in a state of incipient failure. If θ defines the inclination of the failure plane with the horizontal, then the forces acting on the wedge ABC, as shown in Figure 16.4(*b*), and maintaining limiting equilibrium are as follows:

(a) Forces acting on the soil wedge

(i) Total Weight of the soil wedge ABC plus surcharge,

$$\sum W = \frac{1}{2}\gamma H^2 \cot\theta + \frac{1}{2}(\gamma_S - \gamma) H_W^2 \cot\theta + q_s H \cot\theta \qquad (16.1)$$

Effective Weight of soil wedge ABC plus surcharge,

$$\sum W_e = \sum W - \frac{1}{2}\gamma_w H_w^2 \cot\theta = \frac{1}{2}\gamma_e' H^2 \cot\theta \qquad (16.2)$$

where, γ_e' = Equivalent effective unit weight of soil

$$= \left[\gamma - (\gamma - \gamma_{Sub}) \left(\frac{H_W}{H}\right)^2 + 2\left(\frac{q_s}{H}\right) \right] \qquad (16.3)$$

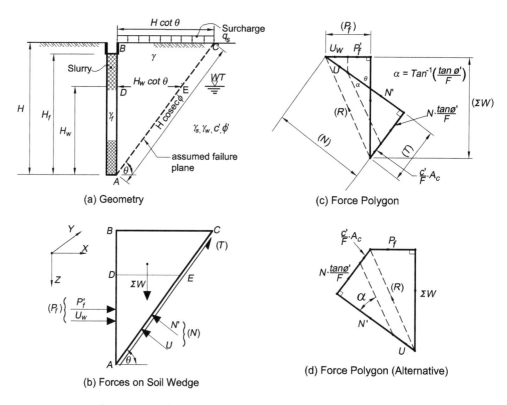

Figure 16.4 Slurry Trench Geometry and Forces in c', ϕ' Soil

and γ_{Sub} = Submerged unit weight of soil

$$= \gamma_s - \gamma_w$$

(ii) Normal Component of the Resultant Force on the Failure Plane,

$$N = N' + U \tag{16.4}$$

where, U = Force from Porewater pressure

$$= \frac{1}{2}\gamma_w H_w^2 \operatorname{cosec} \theta \tag{16.5}$$

(iii) Tangential Component of the Resultant Force on the Failure Plane,

$$T = \text{Mobilized shear resistance}$$

$$= N' \cdot \frac{\tan \phi'}{F} + \frac{c'}{F} \cdot H \cdot \operatorname{cosec} \theta$$

$$= N' \cdot \tan \alpha + \bar{c} \cdot H \cdot \operatorname{cosec} \theta \tag{16.6}$$

where, F = Factor of safety on shear strength of soil

$$\alpha = \tan^{-1}\left(\frac{\tan\phi'}{F}\right) \tag{16.7}$$

$$\bar{c} = \frac{c'}{F}$$

The factor of safety may be defined as the ratio by which available shear strength on the hypothetical shear surface can be reduced before limiting equilibrium occurs. In other words, at limiting equilibrium, i.e. when failure occurs, the factor of safety is unity.

(iv) Force exerted by the Fluid Pressure,

$$P_f = \frac{1}{2}\cdot\gamma_f\cdot H_f^2 \tag{16.8}$$

where, $\gamma_f = $ Unit weight of slurry fluid

(b) *Consider the equilibrium of forces*

- Vertically, $\sum V = 0$:
  $\sum W - T\sin\theta - N\cos\theta = 0$. Substituting values of N, U and T from Equations (16.4), (16.5) and (16.6), respectively, using Equation (16.2) and rearranging, we obtain:

$$N' = \frac{0\cdot5\gamma_e'\cdot H^2\cdot\cot\theta - \bar{c}\cdot H}{\cos\theta + \sin\theta\cdot\tan\alpha} \tag{16.9}$$

- Horizontally, $\sum H = 0$:
  $P_f + T\cos\theta - N\sin\theta = 0$. Substituting for values of N, T and N' and recognizing from the force polygon, Figure 16.4 (c), that $\left(P_f - U\cdot\sin\theta\right) = P_f'$, we obtain:

$$P_f' = 0\cdot5\gamma_e'\cdot H^2\cdot\cot\theta\cdot\tan\left(\theta - \alpha\right) - \bar{c}\cdot H\left[\cot\theta + \tan\left(\theta - \alpha\right)\right] \tag{16.10}$$

(c) *Evaluation of θ*

For angle θ to be a maximum,

$$\frac{\partial P_f'}{\partial\theta} = \frac{\left(0\cdot5\gamma_e'\cdot H^2\cdot\tan\alpha + \bar{c}\cdot H\right)\left(\cos 2\theta + \sin 2\theta\tan\alpha\right)}{[\sin\theta\left(\cos\theta + \sin\theta\tan\alpha\right)]^2} = 0$$

This equation is satisfied when

$$\cos 2\theta + \sin 2\theta\tan\alpha = 0$$

$$\text{Or}\quad \cos\left(2\theta - \alpha\right) = 0 = \cos 90^{\circ}$$

$$\text{whence}\quad \theta = 45° + \frac{\alpha}{2} \tag{16.11}$$

This expression confirms that the inclination of the theoretical failure plane is independent of, and therefore not influenced by, the parameter c'.

(d) Evaluation of factor of safety F

Recalling Equation (16.10) and substituting for θ as per Equation (16.11), the equation can be rearranged as:

$$P'_f = 0\cdot5\gamma'_e \cdot H^2 \cdot \left(\frac{1}{N_\alpha}\right) - 2\bar{c} \cdot H \cdot \left(\frac{1}{\sqrt{N_\alpha}}\right) \tag{16.12}$$

$$\text{where}\quad N_\alpha = \tan^2\left(45° + \frac{\alpha}{2}\right)$$

$$= \tan^2\left[45° + \frac{1}{2} \cdot \tan^{-1}\left(\frac{\tan\phi'}{F}\right)\right] \tag{16.13}$$

Equation (16.13) may also be written in the form:

$$F = \frac{\tan\phi'}{\tan\{2[\tan^{-1}\sqrt{N_\alpha} - 45°]\}} \tag{16.14}$$

Substituting c'/F for $\bar{c}$ and rearranging Equation (16.12), we obtain:

$$F = \frac{2c' \cdot H \cdot \sqrt{N_\alpha}}{0\cdot5\gamma'_e \cdot H^2 - P'_f \cdot N_\alpha} \tag{16.15}$$

$$\text{where}\quad P'_f = P_f - 0\cdot5\gamma_w \cdot H^2_{\text{W}}$$

Let the equivalent effective unit weight of fluid for computing lateral pressure be represented by the expression

$$\gamma'_f = \gamma_f \cdot \left(\frac{H_f}{H}\right)^2 - \gamma_w \cdot \left(\frac{H_{\text{W}}}{H}\right)^2 \tag{16.16}$$

$$\text{Then}\quad P'_f = 0\cdot5\gamma'_f \cdot H^2 \tag{16.17}$$

and

$$F = \frac{2c' \cdot H \cdot \sqrt{N_\alpha}}{0\cdot5\gamma'_e \cdot H^2 - 0\cdot5\gamma'_f \cdot H^2 \cdot N_\alpha} = \frac{4c'}{H}\left[\frac{\sqrt{N_\alpha}}{\gamma'_e - \gamma'_f \cdot N_\alpha}\right] \tag{16.18}$$

Note that since N_α is also a function of the factor of safety in its own right, the value of F can be obtained by trial and error. Note also that it is easier to use N_α as the base item for trial rather than F itself directly. For the trial, adopt the following procedure:

- Compute the ratio γ_e'/γ_f'. Values of N_α chosen for trial must always be less than this ratio so as to ensure that F is not negative and therefore meaningless.
- Substitute the trial value of N_α into the Equations (16.14) and (16.18) to obtain the value of F in each case.
- If both equations yield the same value, then that value represents the factor of safety. If not, repeat the trials with different N_α values until convergence is reached.

The depth of the trench for which the value of $F = 1$ is, by definition, the critical depth. By making this substitution in Equation (16.18), the critical depth can be obtained from the following expression.

$$H_{crit} = \frac{4c'}{a \cdot \gamma_e' - b \cdot \gamma_f'} \tag{16.19}$$

$$\text{where} \quad a = \tan\left(45° - \frac{\phi'}{2}\right)$$

$$b = \tan\left(45° + \frac{\phi'}{2}\right),$$

$$\text{and} \quad a = 1/b$$

(B) $c' = 0, \phi'$ soils

Recalling that the angle of inclination θ of the failure plane is independent of the parameter c', the foregoing procedure must apply equally to a $c' = 0, \phi'$ soil, provided that all the c' terms are discretely excluded. Therefore, setting $c' = 0$ in Equation (16.12), we obtain:

$$P_f' = 0 \cdot 5 \gamma_e' \cdot H^2 \cdot \left(\frac{1}{N_\alpha}\right) \tag{16.20}$$

Comparison of Equations (16.17) and (16.20) yields:

$$\sqrt{\frac{\gamma_e'}{\gamma_f'}} = \sqrt{N_\alpha} = \tan\left(45° + \frac{\alpha}{2}\right) = \sqrt{(1 + \tan^2\alpha)} + \tan\alpha$$

Upon squaring and rearranging, we get:

$$\tan\alpha = \frac{\gamma_e' - \gamma_f'}{2 \cdot \sqrt{\gamma_e' \cdot \gamma_f'}} = \frac{\tan\phi'}{F} \quad \text{recalling Equation(16.7)}$$

$$\therefore \text{Factor of Safety,} \quad F = 2\tan\phi' \left[\frac{\sqrt{\gamma'_e \cdot \gamma'_f}}{\left(\gamma'_e - \gamma'_f \right)} \right] \tag{16.21}$$

Equation (16.21) confirms that the stability of a trench primarily depends upon the strength of the soil and the unit weight of the slurry. By inspection of Equations (16.3) and (16.16), it can also be seen that the values of the terms γ'_e and γ'_f at any time are functions of the depth of excavation at that time. While the values of these terms, individually, will vary with the depth of excavation, their difference $\left(\gamma'_e - \gamma'_f \right)$, however, remains invariant. Accordingly, the factor of safety of the slurry-filled trenches in $c' = 0, \phi'$ soils based upon simple wedge failure will also vary with the advancement in the depth of excavation.

Alternatively, the expression for the factor of safety as given by Equation (16.21) can also be expressed in the form:

$$F = \frac{\tan\phi'}{\tan\beta} \tag{16.22}$$

$$\text{where} \quad \beta = 2\tan^{-1}\left[a - \sqrt{(a^2 - 1)} \right]$$

$$\text{and} \quad a = \left[\frac{\gamma - 2\gamma_W \left(\frac{H_W}{H} \right)^2 + \gamma_f \left(\frac{H_f}{H} \right)^2}{\gamma - \gamma_f \left(\frac{H_f}{H} \right)^2} \right]$$

By assigning some values to the various parameters, it can be confirmed, as is to be expected, that both the expressions (Equations 16.21 and 16.22) yield identical values for the factor of safety.

For a given set of parameters, increase in the value of the factor of safety, F, can be achieved by one or more of the following:

- Limiting the surcharge load, q_S;
- Increasing γ_f, the unit weight of the slurry, but only up to the extent that the displacement of the slurry by concrete does not become problematic;
- Increasing the differential slurry head $\left(H_f - H_W \right)$.

It may not be possible to reduce the surcharge load below that generated by the activity of the construction rig. This should not be overlooked.

For slurry, bentonite concentration in the range of 4–8 per cent represents the density in the range of 1·023–1·045 g/ml; for intermediate values of concentration, the density can be obtained by interpolation. Bentonite concentration in the slurry should neither be too high (>8 per cent) to cause an impediment in the desired rate of flow, nor too low (<4 per cent) to cause flocculation or prevent formation of a filter cake. A concentration in the range of 5–6 per cent has come to be recognized as the accepted norm. However, in order to reduce costs, part of bentonite is often replaced with an appropriate clay admixture. Where relatively higher concentrations are required to ensure stability, special weighting agents which do not increase the initial gel strength can also be added.

Increasing the differential slurry head can be achieved by the following:

- Reducing H_W by lowering the groundwater level;
- Increasing H_f by raising the slurry level;
- Some combination of the two.

Lowering the groundwater level, if accompanied by an unacceptable level of settlements of the surrounding ground and structures, may not be feasible. As far as raising the slurry head is concerned, it is essential that the head is maintained at the very least 1*m* higher than the groundwater or the piezometric level outside. However, if the lateral ground pressures are high enough to demand differential head to be well in excess of the 1*m* or that the piezometric head is close to the existing ground level, there may arise the need to raise the slurry level above the ground surface. This would call for a commensurate extension of the guide trench above the ground level. Practicalities of such a measure, especially in areas where any obstruction to the free flow of traffic, etc., could cause problems and must be duly taken into account.

In the case of layered soils, the potential implications of an underlying aquifer being under artesian pressure must not be ignored unless it is possible to satisfactorily vent the excess pressures to atmosphere through 'bleeder' wells. However, if ignored, it could cause collapse of the trench as the excavation reaches the artesian layer. For the successful use of the slurry trench technique, it is therefore essential to establish the porewater pressure profile with depth as accurately as possible.

(C) '$\phi = 0$' soils

For saturated clays, the short-term stability of long, unbraced cuts filled with slurry may be obtained by using the '$\phi = 0$' concept. In this case, for $\phi = 0$, c' is replaced by s_u, the undrained shear strength.

In his approach, Meyerhof (1972) considered the horizontal equilibrium of active earth and slurry pressures based on plastic theory, assuming isotropic and homogeneous clay medium. Gel strength was considered negligible, and tension cracks were ignored. It is believed (Tschebotarioff, 1967) that the presence of the slurry reduces the tendency for tension cracks and fissures to open in the upper parts of the cut, and that this adds greatly to the stability. In what follows, this approach has been extended to a general case of soil with groundwater and surcharge load.

Consider the geometry of a generalized problem as shown in Figure 16.5(*a*). Figure (*b*) gives a breakdown of the pressures acting on the 'slurry–soil' interface of a long cut.

Active Thrust

$$= \frac{1}{2}\gamma\left(H - H_w\right)^2 + \gamma\left(H - H_w\right)H_w + \frac{1}{2}\gamma_{Sub} \cdot H_W^2 + q_s \cdot H - 2s_u \cdot H$$

$$= \frac{1}{2}\gamma_e' \cdot H^2 - 2s_u \cdot H$$

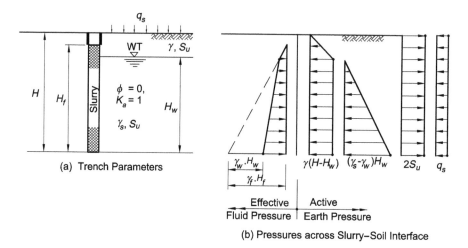

(a) Trench Parameters

(b) Pressures across Slurry–Soil Interface

Figure 16.5 Slurry Trench Pressures in '$\phi' = 0$' Soil

$$\text{where, } \gamma'_e = \gamma - (\gamma - \gamma_{Sub}) \left(\frac{H_W}{H}\right)^2 + 2 \cdot \frac{q_S}{H},$$

$$\text{Effective Fluid Resistance} = \frac{1}{2}\gamma_f \cdot H_f^2 - \frac{1}{2}\gamma_W \cdot H_W^2 = \frac{1}{2}\gamma'_f \cdot H^2$$

$$\text{and, } \gamma'_f = \gamma_f \cdot \left(\frac{H_f}{H}\right)^2 - \gamma_W \cdot \left(\frac{H_W}{H}\right)^2,$$

By introducing a factor of safety F on the undrained shear strength, its value can be obtained by equating the resistance and the thrust.

$$\text{So } \frac{1}{2}\gamma'_f \cdot H^2 = \frac{1}{2}\gamma'_e \cdot H^2 - 2\frac{s_u}{F} \cdot H$$

$$\text{whence, } F = \frac{4s_u}{H\left(\gamma'_e - \gamma'_f\right)} \tag{16.23}$$

Expressions for the various factors of safety for trench stability represented by the three Equations (16.18), (16.21) and (16.23) for the three types of soils are likely to yield conservative results since they presuppose cuts of infinite lengths and assume plane strain conditions. They also ignore the geometry and shape of the trenches. However, in reality, trenches are excavated in finite panel lengths that are also considerably smaller than their respective depths. These aspects can result in significant enhancements in the factors of safety, as will become evident later on.

The foregoing expressions also disregard the inherent strength of the gel or the slurry. However, it is believed (Elson, 1968) that the hydrostatic slurry pressure can account for, at the most, up to 90 per cent, of the resistance against trench stability and the shearing resistance of the slurry saturated soil up to 25 per cent. It is considered unlikely that the plastic resistance of the slurry will exceed 5 per cent of the stabilizing force and could, in fact, be much less during the panel excavation when the slurry is likely to be disturbed. However, it is worth remembering that the potential contribution to stability of gel strength can be usefully employed in construction works using prefabricated wall panels where the problems of displacement of the slurry by the fluid concrete do not arise. In certain other applications, such as the formation of slurry curtains and cut off walls, slurry strengths even as small as $0.5-0.75 kN/m^2$ can help stability appreciably.

16.7 Effect of gel strength

Slurry in the trench may possess a certain shear (gel) strength that can help resist the tendency of the soil faces of the trench to close in (Xanthakos, 1974). The mechanism can be likened to the compression of a perfectly plastic material between two rough rigid plates as analysed by Prandtl (1923) and extended by Bishop (1952) in his earth dam puddle core problem.

Consider an infinitesimal element dx, dy of slurry of unit weight γ_f, per m run of the trench acted upon by stresses as shown in Figure 16.6.

Static equilibrium $(\sum H = 0 = \sum V)$ requires that:

$$\frac{\partial \sigma_x}{\partial x} + \frac{\partial \tau_{xy}}{\partial y} = 0 \tag{16.24}$$

$$\text{and} \quad \frac{\partial \sigma_y}{\partial y} + \frac{\partial \tau_{xy}}{\partial x} - \gamma_f = 0 \tag{16.25}$$

Plastic flow of the slurry requires that the stresses satisfy the yield criterion:

$$\left(\sigma_x - \sigma_y\right)^2 + 4\tau_{xy}^2 = 4\tau_f^2 \tag{16.26}$$

where, τ_f is the gel strength and τ_{xy} represents the complementary shear stress on, or the distortion of, the gel element. Subject to the appropriate boundary conditions, the stresses can be statically determined from the Equations (16.24) and (16.25). For a trench of width $2a$, it can be shown (Xanthakos, 1974) that the horizontal stress,

$$\sigma_x = \gamma_f \cdot y + \frac{\tau_f \cdot y}{a} + \frac{\pi}{2} \cdot \tau_f \tag{16.27}$$

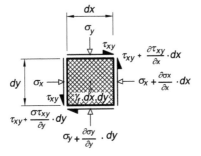

Figure 16.6 Stresses on Gel Element

Summating for the full slurry height H_f in the trench, we obtain the total resistance offered by the slurry as:

$$P_f = \int_o^{H_f} \sigma_x \cdot dy$$

$$= \frac{1}{2} \cdot \gamma_f \cdot H_f^2 + \frac{1}{2a} \cdot \tau_f \cdot H_f^2 + \frac{\pi}{2} \cdot \tau_f \cdot H_f \qquad (16.28)$$

From the equation, it can be seen that, if the gel strength is ignored, expression for the force exerted by the fluid can be recovered.

Factors of safety, including the contribution of gel strength, against trench collapse in the $c' - \phi'$, $c' = 0$, ϕ' soils and saturated ($\phi = 0$) clays can be obtained from Equations (16.18), (16.21) and (16.23), respectively, simply by setting the parameter

$$\gamma_f' = \left[\gamma_f + \frac{\tau_f}{a} + \pi \left(\frac{\tau_f}{H_f} \right) \right] \left(\frac{H_f}{H} \right)^2 - \gamma_W \cdot \left(\frac{H_W}{H} \right)^2 \qquad (16.29)$$

It must be appreciated, however, that the contribution of the gel strength to the trench stability is only appreciable for values in excess of $0.75 kN/m^2$. Although such gel strengths can be achieved, they are neither practical nor desirable in the context of the commonly used slurry trench applications in metro works, e.g. diaphragm wall panel construction.

16.8 Use of artificial slurries

Experience available from the soil drilling industry suggests (Puller, 1996) that polymeric and the mixture of polymeric and bentonite slurries can be successfully used to maintain trench support. Polymer-based artificial slurry behaves as a pseudo-plastic fluid and, unlike bentonite slurry, achieves trench support without forming a filter cake. The molecular lattice structure within the polymeric slurry is more efficient in the suspension and transportation of soil particles and leads to reduced energy costs in pumping from excavation to slurry station on the site. As the fluid loss from polymeric slurries is less than that from bentonite slurries, the former can be used with advantage in weak soils where an increase in soil moisture content could otherwise cause risk of panel instability. However, polymeric slurries cost considerably more than bentonite slurries, but since, with the use of shaker screens and hydrocyclones much improved slurry cleaning is achieved, the slurry is reusable many times over and the disposal costs can, at least partly, compensate for the high initial cost.

16.9 Effect of arching

Nash and Jones (1963) pointed out that the limit equilibrium analysis using a simple Coulomb wedge gave very conservative values of trench stability since the effects of arching around the ends of the trench were ignored.

When a deep vertical cut in the shape of a rectangular trench panel is made in an elastic half-space, the resultant pressure relief will cause the ground on either side to move laterally towards the cut. If the cut is long enough in plan, the strain in the soil mass either side of it could be unrestrained. It would therefore be reasonable to expect that the thrust generated on the sides of the panel will be close to their full active value. However, if the cut is relatively short, the soil will tend to arch across its length relieving part of the thrust on the panel by carrying it in horizontal arch action. Where the penetration of the panel into the ground is also considerably greater than its length, as is often the case, soil pressures realized in practice have been known to be smaller than those predicted by the classical earth pressure theories. Clearly, these factors can subscribe to enhancement in the safety of slurry trenches in a significant way.

In the case of trenches, especially in unfavourable ground conditions, i.e. with high water table and low shear strength, the effects of arching can be particularly beneficial. Full-scale field test of a slurry trench, $1m$ wide, $5m$ long and $28m$ deep, carried out in Studenterlunden in Oslo (Dibiagio, Myrvoll, 1972) confirms the importance of the end restraints on the displacement field around the excavation on account of its geometry. The average lateral displacement near the edges of the panel measured about two-thirds of that at the centre, thus confirming the effect of arch action. A number of investigators (Meyerhof, 1951; Schneebeli, 1964; Piaskowski and Kowalewski, 1965; Huder, 1972) have looked into such helpful ground response in different soil types.

The main difference between the various approaches incorporating the effect of arching for estimating the available factor of safety is, essentially, in the manner in which the effective lateral (driving) stresses are estimated.

(A) c', ϕ' soils

Schneebeli approach

Schneebeli (1964) investigated the effect of arching on short panels on the basis of the theory of lateral earth pressure on silos (Caquot and Kerisel, 1956). For a panel of length L, the following general expression for the lateral earth pressure at depth z modified to incorporate the effect of parameter c' was derived:

$$\sigma_z = \frac{\gamma L}{N_\phi \sin 2\phi'} \cdot \left[1 - e^{-n \sin 2\phi'}\right] - \frac{c'}{\tan \phi'} \cdot \left[\frac{N_\phi - 1}{N_\phi}\right] \qquad (16.30)$$

$$\text{where,} \quad n = z/L, \text{ and } N_\phi = \text{ Flow Factor} = \frac{(1 + \sin \phi')}{(1 - \sin \phi')}$$

For large values of n (i.e. $z >> L$), the maximum value of lateral earth pressure σ_z on the panel is given by:

$$\sigma_{max} = \frac{\gamma L}{N_\phi \sin 2\phi'} - \frac{c'}{\tan \phi'} \cdot \left[\frac{N_\phi - 1}{N_\phi}\right] \qquad (16.31)$$

In both the Equations (16.30) and (16.31), γ is depth-specific; it represents both the dry unit weight (γ_d) of soil above water table as well as the submerged unit weight (γ_{sub}) below the water table to be adopted as appropriate.

An idea of trench stability can be obtained simply by comparing the Schneebeli pressure (i.e. lateral active earth pressure with allowance for arching effect) diagram with the net fluid pressure (slurry pressure minus groundwater pressure) diagram (Schneebeli, 1964; Xanthakos, 1979). The use of this method is illustrated in the solved example towards the end of the next chapter.

Alternatively, factors of safety available at the water table level and at the base of the trench can be obtained directly from the following expressions:

$$\text{At water table level}\quad F = \left[\frac{\gamma_f \cdot D_W}{a(1-b)-d}\right] \tag{16.32}$$

$$\text{At base of cut}\quad F = \frac{\left(\bar{\gamma}_f \cdot H_f + \gamma_W \cdot D_W\right)}{\left(a \times c - d\right)} \tag{16.33}$$

where

$$a = \frac{\gamma_d \cdot L}{N_\phi \cdot \sin 2\phi'}$$

$$b = \left[e^{-n \cdot \sin 2\phi'}\right]$$

$$c = \left[1 - \left(1 - \frac{\gamma_{sub}}{\gamma_d}\right)b\right]$$

$$d = \frac{c'}{\tan\phi'}\left[\frac{N_\phi - 1}{N_\phi}\right]$$

$$\bar{\gamma}_f = \text{Effective unit weight of slurry} = \gamma_f - \gamma_W$$

$$D_W = \text{Slurry–water differential head} = H_f - H_W$$

All other terms are as defined previously in the chapter.

(B) $c' = 0, \phi'$ soils

Schneebeli approach

By setting the parameter $c' = 0$ discretely in Equation (16.30), the lateral earth pressure σ_z at depth z appropriate to $c' = 0, \phi'$ soils can be obtained as:

$$\sigma_z = \frac{\gamma \cdot L}{N_\phi \sin 2\phi'} \cdot \left[1 - e^{-n \sin 2\phi'}\right] \tag{16.34}$$

Similarly, for large values of n (i.e. $z >> L$), the maximum earth pressure σ_z on the panel can be seen to approach the asymptotic value:

$$\sigma_{max} = \frac{\gamma \cdot L}{N_\phi \sin 2\phi'} \tag{16.35}$$

This implies that, for a given length of the panel, the lateral earth pressure does not increase indefinitely with depth. It attains a maximum value by a certain depth below which there is no further increase.

Again, simply by comparing the Schneebeli pressure diagram with the net fluid pressure diagram, an idea of trench stability can be obtained. Alternatively, factors of safety available at the water table level and at the base of the trench can be obtained by setting the term d in Equations (16.32) and (16.33) above to zero yielding the following expressions:

$$\text{At water table level} \quad F = \frac{\gamma_f \cdot D_W}{a(1-b)} \tag{16.36}$$

$$\text{At base of trench} \quad F = \frac{\left(\bar{\gamma}_f \cdot H_f + \gamma_W \cdot D_W\right)}{a \times c} \tag{16.37}$$

All terms are as defined previously.

Stability of cylindrical cuts, for instance in the construction of bored piles under bentonite, in $c' = 0, \phi'$ soils may be analysed, approximately, using the balanced stress concept as before by replacing the length L of the panel by the diameter of the bore. However, this idealization is likely to yield conservative results.

Balanced stress approach

Piaskowski and Kowalewski (1965) examined the stability problem of slurry trenches using the balanced stress concept. They suggested that the caving-in of the sides at any depth could be avoided if the net hydrostatic pressure of the slurry was at least equal to the effective active earth pressure at that depth. In other words, by reference to Figure 16.7, for a factor of safety of 1,

$$p_f - p_w = p'_a \tag{16.38}$$

For trenches of infinite length and $H_x > h_W$, by substituting appropriate values in Equation (16.38), the expression for the critical effective slurry pressure can be

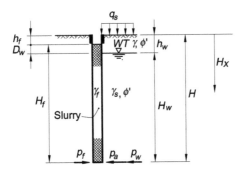

Figure 16.7 Trench of 'Infinite' Length in $c' = 0, \phi'$ Soil

obtained as:

$$\gamma_f^c \cdot H_f - \gamma_W \cdot H_W = \left[\gamma\left(H_x - H_W\right) + \gamma_{Sub} \cdot H_W + q_S\right] \cdot K_a \qquad (16.39)$$

For a given geometry of the trench and ground conditions, all terms on the right hand side of the equation are known. So, by substituting these known values, the expression on the left hand side of the equation representing the value for the critical effective slurry pressure reflecting a factor of safety of 1 can be readily evaluated.

As the excavation advances downwards, the face of the trench would tend to yield transversely towards the cut. However, in short panels this tendency is resisted at the ends of the panel. As a result, horizontal arch action is developed over the panel length. For panels of finite dimensions, this arching is considered similar to that of the ground response over the crown of a tunnel. To allow for the effect of such arch action in relatively short panels, the Coulomb wedge is modified to assume the shape of the segment of a vertical parabolic cylinder $ABC - A'B'C'$ as shown in Figure 16.8 with the failure occurring as sliding along the plane $A'B'C'$ inclined at a known angle α to the horizontal.

The beneficial effect of such arch action is accounted for by using a reduced value of K_a. The stability of a trench with a given aspect ratio H/L, in a soil of given strength and unit weight, and known water table geometry, can be examined by using the modified K'_a-value obtained from nomographs similar to those of Figure 16.9. It can be clearly seen from this figure that, with the shortening of panel length (i.e. increase in the H/L ratio), significant reductions in the values of active earth pressure coefficient are indicated suggesting greater inherent reserves of safety. Beneficial effects of lowering the water table are also in evidence.

The balanced stress concept can also be applied to a slurry trench in a multi-layered soil with different values of K_a and γ for the different layers. For stratified soil, with m different layers above and n below the water table, the expression corresponding to

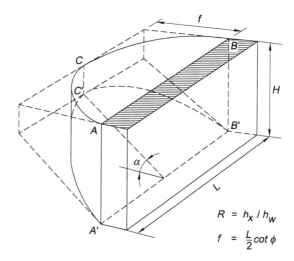

$$R = h_x / h_w$$
$$f = \tfrac{L}{2}\cot\phi$$

Figure 16.8 Failure Wedge (after Piaskowski and Kowalewski, 1965)

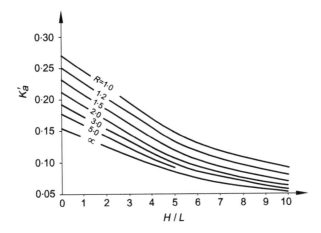

Figure 16.9 Active Earth Pressure Coefficient vs. Trench Size
(adapted from Piaskowski and Kowalewski, 1965)

that of Equation (16.39) can be written as follows:

$$\gamma_f^c \cdot H_f - \gamma_W \cdot H_W = \left[\sum_{i=1}^{m} (K_a \cdot \Delta H \cdot \gamma)_i + \sum_{i=1}^{n} (K_a \cdot \Delta H \cdot \gamma)_i + K_{a(i)} \cdot q_S \right] \quad (16.40)$$

where, $K_{a(i)}$ = coefficient of active earth pressure in the ith layer duly modified if arch action is to be taken into account, and ΔH = thickness of a typical layer. All other terms are as defined previously.

The foregoing equation gives an indication of the minimum unit weight of slurry and the level of slurry in the trench necessary to ensure a factor of safety of at least 1. However, in case the earth pressure at the base does not, for some reason, represent the maximum soil stress, this expression could lead to an underestimation of the value of γ_f^c. It should also be remembered that in the application of this method incorporating the beneficial contribution of the arch action, nomographs for the specific ground conditions need to be plotted to help obtain the appropriate K_a values.

The factor of safety against trench collapse is estimated, by comparing the actual effective slurry pressure with the critical effective slurry pressure, using the following expression:

$$F = \left[\frac{\gamma_f \cdot H_f - \gamma_W \cdot H_W}{\gamma_f^c \cdot H_f - \gamma_W \cdot H_W} \right] \quad (16.41)$$

where, γ_f = unit weight of slurry as actually used in the trench.

Huder approach

Huder (1972) also uses the balanced stress concept albeit with a slightly different approach to the modification of the active earth pressure coefficient. To allow for the

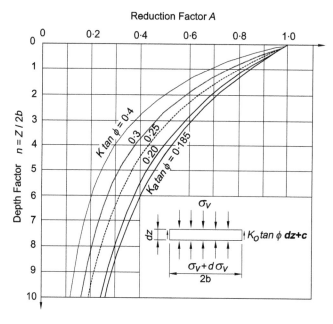

Figure 16.10 Reduction Factor A for Earth Pressure in Trench of Length 2b
(adapted from Huder, 1972)

effect of arching, factor 'A' is introduced by which to reduce the active earth pressure coefficient. In the light of this, the expression for the critical effective slurry pressure modifies to:

$$\left(\gamma_f^c \cdot H_f - \gamma_W \cdot H_W\right) = \left[\gamma\left(H_x - H_W\right) + \gamma_{Sub} \cdot H_W + q_S\right] \cdot A \cdot K_a \qquad (16.42)$$

Value of A is given by : $\quad A = \dfrac{1 - e^{-2nK_a \tan\phi'}}{2nK_a \tan\phi'}$ where, $\quad n = H_x/L \qquad (16.43)$

It can be seen that whereas the reduction factor A tends to approach a nearly limiting value, the horizontal earth pressure continues to increase with depth. The values for the reduction factor for various values of n and $K_a \tan\phi'$, can be directly obtained from Figure 16.10. After obtaining the critical effective slurry pressure from Equation (16.42), factor of safety can be estimated from Equation (16.41) as before.

(C) '$\phi = 0$' soils

Cylindrical cut: general

In a cylindrical cut, hoop stresses act normal to the radial (vertical) planes. Near the face of the cut, the hoop stresses are nearly equal to the vertical stresses, whereas, away from the cut, both hoop and radial stresses approach the earth pressure at rest (Meyerhof, 1972). At failure of the cut, plastic flow of the soil occurs in both horizontal and vertical planes.

Cylindrical cut: shallow

For cylindrical cuts with depth of penetration to diameter ratio less than 12, it has been shown (Hencky, 1923) that, in the radial planes, the plastic zones and the failure surface are similar in shape to, although somewhat smaller in size than, those in the transverse planes of a corresponding long cut. Using the same approach as for long cuts, Meyerhof (1972) has presented an approximate solution to the problem by introducing the coefficient K in place of 2 associated with the undrained shear strength term. The generalized Equation (16.23) can now be expressed as:

$$F = \frac{2 \cdot K \cdot s_u}{\left(\gamma'_e - \gamma'_f\right) \cdot H} \tag{16.44}$$

$$\text{where} \quad K = 2\left[1 + \log_e\left(\frac{2H}{B} + 1\right)\right] \tag{16.45}$$

$$H = \text{Depth of penetration of cut } (= D)$$
$$B = \text{Width (or diameter) of cut}$$

For a cylindrical (or square) cut, the value of the coefficient K may be read directly off the top curve in Figure 16.11.

It is obvious from the curve that the value of K increases at a reducing rate with the increase in the H/B ratio. This implies that, for a given size $(L \times B)$ of cut in a clay soil, although the critical depth H_c increases with the increase in depth, the value of F actually decreases.

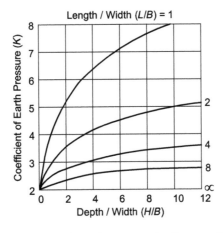

Figure 16.11 Coefficients of Earth Pressure for Rectangular Cuts in Clay
(adapted from Meyerhof, 1972)

Cylindrical cut: deep

For H/B ratio in excess of 12, the earth pressure and stability problem of a circular (or square) cut approaches that of the bearing capacity of a deep (vertical) strip foundation with an overburden pressure equal to earth-at-rest pressure (Meyerhof, 1972). Applying the factor of safety, F, on the undrained shear strength of the soil and ignoring the lateral shearing resistance of the base of the cut and the shearing strength of the slurry, the factor of safety may be expressed as:

$$F = \frac{N \cdot s_u}{\left(K_o \cdot \gamma_e' - \gamma_f'\right) \cdot H} \tag{16.46}$$

where K_o = Coefficient of earth pressure at rest

N = Bearing capacity factor for deep strip foundations

For a rigid-plastic material, Meyerhof (1951) has shown that $N = 2\pi + 2 = 8 \cdot 28$. This represents an upper limit for both the coefficient K and the stability number N for saturated clays.

For an elastic–plastic material, factor N is given (Bishop *et al.*, 1945) by the expression:

$$N = \left[\log_e \frac{E_0}{3s_u} + 1\right] \tag{16.47}$$

where, E_o = Initial tangent modulus of clay

For saturated clays, E_o/s_u ratios typically range from 500 to 2,000; the corresponding N-values range from 6 to 7·5. However, for practical purposes, both the value of N and the maximum value of K are frequently taken as 8. This ignores the lateral shearing resistance of soil at the base of the cut. It has been shown (Meyerhof, 1951) that, with the inclusion of the base resistance, the bearing capacity factor N_c for a rigid-plastic material increases to 9·34. This represents an increase of about 12 per cent in the factor of safety against collapse of a deep cut, circular or square in cross-section. In other words, if the friction at the base of the cut is ignored the critical height is underestimated by about 12 per cent. In view of this, the aforementioned approach is considered to be safe. It should be noted that the percentage difference decreases commensurate with the decrease in the depth of the cut and becomes zero for a shallow cut.

Short rectangular cut

Stability of a short rectangular cut, with its depth of penetration greater than its plan length, is increased by the resistance of the clay on the vertical planes through the end edges of the cut. An approximate but conservative analysis of the stability of a rectangular cut of width B, length L and depth of penetration H, for $H/L > 1$, can be obtained by analogy to the bearing capacity problem (Meyerhof, 1951). Figure 16.12 shows the plan of such a cut.

From the considerations of the earth pressure, the perimeter of the cut may by split up into two parts. The two end portions of width B and the return lengths $B/2$ adjacent to each end are, together, assumed as being the equivalent of a square (or circular)

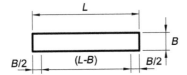

Figure 16.12 Geometry of Short Trench

cut for which the stability factor N increases from a value of 4 for $H/B = 0$ to a maximum value of about:

$$N = 4(1 + B/L) \text{ at great depth} \tag{16.48}$$

The earth pressure on the remaining length $(L-B)$ of the cut is taken as that for a long cut for which the earth pressure coefficient K increases from a value of 2 for $H/B = 0$ to a maximum value of, approximately,

$$K = 2(1 + 3B/L) \text{ at great depth} \tag{16.49}$$

For intermediate depths, the appropriate values of K and N to be used may be obtained by interpolation (Brinch Hansen, 1961). For ease of use, the values, ignoring the base shear, have been plotted graphically in Figures 16.11 and 16.13.

While the coefficient K in Figure 16.11 refers to unit earth pressure, the stability factor N in Figure 16.13 was based on bearing capacity considerations. In the latter case, an interpolation between the upper limit of the zero net thrust criteria (represented by the top curve in Figure 16.13) and the lower limit of the zero net earth pressure criterion based on the top curve of Figure 16.11 (represented by the dashed curve in Figure 16.13) was used (Meyerhof, 1985) for the value of N to avoid local stability failure.

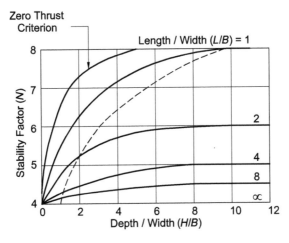

Figure 16.13 Stability Factors for Rectangular Cuts in Clay (adapted from Meyerhof, 1985)

For deep cuts at failure, the net active pressure at depth H is given by:

$$p_a = p_e - p_s - Ks_u \qquad (16.50)$$

The terms p_e and p_s represent the total earth and the slurry pressures at depth H across the interface. With the maximum values for these pressures suffixed appropriately, the factor of safety against failure is given by:

$$F = \frac{N \cdot s_u}{\left(p_{e(\text{max})} - p_{s(\text{max})}\right)} \qquad (16.51)$$

Huder approach

Huder (1972) also uses the bearing capacity approach by comparing the overburden pressures on either side of the soil–slurry interface. In accordance with this approach, by applying a factor of safety on the undrained shear strength, the expression for the factor of safety for a short trench in clay with groundwater and surcharge loading may be expressed as:

$$\gamma \left(H - H_W\right) + \gamma_s \cdot H_W + q_S = N \cdot \frac{s_u}{F} + \gamma_f \cdot H_f$$

$$\text{Or,} \quad F = \frac{N \cdot s_u}{\left[\gamma \left(H - H_W\right) + \gamma_S \cdot H_W - \gamma_f \cdot H_f + q_S\right]} \qquad (16.52)$$

Stability number N is used to allow for the arching effect. Its value, in the case of circular or square cuts (i.e. $L/B = 1$), may be taken as 4 at the surface, i.e. for shallow trenches, and increasing as a function of H/L ratio approaching a maximum of 8 at great depth. For saturated clays, N value typically ranges from 6 to 7·5. All parameters are as defined previously.

16.10 Approach for soft clay

A number of full-scale field tests were carried out in the seventies and eighties on the Norwegian soft clays at a number of different sites in Oslo and subsequently reported on by a number of investigators (DiBiagio and Myrvoll, 1972; Eide *et al.*, 1972; Karlsrud *et al.*, 1980, Karlsrud, 1983). On the basis of experience gained from some of these tests, a semiempirical method of evaluating the stability of slurry trenches in such clays was developed (Aas, 1976). The failure condition assumed is, as also confirmed by the observed displacement patterns in the test trenches, as shown in Figure 16.14. It comprises two distinct blocks – a lower wedge-shaped free body which is assumed to slide down and sideways into the trench along the two 45° inclined planes, and an upper free body which is assumed to move only downwards.

Taking into account its anisotropic nature, the use of the undrained shear strength determined by triaxial (or plane strain) compression tests on the two 45° planes in the lower free body, and the use of *in situ* vane tests (or possibly direct simple shear tests) on the three vertical planes in the free body above, were proposed as being appropriate.

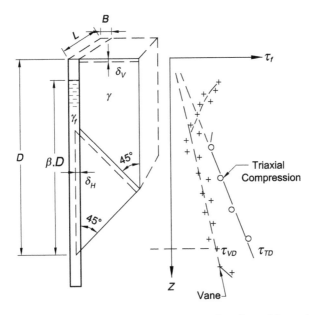

Figure 16.14 Assumed Failure Condition in Trench (adapted from Aas, 1976)

With the help of some simplifying assumptions particularly for the Oslo conditions (Aas, 1976), the following semiempirical formula for the factor of safety for trench stability, allowing for trench geometry and anisotropy of clay, was evolved as given by:

$$F = \frac{\tau_{VD}}{D\left(\gamma - \beta^2 \cdot \gamma_f\right)} \cdot \left[2\frac{\tau_{TD}}{\tau_{VD}} + 0\cdot86\frac{D}{L} + 0\cdot6\right] \qquad (16.53)$$

where D = Depth to lowest point of failure surfaces,
$\quad\quad\quad L$ = Length of trench panel,
$\quad\quad\quad \beta$ = Slurry depth as a fraction of failure depth,
$\quad\quad\quad \gamma$ = Unit weight of clay,
$\quad\quad\quad \gamma_f$ = Unit weight of trench slurry,
$\quad\quad\quad \tau_{TD}$ = Triaxial compression strength at depth D, and
$\quad\quad\quad \tau_{VD}$ = Vane shear strength at depth D.

The foregoing equation as it stands does not include the effect of external surcharge load. However, this can be easily allowed for by replacing the 'γ' term in the denominator by '$\gamma + q/D$'.

It is not surprising to note that, in all the cases referred to above, lateral earth pressures can be seen to increase with time after the construction was completed suggesting continued progress towards the original at-rest conditions (Figure 16.15).

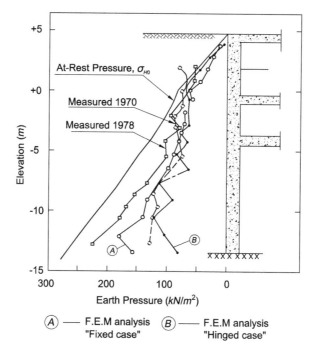

Figure 16.15 Comparison between Measured and Computed Earth Pressures at end of Excavation, Kongensgt (adapted from Karlsrud, 1983)

16.11 Displacement of trench faces

Trench faces can be exposed to two types of lateral deformations – inward movement during excavation and outward movement during the replacement of slurry by wet concrete. The former, depending upon its magnitude, can have an impact on the stiffness and strength of the wall and the latter on the extent of the breakout of the concrete that may become necessary when the face is exposed. Some idea of the extent of both of these can be useful. An indication of the initial, short-term deformation in the case of the former phenomenon can be obtained using the principles of elastic theory whereas that of the latter, by treating the problem as one of trench stability from the perspective of bearing capacity.

Inward displacement

During trench excavation the stress history around the cut is lost and the disturbance causes a change in the ambient *in situ* stresses at its faces. As the soil face undergoes strain inwards into the cut, there is a drop in the earth pressure from its at-rest value. Since the presence of the slurry in the trench is bound to prevent full and free lateral strain, the earth pressure at the face is unlikely to attain its active failure value. While the precise magnitude is not easy to quantify, depending upon the extent of the strain undergone, the earth pressure will lie somewhere between the at-rest and the active value. What is beyond doubt, however, is that the stresses will no longer be those at rest for quite some time to come.

From the considerations of design, it is useful to have some feel for the inward displacement of the faces not only for the estimation of the design pressures to be anticipated but also for the extent of its potential impact on wall thickness. However, achieving these objectives in anisotropic, heterogeneous soil is rather difficult. Of necessity, therefore, and as an approximation, equivalent effective unit weights are used with the half-space assumed to be fully saturated (Poisson's ratio, $\mu = 0\cdot 5$), elastic, homogeneous, isotropic, and under geostatic stresses. Under these idealized conditions, the inward displacement at the centre of the long face of a deep, rectangular trench panel, for a generalized case, is given (Xanthakos, 1979) by:

$$\Delta = 0\cdot 75 \left(K_o \cdot \gamma'_e - \gamma'_f \right) \cdot \frac{zL}{E_o} \tag{16.54}$$

where, γ'_e and γ'_f are the equivalent effective unit weights for the soil and slurry, respectively, as defined previously in Equations (16.3) and (16.16), z is the depth under consideration, L is the panel length and E_o the initial tangent modulus for the soil.

Outward displacement

This is of particular importance in soft clays. The question as to whether the soft clay medium can generate enough resistance and mobilize sufficient strength to withstand the pressure from wet concrete and prevent its intrusion into the soil can be analysed (Xanthakos, 1979) by its analogy to the bearing capacity problem. It is possible to prevent such intrusion as long as

$$\gamma \cdot H + N_C \cdot s_u + q_s (= 0) \geq \gamma_c \cdot H \cdot K_C. \tag{16.55}$$

In the equation, γ_c is the unit weight of wet concrete, K_C its pressure coefficient $(= 0\cdot 8)$ and the value of the N_C factor varies from 4 at the surface and approaching a maximum of 8 at depth; other terms are as defined previously. In order to maximize the driving effect of concrete, the contribution of the surcharge load, which would oppose this effect, should be disregarded.

Illustrative examples amplifying the design principles discussed in this chapter are presented at the end of the next chapter.

17 Slurry Trench Stability: II

17.1 Introduction

Principles of limit equilibrium and trench stability have been discussed in the previous chapter. In this chapter, some recent advances made in the analytical approach to the trench stability problems in different ground conditions are presented.

Examples illustrating the use of limit equilibrium principles, arching, etc., have been worked out at the end of this chapter. Solutions using different approaches have been examined and, based on the comparative studies carried out, useful conclusions drawn.

17.2 Recent analytical advances

During the past decade or so, interesting and useful experimental works have been carried out in relation to the slurry trench stability in different ground conditions including soft clays (Tamano *et al.*, 1996). A three-dimensional approach based on limit equilibrium theory (Tsai and Chang, 1996) and using elasto-plastic finite element method (Oblozinsky *et al.*, 2001) for the stability of trenches in cohesionless soils has been proposed by some investigators. The stability of layered soils, in particular, against the lateral extrusion of a weak cohesive layer sandwiched between stronger layers (Tsai, 1997) is also examined. The principles and the procedures for examining the stability are presented hereunder.

Cohesionless soils: three-dimensional approach

Tsai and Chang (1996) have proposed a three-dimensional analysis for slurry trenches in cohesionless soils based on limit equilibrium theory, incorporating the effects of both horizontal and vertical arching. The analytical approach is developed by considering the stability of a shell-shaped sliding wedge of soil subject to potential movement within a fictitious half-silo surrounded by a rough wall (Figure 17.1). Identification of the slip surface is based on two major assumptions:

- The possible horizontal movement of soil mass induced by the net lateral earth pressure is confined within a horizontal compression arch, i.e. a half-silo, supported at the ends of the trench.
- The possible vertical movement of the soil mass within the half-silo is equivalent to the downward movement of a vertical extension arch, i.e. a catenary, hung from the silo wall.

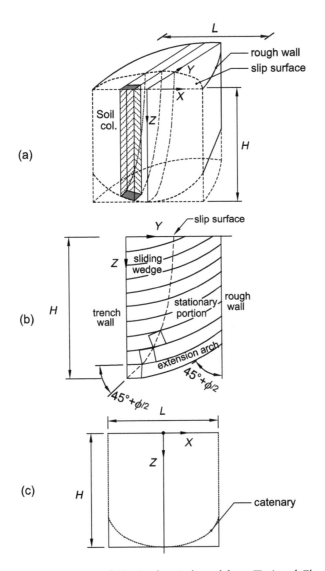

Figure 17.1 Determination of Slip Surface (adapted from Tsai and Chang, 1996)

The sliding wedge is divided into a number of vertical soil columns transverse to the trench. The slip surface is defined using the Mohr-Coulomb criterion on the catenaries. The force equilibrium on the slip surface is checked by summing up the external forces on each discrete soil column. The factor of safety is defined as the ratio of the resisting force generated by the slurry pressure and the lateral driving force required to maintain the stability of the sliding soil wedge, i.e.

$$F = P_f/P_H \tag{17.1}$$

The effective vertical pressure of a typical column of soil for force equilibrium in the vertical direction is obtained from Huder's formula (1972) based on Terzaghi's arch theory (1943) for granular soils, i.e.

$$\sigma'_V = \frac{L \cdot \gamma'}{2K \cdot \tan\phi} \left[1 - e^{-2K \cdot n \cdot \tan\phi} \right] + q \cdot e^{-2K \cdot n \cdot \tan\phi} \quad \text{for } (z < 2 \cdot 5L) \tag{17.2}$$

$$= \frac{L \cdot \gamma'}{2K \cdot \tan\phi} \left[1 - e^{-2K \cdot n \cdot \tan\phi} \right] + (\gamma' \cdot z' + q) e^{-2K \cdot n \cdot \tan\phi} \quad \text{for } (z \geq 2 \cdot 5L) \tag{17.3}$$

where, $L =$ Length of trench panel

$z =$ Depth to the base of soil column

$z' =$ Effective depth for deep soil column

$= z - 2 \cdot 5L$

$n = z / L$

$\phi =$ Undrained friction angle

$\gamma' =$ Effective unit weight of soil

$q =$ Surcharge load on ground surface

Coefficient $K = \dfrac{1 + \sin^2\phi}{\cos^2\phi + 4\tan^2\phi}$ as suggested by Jacobson (1958).

In solving the force equilibrium, mobilized friction angle ϕ_m is required. This is defined by the expression:

$$\phi_m = \tan^{-1}\left(\frac{\tan\phi'}{F} \right). \tag{17.4}$$

The factor of safety is obtained by iteration when the F-values in Equations (17.1) and (17.4) converge.

c′, φ′ soils: elasto-plastic FEM approach

Oblozinsky *et al.* (2001) have proposed a two-stage approach for examining the stability of a slurry trench panel based on the elasto-plastic finite-element method (*FEM*). The first stage is concerned with the evaluation of the trench stability in terms of the factor of safety and the second suggests a method of choosing the key preliminary slurry parameters, i.e. the unit weight of slurry and its height in the trench.

The *FEM* analysis, in the first stage, makes use of the shear strength reduction (*SSR*) technique. The soil strength parameters used are defined as follows:

$$c_F = \frac{c'}{F}; \quad \phi_F = \tan^{-1}\left(\frac{\tan\phi'}{F} \right) \tag{17.5}$$

In the above equation, c' and $\tan\phi'$ are the usual effective stress shear strength parameters and F is the parameter which reduces these. Start is made with a very small initial value of F equal to, say, 0·01, for use in the Newton-Raphson iterations.

This, naturally, exaggerates the soil strength parameters so as to ensure that the domain remains in the elastic range. The parameter F is then incrementally raised by similar amounts until failure occurs. At this point, the value of F represents the global minimal factor of safety, which is the equivalent of the safety factor defined in the limit equilibrium method. The convergence criterion in the iterations is assumed to be satisfied if the displacement increment between two successive steps, such as F_i and F_{i+1}, divided by the total displacement is less than 10^{-5} within 1000 iterations. If this is not achieved, it is assumed that the solution diverges and the system has collapsed.

In the second stage, the unit weight of the slurry or its height, or both, are found from an active earth pressure *FEM* analysis. The face of the trench is subjected to a prescribed uniform displacement towards the cut at a constant rate, reducing the soil stress against the trench from its initial K_o state to its final active state. Stability of the trench requires the stabilizing pressure from the slurry to be at least equal to or higher than the driving active pressure from soil. The effective stabilizing pressure is given by the expression:

$$\bar{p}_f = H \cdot \bar{\gamma}_f \tag{17.6}$$

where, H = depth under consideration, $\bar{\gamma}_f$ = effective unit weight of slurry, $= \gamma_f - \gamma_w$ as appropriate.

The interesting results of the *FEM* study carried out in sandy ground on 3*m*-, 6*m*- and 9*m*-long panels, 1*m* thick and 15*m* deep are shown in Figures 17.2–17.6 inclusive. The results have been verified by comparison with the available results of the centrifuge experiments reported by Katagiri *et al.* (1997, 1998). Water table is assumed to be at ground level and the trench completely filled with slurry. The soil parameters used in the study throughout the *FEM* analysis were determined from the triaxial compression test conducted on the sand sample which was used for the centrifuge tests. The parameters are as follows:

$$\gamma_{Sub} = 8 \cdot 7 kN/m^3, c' = 0, \gamma_f = 10 \cdot 5 kN/m^3, \phi' = 39^\circ, E = 2 \times 10^4 kN/m^2,$$

$$\psi' = 39^\circ, \nu = 0 \cdot 3, \text{ where } \psi' \text{ is the Dilatancy Angle}$$

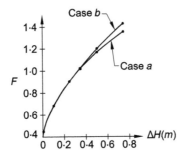

Figure 17.2 F vs. ΔH (adapted from Oblozinsky *et al.*, 2001)

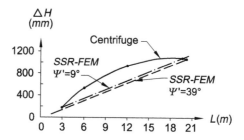

Figure 17.3 Centrifuge vs. *SSR-FEM* Results (adapted from Oblozinsky *et al.*, 2001)

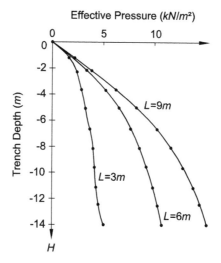

Figure 17.4 Active Earth Pressure Profiles (adapted from Oblozinsky *et al.*, 2001)

The value for the angle of dilatancy, ψ', was assumed to be equal to ϕ'. However, since this value appeared to be too high, ψ' of $9°$ was also used in order to investigate the sensitivity of the *SSR-FEM* analysis against this parameter.

For simplicity, soil beyond the ends of the trench was ignored and only that behind the face was considered. However, a set of analyses was carried out to investigate the extent of the likely error caused by such simplification. Accordingly, Case (*a*) assuming rigid boundaries at the ends of the trench, and Case (*b*), with soil extending beyond the ends of the trench were investigated. As indicated in Figure 17.2, this simplification proved to have only a marginal influence on the values of safety factors obtained using the *SSR-FEM* analysis. Besides, a reasonably good agreement was achieved between the centrifuge experiments and the *FEM* analysis.

The *SSR-FEM* analysis using the Mohr-Coulomb soil model does not appear to be sensitive (<2 per cent difference) to variation in the angle of dilatancy as shown in Figure 17.3. Also, as is to be expected, with the increase in the panel length, increase in the slurry head is required to maintain stability. It is particularly interesting to find that, unlike that obtained from the centrifuge test, the relationship in the case of the 3D-FEM analysis is almost linear.

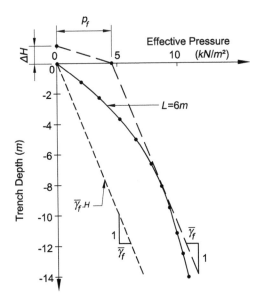

Figure 17.5 Procedure for Estimating ΔH (adapted from Oblozinsky *et al.*, 2001)

The graphs shown in Figure 17.4 represent the active earth pressure profiles (without the hydrostatic pressures) obtained by performing the *FEM* analyses for trench lengths of 3, 6 and 9*m*. The effect of increase in soil pressures with the increase in panel length is obvious. However, the pressures increase on a reducing rate. This suggests that the influence of the three-dimensional effect is likely to be felt only up to a certain limit beyond which the two-dimensional plane strain conditions can be expected to prevail.

Figure 17.5 illustrates, for the 6*m*-long panel as an example, the procedure for estimating the height of slurry (ΔH) above ground level corresponding to the unit weight of the slurry in the trench or, for a specified head of slurry above ground, the required unit weight of the slurry in the trench. The dashed line represents the effective slurry pressure ($H \cdot \bar{\gamma}_f$) profile and its slope, the parameter $\bar{\gamma}_f$. If a line is drawn tangential to the *FEM* active pressure profile in such a manner that it is also parallel to the slurry pressure line, then the point where it meets the horizontal coordinate axis at ground surface represents the slurry pressure (p_f) required at the ground level. With the value of the required slurry pressure thus obtained, the height of the slurry above ground (ΔH), corresponding to the unit weight of slurry (γ_f) in the trench, can be simply obtained from the expression: $p_f = \gamma_f \cdot \Delta H$.

If, owing to a site-specific constraint, ΔH cannot exceed a certain limit, it may become necessary to modify the unit weight of the slurry to be used in the trench to ensure conformance. Then, for the stipulated value of ΔH, the modified unit weight may be calculated from the expression:

$$\gamma_f = \left[\frac{p'_{a(z)} + \gamma_W \cdot z}{z + \Delta H} \right] \tag{17.7}$$

In the expression, $p'_{a(z)}$ represents the active earth pressure at depth z below the surface as per the relevant *FEM* active pressure profile.

There are two unknowns, γ_f and z, in the equation. Obtaining the value for γ_f, therefore, becomes a function of the choice of the parameter z, i.e. the location of the tangent point T on the *FEM* active pressure profile. A number of different values of z and, therefore, also of the parameter γ_f are possible. For example, for $(p'_{a(z)}/z)$ in the range of 0·4–2·0 and $(\Delta H/z)$ in the range of 0·002–0·022, γ_f would fall in the range of 10·2–12·0. In selecting the tangent point, therefore, care should be exercised to ensure that the slope of the tangent line is sensible.

Table 17.1 Comparison of ΔH and F

Predictions made by	$L = 3m$		$L = 6m$		$L = 9m$	
	ΔH	F	ΔH	F	ΔH	F
SSR-FEM	155	1·00	325	1.00	500	1·00
Active Pressure FEM	142	0·98	381	1·06	770	1.10

(adapted from Oblozinsky *et al.*, 2001)

Table 17.1 compares the slurry head ΔH (*mm*) above ground and the factors of safety as predicted by the *SSR-FEM* analysis with those predicted by the active earth pressure *FEM* analysis conducted on *3m*-, *6m*- and *9m*-long trench panels using γ_f of 10·5kN/m^3. The factors of safety (F) in the second row, in the range of 0·98–1·10 have been evaluated by the *SSR-FEM* for the ΔH obtained from the active pressure *FEM* analysis. It is obvious that if this method of analysis is chosen for determining the preliminary parameters (γ_f and ΔH) for the slurry, factor of safety would be expected to be around 1·00. The slurry parameters would then need to be adjusted in accordance with the required minimum factor of safety.

Figure 17.6 shows the effect of the rigidity of guide walls in improving the overall stability of the trench. Although it appears to be marginal, it becomes relatively more pronounced as the panel length becomes shorter, especially when the slurry level is close to the ground surface. This may be taken into account by the designers where appropriate.

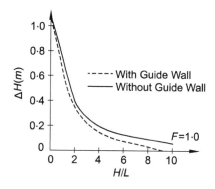

Figure 17.6 Influence of Guide Wall (adapted from Oblozinsky *et al.*, 2001)

The procedure for the design of a stable slurry trench as suggested by the authors can be summed up as follows:

- Perform active earth pressure *FEM* analysis on the idealized mesh and obtain the 'preliminary' slurry parameters, i.e. slurry height and unit weight.
- Based on the 'preliminary' information, choose the preferred design slurry parameters and, using the same mesh as before, perform the *SSR-FEM* analysis for stability. A factor of safety of 1·2 is recommended for design.

The study discussed in the foregoing section has been carried out assuming the water table to be at ground level and the trench completely filled with slurry. However, the procedure can be extended to trenches with water table below ground level and the slurry filled up to the water table level. This can be achieved by assuming the depth of the trench to be, effectively, the same as its depth below the water table and treating the effect of the soil above this level as the surcharge at the water table level. The rest of the procedure then is the same as before with the exception that the surcharge load will also be reflected, as appropriate, in the computations of the active pressure profile.

Soft cohesive soil

Tamano *et al.* (1996) studied the trench wall stability in soft ground. The experimental trench was $1m$ wide, $9·5m$ long and $21m$ deep. The $12m$ deep soft alluvial clay layer, located $9·8m$ below ground level comprised normally consolidated clay with a natural water content of 40–50 per cent, liquid limit of 50–70 per cent, plastic limit of 20–27 per cent and undrained shear strength ranging from 29 to $73kN/m^2$. In order to exclude the influence on the trench wall integrity of the overlying sand and the reclaimed layers during the slurry level lowering test, the trench was excavated inside a perimeter cofferdam formed by $0·46m$ diameter soil-cement piles installed to a depth of $11m$ below ground level. The unit weight of the slurry used was $10·2kN/m^3$ and its level was maintained at $300mm$ below the ground level.

On the basis of the readings available from the inclinometers located $1m$ and $5m$ away from the test panel, the authors estimated a maximum inward displacement of the trench face of about $20mm$. It was observed that the extent of the movement undergone was not significant enough to cause collapse. However, it was sufficient enough to reduce the lateral earth pressure to such an extent as to achieve stability by balancing out the lateral pressures across the slurry–soil interface. It was also interesting to note that by the time the trench face suffered the maximum inward displacement and achieved stability the earth pressure had dropped down to its active value.

Soft cohesive soil: weak sub-layers

Tsai (1997) proposed an analytical method for examining the stability against lateral extrusion of weak sublayers in slurry supported trench panels. A weak sublayer is defined as a soft cohesive layer with strength much lower than those of the sandwiching layers above and below. The proposed method makes use of two theories – one for finding the bearing capacity of the weak sublayer, and the other, for determining the driving lateral pressure. In the first theory, the bearing capacity is obtained using a theoretical model of a weightless soft sublayer compressed between rough parallel plates in an undrained mode. The loss of weight of the weak layer assumed is subsequently compensated for by including it in the second theory in calculating the distribution of the vertical effective stress that mobilizes the driving lateral pressure. The effect of arching induced by the finite length of the panel is also allowed for.

The effective vertical stress in the stronger sandwiching layers is obtained, once again, from the following modified Huder formula based on Terzaghi's arch theory:

$$\sigma_V' = \frac{L\left(\gamma' - \frac{2c}{L}\right)}{2K \cdot \tan\phi}\left[1 - e^{-2K \cdot n \cdot \tan\phi}\right] + q \cdot e^{-2K \cdot n \cdot \tan\phi} \tag{17.8}$$

where, z = depth to the top boundary of weak sublayer, and c, ϕ = undrained shear strength parameters. Other parameters are as defined previously.

Equation (17.8) can also be applied to stratified soils. In such soils, each layer is treated discretely; the effective vertical stress obtained for an upper layer is treated as the surface surcharge in the computation for the effective vertical stress in the layer immediately below it.

In the case of the weak sublayers, the effective vertical stress is computed from the expression:

$$\sigma_V' = z\left(\gamma' - \frac{2s_u}{L}\right) + q \tag{17.9}$$

The expression on the right hand side of the above equation excluding the term 'q' represents the variation in the effective vertical stress locally within the weak sublayer, where s_u = undrained shear strength and γ' = effective unit weight of the weak sub-layer.

The factor of stability against the lateral extrusion of the weak sublayer is defined as the ratio between the net applied horizontal slurry pressure and the horizontal driving pressure due to the vertical effective stress at the top boundary of the trench end of the sublayer and is given by the expression:

$$F = \frac{P_f - P_W}{K_h\left(\sigma_V' - 1 \cdot 57 s_u\right) + 0 \cdot 43 s_u} \quad \text{for } \sigma_V' > 1 \cdot 57 s_u \tag{17.10}$$

$$\text{and} \quad F = \frac{P_f - P_W}{0 \cdot 43 s_u} \quad \text{for } \sigma_V' \leq 1 \cdot 57 s_u \tag{17.11}$$

The earth pressure coefficient K_h is taken, conservatively, as equal to K_o.

$$K_o = \text{coefficient of earth pressure at rest for the sublayer}$$
$$= 1 - \sin\phi' \text{ (For normally consolidated clays)}$$
$$\sin\phi' = 0 \cdot 82 - 0 \cdot 24\log(PI) \text{ (Kenney, 1959)}$$
$$PI = \text{Plasticity index}$$
$$1 \cdot 57 s_u = \text{Bearing capacity of sub-layer at trench end}$$

Other terms are as defined previously.

Field experiment was carried out on two slurry trench panels *A* and *B* of 2·5 and 6·5*m* lengths, respectively, each penetrating 26*m* into the ground. Table 17.2 shows the simplified soil profile of the test site and lists the soil parameters used.

The groundwater level was at 1*m* below the ground level, and the slurry head was maintained at 300*mm* above the surface. The weak soils comprised moderate plastic clay (*CL*) with undrained shear strength ranging from 11 to 28*kN/m²*.

With $c = 0$, the expression for the effective vertical stress in Equation (17.8) reduces to that in Equation (17.2), i.e.

$$\sigma_V' = \frac{L \cdot \gamma'}{2K \tan \phi} \left[1 - e^{-2Kn \tan \phi} \right] + q \cdot e^{-2Kn \tan \phi}$$

The values of ϕ are as listed in Table 17.2.

Using this equation for the strong layers and Equation (17.9) for the week sub-layers, effective vertical stress in each layer can be worked out by treating each layer discretely. The effective vertical stress at the bottom of the layer above is treated as the new surcharge load on the layer immediately below it. This can be clearly seen by comparing the last two columns of the tables. The results for the two panel lengths are given in Tables 17.3 and 17.4.

Table 17.2 Simplified Soil Profile and Parameters

Layer	Depth (m)	USCS*	γ_d (kN/m³)	γ_{sat} (kN/m³)	ϕ (°)	PI	s_u (kN/m²)
1	0·0~3·0	SM/FILL	18·3	19.0	30	–	–
2	3·0~4·2	CL	–	19·2	–	32	11
3	4·2~7·5	SM	–	19·9	33	–	–
4	7·5~8·5	CL	–	19·4	–	25	21
5	8·5~10·6	SM	–	19·8	35	–	–
6	10·6~11·8	CL	–	19·2	–	16	28
7	11·8~	SM	–	20·1	37	–	–

USCS*: *Unified Soil Classification System;*
ϕ: *Angle of internal friction from consolidated, undrained triaxial tests.*
(after Tsai, 1997)

Table 17.3 Effective Stresses at Different Layers

			Panel Length, L = 2·5m				
Layer	z (m)	γ' (kN/m³)	ϕ (°)	K	s_u (kN/m²)	q (kN/m²)	σ_V' (kN/m²)
1a	1·0	18·3	30	0·60	–	15	27·35
1b	2·0	9·0	30	0·60	–	27·35	29·53
2	1·2	9·2	–	–	11	29·53	30·01
3	3·3	9·9	33	0·54	–	30·01	33·20
4	1·0	9·4	–	–	21	33·20	25·80
5	2·1	9·8	35	0·50	–	25·80	29·89
6	1·2	9·2	–	–	28	29·89	14·05

(adapted from Tsai, 1997)

Table 17.4 Effective Stresses at Different Layers

			Panel Length, $L = 6 \cdot 5m$				
Layer	z (m)	γ' (kN/m^3)	ϕ (°)	K	s_u (kN/m^2)	q (kN/m^2)	σ_V' (kN/m^2)
1a	1·0	18·3	30	0·60	–	15	30·84
1b	2·0	9·0	30	0·60	–	30·84	41·13
2	1·2	9·2	–	–	11	41·13	48·11
3	3·3	9·9	33	0·54	–	48·11	61·18
4	1·0	9·4	–	–	21	61·18	64·12
5	2·1	9·8	35	0·50	–	64·12	69·56
6	1·2	9·2	–	–	28	69·56	70·26

(adapted from Tsai, 1997)

Table 17.5 Factors of Stability of Sub-Layers

L(m)	Layer	$K_h = K_o$	P_f (kN/m^2)	P_W (kN/m^2)	Ave.σ_V' (kN/m^2)	F	Remarks
	2	0·54	347	20·0	29·8	1·28	Stable
2·5	4	0·52	81·9	65·0	29·5	1·87	Stable
	6	0·47	114·5	96·0	22·0	1·54	Stable
	2	0·54	34·7	20·0	44·6	0·75	Fall-off
6·5	4	0·52	81·9	65·0	62·6	0·69	Fall-off
	6	0·47	114·5	96·0	69·9	0·76	Fall-off

(adapted from Tsai, 1997)

In Tables 17.3 and 17.4, z is not unique but case-specific and is taken as the depth of a discrete layer under consideration.

Average of the values at the top and bottom of the weak sub-layers may be taken to represent the effective vertical stresses in these layers. For the 2·5m panel, the average σ_V' values in the various weak sublayers are as follows:

$$\text{In sub-layer 2, } (29 \cdot 53 + 30 \cdot 01)/2 = 29 \cdot 77 kN/m^2$$

$$\text{In sub-layer 4, } (33 \cdot 20 + 25 \cdot 80)/2 = 29 \cdot 50 kN/m^2$$

$$\text{In sub-layer 6, } (29 \cdot 89 + 14 \cdot 05)/2 = 21 \cdot 97 kN/m^2$$

The corresponding values for the 6·5m panel from Table 17.4 above work out to: 44·62, 62·65 and 69·91kN/m^2, respectively. Factors of stability for the weak sublayers as evaluated are listed in Table 17.5.

Figure 17.7 shows the effective vertical stress profiles over the depth of the panels clearly highlighting the beneficial effect of arching due to the finite panel lengths.

Ultrasonic investigations revealed no evidence of lateral extrusion of weak sublayers in panel A. However, fall-offs down the trench face were detected in the longer panel B, as indicated in Figure 17.8. The analytical results, as listed in Table 17.5, obtained using the proposed approach appear to be in good agreement with the findings of the ultrasonic investigations in the field.

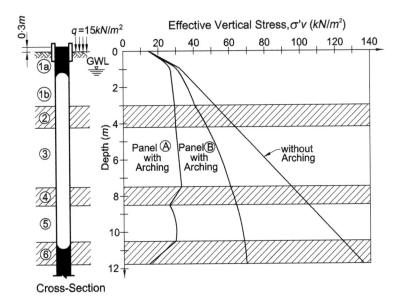

Figure 17.7 Effective Vertical Stress Profiles (adapted from Tsai, 1997)

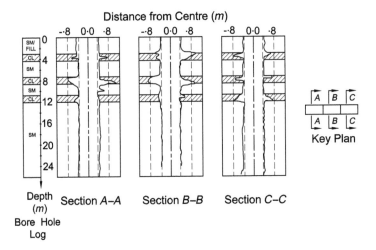

Figure 17.8 Ultrasonic Profiles (Panel B) (adapted from Tsai, 1997)

Example 17.1

Given data: $\gamma = 19{\cdot}6 kN/m^3$ (dry soil)

$\gamma_S = 20 kN/m^3$ (Saturated soil)

$\gamma_f = 11{\cdot}2 kN/m^3$ (Slurry fluid)

Analyse the stability of a long, 1·2m-wide and 30m-deep trench filled with slurry up to within 1m of ground level and 1m above the ambient water table, exposed to a surcharge load of $5kN/m^2$, in the three ground conditions with the following parameters:

- $c' = 2kN/m^2, \phi' = 32°$
- $c' = 0, \phi' = 35°$
- $s_u = 50kN/m^2$
- $K_o = 0·5$.

A minimum factor of safety of 1·3 is required.

Solution

From trench geometry, $H = 30m, H_f = 29m$ and $H_W = 28m$. Also, $q_S = 5kN/m^2$.

(A) c', ϕ' *Soil*

Recalling relevant equations from Chapter 16,

$$F = \frac{4c'}{H}\left[\frac{\sqrt{N_\alpha}}{\gamma_e' - \gamma_f' \cdot N_\alpha}\right]$$

$$\gamma_e' = \left[\gamma - (\gamma - \gamma_{Sub})\left(\frac{H_W}{H}\right)^2 + 2\left(\frac{q_S}{H}\right)\right]$$

$$= \left[19·6 - (19·6 - 10)\left(\frac{28}{30}\right)^2 + 2\left(\frac{5}{30}\right)\right] = 11·57$$

$$\gamma_f' = \gamma_f \cdot \left(\frac{H_f}{H}\right)^2 - \gamma_W \cdot \left(\frac{H_W}{H}\right)^2$$

$$= 11·2\left(\frac{29}{30}\right)^2 - 10\left(\frac{28}{30}\right)^2 = 1·75$$

On substitution, we get: $F = \dfrac{\sqrt{N_\alpha}}{43·39 - 6·56N_\alpha}$

where, $N_\alpha = \tan^2\left[45° + \dfrac{1}{2} \cdot \tan^{-1}\left(\dfrac{\tan\phi'}{F}\right)\right]$

Trial and error solution of the preceding two equations yields a value of $F = 0·61$. This is not adequate. For the desired value of $F = 1·3$, the required value of N_α works out to 2·529. On substituting this in the factor of safety equation, we get

$$1·3 = \frac{8}{30}\left[\frac{\sqrt{2·529}}{11·57 - \gamma_f' \times 2·529}\right]$$

So, required $\gamma'_f = 4\cdot46$ which is, in turn,

$$= 11\cdot2 \left(\frac{H_f}{30}\right)^2 - 10 \left(\frac{28}{30}\right)^2$$

whence, the required $H_f = 32\cdot6m$. This would necessitate extending the guide trench by about $3m$ above ground level, which is not feasible. If the gel strength of, say, $\tau = 1kN/m^2$ is included, then recalling Equation (16.29), Chapter 16, and setting $\gamma'_f = 4\cdot46$, we get

$$\gamma'_f = 4\cdot46 = \left[11\cdot2 + \frac{1}{0\cdot6} + \pi \left(\frac{1}{H_f}\right)\right] \left(\frac{H_f}{30}\right)^2 - 10\cdot\left(\frac{28}{30}\right)^2$$

whence, the required $H_f = 30\cdot22m$. This is more likely to be acceptable.

However, it is possible to opt for this measure to achieve the desired factor of safety only if the following two questions can be answered in the affirmative:

- In achieving the gel strength of $1kN/m^2$, is the flowability of the slurry likely to be maintained unaffected?
- Given the thixotropic characteristic of the gel, is it possible to maintain the state of gel that is capable of providing the requisite strength during the disturbance that may be occasioned by the insertion of the rebar cage and pumping of concrete?

If the answer to either of the above questions is in the negative, then other measures of improving the factor of safety must be considered. In any case, in the face of even the least uncertainty, it is advisable to tread carefully. In computing stability, it may not be prudent to rely on the gel strength of the slurry in the trench.

(B) $c' = 0, \phi'$ soil

For this soil, $\gamma'_e = 11\cdot57$ and $\gamma'_f = 1\cdot75$ as before. By substituting appropriate values into Equation (16.21), Chapter 16, we get

$$F = 2\cdot\tan 35° \cdot \left[\frac{\sqrt{11\cdot57 \times 1\cdot75}}{(11\cdot57 - 1\cdot75)}\right] = 0\cdot64 \text{ This is unsafe!}$$

For the desired value of $F = 1\cdot3$, the required value of γ'_f can be obtained from the equation:

$$1\cdot3 = 2\cdot\tan 35° \cdot \left[\frac{\sqrt{11\cdot57 \times \gamma'_f}}{\left(11\cdot57 - \gamma'_f\right)}\right]$$

whence, $\gamma'_f = 4\cdot12$ which is, in turn $= 11\cdot2 \left(\frac{H_f}{30}\right)^2 - 10 \left(\frac{28}{30}\right)^2$

whence, the required $H_f = 32 \cdot 17m$. This would necessitate extending the guide trench by over $2m$ above ground level. Again, this is not feasible. If the gel strength of, say, $\tau = 1kN/m^2$ were to be included, then recalling Equation (16.29), Chapter 16, and setting $\gamma_f' = 4 \cdot 12$, we get

$$\gamma_f' = 4 \cdot 12 = \left[11 \cdot 2 + \frac{1}{0 \cdot 6} + \pi \left(\frac{1}{H_f} \right) \right] \left(\frac{H_f}{30} \right)^2 - 10 \cdot \left(\frac{28}{30} \right)^2$$

whence required $H_f = 29 \cdot 83m$.

Balanced stress approach

Using Equation (16.39), Chapter 16, we obtain

$$\gamma_f^c \cdot H_f - \gamma_W \cdot H_W = [19 \cdot 6 \times 2 + 10 \times 28 + 5] \tan^2 \left(45° - 35°/2 \right) = 87 \cdot 85 kN/m^2$$

$$\gamma_f \cdot H_f - \gamma_W \cdot H_W = 11 \cdot 2 \times 29 - 10 \times 28 = 44 \cdot 8 kN/m^2$$

Factor of safety, $F = 44 \cdot 8/87 \cdot 85 = 0 \cdot 51$ – a failure situation!

(C) '$\phi = 0$' soil

By substituting appropriate values into the Equation (16.23), Chapter 16, we obtain: $F = \dfrac{4 \times 50}{30} \cdot \left[\dfrac{1}{(11 \cdot 57 - 1 \cdot 75)} \right] = 0 \cdot 68$ – a failure situation!

For the desired value of $F = 1 \cdot 3$, the required value of γ_f' can be obtained from the equation:

$$1 \cdot 3 = \frac{4 \times 50}{\left(11 \cdot 57 - \gamma_f' \right) \times 30}$$

whence $\gamma_f' = 6 \cdot 44$ which is, as before $= 11 \cdot 2 \left(\dfrac{H_f}{30} \right)^2 - 10 \left(\dfrac{28}{30} \right)^2$

From the above, the required $H_f = 35 \cdot 76m$. Again, this would necessitate extending the guide trench by nearly $6m$ above ground level, which is not feasible at all. If the gel strength of, say, $\tau = 1kN/m^2$ were to be included, then recalling Equation (16.29), Chapter 16, and setting $\gamma_f' = 6 \cdot 44$, we get

$$\gamma_f' = 6 \cdot 44 = \left[11 \cdot 2 + \frac{1}{0 \cdot 6} + \pi \left(\frac{1}{H_f} \right) \right] \left(\frac{H_f}{30} \right)^2 - 10 \cdot \left(\frac{28}{30} \right)^2$$

$\therefore$ Required $H_f = 32 \cdot 42m$. Once again, this is not acceptable.

Conclusions

- In each of the three cases considered, the available factor of safety is less than 1 implying failure situation.
- Also, the extent of the increase in the slurry head needed to achieve the required factor of safety in each case is not feasible.

Example 17.2

Examine the improvement in the factors of safety of all the above three cases if the trench is *5m* long and arch action is taken into account. This time assume full height of slurry in the trench, i.e. $H_f = 30m$ and ignore the surcharge load on the surface.

Solution

(A) c', ϕ' soil

Schneebeli approach

Recalling Equation (16.30), Chapter 16,

$$\sigma_z = \frac{\gamma \cdot L}{N_\phi \cdot \sin 2\phi'} \left[1 - e^{-n \cdot \sin 2\phi'} \right] - \frac{c'}{\tan \phi'} \left[\frac{N_\phi - 1}{N_\phi} \right]$$

$$N_\phi = \frac{(1 + \sin 32°)}{(1 - \sin 32°)} = 3.25$$

$$n = 2/5 = 0.4 \text{ at water table level}$$

$$= 30/5 = 6 \text{ at base of cut}$$

Net fluid pressure:

(i) at water table level $= 11.2 \times 2 = 22.4 kN/m^2$
(ii) at base of cut $= 22.4 + (11.2 - 10) \times 28 = 56 kN/m^2$

Earth pressure without arching:

(i) at ground level $= -2 \times 2 = -4 kN/m^2$
(ii) at water table level $= 19.6 \times 2 \times 0.307 - 2 \times 2 = 8 kN/m^2$
(iii) at base of cut $= 8 + (20 - 10) \times 28 \times 0.307 = 93.96 kN/m^2$

Earth pressure with arching: (Using dry unit weight of soil)

(i) at ground level $= -\dfrac{2}{\tan 32°} \left[\dfrac{3.25 - 1}{3.25} \right] = -2.2 kN/m^2$

(ii) at water table level $= \dfrac{19.6 \times 5}{3.25 \sin 64°} \left[1 - e^{-0.4 \sin 64°} \right] - \dfrac{2}{\tan 32°} \left[\dfrac{3.25 - 1}{3.25} \right]$

$$= 7.92 kN/m^2$$

(iii) at base of cut $\quad = \dfrac{19 \cdot 6 \times 5}{3 \cdot 25 \sin 64°} - \dfrac{2}{\tan 32°}\left[\dfrac{3 \cdot 25 - 1}{3 \cdot 25}\right]$ approx.

$$= 31 \cdot 33 kN/m^2$$

Earth pressure with arching: (Using submerged unit weight of soil)

(i) at ground level $\quad = -2 \cdot 2 kN/m^2$ as before

(ii) at water table level $= \dfrac{10 \times 5}{3 \cdot 25 \sin 64°}\left[1 - e^{-0 \cdot 4 \sin 64°}\right] - \dfrac{2}{\tan 32°}\left[\dfrac{3 \cdot 25 - 1}{3 \cdot 25}\right]$

$$= 2 \cdot 95 kN/m^2$$

(iii) at base of cut $\quad = \dfrac{10 \times 5}{3 \cdot 25 \sin 64°} - \dfrac{2}{\tan 32°}\left[\dfrac{3 \cdot 25 - 1}{3 \cdot 25}\right]$ approx.

$$= 14 \cdot 90 kN/m^2$$

Above results plotted in Figure 17.9 show significant reserves of safety. Available factor of safety ranges from 2·83 at the water table level to 2·82 at the base of the cut.

These values can also be confirmed with the use of Equations (16.32) and (16.33), Chapter 16,

$$a = \frac{\gamma_d \cdot L}{N_\phi \cdot \sin 2\phi'} = \frac{19 \cdot 6 \times 5}{3 \cdot 25 \sin 64°} = 33 \cdot 55$$

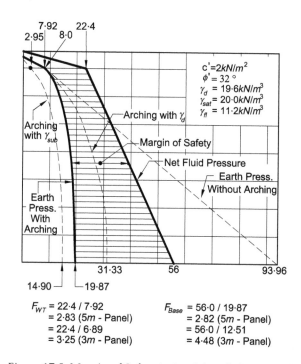

Figure 17.9 Margin of Safety (using Schneebeli approach)

$$b = \left[e^{-n \cdot \sin 2\phi'} \right] = e^{-0.4 \sin 64^\circ} = 0.698$$

$$c = \left[1 - \left(1 - \frac{\gamma_{sub}}{\gamma_d} \right) b \right] = \left[1 - \left(1 - \frac{10}{19.6} \right) 0.698 \right] = 0.66$$

$$d = \frac{c'}{\tan \phi'} \left[\frac{N_\phi - 1}{N_\phi} \right] = \frac{2}{\tan 32^\circ} \left[\frac{3.25 - 1}{3.25} \right] = 2.22$$

At water table level, $F = \dfrac{\gamma_f \cdot D_W}{a(1-b) - d} = \dfrac{11.2 \times 2}{33.55(1 - 0.698) - 2.22} = 2.83$

At base of cut,

$$F = \frac{\left(\bar{\gamma}_f \cdot H_F + \gamma_W \cdot D_W \right)}{(a \times c - d)} = \frac{(1.2 \times 30 + 10 \times 2)}{(33.55 \times 0.66 - 2.22)} = 2.82$$

Factors of safety corresponding to a different panel length can be obtained by replacing the '*L*' and the '*n*' terms appropriately. For example, for a *3m* panel, $L = 3$ and $n = 2/3 = 0.67$; the corresponding values for the factors of safety work out to 3.25 and 4.48, respectively.

(B) $c' = 0, \phi'$ *soil*

Schneebeli approach

Recalling Equation (16.34), Chapter 16:

$$\sigma_z = \frac{\gamma L}{N_\phi \sin 2\phi'} \left[1 - e^{-n \sin 2\phi'} \right]$$

$$N_\phi = \frac{(1 + \sin 35^\circ)}{(1 - \sin 35^\circ)} = 3.69$$

$$n = 0.4 \text{ at water table level as before}$$

$$= 6 \text{ at base of cut as before}$$

Net fluid pressure:

    (i) at water table level $= 22.4 kN/m^2$ as before
    (ii) at base of cut $= 56 kN/m^2$ as before

Earth pressure without arching:

    (i) at water table level $= 19.6 \times 2 \times 0.271 = 10.62 kN/m^2$
    (ii) at base of cut $= 10.62 + (19.6 - 10) \times 28 \times 0.271 = 83.47 kN/m^2$

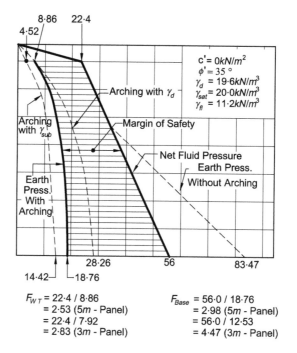

F_{WT} = 22·4 / 8·86
= 2·53 (5m - Panel)
= 22·4 / 7·92
= 2·83 (3m - Panel)

F_{Base} = 56·0 / 18·76
= 2·98 (5m - Panel)
= 56·0 / 12·53
= 4·47 (3m - Panel)

Figure 17.10 Margin of Safety (using Schneebeli approach)

Earth pressure with arching: (Using dry unit weight of soil)

(i) at water table level $= \dfrac{19 \cdot 6 \times 5}{3 \cdot 69 \sin 70°}\left[1 - e^{-0 \cdot 4 \sin 70°}\right] = 8 \cdot 86 kN/m^2$

(ii) at base of cut $= \dfrac{19 \cdot 6 \times 5}{3 \cdot 69 \sin 70°}$ approx. $= 28 \cdot 26 kN/m^2$

Earth pressure with arching: (Using submerged unit weight of soil)

(i) at water table level $= \dfrac{10 \times 5}{3 \cdot 69 \sin 70°}\left[1 - e^{-0 \cdot 4 \sin 70°}\right] = 4 \cdot 52 kN/m^2$

(ii) at base of cut $= \dfrac{10 \times 5}{3 \cdot 69 \sin 70°}$ approx. $= 14 \cdot 42 kN/m^2$

The results are plotted in Figure 17.10. Once again, significant reserves of inherent safety are indicated. Available factor of safety ranges from 2·53 at the water table level to 2·98 at the base of the cut.

For 3m long panel, the corresponding factors of safety work out to 2·83 and 4·47, respectively.

Balanced stress approach (Piaskowski and Kowalewski)

To examine the stability at the bottom of the trench, we have: $H_x/L = 30/5 = 6$, $H_x/D_w = 30/2 = 15$ and, for ϕ' value of $35°$, an unfactored $K_a = 0 \cdot 27$. To reflect the effect of arching, appropriately modified *K*-value is required. However, in the absence of the appropriate nomogram for the given soil, and for illustrating the use of the method, the nomogram in Figure 16.9 is used. This gives a factored value for the active earth pressure coefficient of $K_a = 0.075$. Incidentally, because of only a marginal difference in the γ values of the soil in the two cases, this value should be reasonably close.

Using Equation (16.39), Chapter 16, we then obtain at the base of the trench:

$$\gamma_f^c \cdot H_f - \gamma_W \cdot H_W = [19.6 \times 2 + 10 \times 28] \times 0.075 = 23.94 kN/m^2$$

$$\gamma_f \cdot H_f - \gamma_W \cdot H_W = 11.2 \times 30 - 10 \times 28 = 56 kN/m^2$$

Factor of safety, $F = 56/23.94 = 2.34$ (approx.)

At 2m below ground level, i.e. at the water table level, $H_x/L = 2/5 = 0.4, H_x/D_w = 2/2 = 1$ and $H_W = 0$. By reference to Figure 16.9, a corresponding factored K_a value of 0.26 is indicated and the corresponding

Factor of safety, $F = 11.2 \times 2 \div [(19.6 \times 2) \times 0.26] = 2.20$ (approx.)

Balanced stress approach (Huder)

For $n = H_x/L = 6$ as before, and $K_a \tan\phi' = 0.27 \times \tan 35° = 0.19$, the reduction factor *A* can be obtained from Equation (16.43) or read directly from the graph of Figure 16.10 as 0.395.

From Equation (16.42), Chapter 16, we then obtain at the base of the trench:

$$\gamma_f^c \cdot H_f - \gamma_W \cdot H_W = [19.6 \times 2 + 10 \times 28] \times 0.395 \times 0.27 = 34.06 kN/m^2$$

$$\gamma_f \cdot H_f - \gamma_W \cdot H_W = 56 kN/m^2 \text{ as before}$$

Factor of safety, $F = 56/34.06 = 1.64$

At 2m below ground level, for $n = 0.4$ and $K_a \tan\phi' = 0.19$ (as before), reduction factor *A* of 0.93 is indicated. Then, the corresponding

Factor of safety, $F = 11.2 \times 2 \div [(19.6 \times 2) \times 0.93 \times 0.27] = 2.28$

(C) '$\phi = 0$' soil

Meyerhof approach

Recalling Equation (16.51), Chapter 16, the factor of safety against failure is given by:

$$F = \frac{N \cdot s_u}{\left(p_{e(max)} - p_{s(max)}\right)}$$

For $H/B = 30/1.2 = 25$, the stability number may be obtained from the graph of Figure 16.13 or from Equation (16.49) as follows:

$$\text{Stability number, } N = 4(1 + B/L) = 4(1 + 1 \cdot 2/5) = 4 \cdot 96$$

With water table at $2m$ below ground level, the maximum total lateral earth pressure at the bottom of the trench,

$$p_{e(max)} = K_o \cdot \gamma \left(H - H_w \right) + K_o \left(\gamma_s - \gamma_w \right) H_w + \gamma_w \cdot H_w$$

$$= 0 \cdot 5 \times 19 \cdot 6 (30 - 28) + 0 \cdot 5 (20 - 10) 28 + 10 \times 28 = 439 \cdot 6 kN/m^2$$

For the completely slurry-filled trench, i.e. $H_f = H$, maximum total slurry pressure at the bottom,

$$p_{s(max)} = \gamma_f \cdot H_f = 11 \cdot 2 \times 30 = 336 kN/m^2$$

Factor of safety against failure,

$$F = \frac{4 \cdot 96 \times 50}{(439 \cdot 6 - 336)} = 2 \cdot 39$$

Huder approach

The available factor of safety at the bottom of the trench can be obtained with the help of Equation (16.52), Chapter 16. For saturated clays, N lies, typically, in the range of

Table 17.6 Comparison of Results from Examples 17.1 and 17.2

	Analysis	Factor of safety (F)		
		c', ϕ'	$c' = 0, \phi'$	'$\phi = 0$'
Without Arching	LEM	0·78 (0·61)	0·83 (0·64)	
	Balanced Stress		0·64 (0·51)	
	Meyerhof			0·73 (0·68)
With Arching (L = 5m)	*Schneebeli*	2·83–2·82	2·53–2·98	
	Balanced Stress (a)		2·20–2·34	
	Balanced Stress (b)		2·28–1·64	
	Meyerhof			2·39
	Huder			1·14
With Arching (L = 3m)	*Schneebeli*	3·25–4·48	2·83–4·47	
	Balanced Stress (a)		2·25–3·25	
	Balanced Stress (b)		2·51–2·40	
	Meyerhof			2·70
	Huder			1·42

(a) *Piaskowski and Kowalewski approach*; (b) *Huder approach (figures in parentheses are F-values with slurry level 1m below ground)*

6–7.5. Taking $N = 6$, $q_S = 0$ and substituting other values as appropriate, the factor of safety works out as:

$$F = \frac{6 \times 50}{[19{\cdot}6\,(30-28)+20 \times 28 - 11{\cdot}2 \times 30]} = 1{\cdot}14$$

17.3 Comparative study of the results from Examples 17.1 and 17.2

Results from Examples 17.1 and 17.2 are presented together in Table 17.6. It shows how the various theories in use for examining stability of slurry trenches compare with one another. It also compares, by implication, the significance of panel geometry in short and deep trench panels and brings into focus the positive impact of arch action.

Conclusions
From the above comparison, following broad based conclusions can be drawn:

- As is to be expected, even small rise in the slurry head (about 3·5 per cent) shows a marked gain in the factor of safety. In the examples considered, the improvement using limit equilibrium approach is of the order of 28–30 per cent. Using balanced stress approach, it is of the order of 25 per cent, whereas, with the Meyerhof approach it is much smaller, only of the order of around 7 per cent.
- Disregarding the benefit of arching in short deep trenches can lead to substantial underestimation of the factor of safety to the extent that the trenches which would inherently be very safe could appear to be most unsafe.
- From the downward progression of the panel lengths considered, i.e. long to 5 to 3m, it is abundantly clear that the reduction in panel lengths leads to significant gains in the factors of safety. This is attributable to the beneficial effects of arching.
- In the case of '$\phi = 0$' soils, improvement in the factors of safety according to the Meyerhof and Huder approaches based, essentially, on bearing capacity problem is directly proportional to and limited to the enhancement in the respective stability numbers that can be achieved due to favourable modification in the panel geometry.

Example 17.3
Estimate the lateral deformation of the trench face for the following two conditions:

(a) Trench geometry and Soil Type (C) of Example 17.2
(b) Trench in soft clay medium with undrained shear strength s_u of $20kN/m^2$

Solution
(a) *Inward displacement*

With this type of soil, there is the likelihood for the soil to move inwards into the slurry. To estimate the lateral deformation in the centre of the panel at the base of the cut, recall Equation (16.54), Chapter 16:

$$\Delta = 0 \cdot 75 \left(K_o \cdot \gamma_e' - \gamma_f' \right) \cdot \frac{zL}{E_o}$$

For fully saturated, perfectly elastic soil, i.e. with $\mu = 0 \cdot 5$, assume $K_o = \mu/(1 - \mu) = 1 \cdot 0$. Take $\gamma_e' = 11 \cdot 57, \gamma_f' = 2 \cdot 49, z = 30m$ and $L = 5m$ as used before in Example 17.2. Also assume $E_o = 1500 s_u = 75000 kN/m^2$. Then, the inward displacement in the centre of the panel at the base of the cut is:

$$\Delta = 0 \cdot 75 (1 \times 11 \cdot 57 - 2 \cdot 49) \cdot \frac{30 \times 5}{75000} = 0 \cdot 014m = 14mm$$

For $K_o = 0 \cdot 5$ as stipulated in Example 17.2, 'Δ' works out to $5mm$ which is not significant.

(b) Outward displacement

With the surrounding soft clay medium it is possible that it may not offer enough resistance to the outward movement of wet concrete and which may intrude into the clay during concreting. This possibility can be examined with the use of Equation (16.55), Chapter 16. Assuming the unit weight of concrete to be $24kN/m^3$, the pressure coefficient for concrete to be $0 \cdot 8$ and the bearing capacity factor for the soft clay medium of 6, then

$$\text{Pressure from wet concrete} = \gamma_C \cdot H \cdot K_C = 24 \times 30 \times 0 \cdot 8 = 576kN/m^2$$
$$\text{Resistance from the soil} = \gamma \cdot H + N_C \cdot s_u = 18 \times 30 + 6 \times 20 = 660kN/m^2$$

In view of the likely pressure exerted by the wet concrete being less than the likely resistance available from the soil as indicated, intrusion of wet concrete into the soft clay medium would appear to be unlikely. The existence of and allowance for groundwater will not alter the nature of the difference between the pressure and the resistance.

If the pressure coefficient of the liquid mix concrete in the above example were to be close to unity, then the pressure from wet concrete would be of the order of $720kN/m^2$ which would be in excess of the resistance from the soil and the outward movement of concrete into the soil would then be very likely.

Trench face movements resulting in the increase of wall thickness of the order of $300mm$ in loose silt have been known to have taken place. If the rise in the concrete level is unreasonably slower than expected, movement of concrete into the soil could well be the likely cause. Close monitoring of the anticipated rise of concrete level during pouring, particularly in soft ground, is therefore very essential.

Example 17.4

Consider a $20m$ deep, completely filled slurry trench in a $c' = 0, \phi'$ soil medium with the following parameters: $\phi' = 32 \cdot 5°$ and $\gamma = 18kN/m^3$. Examine the stability of the trench for the following cases:

- Unit weight of slurry: $10.3kN/m^3$ and $12kN/m^3$;
- Length of panel: $3m$, $5m$ and $10m$;
- Water table below ground level at: $2m$ and $5m$.

Compare the results using the Schneebeli approach with those of the balanced stress concepts of Piaskowski *et al.* and Huder. Compare also the 'factor of safety – depth' profiles obtained using these three approaches with those of Morgenstern-Amir-Tahmasseb, Tsai-Chang and Walz-Prager approaches (Tsai and Chang, 1996).

Solution
To outline the procedure, the factors of safety for only '$\gamma_f = 10.3kN/m^3$, $L = 5m$, water table $2m$ below ground level' case at the water table level and the base of the cut are worked out in detail. Factors of safety corresponding to other cases and at other levels between the water table and the base can be similarly worked out by using the parameters appropriate to those cases and the 'n' values appropriate to those levels.

Schneebeli approach

Recalling Equation (16.34), Chapter 16:

$$\sigma_z = \frac{\gamma \cdot L}{N_\phi \sin 2\phi'} \cdot \left[1 - e^{-n \sin 2\phi'}\right]$$

$$N_\phi = \frac{\left(1 + \sin 32.5°\right)}{\left(1 - \sin 32.5°\right)} = 3.32$$

$$n = 2/5 = 0.4 \text{ at water table level}$$

$$= 20/5 = 4 \text{ at base of cut}$$

Net fluid pressure:

- (i) At water table level $= 10.3 \times 2 = 20.6 kN/m^2$
- (ii) At base of cut $\quad = 20.6 + (10.3 - 10) 18 = 26 kN/m^2$

Earth pressure without arching:

- (i) At water table level $= 18 \times 2 \times 0.30 = 10.80 kN/m^2$
- (ii) At base of cut $\quad = 10.80 + (18 - 10) \times 18 \times 0.30 = 54.00 kN/m^2$

Earth pressure with arching: (Using dry unit weight of soil)

- (i) At water table level $= \dfrac{18 \times 5}{3.32 \sin 65°} \left[1 - e^{-0.4 \sin 65°}\right] = 9.09 kN/m^2$
- (ii) at base of cut $\quad = \dfrac{18 \times 5}{3.32 \sin 65°} \text{approx.} = 29.91 kN/m^2$

Earth pressure with arching: (Using submerged unit weight of soil)

(i) At water table level $= \dfrac{8 \times 5}{3 \cdot 32 \sin 65^\circ} \left[1 - e^{-0 \cdot 4 \sin 65^\circ} \right] = 4 \cdot 04 kN/m^2$

(ii) At base of cut $\quad = \dfrac{8 \times 5}{3 \cdot 32 \sin 65^\circ} \text{approx.} = 13 \cdot 29 kN/m^2$

On the basis of the above values, the available factor of safety ranges from 2·26 at the water table level to 1·42 at the base of the cut.

   These values will now be confirmed with the use of formulae. Recalling Equations (16.36) and (16.37), Chapter 16, and substituting the appropriate values, the factors of safety are:

$$\text{At water table level, } F = \frac{10 \cdot 3 \times 0 \cdot 4 \times 3 \cdot 32 \times \sin 65^\circ}{18 \left[1 - e^{-0 \cdot 4 \sin 65^\circ} \right]} = 2 \cdot 26$$

$$\text{At base of cut, } F = \frac{(4 \times 0 \cdot 3 + 0 \cdot 4 \times 10) \, 3 \cdot 32 \sin 65^\circ}{\left[18 - 10 \times e^{-0 \cdot 4 \sin 65^\circ} \right]} = 1 \cdot 42$$

Balanced stress approach (**Piaskowski and Kowalewski**)

At water table level, $H_x/L = 2/5 = 0 \cdot 4$. Unfactored $K_a = 0 \cdot 30$. For $R = 2/2 = 1$, from Figure 16.10, factored value of the active earth pressure coefficient $K_a = 0 \cdot 26$. Using Equation (16.39), Chapter 16, at water table level, we obtain:

$$\gamma_f^c \cdot H_f - \gamma_W \cdot H_W = 18 \times 2 \times 0 \cdot 26 = 9 \cdot 36 kN/m^2$$

$$\text{Also, } \gamma_f \cdot H_f - \gamma_W \cdot H_W = 10 \cdot 3 \times 2 - 0 = 20 \cdot 6 kN/m^2$$

$$\text{Factor of safety, } F = 20 \cdot 6/9 \cdot 36 = 2 \cdot 20 \text{ at water table level}$$

At base of cut, $H_x/L = 20/5 = 4$. For $R = 20/2 = 10$, from Figure 16.9, factored value of the active earth pressure coefficient $K_a = 0 \cdot 104$. Again, using Equation (16.39), Chapter 16, we obtain:

$$\gamma_f^c \cdot H_f - \gamma_W \cdot H_W = [18 \times 2 + 8 \times 18] \times 0 \cdot 104 = 18 \cdot 72 kN/m^2$$

$$\gamma_f \cdot H_f - \gamma_W \cdot H_W = 10 \cdot 3 \times 20 - 10 \times 18^2 = 26 kN/m$$

$$\text{Factor of safety, } F = 26/18.72 = 1 \cdot 39 \text{ at base of cut}$$

Balanced stress approach (**Huder**)

At water table level, $n = H_x/L = 0 \cdot 4$ as before, and $K_a \tan \phi' = 0 \cdot 30 \times \tan 32 \cdot 5^\circ = 0 \cdot 19$, the reduction factor A can be obtained from Equation (16.43), Chapter 16, or read directly from the graph of Figure 16.10 as 0·93.

From Equation (16.42), Chapter 16, we obtain:

$$\gamma_f^c \cdot H_f - \gamma_W \cdot H_W = 18 \times 2 \times 0.93 \times 0.30 = 10.04 kN/m^2$$

$$\gamma_f \cdot H_f - \gamma_W \cdot H_W = 20.6 kN/m^2 \text{ as before}$$

Factor of safety, $F = 20.6/10.04 = 2.05$ at water table level

At the base of cut, $n = H_x/L = 4$, and $K_a \tan \phi' = 0.19$ as before, the reduction factor A can be obtained from Equation (16.43), Chapter 16, or read directly from the graph of Figure 16.10 as 0.51.

From Equation (16.42), Chapter 16, we obtain:

$$\gamma_f^c \cdot H_f - \gamma_W \cdot H_W = [18 \times 2 + 8 \times 18] \times 0.51 \times 0.30 = 27.54 kN/m^2$$

$$\gamma_f \cdot H_f - \gamma_W \cdot H_W = 26 kN/m^2 \text{ as before}$$

Factor of safety, $F = 26/27.54 = 0.94$ at base of cut

This, potentially, suggests instability deeper down the trench.

For ready comparison, F-values for all the cases using the different approaches are listed in Table 17.7.

Table 17.7 Comparison of F-values using Different Approaches ($c' = 0, \phi'$ Soil)

		Factor of safety (F)		
D_W (m)	γ_f (kN/m³)	$L = 3m$	$L = 5m$	$L = 10m$
(a) *Results using Schneebeli approach*				
2	10.3	2.55–2.09	2.26–1.42	2.08–1.00
2	12.0	2.97–4.81	2.64–3.27	2.42–1.87
5	10.3	3.68–3.56	2.89–2.41	2.36–1.45
5	12.0	4.29–5.71	3.37–3.88	2.75–2.33
(b) *Results using Balanced stress approach (P and K)*				
2	10.3	2.24–2.06	2.20–1.39	2.15–1.06
2	12.0	2.61–4.76	2.56–3.21	2.51–2.45
5	10.3	2.55–3.42	2.35–2.38	2.18–1.80
5	12.0	2.98–5.49	2.73–3.83	2.58–2.90
(c) *Results using Balanced stress approach (Huder)*				
2	10.3	2.17–1.34	2.05–0.94	1.99–0.69
2	12.0	2.53–3.09	2.40–2.17	2.31–1.59
5	10.3	2.58–2.47	2.30–1.74	2.10–1.27
5	12.0	3.00–3.97	2.68–2.80	2.44–2.04

Observations

Comparative study of the above results leads to the following observations:

- The trends of effects on the F-value at the base of the cut following the three approaches are similar.

- For short panels, F-values at base of the cut as given by the Schneebeli and the Piaskowski and Kowalewski approaches are practically the same.
- For short panels, F-values at the water table as given by the Piaskowski and Kowalewski and the Huder approaches are nearly the same.
- F at base of the cut increases with the decrease in the panel length.
- Increase in the value of γ_f from $10\cdot3kN/m^3$ to $12kN/m^3$ (i.e. $16\cdot5$ per cent), leads to an increase in the value of F at base of the cut in all three approaches of the order of:

  - 130% for $D_W = 2m$ (i.e. $H_W = 18m$)
  - 61% for $D_W = 5m$ (i.e. $H_W = 15m$).

- Lowering of the water table from 2 to $5m$ (i.e. $16\cdot7$ per cent drop), leads to an increase in F at base of the cut (with Huder at the higher end) of the order of:

  - 70–85% for $\gamma_f = 10\cdot3kN/m^3$
  - 18–28% for $\gamma_f = 12kN/m^3$

- Combination of increase of $16\cdot5$ per cent in γ_f and drop of $16\cdot7$ per cent in the water table leads to an increase in F at base of the cut of the order of 170 per cent using the Schneebeli and the Piaskowski approaches and nearly 200 per cent using the Huder approach.
- Combination of decrease of 14 per cent in γ_f (i.e. from $12kN/m^3$ to $10\cdot3kN/m^3$) and drop of $16\cdot7$ per cent in H_W leads to a decrease in F at base of the cut of the order of 20–26 per cent with Huder at the lower end of the scale.

Graphs for the parametric study are shown in Figure 17.11.

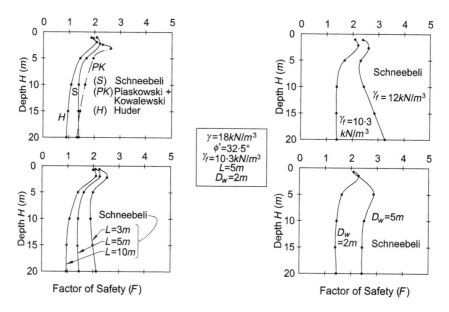

Figure 17.11 Parametric Study (A)

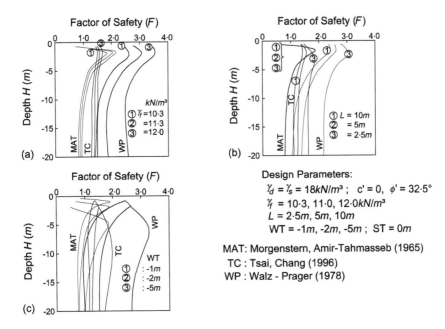

Figure 17.12 Parametric Study (B) (adapted from Tsai, Chang, 1996)

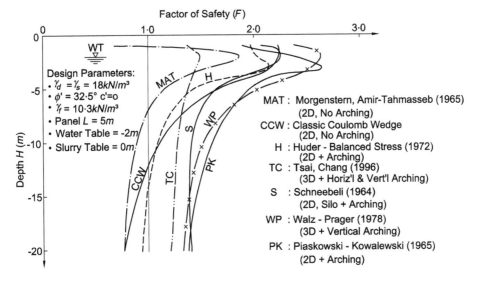

Figure 17.13 Comparison of *F* vs. Depth Profiles

Conclusions

From the foregoing observations, following conclusions may be drawn:

- Arch action due to the geometry of the cut makes a significant improvement in the stability.
- Where water table is high, increase in the unit weight of slurry makes a considerable improvement in the factor of safety. If the water table is low, the improvement is still there; however, as is to be expected, it is much reduced because of the compensating effect due to a corresponding increase in the lateral earth pressure.
- Increase in the factor of safety as a result of lowering the water table is more pronounced in the case of slurry unit weight at the lower end of its scale; the increase is relatively modest with the slurry unit weight at the higher end of its scale.
- Decrease in the factor of safety due to a decrease in the slurry unit weight even when accompanied by a significant drop in the water table, cannot be ruled out.

For comparison, graphs showing the results of the parametric study carried out by Tsai and Chang (1996) using two- and three-dimensional limit equilibrium approaches and similar parameters as above are reproduced in Figure 17.12.

Figure 17.13 compares the 'factor of safety - depth' profiles based on approaches proposed by a number of investigators using different concepts, e.g. classical two-dimensional coulomb wedge without arching, two- and three-dimensional limit equilibrium models incorporating the effects of arching, balanced stress concepts, etc. While the variations in the F-values are obvious, the significant beneficial effect of arching is clearly in evidence.

18 Groundwater Flow

18.1 Introduction

Presence of subsurface water and the rate of its movement during the various stages of excavation and construction play a major role in the design and construction of cut-and-cover structures. Clear understanding of the phenomenon of groundwater movement and the appreciation of the factors influencing it are therefore essential. In this chapter, the principles and assumptions adopted in the solution of groundwater flow problems are outlined, highlighting, where appropriate, their inherent limitations so as to enable reasoned judgments to be made in their application to design.

In ground, water occurs essentially in three basic forms: absorbed, adsorbed and free water. Absorbed water is chemically combined with the soil particles and is incapable of flowing. Adsorbed water forms a thin film surrounding the soil particles and is bonded to the surface of the particles by electrical forces. It depends for its existence upon its attraction to the soil particles and is also, therefore, incapable of movement. The water in the voids between the adsorbed layers, however, is not subject to any forces of attraction and is capable of flow when the equilibrium is disturbed as a result of an induced stress imbalance. The implication of the movement of this water, known as the free or porewater, on and during the construction of cut-and-cover structures is the subject of this chapter.

In order to appreciate the implications of the groundwater movement on the various stages of design and construction of cut-and-cover structures, the problems likely to arise, given the particular soil environment and sequence of construction, must be identified and quantified beforehand. Such investigation may reveal that it is possible to overcome these problems with the help of some practical and well-thought-out measures which can be immediately implemented should the need arise, or that some construction and design modifications may become necessary. In the extreme case, it may point towards a complete rethink of the whole scheme. In order to assess the potential scale of the problem and take the appropriate decision, it is necessary to carry out the estimates of the following:

- Leakage into excavation;
- Stability of base of excavation;
- Lateral pressures on the cofferdam;
- Change in the porewater pressure regime;
- Change in the effective stresses in the soil mass.

These aspects are dealt with later on in the chapter.

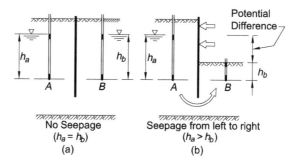

Figure 18.1 Typical Cross-Section

18.2 Groundwater movement

Porewater will tend to flow if the soil structure, including the subsoil water, is subjected to stresses, other than those maintaining *in situ* equilibrium, in a manner setting up a potential difference. If the potential of porewater is not the same everywhere, water will flow from a higher to a lower potential provided that the pores within the soil mass are so communicating as to allow that movement to take place.

Figure 18.1 shows a typical cross-section through a cut-and-cover construction with the perimeter wall in place. If piezometers or standpipes are inserted in the soil on either side of the wall at *A* and *B* (figure a), identical porewater pressures are likely to be registered at these locations. Therefore, no porewater movement is likely to take place between *A* and *B*. However, if the water level were to be lowered and the excavation taken down to the formation level on one side as shown in figure (b), porewater pressure at *A* would be higher than that at *B*. This potential difference would cause seepage flow of water under the wall in the direction as shown. The extent between the entry and the exit horizons over which movement of water takes place is known as the flow (or seepage) region.

In the figure, excavation and draw-down of water to the formation level on the right side of the wall are assumed to be instantaneous. However, in reality, it is likely that by the time the water level is drawn down to the formation level, the level to the left of the wall would also suffer some drop; the amount of this drop being a function of the prevailing boundary conditions and the drainage characteristics of the soil medium. For instance, if the soil medium were relatively free draining, the rate of drop would be faster than that expected of fine-grained soils, in which case the viscous retardation would significantly slow down the rate of flow.

If the porewater pressure dissipation is slow enough to cause the least change in the effective stresses in the very fine-grained soils at least over the duration of construction, then construction-related problems would be minimal. However, the implications of long-term time-dependent changes to be expected in the ground and their potential effects on the structure would need to be adequately addressed. If, on the other hand, the dissipation of excess porewater pressures is almost instantaneous, as would be the case, for instance, in highly granular, free-draining soils, most ground movements would be expected to occur during the period of construction itself. In such a case, the changes to be expected in the ground and their implications can be anticipated reasonably accurately and allowed for in the design and construction. However, in the case of soils, such as silts, which fall within the intermediate range, ground response can be as complex and varied as the complexity of the prevalent ground conditions. In such cases, use of instrumentation and close monitoring become all the more important.

Bernoulli's theorem and total head

In order to understand the effects of construction activity on the surrounding ground and structures, Bernoulli's theorem of conservation of energy is generally applied, with certain modifications, to the flow of water through soils. In order to apply it sensibly, it is important to understand the principles and the limitations of the theorem in its application to flow of water through soils.

According to Bernoulli's equation, the total head 'h' at any point in water under flow is constant and is given by:

$$h = \frac{p}{\gamma_w} + \frac{v^2}{2g} + z \qquad (18.1)$$

where, p/γ_w = pressure head, h_P, $v^2/2g$ = velocity or kinetic head, h_V,
z = elevation head, (h_E,) measured from an arbitrary datum,
γ_w = unit weight of water, and g = acceleration due to gravity.

In other words, total head, $h = h_P + h_V + h_E$. Assuming a velocity of flow, $v = 10\,cms/sec = 0 \cdot 1\,m/sec$, which is fairly high for flow through soils, then,

$$\text{Velocity head}, h_V = \frac{(0 \cdot 1)^2}{2 \times 9 \cdot 81} = 0 \cdot 0005m, \text{ which is negligible!}$$

So, for all practical purposes, in soils, $h_V = 0$ and, with little loss of accuracy, Bernoulli's expression, as applied to soils, can be reduced to:

$$h = h_P + h_E$$

Figure 18.2 shows a piezometer installed in the soil mass. Porewater pressure at A, assuming it to be static, will be given by the expression $u = \gamma_W \cdot h_P$, such that the pressure head, $h_P = u/\gamma_W$. So, for flow in soils, expression for the total head or potential can be safely expressed as:

$$h = z + \frac{u}{\gamma_W} \qquad (18.2)$$

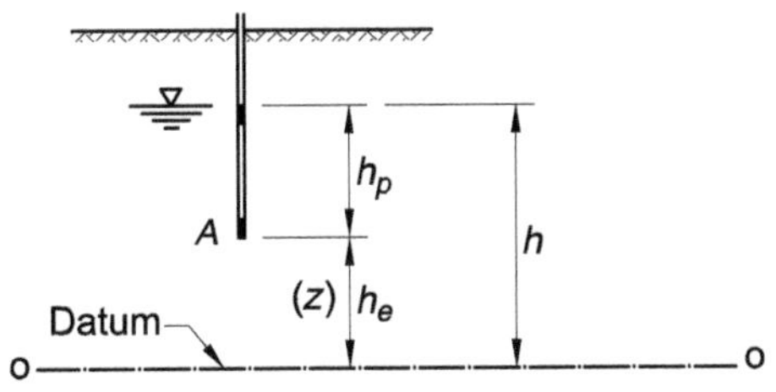

Figure 18.2 Total Head

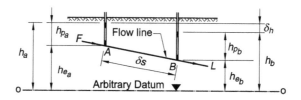

Figure 18.3 Hydraulic Gradient

Hydraulic gradient

In the flow of water through soils, energy is expended in creating the flow. In Figure 18.3, *FL* represents a flow line. Water flows from point *A*, which is at a higher potential (h_a) to point *B*, which is at a lower potential (h_b); δh represents the potential drop or loss of head due to friction between these two points. If the two points, *A* and *B*, are δ_S apart, then δ_h/δ_s is the loss of head per unit distance of flow, or the rate of loss of head, and is known as the hydraulic gradient.

Expressed mathematically, hydraulic gradient,

$$i = - \lim_{\delta_s \to 0} \frac{\delta_h}{\delta_s} = - \frac{dh}{ds}$$

Gradient '*i*' is a dimensionless vector. The negative sign is introduced only by way of convention so as to assign the gradient in the direction of flow a positive value. In Figure 18.3, $dh = h_b - h_a = -\Delta h$

$$\therefore \ i = -\frac{dh}{ds} = -\frac{(-\Delta h)}{\Delta s} = \frac{\Delta h}{\Delta s}$$

Darcy's law

Darcy's law (1856) relates the superficial or discharge velocity of flow to the hydraulic gradient causing flow, by the relation $v = k \cdot i$, where *k* represents the hydraulic conductivity, commonly known as the coefficient of permeability. For $i = 1$, $k = v$, therefore, the coefficient of permeability may be defined as the velocity of flow generated by a unit hydraulic gradient.

Darcy's law is based on 'laminar' flow through soil. Such a flow assumes that the different particles of water travel along discrete individual paths that do not cross each other. The law is based on the observations of flow of water through clean sands. This implies that while the law is valid at least up to 0·5*mm* particle size, representative of medium to coarse sand, and would possibly hold for up to the size of gravels, it would certainly fail while encountering boulders where the flow is likely to be turbulent. However, that still constitutes a wide range of soils generally encountered during the construction of metros and related structures where the law can be safely applied.

18.3 Three-dimensional seepage

For most naturally occurring soils, the behaviour of porewater pressure movement under an imposed imbalance is a function of a number of factors, e.g. soil structure, drainage characteristics, stress history, boundary conditions, threshold effects, etc. In many instances, the flow may not be unidirectional, nor need it necessarily be uniform over the entire area transverse to the flow. This makes an exact flow analysis rather complex and the need for the use of certain simplifying assumptions inevitable. Most of the simplifying assumptions may, however, be gross idealizations of the reality; nevertheless, the mathematical relationships derived for these highly idealized conditions provide the bases for an appreciation of the behaviour of real soils *vis-à-vis* the porewater movement. It is against this backdrop that some of the potential problems faced during the construction of deep cut-and-cover structures can be identified and appropriate safeguards incorporated at the design stage.

Main assumptions

For seepage flow, the following main assumptions are invoked:

- Steady-state of seepage prevails. Steady-state, as defined in Chapter 11, implies a constant rate of flow such that the rate of change of velocity is zero (Figure 18.4). During steady-state flow, porewater pressures remain constant so that there is no change in volume or effective stress and hence no deformation of the soil element is likely to take place.
- Soil is saturated, i.e. the presence and the compressibility of air bubbles are ignored. Air bubbles are highly compressible. If the soil is not saturated, depending upon its porosity, flow of water could be exaggerated.
- Soil particles and porewater are incompressible. Compressibility of water is of the order of $5 \times 10^{-10} m^2/N$, which, in the context of typical civil engineering parameters encountered in practice, is practically zero.
- Temperature is constant throughout. There is a marked change in viscosity with variation in temperature and since permeability is a function of viscosity, its value is likely to be different at different temperatures. However, assumption of constant temperature in subsoil conditions is not unreasonable.
- No soil is displaced during flow, i.e. porewater alone moves.
- Darcy's law is valid, i.e. flow is laminar.

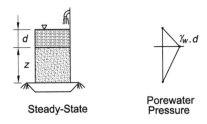

Steady-State   Porewater Pressure

Figure 18.4 Steady-State Flow

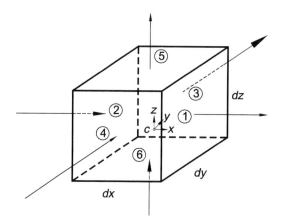

Figure 18.5 Flow through an Element

Steady-state seepage

Consider an elemental block $dx \cdot dy \cdot dz$ in the flow region of an isotropic soil medium as shown in Figure 18.5.

At the centre '*c*' of the element, let the components of discharge velocity in the principal coordinate directions x, y and z be v_x, v_y and v_z, respectively. Then the average discharge velocity through plane 1 of the element, in positive x-direction is given by:

$$v_1 = v_x + \frac{\partial v_x}{\partial x} \cdot \frac{dx}{2}$$

where $dx/2$ is the distance of plane 1 measured from the centre '*c*'.

Volumetric rate of flow through plane 1:

$$= v_1 \times \text{Area of plane 1} = \left(v_x + \frac{\partial v_x}{\partial x} \cdot \frac{dx}{2} \right) \cdot dy \cdot dz$$

$$= \text{'Outflow' in direction } x.$$

Volumetric rate of flow through plane 2, likewise:

$$= \left(v_x - \frac{\partial v_x}{\partial x} \cdot \frac{dx}{2} \right) \cdot dy \cdot dz$$

$$= \text{'Inflow' in direction } x.$$

Net inflow in x-direction per unit of time

$$= \text{Inflow–Outflow} = -\frac{\partial v_x}{\partial x} \cdot dx \cdot dy \cdot dz$$

Similarly, net inflows per unit of time in the y and z-directions will be

$$-\frac{\partial v_y}{\partial y} \cdot dx \cdot dy \cdot dz \quad \text{and} \quad -\frac{\partial v_z}{\partial z} \cdot dx \cdot dy \cdot dz, \text{ respectively.}$$

By invoking the various assumptions listed previously, we know that the soil element suffers no change in volume since the volumes of water per unit time entering and leaving the element are the same. Therefore Net Inflow = 0, i.e.

$$\frac{\partial v_x}{\partial x} \cdot dx \cdot dy \cdot dz + \frac{\partial v_y}{\partial y} \cdot dx \cdot dy \cdot dz + \frac{\partial v_z}{\partial z} \cdot dx \cdot dy \cdot dz = 0$$

$$\text{or} \quad \frac{\partial v_x}{\partial x} + \frac{\partial v_y}{\partial y} + \frac{\partial v_z}{\partial z} = 0 \tag{18.3}$$

This is the so-called steady-state continuity equation, which confirms the invariability of the velocity components such that the change of velocity per unit distance in any direction is zero. Furthermore, from Darcy's law:

$$v = k \cdot i = -k \cdot \frac{\partial h}{\partial s}; \quad v_x = -k_x \cdot \frac{\partial h}{\partial x}; \quad \therefore \frac{\partial v_x}{\partial x} = -k_x \cdot \frac{\partial^2 h}{\partial x^2};$$

$$v_y = -k_y \cdot \frac{\partial h}{\partial y}; \quad \frac{\partial v_y}{\partial y} = -k_y \cdot \frac{\partial^2 h}{\partial y^2};$$

$$v_z = -k_z \cdot \frac{\partial h}{\partial z}; \quad \frac{\partial v_z}{\partial z} = -k_z \cdot \frac{\partial^2 h}{\partial z^2}.$$

On substituting these values, Equation (18.3) assumes the form:

$$k_x \cdot \frac{\partial^2 h}{\partial x^2} + k_y \cdot \frac{\partial^2 h}{\partial y^2} + k_z \cdot \frac{\partial^2 h}{\partial z^2} = 0 \tag{18.4}$$

This represents the continuity equation for anisotropic soil. In the case of isotropic soils, $k_x = k_y = k_z$. For such soils, Equation (18.4) reduces to the Laplace form:

$$\frac{\partial^2 h}{\partial x^2} + \frac{\partial^2 h}{\partial y^2} + \frac{\partial^2 h}{\partial z^2} = \nabla^2 h = 0 \tag{18.5}$$

where ∇^2 ('del' squared) is the Laplacian Operator.

The key to the estimation of groundwater flow lies, essentially, in the solution of the Laplace equation of continuity, which assumes the steady-state flow condition for a given point within the soil mass. The concept of flow-net sketching that is employed in the calculation of groundwater flow is based on this equation.

18.4 Two-dimensional seepage

Physically, and in reality, most flow systems operate three-dimensionally. However, generally, cut-and-cover metro structures are relatively long linear structures; as such, the vast majority of flow problems encountered during the excavation for such structures and over most of their lengths tend to be two-dimensional which simplifies the solution considerably. Thus, for a homogeneous, isotropic soil medium, Laplace equation for two-dimensional seepage is given by:

$$\nabla^2 h = \frac{\partial^2 h}{\partial x^2} + \frac{\partial^2 h}{\partial z^2} = 0 \qquad (18.6)$$

Solution of this equation under appropriate boundary conditions should yield a general value of h, the total head, throughout the region of seepage. This, in turn, makes it possible to estimate the rate of leakage into the excavation, porewater pressures on the cofferdam and the state of effective stress in the soil. Rate of leakage can provide the basis for the design of an appropriate pumping system and the estimate of the effective stresses in the soil helps to assess the strength and stability of the excavation and the lateral pressures on the perimeter walls.

Consider groundwater flow from A to B in the seepage zone within impervious boundaries as shown in Figure 18.6.

Curvilinear lines *1–1*, *2–2* and *3–3* represent the stream lines, also known as the flow lines (*FLs*), and *a–a*, *b–b*, *c–c*, etc. are the 'contours' drawn through the points of equal total (or potential) head across the stream lines; these are commonly known as the equipotential lines (*ELs*). Since, by definition, there is no potential head difference anywhere along an equipotential line, there can be no hydraulic gradient and hence no component of velocity in that direction (Darcy's Law). So, the flow must be at right angles to the equipotential lines. If the intervals between successive flow and equipotential lines are made equal, an orthogonal mesh of elementary curvilinear squares will be formed. Such a family of 'orthogonal' flow and equipotential lines within the flow region, constituting what is referred to as a flow-net, forms the basis for the graphical solution. However, before attempting the flow-net sketching, it will be helpful to identify and understand the characteristics of the various types of boundaries likely to be encountered within the seepage region during the course of the construction of a typical cut-and-cover structure.

Boundary conditions

Excavation for the construction of a typical cut-and-cover metro structure is carried out within a perimeter wall, which forms either an outer temporary cofferdam or an integral part of the permanent structure. In either case, in order to make the groundwater flow problem tractable, deep excavation within the perimeter is

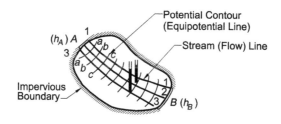

Figure 18.6 Flow and Equipotential Lines

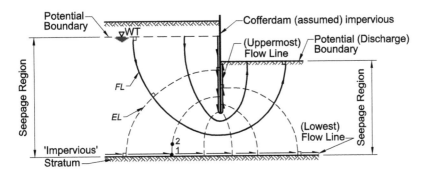

Figure 18.7 Seepage Region – Boundary Conditions

commonly idealized as a confined flow problem with known boundaries as illustrated in Figure 18.7.

Essentially, two types of boundaries are considered; these are the 'impervious' and the 'potential' boundaries. In the figure, an 'impervious' stratum has been assumed at the threshold of the seepage region. There is, by definition, no flow and hence no component of velocity across the impervious boundary. This implies that the flow takes place in the tangential direction, i.e. parallel to the boundary. An impervious boundary can be seen to define the locus of the lowest fixed flow line; and by the same token, any other flow line also satisfies the condition for an impervious boundary and can therefore be accepted as such. The uppermost fixed flow line is defined by the profile of the cofferdam, provided that it is, or can be assumed to be, impervious. This may not always be the case, for instance, if the leakage of water through sheet pile cofferdam cannot be reasonably controlled.

Consider two points 1 and 2 closely spaced apart along the normal to the impervious boundary as shown in the figure. Since there is no flow and no velocity normal to the boundary, there can be no hydraulic gradient, implying thereby the existence of a constant head between the points 2 and 1, i.e. $\partial h/\partial n = 0$, since h is constant in the normal direction n. It can, therefore, be concluded that where the equipotential lines meet the impervious boundary, they must do so at right angles, whereas further away from the boundary, they are free to change direction.

Potential boundary, on the other hand, is the boundary of the seepage region that is in contact with the reservoir of water, i.e. 'water table – seepage region' interface. It is clearly a line of equal total or potential head. Since there is no hydraulic gradient along this boundary, the velocity component tangential to the boundary must be zero. Therefore, at the potential boundary, flow is normal to it.

Besides the impervious and potential boundaries, which constitute the actual boundaries, there can also be 'virtual' boundaries defined by the geometry of the particular problem. For example, in the case of twin parallel cofferdams, where the depth of the seepage region outside the cofferdams is greater than that between them, the axis of symmetry halfway between the cofferdams can be looked upon as a virtual impervious boundary with the equipotential lines meeting it orthogonally. For this reason, the flow problem can be fully addressed by considering only half the seepage region, i.e. the region to one side of the axis of symmetry.

Flow-net sketching

The procedure for flow-net sketching is shown in Figure 18.8 and outlined as follows:

- From the geometry of the problem, identify the seepage region. Seepage region is the soil area of seepage from the entry (*AB*) to the exit (*CD*) horizons (Sketch *a*).
- Identify the boundary conditions of the seepage region establishing the fixed uppermost (*BEC*) and the lowermost (*FG*) flow lines as appropriate (Sketch *b*).
- Lightly sketch a flow line *HJ* near the cofferdam defining a smooth curve around the pile toe (Sketch *c*).
- Sketch trial equipotential lines (*ELs*) between *HJ* and *BEC* forming curvilinear squares ensuring bigger squares in zones of lesser movement (Sketch *d*).
- Adjust *HJ*, if need be, to achieve whole number of squares in the flow channel.
- Sketch the next flow line *KL* and extend the *ELs* to describe similar squares in the second flow channel (Sketch *e*).
- Repeat the procedure until the last flow channel with the fixed flow line *FG* is defined (Sketch *f*). For most problems, two or three flow lines are enough.
- Any closing discrepancy of the net at the boundary *FG* can be corrected by adjusting the flow line *HJ* and following it through the whole net. Otherwise, ensure that, if the flow channel is not a whole square, it represents a uniform fraction of a constant *L/B* ratio.

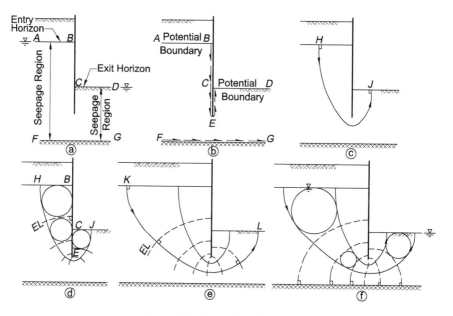

Figure 18.8 Flow-Net Sketching

During flow-net sketching, the following rules must be carefully observed.

- *FLs* do not meet.
- Boundary conditions are obeyed.

- *FLs* and *ELs* cross orthogonally forming elementary 'squares'.
- For every element of the fractional flow channel, the *L/B* ratio is constant.

The observance of these rules can be clearly seen in the construction of the flow-net as outlined in Figure 18.8.

Properties of flow-nets

The main characteristics of flow-nets are as follows:

(1) *Head loss across every square is the same*: Figure 18.9 shows a typical flow channel of unit thickness with two successive squares. Let Δq be the flow or seepage through the channel and Δh_1 and Δh_2 the head drops in the two squares.
   Since flow cannot cross flow lines,

Rate of flow through square 1 = rate of flow through square 2

So, flow through the first square,

$$\Delta q = v_1 \cdot A_1 = k \cdot i_1 \cdot A_1 = k \cdot \left(\frac{\Delta h_1}{a_1}\right)(a_1 \cdot 1) = k \cdot \Delta h_1$$

Similarly, flow through the second square, $\Delta q = k \cdot \Delta h_2$

$$\therefore \Delta h_1 = \Delta h_2$$

So, it is clear that, in the direction of flow, each square represents an equal head drop. It follows, therefore, that the total head loss between entry and exit horizons of the seepage region will be shared out as equal head drops by all the squares in a flow channel irrespective of the disparity in their sizes, i.e.

$$h = n_d \times \Delta h, \quad \text{or} \quad \Delta h = \frac{h}{n_d}$$

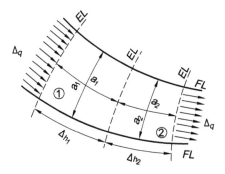

Figure 18.9 Flow Channel

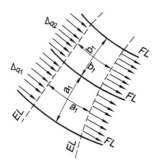

Figure 18.10 Adjoining Flow Channels

where, h = total head loss,

n_d = number of head drops

= number of squares per flow channel

(2) *Flow along each flow channel is the same*:   Consider one square in each of the two adjoining flow channels of unit thickness as shown in Figure 18.10. Let the flows in the two channels be Δq_1 and Δq_2.
Flow through channel 1,

$$\Delta q_1 = k \cdot i_1 \cdot A_1 = k \left(\frac{\Delta h}{a_1} \right) (a_1 \cdot 1) = k \cdot \Delta h$$

Flow through channel 2,

$$\Delta q_2 = k \cdot i_2 \cdot A_2 = k \left(\frac{\Delta h}{b_1} \right) (b_1 \cdot 1) = k \cdot \Delta h$$

$$\therefore \Delta q_1 = \Delta q_2$$

So, total flow is shared equally by all the flow channels with the exception of a fractional flow channel adjacent to a boundary which will carry, correspondingly, the fractional flow.

$$\text{Total flow or seepage,} \quad q = \sum \Delta q = n_f \cdot \Delta q$$

$$\text{or} \quad \Delta q = \frac{q}{n_f},$$

where n_f is the number of flow channels.

Estimation of rate of flow

Consider a typical elementary square of unit thickness as shown in Figure 18.11 in a flow channel of a complete net.

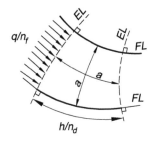

Figure 18.11 Typical Elemental Square

Let q = rate of flow per unit run, h = total head loss across seepage region, n_f = number of flow channels, n_d = number of equipotential drops, k = coefficient of permeability of soil. Flow through a typical flow channel square,

$$\frac{q}{n_f} = k \cdot i \cdot A = k \left(\frac{h}{n_d} \times \frac{1}{a} \right) (a \times 1) = k \cdot \frac{h}{n_d}$$

$$\text{or} \quad q = k \cdot h \cdot \frac{n_f}{n_d} = k \cdot h \cdot S \qquad (18.7)$$

In the above equation, S is the ratio between the number of flow channels and the number of drops and constitutes the 'shape factor' of the net. If the number of flow channels were to be doubled by halving the width of each channel, each drop would also need to be halved in order to restore the 'square-ness' of each element. In other words, if the net is subdivided the shape factor S does not change. It is therefore clear that, with regard to flow, there is no advantage to be gained in going for a finer mesh. Two to three flow lines can be more than enough. However, this is not so for porewater pressure distribution.

Estimation of porewater pressures

Recalling Equation (18.2) for total head and rearranging it, the expression for porewater pressure at any point in the seepage region can be obtained as follows.

$$u = (h - z) \cdot \gamma_W \qquad (18.8)$$

Before listing the steps for computation, it is worth examining the inter-relationship between the parameters, total head (h), elevation head (z), pressure head (u/γ_W) and the location of datum, as illustrated in Figure 18.12.

In the figure, let us assume that the level of water in a piezometer installed at a point P in a seepage region rises to a level L. The pressure head at this point is then represented by PL. Also let 'D' represent an arbitrary datum line which, for a given case, may be fixed at any convenient elevation. Whatever the position of the datum, pressure head, being exclusively a function of the porewater pressure u, carries a unique value at point P. Total head h and elevation head z, on the other hand, are both measured

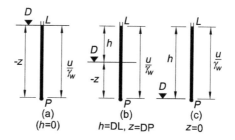

Figure 18.12 Relationship between h, z and U_w

from the datum and can therefore assume different values for different chosen levels of the datum. However, h and z complement each other in such a manner that the algebraic difference between the two at any point always represents the pressure head at that point to γ_w-scale. Total head at a point is obtained from the flow net whereas the elevation head simply from the geometry of the section. It is interesting to note that while h and z have no actual physical significance, their concept provides a convenient mathematical expedient that enables estimation of the porewater pressures at various locations to be made easily and without recourse to expensive instrumentation and *in situ* measurements. It should also be noted that whereas the use of a finer mesh makes no difference to the shape factor, and therefore also to the rate of flow, it can certainly influence the porewater pressure distribution on the cofferdam; finer the mesh, more linear the distribution of porewater pressure.

The procedure for evaluation of porewater pressures at various points in the seepage region may now be outlined as follows:

- Draw the flow-net and obtain head drops based on the net head causing flow.
- Fix an arbitrary datum line, preferably soil horizon, defined by the entry or the exit points of the flow lines.
- Starting off at a point of known porewater pressure in the seepage region, such as the water table, obtain the total head, h, at that point.
- Adjust the starting off total head by the respective head drops to obtain total heads at the successive points.
- Obtain elevation head, z, from geometry of the section, i.e. by scaling directly from the datum, measuring +ve upwards and −ve downwards.
- Expression $(h\text{-}z)$ at a point gives the porewater pressure at that point to γ_w-scale.

It is important to remember that the magnitude of porewater pressure at any point is direction-invariant, i.e. it is constant in all directions. However, if there happens to be a curvilinear square at the tip of the toe of a thin cofferdam, then the porewater pressure on either side of the tip will not be the same owing to the presence of, and drop in the pressure along, the intervening square. In such a case, average value should be taken. Besides, the porewater pressure distribution is also insensitive to the differences in the sizes of the squares which is an approximation; however, finer the mesh, smaller the approximation.

Padfield and Mair approach

Alternatively, porewater pressure distribution across a cofferdam for steady-state seepage condition can be estimated by reference to Figure 18.13. It is assumed that the seepage medium is homogeneous and that the head differential is uniformly dissipated (Padfield and Mair, 1984) along the entire length of the flow path, i.e. down the back and up the front of the cofferdam. The difference in the porewater pressure distributions estimated by the flow-net analysis and this approach lies in the quality of the distribution achieved. Whereas the distribution in the former is a function of the number of 'squares' in a flow channel irrespective of the differences in their sizes, the latter simply follows a linear variation along the flow path. Accordingly, as the mesh is made finer, the results from the former will progressively approach those of the latter.

Assuming the water table levels to be '*m*' below the formation and '*n*' below the surface on the two sides of the cofferdam as shown in Figure 18.13, the head difference causing flow $= (h + m - n)$ and the length of the flow path $= (2d + h - m - n)$; the ratio between the two represents the rate of head drop or the gradient. If the seepage entry horizon is assumed as the datum, then the total head (h_e) at this level $= 0$. Typically, head 'drop' at depth '*x*' below the entry horizon, assuming it to vary linearly along the flow path, is given by:

$$h_{d(x)} = \left[x / (2d + h - m - n) \right] (h + m - n) \tag{18.9}$$

Total head at the toe, h_t = head at entry horizon – drop in head from entry down to toe

$$= h_e - h_{d(t)} = 0 \cdot 0 - \left[(h + d - n) / (2d + h - m - n) \right] (h + m - n)$$

$$= - (h + m - n)(h + d - n) / (2d + h - m - n)$$

Elevation head at toe, $z_t = - (h + d - n)$

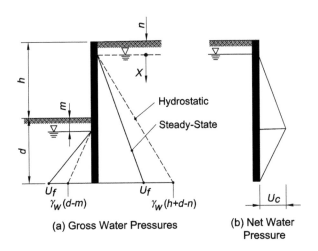

Hydrostatic

Steady-State

U_f
$\gamma_w (d-m)$

U_f
$\gamma_w (h+d-n)$

U_c

(a) Gross Water Pressures

(b) Net Water Pressure

Figure 18.13 Water Pressure across Cofferdam in Homogeneous Soil Medium
(adapted from Padfield and Mair, 1984)

Then the water pressure at the toe of the Cofferdam at steady-state seepage works out to:

$$u_t = (h_t - z_t)\gamma_w = \frac{2(h+d-n)(d-m)}{(2d+h-m-n)} \cdot \gamma_w \qquad (18.10)$$

Water pressures at any depth 'x' below the water table are given by the expressions:

$$u_x = \frac{'x'}{(h+d-n)} \cdot u_t \quad \text{(behind cofferdam)} \qquad (18.11)$$

$$= \frac{'x'}{(d-m)} \cdot u_t \quad \text{(in front of cofferdam)} \qquad (18.12)$$

Water pressures at steady-state seepage on either side of the cofferdam according to Equations (18.10, 11 and 12) are presented in Figure 18.13 with static pressures superimposed for comparison.

Note that in the expressions for u_x, 'x' is not unique but side-specific. Note also that it is enough to work out the maximum porewater pressure u_t at the toe; the variation being linear, pressures at other levels on the respective sides can then be obtained simply by interpolation. Accordingly, by substituting the value of $x = (h+m-n)$ in Equation (18.11), maximum net water pressure acting at the level corresponding to the water table below the formation works out to:

$$u_c = \frac{2(h+m-n)(d-m)}{(2d+h-m-n)} \cdot \gamma_w \qquad (18.13)$$

In the case of nonhomogeneous soil medium such as clay, where pervious interlayers of silt or sand partings are so disposed within the soil mass as to convey water to the toe of cofferdam at static pressure, the net pressures on the cofferdam would be substantially greater than those for the homogeneous medium.

18.5 Stability of excavation

Consider a cross-section through a supported excavation in granular soil as shown in Figure 18.14.

In the figure, h is the head causing seepage from the entry to the exit or the discharge horizon. By the time the flow reaches the surface *A–A*, let us assume that all but Δh

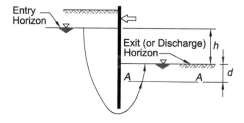

Figure 18.14 Typical Cross-Section of Support Excavation

of the net head has been lost. Let γ be the unit weight of soil and d the depth of surface A–A below the exit horizon. Let i be the hydraulic gradient caused by Δh so that $i = \Delta h/d$.

At Section A–A, hydrostatic pressure $= \gamma_W \cdot d$

Hydro-dynamic pressure due to seepage $= \gamma_W \cdot \Delta h = \gamma_W \cdot i \cdot d$

$\therefore$ Porewater pressure, $u = \gamma_W \cdot d + \gamma_W \cdot i \cdot d$

Vertical total stress, $\sigma_V = \gamma \cdot d$

Vertical effective stress, $\sigma_V' = \sigma_V - u = \gamma \cdot d - \gamma_W \cdot d - \gamma_W \cdot i \cdot d$

$= (\gamma - \gamma_W) \cdot d - \gamma_W \cdot i \cdot d$

If i attains the critical value i_c such that $\sigma_V' = 0$, a state of instability known as quicksand condition, blow-up or piping failure, becomes incipient. Making these substitutions in the above equation, expression for the critical hydraulic gradient can be obtained as:

$$i_c = \frac{\gamma - \gamma_W}{\gamma_W} = \frac{\gamma_{Sub}}{\gamma_W} \simeq 1, \quad \text{generally} \tag{18.14}$$

where, $\gamma_{Sub} =$ Submerged unit weight.

The factor of safety against quicksand condition is ascertained by comparing the maximum hydraulic gradient along the discharge boundary, called the exit gradient, with the critical gradient, i.e.

$$F_C = i_c/i_e \tag{18.15}$$

While the mechanics of piping failure by heave can be understood with the help of theoretical flow analysis, the actual *in situ* flow gradients required to initiate piping in natural soil deposits are, to a large extent, influenced by geological irregularities and accidental features of construction methods. Unfortunately, these factors cannot generally be known prior to construction. Also, in practice, piping is usually initiated at a smaller average gradient than the critical gradient predicted by theory (D'Appolonia, 1971). Theoretical analysis, therefore, serves primarily as an aid to judgment in design and for planning surveillance during construction. To allow for the various uncertainties, a high factor of safety, of the order of four is generally recommended. In other words, the exit gradient is limited to 0·25 or less.

Terzaghi (1922) conducted several model tests to investigate the cause of failure due to soil heave of sheet piles driven into a permeable layer on the downstream side. He found that the failure generally occurred within a distance of $D/2$ from the sheet piles where D is the depth of embedment of the sheet piles. By considering the stability of the submerged weight, W', of the critical prism of the size $(D/2) \times D$, against the uplifting force, U, due to seepage on the prism, the factor of safety against heave can be given by:

$$F = \frac{W'}{U}, \text{ which may also be represented as } \quad F_T = \frac{i_c}{i_T}$$

where $i_T = \dfrac{h_a}{D}$ and may be looked upon as the 'Terzaghi' gradient.

The term h_a is the average total head across the bottom of the prism. It can be obtained from the flow net. Using this approach, a value of F_T of 2·5 to 3 is recommended.

In the event that the available factor of safety is inadequate, the required shortfall can be made up by placing filter material of unit weight γ_F and of depth D_1 on the downstream side of the sheet piles. The required increase in the factor of safety is given by:

$$\Delta F_T = \frac{i_{fc}}{i_{fT}} \qquad (18.16)$$

where, $\quad i_{fc} = $ critical gradient for filter material

$$= \frac{\gamma_F - \gamma_W}{\gamma_W} = \frac{\gamma_F'}{\gamma_W} = \frac{G_s - 1}{1 + e},$$

$$i_{fT} = \text{exit gradient for filter material} = \frac{h_a}{D_1},$$

$$G_s = \text{specific gravity of soil particles, and}$$

$$e = \text{void ratio.}$$

Alternatively, the thickness of the filter material required to make up for the shortfall in the factor of safety is given by:

$$D_1 = \frac{h_a}{i_{fc}} \cdot \Delta F_T = \frac{\gamma_W}{\gamma_F'} \cdot h_a \cdot \Delta F_T = \left(\frac{1+e}{G_s - 1}\right) \cdot h_a \cdot \Delta F_T \qquad (18.17)$$

Potential for instability of formation can also exist if, during the course of excavation, a thin layer of relatively impermeable soil overlying a permeable layer is encountered as shown in Figure 18.15.

In the figure, H_W is the porewater pressure head in the sand immediately underlying the impermeable clay layer, which may or may not be hydrostatic. In such a situation, the bottom of excavation may 'blow-up' if the porewater pressure immediately below the impermeable layer is too great, i.e. $\gamma_W \cdot H_W > \gamma \cdot h$. Factor of safety against

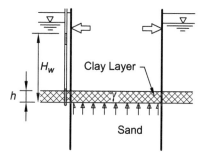

Figure 18.15 Susceptibility to Blow-Up

'blow-up' is given by:

$$F = \frac{\gamma \cdot h}{\gamma_W \cdot H_W} \qquad (18.18)$$

In order to compare it with Equation (18.15), it may be written in the form:

$$F = \frac{\gamma / \gamma_W}{H_W / h} \simeq \frac{2}{i_e}$$

Here H_W/h has been approximated to the Exit Gradient. At the point of incipient instability, $F = 1$ or $H_W \simeq 2h$. This is on the conservative side since adhesion at the soil–wall interface is ignored.

Measures to improve factor of safety

Failure from piping can be avoided either by reducing the exit gradient or increasing the critical gradient, or both; or by employing any other means which can ensure that the excess porewater pressures are released without disturbing the soil. Commonly available methods of preventing piping are:

- *Use of well points.* Porewater pressures can be relieved through well points. In this way, exit gradients can be controlled eliminating the possibility of springs and boiling at the formation.
- *Increasing penetration of perimeter wall below formation.* This measure lengthens the flow path and thereby reduces the exit gradient. As a cheaper alternative to increasing the depth of the structural wall, grout curtain can be constructed below the cut-off level of the wall forming, in effect, an extension of the wall.
- *Use of ballast.* Surcharging the formation by ballasting with sand bags, etc. can help raise the critical gradient.
- *Use of graded filter.* Introduction of a graded filter blanket on formation can help in two ways – first, by providing an easy vent allowing the seepage pressures to be relieved without disturbing the soil and second, by increasing the effective stress by its weight.
- *Flooding.* Maintaining, temporarily, a certain head of water within the excavation through flooding can reduce the net head causing seepage and therefore also the exit gradient. However, this measure is likely to interfere with the construction activity and may therefore not always be practical.
- *Allowing controlled leakage through wall.* This principle was successfully used on a section of a subway excavation in New York (Prentis and White, 1950). The retaining wall consisted of soldier piles and timber lagging with spaces left between the lagging boards. The spaces were packed with hay to allow flow through the lagging while, at the same time, preventing erosion of the sand. No noticeable settlements were recorded.

In the case of cut-and-cover metro works, the depth of penetration of the perimeter wall generally dictated by other overriding considerations may provide an adequate factor of safety against quick condition. However, in circumstances where the theoretical factor of safety with regard to the exit gradient using the conventional approach, i.e. F_C, may fall short of the recommended value, it is feasible to work to a reduced value of 2·5 to 3·0, provided always that arrangements have been made such that, at the first signs of any distress, appropriate corrective measures can be put into operation straightaway.

18.6 Pavlovsky's method of fragments

For a confined, homogeneous and isotropic flow system of finite depth, Pavlovsky (1956) evolved an approximate analytical method in 1935 that lends itself readily to design. The method approximates the equipotential lines at the critical points of the seepage region, such as under the barrier wall, as straight vertical dotted lines. These lines divide the region into discrete fragments as shown in Figure 18.16.

On the basis of different combinations of seepage geometry and boundary conditions of flow regions, Pavlovsky classified six different types (I–VI) of fragments. Both the upstream fragment 1 and the downstream fragment 2 in Figure 18.16 are of similar geometry and boundary conditions and can be identified as Pavlovsky's Type II Fragments. Since this is typically the type of fragment encountered during the construction of cut-and-cover metro structures, treatment here will be confined to Pavlovsky's Type II Fragments only.

Let k be the isotropic permeability of the seepage region and h the total head lost. For a typical fragment m, let Φ_m be the dimensionless form factor and h_m the head loss across the fragment. Factor Φ_m is a function of the geometry of the seepage region and the boundary conditions. If the discharge through the mth fragment is represented (Harr, 1962) by

$$q = \frac{k \cdot h_m}{\Phi_m}$$

then, since the discharge through all the fragments must be the same, the ratios $\frac{q}{k}$ and $\frac{h_m}{\Phi_m}$ must be constant. Thus,

$$\frac{q}{k} = \frac{h_1}{\Phi_1} = \frac{h_2}{\Phi_2} = \cdots \cdots = \frac{h_n}{\Phi_n} = \frac{\sum h_m}{\sum \Phi_m} = \frac{h}{\sum \Phi_m}$$

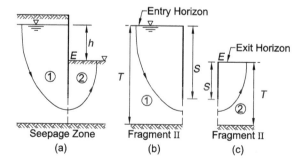

Figure 18.16 Identification of Pavlovsky Fragments

$$\therefore \text{ Discharge, } q = \frac{k \cdot h}{\sum\limits_{m=1}^{n} \Phi_m} \tag{18.19}$$

$$\text{Head loss across } m^{\text{th}} \text{ fragment, } h_m = \frac{\Phi_m}{\sum \Phi_m} \cdot h \tag{18.20}$$

The dimensionless form factor Φ is derived from an analytical solution using complex variable technique. In the case of fragment type II,

$$\Phi = K/K', \quad m = \sin \frac{\pi \cdot s}{2T}, \quad i_e = \frac{h \cdot \pi}{2K \cdot T \cdot m}$$

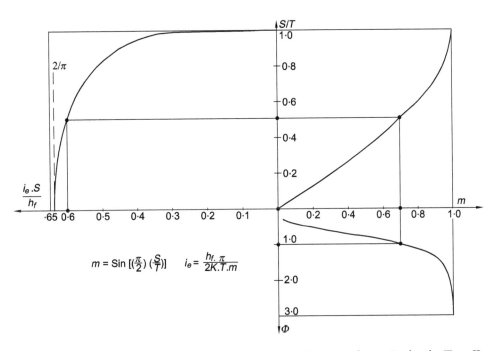

Figure 18.17 Charts for Estimation of Form Factors and Exit Gradients (Pavlovsky Type II Fragments) (adapted from Pavlovsky, 1956)

However, the modulus m (as distinct from a typical fragment denoted by the subscript m), the dimensionless form factors Φ and the exit gradient i_E at point E in Figure 18.16 can be directly obtained after entering the graphs in Figure 18.17 with the respective S/T parameter.

18.7 Layered and anisotropic soils

The treatment so far has been confined to homogeneous and isotropic soils with a single value of the coefficient of permeability k. In nature, however, most soils are anisotropic to varying degrees and may often exhibit nonhomogeneity or layering resulting from preferred particle orientation. This leads to higher coefficient of permeability in the

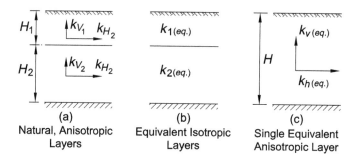

Figure 18.18 Layered Soil

direction of stratification and lower in the direction normal to it. This complicates the solution to problems of seepage in such complex formations. However, the problems can be simplified with little loss of accuracy by converting the complex, anisotropic seepage region into an equivalent, isotropic flow domain in order to enable the potential energy problems of design and construction of works to be reasonably quantified.

Keeping in perspective the degree of accuracy obtainable in the magnitude of the various design parameters and the approximations inherent in the design assumptions, reasonable answers can be obtained by employing certain sensible idealizations. Nonhomogeneous soils, for example, can be looked upon as an aggregate of thin, homogeneous and isometric layers of different permeabilities. Also, in defining the boundaries of a layered seepage region, if the ratio of the coefficients of permeabilities of touching layers is greater than 10 to 1, the layer with the lesser permeability may, for all practical purposes, be regarded as an impervious boundary (Harr, 1962).

Figure 18.18(*a*) shows two anisotropic layers of different thicknesses $H1$ and $H2$ with their respective permeabilities as indicated. For the equivalent isotropic layers of thicknesses $H1$ and $H2$ as shown in figure (*b*), the respective permeabilities are given by:

$$k_{1(eq)} = \sqrt{(k_{H1} \cdot k_{V1})} \quad \text{and} \quad k_{2(eq)} = \sqrt{(k_{H2} \cdot k_{V2})}$$

The two isotropic layers can, in turn, be approximated to an equivalent single anisotropic layer of thickness $H(= H1 + H2)$ as shown in figure (*c*), with the horizontal and vertical coefficients of permeability given by the unidirectional flow equations as follows:

$$k_{h(eq)} = \frac{\sum H_i \cdot k_{i(eq)}}{\sum H_i} = \frac{\sum H_i \cdot k_{i(eq)}}{H}$$

$$k_{v(eq)} = \frac{\sum H_i}{\sum \left(\dfrac{H_i}{k_{i(eq)}}\right)} = \frac{H}{\sum \left(\dfrac{H_i}{k_{i(eq)}}\right)}$$

where subscript '*i*' denotes a typical layer.

It should be noted that the layers $H1$ and $H2$ could be made up of a number of sublayers of different thicknesses themselves, provided that their respective permeabilities are identical. In other words, for the purpose of flow calculations, various sublayers of identical permeabilities can be grouped together as single equivalent layers.

Alternatively, for an equivalent anisotropic layer assumed in place of multilayered soil, equivalent horizontal and vertical permeabilities can be expressed as parts of the following expressions:

$$k_{H(eq)} \cdot H = k_{H1} \cdot H_1 + k_{H2} \cdot H_2 + \cdots\cdots + k_{Hn} \cdot H_n = \sum k_{Hn} \cdot H_n \qquad (18.21)$$

$$\frac{H}{k_{V(eq)}} = \frac{H_1}{k_{V1}} + \frac{H_2}{k_{V2}} + \cdots\cdots + \frac{H_n}{k_{Vn}} = \sum \frac{H_n}{k_{Vn}} \qquad (18.22)$$

18.8 Two-dimensional flow in anisotropic medium

Having obtained the components of permeability in the two orthogonal directions, general steady-state continuity equation for the three-dimensional anisotropic soil, Equation (18.4), can now be rewritten to represent the two-dimensional seepage case as follows:

$$k_x \cdot \frac{\partial^2 h}{\partial x^2} + k_z \cdot \frac{\partial^2 h}{\partial z^2} = 0$$

$$\text{or} \quad \frac{\partial^2 h}{\partial x^2} + \frac{k_z}{k_x} \cdot \frac{\partial^2 h}{\partial z^2} = 0 \qquad (18.23)$$

This is not in the Laplace form. In this form, therefore, the principles of flow-net sketching cannot be applied to obtain the solution. However, it is possible to transform it into the Laplacian form and thus enable the solution to be obtained.

Scale transformation

Equation (18.23) can be transformed into the Laplacian form by using the scale factor '*T*' based on the two orthogonal permeabilities.

$$\text{Let} \quad T = \left(\frac{k_x}{k_z}\right)^{0.5} \cdot z; \quad \text{then} \quad \frac{\partial T}{\partial z} = \left(\frac{k_x}{k_z}\right)^{0.5}$$

$$\text{Also} \quad \frac{\partial h}{\partial z} = \frac{\partial h}{\partial T} \cdot \frac{\partial T}{\partial z} = \frac{\partial h}{\partial T} \cdot \left(\frac{k_x}{k_z}\right)^{0.5}$$

$$\therefore \frac{\partial^2 h}{\partial z^2} = \frac{\partial^2 h}{\partial T^2} \cdot \frac{\partial T}{\partial z} \cdot \left(\frac{k_x}{k_z}\right)^{0.5} = \frac{\partial^2 h}{\partial T^2} \cdot \left(\frac{k_x}{k_z}\right)$$

which rearranged gives: $\left(\dfrac{k_z}{k_x}\right) \cdot \dfrac{\partial^2 h}{\partial z^2} = \dfrac{\partial^2 h}{\partial T^2}$. Substitution in Eq. (18.23),

yields the Laplace form: $\dfrac{\partial^2 h}{\partial x^2} + \dfrac{\partial^2 h}{\partial T^2} = 0$.

After the flow cross-section is transformed by the above scale factor, flow-net can be drawn in the usual way. To understand the implications of the scale factor, let us assign some values to the permeabilities.

Generally, in homogeneous natural deposits, $k_x > k_z$, although, in the case of Loess, because of its vertical structure, $k_z > k_x$. However, if it is assumed that $k_x = 16k_z$, then the Scale Factor $T = (k_x/k_z)^{0.5} \cdot z = 4z$, or $z = T/4$. This implies that, before sketching the flow-net, the flow cross-section should either be magnified fourfold in the z-direction while maintaining the x-direction to normal scale, or be reduced to quarter the size in the x-direction while maintaining the z-direction to normal scale. With the former, the geometry of the problem may sometimes become too large to handle; so, the requirement of transformation for an anisotropic flow medium may be better served by: '*shrinking the section in the direction of the greater permeability*.'

The rate of leakage into the excavation can be estimated by using the coefficient of permeability that is invariant with the direction, i.e. the equivalent k for the isotropic medium. This is given by:

$$k_t = \sqrt{(k_x \cdot k_z)} \qquad (18.24)$$

18.9 Permeable medium of infinite depth

In the case of permeable soil medium of 'infinite' depth, where excavation and dewatering preparatory to the construction of a metro structure is to be carried out inside a perimeter cofferdam, quantity of discharge into the excavation and the factor of safety against piping can be readily obtained from the graphs (Harr, 1962) in Figure 18.19.

In Figure 18.20, the net head causing seepage is equal to the depth of the exit horizon below that of the entry horizon, i.e. $h = d$. In off-shore structures, where the water level would be above entry horizon, dimensions h and d would not be the same.

The only parameters required for entry into the graphs are those provided by the geometry of the problem, i.e. s/b and d/b. While the maximum exit gradient at E is available from the graph, the average exit gradient along the discharge horizon is given by the expression:

$$i_{ave} = \frac{1}{2b}\left(\frac{q}{k}\right) \qquad (18.25)$$

The applications of the principles discussed in this chapter are illustrated in the following worked example.

Example 18.1
The typical cross-section in Figure 18.21 shows the extent of excavation and dewatering necessary to enable the construction of a cut-and-cover metro structure to be carried out in the dry.

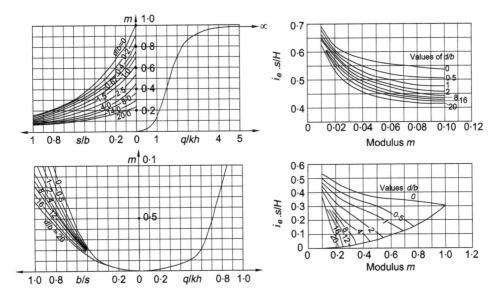

Figure 18.19 Charts for Estimation of Leakage and Exit Gradient
(Seepage medium of Infinite Depth) (adapted from Harr, 1962)

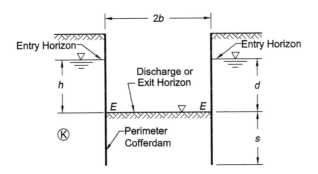

Figure 18.20 Typical Excavation Geometry

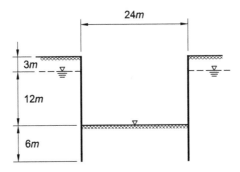

Figure 18.21 Typical Cross-Section of Excavation

Assuming the unit weight 'γ' for the soil of $20kN/m^3$, estimate the discharge into the excavation, factor of safety against piping and the distribution of porewater pressure on the cofferdam, for the following cases:

(A) Homogeneous, isotropic seepage zone 24m deep, $k = 6 \times 10^{-6} m/s$, using

   (a) Flow-net
   (b) Padfield–Mair approach
   (c) Pavlovsky's method of fragments

   Compare the total and the effective lateral pressures on the cofferdam using the methods (a) and (c) assuming, for illustrative purposes only, a value for the coefficient of earth pressure of 0·3.
(B) Anisotropic seepage zone 24m deep, $k_H = 1·8 \times 10^{-5} m/s, k_V = 2 \times 10^{-6} m/s$ using flow-net.
(C) Two homogeneous, isotropic layers of 15m and 9m thickness with respective permeabilities of $k_1 = 6 \times 10^{-6} m/s$ and $k_2 = 4 \times 10^{-7} m/s$ using flow-net.
(D) Homogeneous, isotropic seepage zone of infinite depth and $k = 6 \times 10^{-6} m/s$.

Solution:
(A) *Homogeneous, isotropic seepage medium of finite depth*

(a) Flow-net method

(i)   For flow-net, see Figure 18.22. From the flow-net,

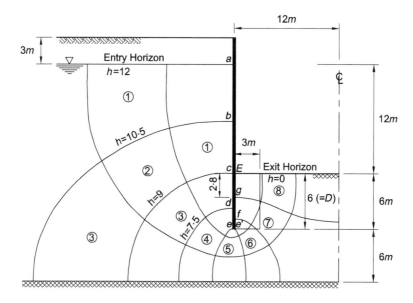

Figure 18.22 Flow-Net

No. of flow channels, $n_f = 3$

No. of head drops, $n_d = 8$

$\therefore$ Shape Factor, $S = n_f/n_d = 3/8$

Head causing seepage, $h = 12m$

(ii) Leakage into excavation per m run per side:

$$q = k \cdot h \cdot S = 6 \times 10^{-6} \times 12 \times 3/8 = 27 \times 10^{-6} m^3/s$$

Total leakage/m run, $\sum q = 2 \times 27 \times 10^{-6} = 54 \times 10^{-6} m^3/s = 4 \cdot 67 m^3/day$

(iii) Maximum hydraulic gradient at exit E:

At exit, head loss
across elemental square, $\Delta h = 12/8 = 1 \cdot 5m$

Length – elemental square, $\Delta l = 2 \cdot 8m$ (By scaling)

$\therefore$ Exit gradient, $i_e = \Delta h / \Delta l = 1 \cdot 5/2 \cdot 8 = 0 \cdot 53$

(iv) Factor of safety against piping:
Critical hydraulic gradient,

$$i_c = \frac{\gamma_{Sub}}{\gamma_W} = \frac{20 - 10}{10} = 1$$

$$\therefore F = \frac{i_c}{i_e} = \frac{1}{0 \cdot 53} = 1 \cdot 87$$

or, $F_T = \dfrac{i_c}{i_T} = \dfrac{1 \times 6}{3 \cdot 55} = 1 \cdot 7$

If it is desired to raise the factor of safety by a margin of $\Delta F_T = 0 \cdot 8$ by placing temporary filter blanket on the formation then, assuming $e = 0 \cdot 46$ and $G_s = 2 \cdot 68$, the required thickness of the blanket is given by the Equation (18.17):

$$D_1 = \left(\frac{1+e}{G_s - 1}\right) \cdot h_a \cdot \Delta F_T = \left(\frac{1+0 \cdot 46}{2 \cdot 68 - 1}\right) \times 3 \cdot 55 \times \cdot 8 = 2 \cdot 5m$$

(v) Porewater pressure distribution on cofferdam:
Consider points a to E as shown on the flow-net sketch. Assume entry horizon as the datum. The elevation heads (z-values) measured below this line will be negative in accordance with the convention assumed. The total head at entry is equal to zero. Total head at any point, $h =$ total head at entry - total head drop up to the point under consideration. Then porewater pressure at any point is given by: $u = (h - z)\gamma_W$

Table 18.1 Porewater Pressure Distribution on Cofferdam

Ref. point	$h =$ (total head at entry – total head drop to point) (m)	z (m)	$u = (h - z)\, \gamma_W$ (kN/m²)	Padfield–Mair approach u(kN/m²)
a	0·0	0·0	0	0
b	−1·5	−6·2	47	31
c	−3·0	−12·0	90	60
d	−4·5	−15·8	113	79
e	−6·0	−18·0	120	90
e′	−7·5	−18·0	105	90
f	−9·0	−17·0	80	75
g	−10·5	−14·8	43	42
E	−12·0	−12·0	0	0

Remember that the value of u at any point is constant in all directions at that point. For the values of porewater pressure, u, at different locations, see Table 18.1 and for their distribution, see Figure 18.26(a).

(b) Padfield–Mair approach

In the example, $h = 15$, $d = 6$, $n = 3$ and $m = 0$. Using Equation (18.10), the maximum porewater pressure at the tip of the cofferdam toe works out as:

$$u_t = \frac{2(d+h-n)(d-m)}{(2d+h-m-n)}\gamma_w = \frac{2(6+15-3)(6-0)}{(12+15-0-3)} \times 10 = 90 kN/m^2$$

Pressures at other levels can be obtained by linear interpolation as per Equations (18.11 and 18.12). For comparison, the pressures are listed in the Table 18.1.

(c) Pavlovsky's method of fragments

Both the upstream fragment 1 and downstream fragment 2 as shown in Figure 18.23 can be identified as Pavlovsky's Type II Fragments.

(i) Form factors:

For fragment 1, $S = 18m$, & $T = 24m$; $\therefore S/T = 18/24 = 0.75$

From graph of Figure 18.17, we obtain $m = 0.922$ and $\Phi_1 = 1.45$

For fragment 2, $S = 6m$, & $T = 12m$; $\therefore S/T = 6/12 = 0.5$

From graph of Figure 18.17, we obtain $m = 0.71$ and $\Phi_2 = 1.0$

$\therefore \sum \Phi = \Phi_1 + \Phi_2 = 1.45 + 1.0 = 2.45$

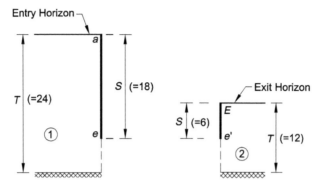

Figure 18.23 Geometry of Fragments

(ii) Leakage into excavation per m run from one side:

$$q = \frac{k \cdot h}{\sum \Phi} = \frac{6 \times 10^{-6} \times 12}{2 \cdot 45} = 29 \cdot 4 \times 10^{-6} m^3/s$$

Total leakage/m run, $\sum q = 2 \times 29 \cdot 4 \times 10^{-6} = 58 \cdot 8 \times 10^{-6} m^3/s$

$$= 5 \cdot 08 m^3/day$$

(iii) Minimum exit gradient:

Head loss in fragment 2, $h_{f2} = \dfrac{\Phi_2}{\sum \Phi} \times h = \dfrac{1}{2 \cdot 45} \times 12$

$$= 4 \cdot 9 m$$

From graph, for $S/T = 0 \cdot 5$, $i_e \cdot \dfrac{S}{h_f} = 0 \cdot 6$

$$\therefore i_e = 0 \cdot 6 \times 4 \cdot 9/6 = 0 \cdot 49$$

(iv) Factor of safety against piping:

$$F = \frac{i_c}{i_e} = \frac{1}{0 \cdot 49} = 2 \cdot 04$$

(v) Pore pressure distribution on cofferdam:
Head loss within a fragment is assumed to be linearly distributed along the upper 'impervious' boundary. Thus, the 'rate of head loss' along a−e and e′−E is constant. Head loss across fragment 1,

$$h_{f1} = \frac{\Phi_1}{\sum \Phi} \times h = \frac{1 \cdot 45}{2 \cdot 45} \times 12 = 7 \cdot 10 m$$

Recall also that $h_{f2} = 4 \cdot 9 m$

Table 18.2 Porewater Pressure Distribution on Cofferdam

Ref. point	h = (total head at entry − total head drop to point)(m)	z (m)	$(h-z)$ (m)	$u = (h-z)\gamma_W$ (kN/m^2)
a	0·0	0·0	0·0	0
e/e'	−7·1	−18·0	10·9	109
E	−12·0	−12·0	0·0	0

Assuming the entry horizon as the datum, porewater pressures are presented in Table 18.2. For porewater pressure diagram, see Figure 18.26 (*b*).

For the comparison of the total and the effective lateral pressures on the cofferdam, see Example 11.3 Chapter 11.

(B) *Anisotropic seepage medium of finite depth*

(i)   Scale transformation:

$$\text{Scale factor} = \left(\frac{k_H}{k_V}\right)^{0·5} z = \left(\frac{18}{2}\right)^{0·5} z = 3z$$

So, shrink the cross-section horizontally to a third. For the transformed section then, the equivalent isotropic permeability,

$$k_t = \sqrt{(k_H \cdot k_V)} = 6 \times 10^{-6} m/s$$

(ii)   Shape factor: (For flow-net, see Figure 18.24)

$$\text{No. of flow channels, } n_f = 2$$

$$\text{No. of head drops, } n_d = 6$$

$$\therefore \text{ Shape factor, } S = n_f/n_d = 1/3$$

(iii)   Leakage into excavation per m run per side:

$$q = k_t \cdot h \cdot S = 6 \times 10^{-6} \times 12 \times 1/3 = 24 \times 10^{-6} m^3/s$$

$$\text{Total leakage/m run, } \sum q = 2 \times 24 \times 10^{-6} \times 60 \times 60 \times 24 = 4.15 m^3/day$$

(iv)   Hydraulic gradient at exit g:

$$\text{Head loss in square 6, } \Delta h = 12/6 = 2$$

$$\text{Length of square 6, } \Delta l = 2$$

$$\text{Exit gradient, } i_e = \Delta h/\Delta l = 1$$

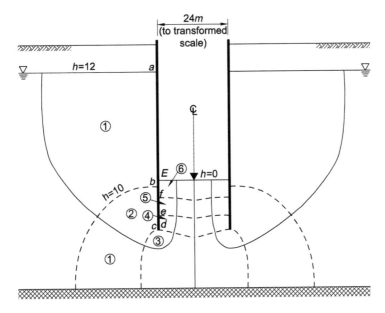

Figure 18.24 Flow Net to Transformed Scale

(v) Factor of safety against piping:

$$\text{As before, } i_c = 1$$

$$\text{Factor of safety, } F = i_c/i_e = 1$$

This implies, potentially, an incipient failure condition.

(vi) Porewater pressure distribution on cofferdam:

Assume the entry horizon as the datum. Porewater pressures are presented in Table 18.3. For porewater pressure diagram, see Figure 18.26(*c*).

Table 18.3 Porewater Pressure Distribution on Cofferdam

Ref. point	$h = $ (total head at entry – total head drop to point) (*m*)	z (*m*)	$(h - z)$ (*m*)	$u = (h - z)\gamma_W$ (kN/m^2)
a	0·0	0·0	0·0	0
b	−2·0	−12·8	10·8	108
c	−4·0	−18·0	14·0	140
d	−6·0	−18·0	12·0	120
e	−8·0	−16·0	8·0	80
f	−10·0	−14·0	4·0	40
E	−12·0	−12·0	0·0	0

(C) *Two homogeneous isotropic layers*

(i) Directional permeabilities for the equivalent anisotropic layer:

$$\text{Horizontal, } k_H = \frac{1}{H}\left[k_1 \cdot H_1 + k_2 \cdot H_2\right] = \frac{1}{24}\left[6 \times 10^{-6} \times 15 + 0.4 \times 10^{-6} \times 9\right]$$

$$= 3.9 \times 10^{-6} m/s$$

$$\text{Vertical, } k_V = \frac{H}{\left[\dfrac{H_1}{k_1} + \dfrac{H_2}{k_2}\right]} = \frac{24 \times 10^{-6}}{\left[\dfrac{15}{6} + \dfrac{9}{0.4}\right]} = 0.96 \times 10^{-6} m/s$$

(ii) Scale transformation:

$$\text{Scale factor } = \left(\frac{k_H}{k_V}\right)^{0.5} z = \left(\frac{3.9}{0.96}\right)^{0.5} z = 2z$$

So, soil flow profile will be shrunk to half the size horizontally maintaining the vertical size to normal scale. For the 'transformed' section, the equivalent isotropic permeability,

$$k_t = \left(k_H \cdot k_V\right)^{0.5} = (3.9 \times 0.96)^{0.5} \times 10^{-6} = 1.93 \times 10^{-6} m/s$$

(iii) For flow-net, see Figure 18.25.

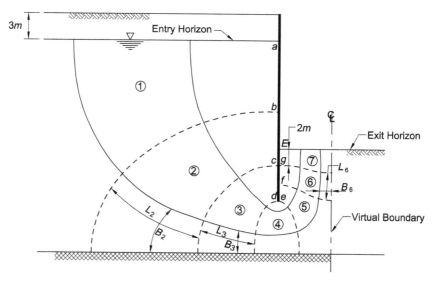

Figure 18.25 Flow Net with Fractional Channel

No. of flow channels, $n_f = 2 \cdot 5$

No. of head drops, $n_d = 7$

$\therefore$ Shape Factor, $S = n_f/n_d = 2 \cdot 5/7 = 0 \cdot 36$

(iii) Leakage into excavation per m run per side:

$$q = k_t \cdot h \cdot S = 1 \cdot 93 \times 10^{-6} \times 12 \times 0 \cdot 36 = 8 \cdot 34 \times 10^{-6} m^3/s$$

Total leakage/m run, $\sum q = 2 \times 8 \cdot 34 \times 10^{-6} \times 60 \times 60 \times 24 = 1 \cdot 44 m^3/day$

(vi) Factor of safety against piping:

Critical gradient, $i_c = 1$

Exit gradient, $i_e = \Delta h/\Delta l = \dfrac{12}{7} \cdot \dfrac{1}{2} = 0 \cdot 85$

Factor of safety, $F = i_c/i_e = 1/0 \cdot 85 = 1 \cdot 18$

Clearly, the margin is not adequate and so appropriate measures need to be put in place to ensure the desired factor of safety.

(v) Porewater pressure distribution on cofferdam:

Assume the entry horizon as the datum. Porewater pressures are presented in Table 18.4. For porewater pressure diagram, see Figure 18.26(d).

(D) Isotropic seepage medium of infinite depth

(i) Factor m:

By reference to flow geometry in Figures 18.20 and 18.21, we obtain:

$$S/b = 6/12 = 0 \cdot 5; \quad d/b = 12/12 = 1$$

Table 18.4 Porewater Pressure Distribution on Cofferdam

Ref. point	$h = $ (total head at entry – total head drop to point)(m)	z (m)	$(h-z)$ (m)	$u = (h-z)$ $\gamma_W (kN/m^2)$
a	0·0	0·0	0·0	0
b	−1·71	−8·0	6·29	63
c	−3·43	−14·0	10·57	106
d	−5·14	−18·0	12·86	129
e	−6·86	−18·0	11·14	111
f	−8·57	−16·0	7·43	74
g	−10·29	−14·0	3·71	37
E	−12·0	−12·0	0·0	0

Corresponding to these values, from graph, $m = 0.27$.

(ii) Leakage into excavation per metre run:

From graph, Figure 18.19, for $m = 0.27$,

$$q/(k \cdot h) = N_c 1.25, \text{ where q is the leakage from both sides}$$

$$\therefore q = 1.25 \times 6 \times 10^{-6} \times 12 = 90 \times 10^{-6} m^3/s$$

$$= 7.8 m^3/day$$

(iii) Hydraulic gradient at exit boundary:

From graph, Figure 18.19, for $m = 0.27$ and $d/b = 1$,

$$i_e \cdot \frac{S}{h} = 0.345$$

$$\therefore i_e = 0.345 \times \frac{12}{6} = 0.69$$

Also, average hydraulic gradient along the exit boundary,

$$i_{ave} = \frac{1}{2b} \cdot \left(\frac{q}{k}\right) = \frac{1}{24} \times 1.25 \times 12 = 0.625$$

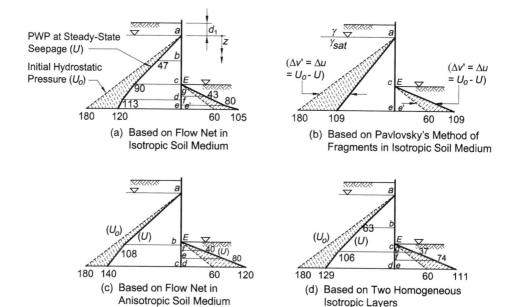

Figure 18.26 Porewater Pressure Distributions on Cofferdam

(iv) Factor of safety against piping:

$$F_{crit} = 1/0 \cdot 69 = 1 \cdot 45$$
$$F_{ave} = 1/0 \cdot 625 = 1 \cdot 6$$

Factor of safety is on the low side.

Porewater pressure distribution

Porewater pressure distributions for cases (A), (B) and (C), are shown in Figure 18.26.

With the porewater pressure distributions thus known, effective and total lateral pressures on the cofferdam can be obtained as outlined in the worked Example 11.3 of Chapter 11.

19 Mechanism of Heave

19.1 Introduction

Construction of a typical cut-and-cover structure requires excavation to be carried out from the surface of the half-space down to the required formation. The excavation may be progressed in stages or carried out *en masse*, depending on whether the construction is to be top-down or bottom-up, or any combination of the two. In any case, a significant mass of soil and volume of water can be expected to be removed from the ground. The resultant release in the *in situ* stresses, both horizontally and vertically, due to the removal of this overburden sets up a state of imbalance along the boundaries of the cut, causing its sides to displace inwards and the bottom to rise upwards. The upward deformation of soil formation in response to the relief of the vertical stress is known as heave.

The horizontally inward movement of the sides of the cut, additionally, causes vertically downward movement (settlement) of the ground surrounding the cut.

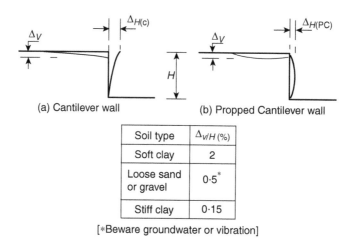

Soil type	Δ_V/H (%)
Soft clay	2
Loose sand or gravel	0.5^*
Stiff clay	0.15

[*Beware groundwater or vibration]

Figure 19.1 Wall Movements due to Horizontal Stress Relief
(adapted from Burland *et al.*, 2004)

As an illustration, Figure 19.1 compares, qualitatively, the movements that result from the relief of horizontal stress against: (*a*) an embedded cantilever wall and, (*b*) a propped cantilever wall, both of identical structural stiffness. As is to be expected, the lateral displacements towards the cut differ appreciably, being much greater in the case of the simple cantilever. The accompanying downward movements of the surrounding ground, on the other hand, are about the same in both the cases. However, their magnitude is a function of the type of the ground as tabulated in the figure (Burland *et al.*, 2004).

During the construction of cut-and-cover metro structures, particularly in soft soils, involving large volumes of excavation and dewatering, effect of heave can present, potentially, serious problems both in design and during construction. For a sound design and successful construction, therefore, proper understanding of the phenomenon of heave, its implications and how to deal with it are important.

Upward deformation or heave can be caused by one or more of the following factors:

- Vertical stress relief on formation due to removal of overburden upon excavation;
- Lowering of water table and the resultant settlement of ground outside the excavation even when the sides of the cut are restrained from lateral movement;
- Lateral inward movement of the sides of the excavation due to the yield of the ground support system; and
- Deep-seated lateral inward movement and the associated rotation of the walls below the formation.

In most soils, the first three factors listed above represent the predominant cause of heave. However, in soft ground, particularly the fourth factor, i.e. deep-seated movement below formation, can prove to be quite significant. If not properly appreciated and adequately addressed, it has the potential to overload the bottom-most brace, cause its failure and even trigger progressive collapse of the cut. Existence of high water table generating high lateral pressures can exacerbate the situation even further.

Upward deformation of soil can also be caused due to frost action in the ground; however, in that case it is referred to as 'frost heave'. Inherent characteristics of certain types of clay mineralogy, such as montmorillonite, upon coming into contact with water, can also lead to upward movement of ground; such 'heave' is commonly referred to as 'swelling'. The treatment of upward deformation of ground in this and the following two chapters is confined exclusively to that aspect of heave which results from the relief of stress upon removal of overburden pressure. However, it should be recognized that heave can also be enhanced by such factors as artesian conditions, pile-driving activity in the surrounding areas, hydrodynamic conditions due to pumping, etc.

19.2 Settlement (or negative footing) analogy

Heave is the deformation of soil resulting from the removal of load; deformation taking place, initially, at constant volume during the short-term, undrained mode, and then continuing as the excess negative porewater pressures dissipate with time. Removal of load upon excavation can be treated as an applied upward load (i.e. negative load); as such, an analogy to loading can be seen to exist in the response of soil to unloading. However, since soil is not a perfectly elastic medium, its quantitative response to these modes (i.e. positive and negative) of load application is not identical. All other things being equal, an upward displacement (i.e. heave) due to an applied negative load (i.e. stress decrement) is always less than the downward displacement (i.e. settlement) under a similar but applied positive load (i.e. stress increment). This is so because soil is generally stiffer in unloading than it is in loading.

19.3 Mechanism of heave

Intuitive model

Consider a potential area of excavation *1–2–3–4* of soil in the half-space as shown in Figure 19.2.

Before the cut is made (Figure *a*), since the area is under static equilibrium, there is no lateral strain across the potential interfaces *1–2* and *3–4*, so that the earth-at-rest pressures prevail. Also, across the potential formation plane *2–3*, $\sum V = 0$.

However, upon excavation, the state of equilibrium is disturbed by the removal of one set of the normal stresses, as implied in Figure(b), that originally maintained equilibrium along the boundaries *1–2, 2–3* and *3–4* of the cut. Modifications in the boundary conditions are thus caused by the following factors:

Vertical stress relief at formation

Vertical stress relief gives rise to upward displacement or heave of the base of excavation. This will, in turn, cause lateral relaxation resulting in some inward movement of the sides *1–2* and *3–4* also. These are shown in Figure(c).

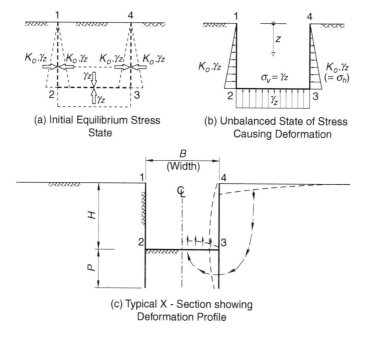

(a) Initial Equilibrium Stress
State

(b) Unbalanced State of Stress
Causing Deformation

(c) Typical X - Section showing
Deformation Profile

Figure 19.2 Intuitive Model

Horizontal stress relief across sides

Horizontal relief of stress causes the sides to move inwards. Even if the sides of the excavation are promptly and rigidly braced, the at-rest pressures existing prior to excavation are unlikely to be maintained, and so certain amount of lateral deformation (i.e. inward displacement) of the sides is inevitable. In the case of the Neasden Lane Underpass, London, UK, in spite of the installation of ground anchors, the diaphragm walls in London Clay have been recorded (Burland *et al.*, 1979) to have moved 50*mm* inwards.

Generation of negative porewater pressure

The change in total stress brought about by the removal of load also alters the boundary porewater pressures around the excavation. The excess negative porewater pressures thus set up subscribe, with subsequent dissipation, to the time-dependent deformations, generally referred to as 'progressive' or 'consolidation' heave. It is important to recognize that the undrained or initial heave and progressive or consolidation heave do not necessarily occur in two distinct phases; the latter can actually commence while the former is still going on.

In the short term, inward movement of the sides and heave at the formation is accompanied by settlement in the surrounding areas in order to maintain constant volume. It is therefore obvious that in the undrained mode, the primary deformations, i.e. heave at the base due to vertical stress relief and inward movement of the sides due to the horizontal stress relief, must generate further complementary secondary deformations enhancing, albeit modestly, the magnitude of the primary deformations. However, owing to the Poisson's ratio effect, the enhancements progressively decay at an exponential rate.

'Realistic' model

As a further aid to understanding the response of soil around an excavation, paths followed by the total and effective stresses in the two elements – one below the formation and the other to one side of the cut – during excavation can be plotted in the p–q stress space (Lambe, 1968, 1970). In Figure 19.3, A_o and A'_o, and B_o and B'_o represent the initial states of total and effective stresses and $A_oA'_o$ and $B_oB'_o$ the corresponding static porewater pressures in elements A and B, respectively. During excavation, there is a drop in the horizontal stress at A, vertical stress remaining constant, and so the total stress point moves 'up' from A_o to A_1. At B_o, on the other hand, the vertical stress reduces while the horizontal stress stays constant, so that the total stress point moves 'down' from B_o to B_1. The effective stresses move from A'_o to A'_1 and B'_o to B'_1. Upon unloading the boundary porewater pressures are altered, resulting in a net reduction $A_1A'_1$ at A and a generation of negative porewater pressures $B_1B'_1$ at B. Element A will settle, whereas B will undergo a net vertical expansion due to unloading.

During the subsequent consolidation phase, element A, in its bid to approach equilibrium condition, experiences a further reduction $A'_1A'_{ss}$ in the porewater pressure, thereby increasing the settlement, whereas element B, in order to dissipate the excess negative porewater pressures, $B'_1B'_{ss}$, imbibes water, causing further expansion.

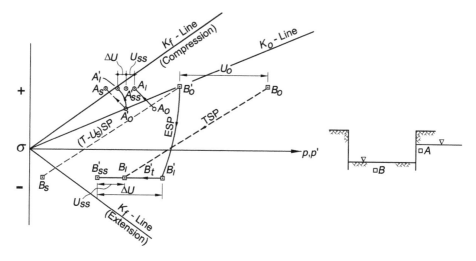

Figure 19.3 Stress Paths for Soil Elements near Excavation (adapted from Lambe, 1970)

The process continues until a steady-state seepage represented by A'_{ss} and B'_{ss} develops, whereat the equilibrium porewater pressure U_{ss} can be obtained from a flow-net.

19.4 Factors influencing heave

Since the base of an excavation, whether braced or unbraced, is not in any sense structurally isolated, it must be appreciated that many factors other than just the shear strength of soil immediately underlying the excavation must have a bearing on heave. These factors include:

- Geometry of excavation
- Geotechnical aspects
- Time dependence.

Geometry of excavation

Size of cut

In an excavation, the amount of heave that can be tolerated will dictate, *inter alia*, the size, i.e. length and breadth, of the cut that may be opened and the depth to which it may be safely taken at any given time. In general, deeper the cut and larger its surface area, greater will be the relief in total stress, and so potentially greater the heave of the base.

However, in his examination of displacements around an unbraced cut using finite difference technique, Dibiagio (1966) arrived at the following conclusions:

- For a given depth of the elastic medium, increasing the depth of the cut while maintaining a constant width increases the base heave achieving an optimum at

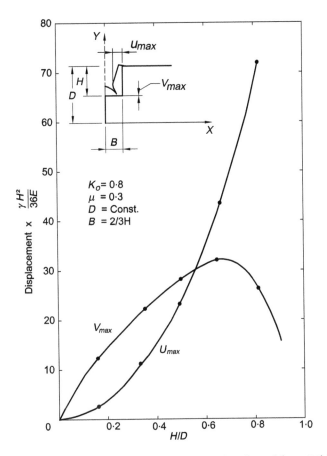

Figure 19.4 Maximum Ground Movements with Depth (adapted from Dibiagio, 1966)

around $H/D = 0.7$. Beyond this point it declines, although the lateral deformation continues to increase rather steeply as shown in Figure 19.4.

- For an increase in the width of the cut beyond twice its depth, increase in heave seems to be insignificant.

Boundary conditions

In nature, soils invariably exhibit discontinuities and zones of weakness of various types adding to the complexity of the heave problem. In heavily overconsolidated clays, for instance, forward movement of the sides into an excavation may take place along a horizontal or an inclined plane (Henkel, 1970). During excavation, as the soil is removed, stress fields and drainage boundaries keep changing. This situation is further complicated by the presence of barriers such as impervious cut-offs to porewater movement.

In estimating the amount of heave in excavations, the problem is generally over-simplified and the boundary condition idealized by ignoring the effects of complicating factors such as vertical shearing forces, friction, etc. However, to be truly representative, boundary conditions must represent accurately the sequence of construction and soil–support interaction. For instance, in the case of a rigid model, the condition of no lateral yield, besides requiring that the support element itself is unyielding, must also preclude lateral strain elsewhere in the region under consideration.

In a deep clay layer, relief of vertical stress at the base of an excavation induces deep-seated inward displacements of the sides below the formation, which cannot be controlled even by bracing the excavation above the formation (Lambe *et al.*, 1970; Burland *et al.*, 1979). These displacements could influence base heave to a certain extent. In the case of soft soils, especially with high water table, the influence can be particularly critical.

Most heave occurs in layers close to the surface of excavation (Lambe, 1968). Skopek (1976) observed the seat of heave to be just *5m* deep. Besides, the presence of silt lenses in an otherwise clayey soil can also alter the drainage characteristics and therefore affect heave.

Geotechnical aspects

Stress history

Deformations around excavations are primarily caused by changes in stresses in relation to those already existing in the ground prior to the commencement of excavation. If the ratio of principal effective stresses, i.e. σ'_h/σ'_v, at zero lateral strain, i.e. the coefficient of earth pressure at rest, K_o, is known, the *in situ* stresses can be fully established. Coefficient K_o is a function of the stress history of the soil. Figure 19.5 gives a typical diagrammatic representation of K_o-values plotted in the normal principal effective stress space.

During the course of initial deposition of soil and its consolidation by sedimentation, K_o has been found to remain relatively constant. This is represented by the straight line OA. However, if subsequent erosion causes a reduction of the overburden, K_o-curve does not retract its path (i.e. AO), but follows a steeper curve AB, exhibiting higher K_o-values over this unloading phase. For the overconsolidated clays, therefore, it is clear that K_o increases with the decrease in overburden pressure.

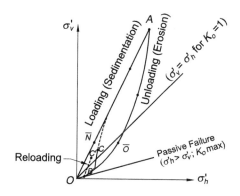

Figure 19.5 Idealized K_o-Consolidation and Swelling Curves

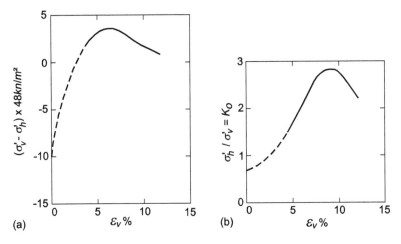

Figure 19.6 Stress–Strain Behaviour under K_o – Unloading (adapted from Henkel, 1970)

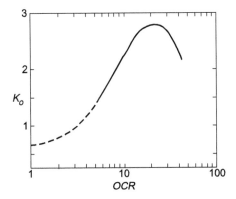

Figure 19.7 Variation of K_o with OCR (London Clay – field data)
(adapted from Skempton, 1961; Henkel, 1970)

If the principal effective stress difference $(\sigma'_h - \sigma'_v)$ and the principal effective stress ratio (σ'_h/σ'_v) are plotted against vertical strain (ε_v) (Figure 19.6), it can be seen (Henkel, 1970) that peak strength may be reached at a strain smaller than that required to generate maximum K_o indicating that, at shallow depths, heavily overconsolidated clays may be at or close to a state of passive failure. So in problems relating to base heave in heavily overconsolidated clays, the initial state of stress is of special significance.

From Figure 19.7, it is clear that for an $OCR = 4$ (for definition of OCR, see Appendix C) the K_o-value is slightly greater than one and progressively, with successive stages of excavation, it will tend to get closer and closer to the passive failure envelope. It is for this reason that the development of plastic zone formation is to be expected in the base of excavations in heavily overconsolidated soils.

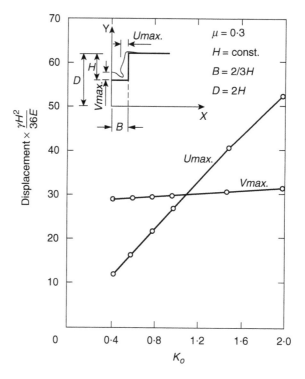

Figure 19.8 Maximum Ground Movements for Different K_o Values
(adapted from Dibiagio, 1966)

Burland *et al.* (1979) observed that, in the case of London Clay, whether σ'_v increased uniformly due to an applied surcharge load or linearly with depth due to lowering of porewater pressures, corresponding increase in σ'_h was very slight. It must be pointed out that the corresponding change in σ'_h will be as much as the change in σ'_v due to the lowering of water table. They were thus able to conclude that distribution of K_o with depth is not unique but particularly sensitive to stress history due to such factors as fill on surface or under-drainage.

Dibiagio (1966) observed that whereas changing the geometry of the problem affected the magnitude, the 'pattern' of the displacements remained primarily a function of the initial stresses in the elastic layer. The maximum heave depends essentially on the magnitude of the initial vertical stress at the formation and the variations in the magnitude of the initial horizontal stresses have only a minor effect. Figure 19.8 shows that although the maximum lateral displacement varies almost linearly, the maximum heave is insensitive to the variation in the value of K_o.

Since heave is a function of the stress history and *in situ* stresses, calculation of the *in situ* stresses and the subsequent stress changes upon load relief must be influenced by the accuracy of K_o-value. It is therefore difficult to reconcile with Dibiagio's findings that maximum heave is insensitive to any variations in K_o. His finding also seems to be directly contradicted by the suggestion (Peck, 1969) that unexpected heave in stiff clays might be due to high lateral at-rest pressures existing in the ground.

Lambe (1970) has attempted to produce a semiempirical plot, relating heave directly to depth D of excavation, by the formula:

$$\rho \uparrow = \frac{D}{100} \times K_o \qquad (19.1)$$

This expression relating displacement with depth may, at best, prove useful where movements are predominantly of an elastic nature and are not influenced by boundary conditions, etc. Since, in reality, the nature of soils is far from elastic, its use may be looked upon as providing no more than just a very rough estimate.

Stress conditions during excavations

The state of stresses that obtains below formation during an excavation is rather complex. However, to get a measure of the potential problem, Henkel (1970) considered the lateral strain compatibility and stress equilibrium along the interface of two elements separated by a 'smooth, impervious, flexible membrane' – one immediately under the excavation and subject to the passive pressure influence, while the other, alongside it under the active zone. Plane strain conditions under an undrained mode are assumed so that any vertical compressive strain in the active element is accompanied by an equal but opposite strain in the passive element. The method is therefore not valid when volume changes take place upon excess porewater pressure dissipation. It also ignores any vertical shearing forces. The models represented by Figures 19.9 and 19.10, although highly idealized, do explain at least qualitatively, the sequence of changes in the stresses with excavation.

Definitions of the normally consolidated and overconsolidated clays are given in Appendix C.

Normally consolidated clays: By tracing the stress paths in the principal effective stress space of the two elements in Figure 19.9, following conclusions are drawn:

- The peak deviator stress in the active and the passive elements are mobilized at different strain levels.
- The depth of excavation to cause actual failure will, in general, be greater than the critical depth implied by this method.
- As excavation approaches the critical path (corresponding to point F), substantial deformation can develop below formation.
- Positive porewater pressures in the active element (i.e. u_a) being greater than those in the passive element (i.e. u_p), porewater pressure will tend to migrate from the active to the passive element.

This approach is likely to yield a conservative value for the critical depth.

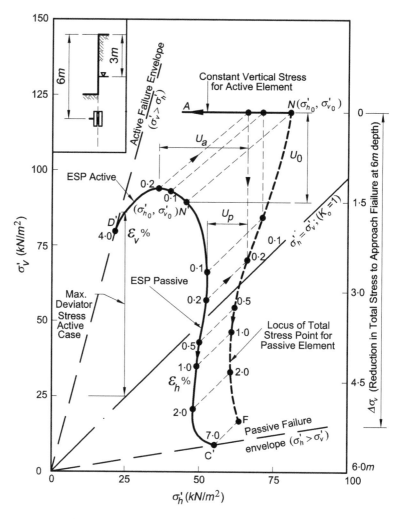

Figure 19.9 Stress Path for Compatible Deformation ($K_o = 0\cdot5$) (adapted from Henkel, 1970)

Overconsolidated clays: Heave in overconsolidated clays has been observed to be greater than expected (Peck, 1969; Henkel, 1970) and so the influence of factors peculiar to these clays and not encountered in the normally consolidated soils is obvious.

Figure 19.10 traces possible stress paths for typical overconsolidated clay with $K_o = 1\cdot5$.

From strain compatibility, the total stress can be seen to follow the path OG, the point G representing the total stress point at zero vertical stress, i.e. when excavation is taken down to the level of the element (i.e. *6m*).

The horizontal projections of the points G and G' along the total stress path (*TSP*) and the effective stress path (*ESP*) represent the porewater pressures. Although the porewater pressures of both the active as well as the passive elements are negative

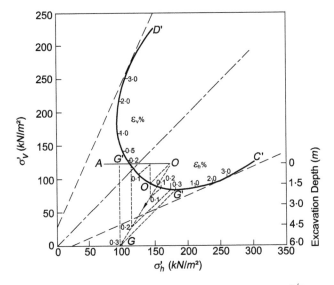

Figure 19.10 Stress Path of Compatible Deformation ($K_0^{o/c} = 1\cdot5$)
(adapted from Henkel, 1970)

(since $\sigma_h' > \sigma_h$), those in the passive element being higher than those in the active element, migration of porewater will take place from the active to the passive element causing, in the process, a reduction in strength and softening of the passive element. Henkel attributes the phenomenon of excessive heave in stiff clays, referred to by Peck (1969), to this softening of the passive element upon porewater imbibition.

Soil properties

How representative an analytical model is of the real soil profile and how accurate the soil parameters used in the computations are, will influence how close the predicted deformations come to the observed values. The effects of nonlinearity, nonhomogeneity, anisotropy, if ignored, can potentially lead to inaccurate stress relief distribution and possibly to wrong predictions of heave. For cross-anisotropic soils, the vertical stress distribution is sensitive to variations in shear modulus in the vertical plane, whereas nonlinearity, heterogeneity and Poisson's ratio have a profound influence on the distribution of horizontal stresses.

Porewater pressure movement

The state of the *in situ* pressures in the soil and the rate of groundwater movement have a profound influence on the deformation around an excavation. Lowering of water table causes an increase in the effective vertical stress, resulting in settlement, whereas with dissipation of excess porewater pressures, time-dependent consolidation takes place. The tendency of the soil to heave due to excess negative porewater pressure dissipation is greater than consolidation due to dewatering (D'Appolonia, 1971).

To enable works to proceed in the dry within an excavation, dewatering either inside or outside is generally resorted to.

Initial heave during excavation is not influenced by dewatering (D'Appolonia and Lambe, 1971). But consolidation heave increases or decreases according as the groundwater level is raised or lowered during excavation. Excavation, on the other hand, tends to set up negative porewater pressures, dissipation of which can only cause swelling. By judicious control of dewatering *vis-à-vis* excavation, it is possible, at least theoretically, to generate mutually compensating deformations, so that there is no net vertical displacement. However, in reality, excavation and dewatering are very often carried out almost simultaneously and it is difficult to separate out the porewater pressure changes, and hence deformation, due to each factor.

Time dependence

When an excavation is made quickly, there is little doubt that the ensuing 'initial heave' occurs without any significant change in volume, i.e. in the undrained mode. However, if the excavation remains open for a length of time, there seems to be some doubt as to whether the additional heave developed can be entirely attributed to the hydrodynamic time lag, i.e. expansion due to the dissipation of excess porewater pressures generated during the 'instantaneous' unloading phase. This 'time-dependent' heave has been variously referred to as 'consolidation', 'deferred', 'delayed' or 'progressive' heave.

Chang and Duncan (1970), in their nonlinear analysis of excavation for Buena Vista Pumping Plant in California, USA, found a remarkable agreement between the total rebound computed from the slow (drained) analysis and observed values and concluded that the significant portion of the rebound was time-dependent, i.e. due to porewater pressure dissipation and 'consolidation'. The authors further stated that apart from few zones of localized failure for which no explanation was offered, the fraction of the average strength mobilized was very low, indicating a high factor of safety against overall failure.

Lambe (1968) ascribes the difference between the total heave and initial heave at constant volume entirely to the time-dependent consolidation. But considerable divergence between his 'predicted' values based on consolidation theory and the 'measured' heave, casts some doubt on this.

Stoll (1969, 1971) expresses grave doubts as to whether appreciable 'consolidation' heave would occur in excavations exposed over the usual construction period of no more than a few months. He suggests that the higher-than-expected rates of heave over a construction period result from soil yield or plastic mass displacements of clay at less than failure shear stresses rather than 'consolidation'.

Stoll also introduces the concept of 'threshold depth' and concludes that:

- Time-dependent heave is negligible for excavations less than the critical depth.
- Adequate sheeting and shoring will effectively increase the threshold depth of excavation and will decrease the rate of heave for any given depth of excavation.
- If the critical depth is exceeded, heave rates will remain relatively constant over normal construction periods unless there is some change in the boundary conditions.

Osaimi and Clough (1979) have shown that the automatic assumption of undrained behaviour of excavation in clay at end-of-construction stage can be widely wrong since porewater pressure dissipation is likely to occur more rapidly than previously accepted and that significant 'consolidation' can in fact occur even during excavation. They underline the importance of the 'permeability' parameter of the soil. Through their one- and two-dimensional finite element analyses of excavation, the authors are further able to demonstrate that even for thick deposits of low permeability ($k = 0.9 \times 10^{-5} m/day$), the classical, fully drained (i.e. 0 per cent 'consolidation') case is unlikely to exist and that for higher field rates of permeability and smaller thickness, significant 'consolidation' could be expected.

Although the evidence to date seems to relate time-dependence of heave to porewater pressure dissipation, further studies in this respect are necessary to explain away satisfactorily the difference between the measured heave and that predicted by the 'consolidation' theory and to establish beyond any doubt in what proportion the 'consolidation' resulting from porewater pressure dissipation, plastic yield of soil or any other factors, contribute to the time-dependence of heave. Not least important in this behalf are also the experimental errors that may creep in while obtaining the requisite soil parameters.

What seems to be widely accepted is that progressive heave occurs very rapidly and that dissipation of excess porewater pressures increases the tendency for soil to yield. There is also no doubt that beyond the excavation and construction period, deformation continues for many years, underlying the time-dependence of heave in the long term. Observations in London Clay at Shell Centre and the underground car park at the Palace of Westminster clearly demonstrate this (Burland *et al.*, 1979).

Depending upon its size, the construction of a typical cut-and-cover metro station structure can take two years or more to complete. This would leave the excavation open for a period significant enough to demand a closer examination of the time-bound implications of heave.

20 Prediction of Heave

20.1 Introduction

In estimating heave at the base of a cut, negative footing analogy is commonly used. In other words, stress relief due to excavation is modelled as a negative, i.e. upward acting, load. Such a load causes upward deformation or heave of the formation. Just as for estimating (positive) settlement, so also for negative settlement or heave, it is important to know the variation in the applied (negative) stress with depth and the corresponding changes in the ambient stresses in the ground. However, for the heave calculations to be meaningful, it is important to ensure, at the very outset, that the base of the cut is not susceptible to failure.

The necessary steps to be gone through preparatory to the prediction of heave may be broadly outlined as follows:

* Checking probability of base failure;
* Establishing the initial soil stresses;
* Evaluating diffusion of stress relief; and
* Evaluating changes in stresses.

20.2 Probability of base failure

Before attempting to estimate actual deformations, it is essential to ensure that, for a given depth, the excavation is safe against the initiation of plastic flow and potential failure of base. As the excavation is advanced downwards, the rate of deformation around the cut increases. If, at any stage, the vertical stress relief at the base is so large that the shear strength of the subsoil is approached, plastic zone formation is initiated around the lower corners of the cut. Then, regardless of the integrity of the bracing system, the deformations, i.e. heave of base and inward displacement of sides, increase rapidly. These deformations can, in turn, cause a progressive extension of the plastic zone and, potentially, lead to instability and base failure. However, if the factor of safety is relatively high, the deformations are likely to be small and essentially elastic, so that a linear elastic analysis may be expected to yield reasonable results. Since metro excavations are by no means instantaneous and with excess porewater pressure dissipation the tendency for soil yield increases, a factor of safety against failure in undrained shear of at least 1·5 is generally aimed at. Stability of the base of a cut can be examined as follows.

Bjerrum and Eide method

Figure 20.1(a) shows the geometry of a braced cut without embedment of ground support elements. Bjerrum and Eide (1956) put forward the negative footing analogy for analysing the stability of such braced excavations. Based on their experience of soft (Norwegian) clays, they proposed the following expression for the critical depth of excavation:

$$H_c = N_c \cdot \frac{s_u}{\gamma} \qquad (20.1)$$

The factor of safety with respect to shear strength against failure is, therefore, given by:

$$F = \frac{N_c \cdot s_u}{\gamma H} \qquad (20.2)$$

where s_u = undrained shear strength underneath and immediately around the base of excavation, γ = unit weight of clay, N_c = coefficient based on Skempton's Bearing Capacity Factors, and H = depth of excavation.

In the event of surcharge load (q) also acting on the surface, 'γH' term in the denominator should be replaced by '$(\gamma H + q)$'.

The graphs in Figure 20.1(b) relate the N_c factors to the geometry of the cut and are, as such, indicators of the factors of safety. It is evident from these graphs, as is

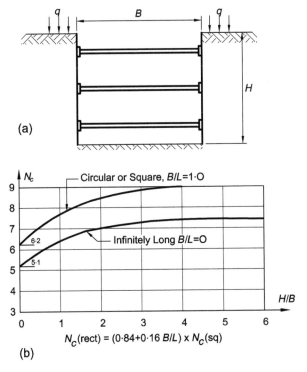

Figure 20.1 Critical Depth of Excavation (adapted from Bjerrum and Eide, 1956)

also to be expected, that, all other things being equal, with the shift in the geometry from the plane strain case (i.e. $B/L = 0$) towards the square cut case (i.e. $B/L = 1$), there is a commensurate increase in the N_c value and, accordingly, in the factor of safety as well.

According to Terzaghi and Peck (1968) and Peck (1969), if the dimensionless stability number $N = \gamma H/s_u$ approaches the value π, the heave is relatively small and essentially elastic. Further increase in the value of N manifests itself by the propagation of plastic zone until, at $N = N_c = \pi + 2$, the base uplift becomes continuous resulting in complete failure.

The expression in Equation (20.2) can also be applied to strutted excavations where the embedment of the ground support below formation, even if present, is disregarded. For most practical situations, it gives results which are satisfactory. In the case of cut-and-cover metro works, however, penetration of the ground support walls below formation is too significant for the helpful mobilization of shear strength at the soil–structure interface to be ignored.

For very long excavations relevant to metro structures, where B/L can be treated as being close to zero thus approaching plane strain case, Equation (20.2) can be modified to incorporate the contribution of the longitudinal wall embedment as follows.

$$F = \frac{s_u}{(\gamma H + q)} \left[N_c + (2h/B)\alpha \right] \tag{20.3}$$

where 'h' is the wall embedment below formation, and 'α' the adhesion factor that generally lies between 0·2 and 0·5; although for soft clay, it can be even as high as 1. Also note that, in this case, 'N_c' is based on the depth of the wall, i.e. $D = (H + h)$, and not on the depth of excavation.

The negative footing analogy may also be applied to overconsolidated clays, but due consideration must be given to the time over which the excavation remains open. Consideration must also be given to the possibility of water coming into contact with the soil which could cause it to soften and result in the diminution in its strength.

Treatment of long cuts

Cut-and-cover tunnels are linear structures with lengths considerably greater than the widths, thus approximating to plane strain conditions. Excavating long stretches at a time can pose problems with regard to the base stability. However, the danger of failure by base heave can be alleviated (Eide *et al.*, 1972) by constructing, within transverse trenches, cross-walls of required depth at regular intervals as shown in Figure 20.2, thereby breaking the long stretch into one or more discrete cells. In this way, the potential area of heave is either significantly reduced by limiting excavation to within a single cell at a time or, in the case of excavating multiple cells, the effect contained by enlisting the help from the bearing resistance under the cross-walls. By extending the walls to appropriate depth below formation, the friction mobilized at the soil–wall interfaces may also be sufficient to prevent initiation of plastic flow and thus increase the stability.

Where the excavation is carried out under two or more cells, the contribution from the bearing resistance under the cross-walls can be assumed to be evenly distributed

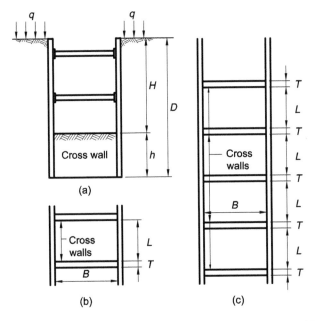

Figure 20.2 Geometry of Long Narrow Cuts for Containment of Heave
(adapted from Xanthakos, 1979)

over the extent of the excavation and included as a stabilizing factor (Eide *et al.*, 1972) against heave. For example, if '*n*' number of cells is excavated at a time, the bearing resistance per unit area against heave offered by the '*n*–1' cross-walls in between, of thickness '*T*' each, is given by:

$$\frac{(n-1)\,BTNs_u}{B\,[nL+(n-1)\,T]} \quad \text{or, simply} \quad \frac{(n-1)\,TNs_u}{[nL+(n-1)\,T]}$$

Accordingly, the expression for *FOS* for cuts without cross-walls as given by Equation (20.3) can be modified to represent the stability of a cut where excavation is carried out under '*n*' cells at a time as follows:

$$F_{nc} = \frac{s_u}{(\gamma H + q)}\left\{N_c + \frac{1}{[nL+(n-1)\,T]}\left[2nh\left(1+\frac{L}{B}\right)\cdot\alpha + (n-1)\,TN\right]\right\} \quad (20.4)$$

In the event of excavation being confined to a single cell at a time, by substituting $n=1$ in the above expression, the corresponding factor of safety for a single-cell case can be obtained as:

$$F_{1c} = \frac{s_u}{(\gamma H + q)}\left[N_c + \frac{2\,(B+L)h}{B\cdot L}\cdot\alpha\right] \quad (20.5)$$

Variation in the value of the adhesion factor between 0·2 and 0·5 is unlikely to yield a variation in the *F*-value in excess of 10 per cent.

Alternatively, in the case of multiple cells excavated at a time and in spite of the presence of the cross-walls, the excavation can be treated as 'infinitely' long (Xanthakos, 1979). The equivalent bearing resistance per unit area against heave offered by the cross-walls may be taken as:

$$\frac{BTNs_u}{B(L+T)} \quad \text{or, simply} \quad \frac{TNs_u}{(L+T)}$$

The factor of safety for the multi-cell case is then given by the expression:

$$F_{mc} = \frac{s_u}{(\gamma H + q)} \left\{ N_c + \frac{1}{(L+T)} \left[2h \left(1 + \frac{L}{B} \right) \cdot \alpha + TN \right] \right\} \tag{20.6}$$

The value of the bearing capacity factor N in the Equations (20.4) and (20.6) can be taken as 7·5. Also, in estimating the values of the N_c factors from the graphs, it should be noted that:

- L and B are case-specific, i.e. B is to be taken as the shorter side and L the longer
- In the ratio D/B, B is to be taken as L or B whichever is the smaller.

20.3 Initial soil stresses

After ruling out the possibility of base failure in the manner outlined above, the next step involves establishing the ambient stress regime in the ground. In establishing the *in situ* state of stress on an element in the soil mass, the following simplified but commonly accepted assumptions are made:

- The stresses are geostatic, i.e. the surface is reasonably level so that the vertical and horizontal stresses represent the principal stresses without any accompanying components of shear stress.
- A condition of radial symmetry exists.
- The vertical stress is obtained from the self-weight of overburden alone.

Variation of porewater pressures with depth can be measured directly through instrumentation and, with its help, the vertical effective stress profile can be established.

The horizontal stresses may be measured *in situ* directly with the use of pressure cells or evaluated indirectly with the help of K_o. The *in situ* horizontal and vertical stresses may then be expressed as outlined in Chapter 13 as follows:

$$\text{Total Stresses: } \sigma_V = \gamma \cdot z$$

$$\sigma_H = K_o \left(\gamma \cdot z - u_o \right) + u_o$$

$$\text{Effective Stresses: } \sigma_V' = \sigma_V - u_o = \gamma \cdot z - u_o$$

$$\sigma_H' = \sigma_H - u_o = K_o \left(\gamma \cdot z - u_o \right)$$

For further details concerning the computations of the *in situ* stresses and the evaluation of K_o, reference should be made to Chapters 11 and 23, respectively.

20.4 Diffusion of stress relief

The extent of potential heave at any depth below formation is, among other things, also a function of the extent of the stress relief due to excavation experienced at that level. Clearly, stress relief as also its effect, i.e. heave, will be the maximum at the formation level. However, with the dissipation of stress relief with depth below the formation, heave will also go on decreasing until it completely tails off at some depth.

The extent of the depth below formation over which heave reduces from its maximum value to zero is referred to as the 'seat of heave'. In order to identify and establish the profile of heave over this depth, it is essential to know the manner in which the influence of relief in the vertical stress is felt through the soil mass. For this, and as an idealization, elastic theory is used assuming the medium to be a linear, homogeneous, isotropic, elastic half-space. Furthermore, the effect of excavation in terms of a negative upward load is modelled in much the same way as a positive downward load in the case of settlement calculations under a footing.

Changes in the stress at various levels due to vertical stress relief are obtained in terms of total stress assuming the undrained soil parameters, i.e. Young's modulus E and Poisson's ratio $\nu\,(=0.5)$, to be constant.

For small deformations, i.e. when the factor of safety is high (>2), elastic theory may well yield reasonable results. Several methods giving the distribution of stresses with depth are available. Of these, the more commonly used are given in Appendix E.

20.5 Changes in the stresses

After establishing the ambient stress regime and the reduction in the stress relief with depth, the next step is to quantify the changes in the former brought about by the latter.

Changes in the porewater pressure

In a fine-grained soil, because of low permeability, if excavation is carried out sufficiently rapidly, total stress relief of the soil mass can occur without any change in volume, simultaneously setting up excess negative porewater pressures that dissipate with time. The excess porewater pressures may either be monitored directly during excavation through instrumentation, or predicted analytically with the help of one of the following methods:

One-dimensional case

In the case of soil that is laterally confined as represented in an odometer test, induced change in the porewater pressure is equal to the applied change in the vertical stress. In other words, $\Delta u = \Delta \sigma_V$.

Skempton's equation

Skempton (1954) pointed out that, in general, the induced porewater pressure is less than the applied change in stress. He introduced the concept of porewater pressure parameters A and B and developed an expression for porewater pressure change in terms of these as represented by the expression:

$$\Delta u = B\left[\Delta\sigma_3 + A\left(\Delta\sigma_1 - \Delta\sigma_3\right)\right] \tag{20.7}$$

where $\Delta\sigma_1$ and $\Delta\sigma_3$ are the changes in the total major and minor principal stresses, respectively. For saturated clays, $B = 1$, and so, the above expression reduces to:

$$\Delta u = \Delta\sigma_3 + A\left(\Delta\sigma_1 - \Delta\sigma_3\right) \tag{20.8}$$

The assumption inherent in this expression is the equality of the principal minor and the intermediate stresses, i.e. $\Delta\sigma_2 = \Delta\sigma_3$, and so it is strictly applicable to axi-symmetric cases. For axi-symmetric case in a perfectly elastic material, the parameter $A = 1/3$.

Henkel's equation

Condition of axi-symmetry, i.e. $\Delta\sigma_2 = \Delta\sigma_3$, is very rarely realized in practice. Henkel (1980) proposed a more general relationship for porewater pressure change in terms of stress invariants, taking into account the principal intermediate stress, as follows:

$$\Delta u = \Delta\sigma_{oct} + \alpha \cdot \Delta\tau_{oct} \tag{20.9}$$

where

$$\Delta\sigma_{oct} = \frac{1}{3}\left(\Delta\sigma_1 + \Delta\sigma_2 + \Delta\sigma_3\right) \tag{20.10}$$

$$\Delta\tau_{oct} = \frac{1}{3}\left[\left(\Delta\sigma_1 - \Delta\sigma_2\right)^2 + \left(\Delta\sigma_2 - \Delta\sigma_3\right)^2 + \left(\Delta\sigma_3 - \Delta\sigma_1\right)^2\right]^{0.5} \tag{20.11}$$

In the preceding expression, σ_{oct} and τ_{oct} are the octahedral total normal and shear stress invariants and $\Delta\sigma_{oct}$ and $\Delta\tau_{oct}$ the respective changes in these, and '$\alpha \cdot \Delta\tau_{oct}$' is the effect of dilation and 'α' is not to be confused with the adhesion factor. The stress parameters the magnitudes of which are independent of the choice of, i.e. do not change with, the variation in the reference axes, are known as stress invariants.

Flow-nets

If a perfectly watertight wall penetrates an impervious formation at the bottom of excavation and if the disturbance caused to the soil mass and porewater pressures due to the insertion of the wall is minimal, hydrostatic conditions may prevail. In reality, since the period of excavation is finite, the mass of soil affected by the excavation follows a path of partial drainage; significant movements can occur below formation and the insertion of the wall causes an increase, albeit slight, in porewater pressures, and so the pattern of flow into the excavation is rather complex.

In order to assess porewater pressures, flow of water may be analysed, graphically, by means of flow-nets or, analytically, by using finite difference or finite element techniques, recognizing, however, that under certain types of excavation and soil structure, the assumptions inherent in these techniques, i.e. two-dimensional flow and steady-state seepage, can represent gross simplifications. For instance, rather than excavating *en masse*, small stretches may be opened out at a time; there may be flow parallel to the wall so that it is far from being two-dimensional; in very fine-grained soils, due to the very low permeability, the state of steady seepage may not be reached

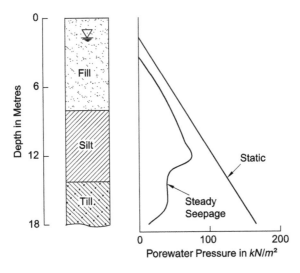

Figure 20.3 Typical Porewater Pressure Change following Excavation
(adapted from Lambe, 1970)

for a long time, and so on. Further uncertainties and complications can also creep in
due to the following factors:

- Leakage through the wall;
- Length of time the excavation stays open;
- Lack of accuracy in the assessment of *in situ* permeability.

Porewater pressures can be estimated with the help of flow-nets or by following the
Padfield and Mair approach as discussed in Chapter 18.

According to Lambe (1970), there seems to be little doubt that, upon excavation,
the porewater pressures fall to less than their static values (Figure 20.3). He even hints
that excavation causes a slight but permanent modification in the ambient porewater
pressures by pointing out that, in spite of extensive recharging, the porewater pressures
could never be restored to their original static values. This could well be due to the
threshold effect.

Stress path method

It must be recognized that even for an elastic soil, porewater pressure parameter A
is not unique. For axi-symmetric, triaxial compression test, with $\sigma_2 = \sigma_3, A = 1/3$;
for triaxial extension test with $\sigma_1 = \sigma_2 > \sigma_3, A = 2/3$ and for plane strain conditions,
$A = 1/2$. It is therefore obvious that the porewater pressure parameters A and B are not
entirely satisfactory as the absolute measures of porewater pressure behaviour under
different conditions. The various factors influencing the parameter A are:

- Strain to which soil has been subjected;
- Initial stress system – whether isotropic or anisotropic;

- Stress history – whether normally or overconsolidated;
- Type of stress change – whether due to loading or unloading.

By subjecting a sample of clay in the laboratory to the same stress regime, total and effective, as existing in the ground, and then unloading along the total stress path according to the mode of excavation, the porewater pressures may be measured directly or evaluated with the help of the porewater pressure parameters.

Changes in the effective stress

The concept of effective stress is essential to the understanding of the phenomenon of heave. This is dealt with in Chapter 11.

Changes in the effective stress or effective stress relief for various stages of excavation can be obtained by deducting the porewater pressure change from the total stress relief, i.e. $\Delta\sigma' = \Delta\sigma - \Delta u$. Estimation of the total stress relief, $\Delta\sigma$, at various depths can be obtained in accordance with the elastic theory and various graphs as outlined in Appendix E and the changes in the porewater pressure in the manner outlined in the foregoing section.

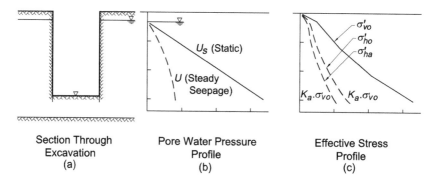

Figure 20.4 Stresses in Soil Adjacent to Excavation

Figure 20.4 shows qualitatively, as an illustration, possible porewater pressures and effective stress profiles adjacent to an excavation.

20.6 Estimation of heave

Type of the problem

Depending on the geometry of an excavation and the position at which the deformation is required, the problem for estimating heave may be considered as:

- Three-dimensional
- Axi-symmetric
- Plane strain.

These are briefly discussed next.

Three-dimensional case ($\varepsilon_1 \neq \varepsilon_2 \neq \varepsilon_3$)

The problem of excavation where heave of formation upon stress relief is accompanied by unequal deformations in the two lateral directions due to the asymmetry in the geometry, such as an area adjacent to the boundary of an excavation, is essentially three-dimensional. However, since heave is greater at centre than near the edges, it may be argued that the three-dimensional approach is rarely necessary.

Axi-symmetric case ($\varepsilon_2 = \varepsilon_3$)

If, upon excavation, the resulting lateral strains are equal, i.e. $\varepsilon_2 = \varepsilon_3$, the problem can be treated as axi-symmetric. Such a case can occur during the excavation of a square, finite area in which an upward heave will be accompanied, due to symmetry, by equal lateral displacements.

Plane strain case ($\varepsilon_2 = 0$)

In the case of an excavation where one side is significantly longer than the other, such as a long stretch of a cut-and-cover metro tunnel or a station box structure, vertical displacement due to unloading can be safely assumed to be accompanied by transverse displacements only. Net displacement in the longitudinal direction can be considered to be zero, i.e. $\varepsilon_2 = 0$, since the uniform stress relief along the length can be relied upon to generate mutually cancelling longitudinal strains; so, the problem can be treated as one of plane strain. Matheson and Koppula (1971) also confirm that rebound due to stress relief in the centre of large excavations is essentially one-dimensional.

It must be recognized that the boundary conditions outlined above are idealized so as to render the problem mathematically tractable. In reality, the excavation problem can be rather complex.

Elastic heave

With the use of elastic theory, undrained elastic heave is commonly estimated either from the elastic displacement theory in which soil is idealized as a uniform layer having an undrained, elastic, unloading (i.e. expansion) modulus, $E_{u(e)}$, or from the summation of elastic strains obtained from elastic stress distribution theory. For details of the various methods, reference may be made to Poulos and Davis (1974). Some of the more commonly used methods are given in the following section.

General solution

Solution by the elastic displacement theory is essentially of the form:

$$\delta_e = \frac{q \cdot B}{E_{u(e)}} \cdot I_\sigma \tag{20.12}$$

where, δ_e = initial undrained heave upon unloading, q = stress relief upon unloading, B = width of excavation, $E_{u(e)}$ = undrained Deformation Modulus for unloading, and I_σ = influence factor based on geometry.

The above solution assumes that the half-space is linear elastic, homogeneous and isotropic. Solutions for soils exhibiting characteristics that show a departure from these are not readily available, although some approximations have been suggested (Lambe, 1970; Davis and Poulos, 1968). Besides the limitations inherent in the above assumptions, there are two chief drawbacks of the elastic displacement theory (D'Appolonia and Lambe, 1970). These are:

- Even at factors of safety as high as three, extensive local yield may develop in soil for which it is difficult to model the redistribution of stresses.
- The accuracy of the result is almost exclusively reliant upon the choice of the undrained deformation parameter for unloading, $E_{u(e)}$; this may not always be easy to quantify accurately.

It is widely accepted that the value of $E_{u(e)}$ is considerably influenced by factors like sample disturbance, variability of soil, stress levels and threshold effects and so the possibility that the prediction of heave based predominantly on this parameter may be widely off the mark cannot be ruled out.

D'Appolonia and Lambe (1970) suggest that a relatively more accurate value of $E_{u(e)}$ can be obtained by following the stress path of a laboratory sample which, after first restoring its *in situ* stresses, is subjected to the same stress changes due to unloading as the corresponding element in the ground. Notwithstanding this, it is useful to remember that the history lost in sampling cannot be restored. In any case, the knowledge of the *in situ* stress changes before the test can be performed is required. And so, unless a more refined technique such as a finite element method is employed, one has to settle for the approximate stress distribution based on elastic theory.

Most soils are anisotropic at least to some degree, and the magnitude of this effect is often enhanced by the degree of overconsolidation. The stress path method is, therefore, more appropriate, particularly for overconsolidated soils.

Boussinesq solution for Influence Factor I_σ is available from Butler's graphs presented in Appendix E.

Adaptation of Butler's type solution

Estimates of immediate undrained deformation of the corner of a uniformly loaded, flexible rectangular area on the surface of a heterogeneous elastic layer with the modulus increasing linearly with depth can be made with the use of Butler (1975) type approach.

The principle adopted in this approach is similar to that which involves a simple extrapolation of Steinbrenner's (1934) approximation for settlement calculations. Boussinesq displacements for an infinite depth are used, and the effect of the rigid base is simulated by the approximation that the compression of a finite layer of z metres thickness on a rigid base is equivalent to the compression within the top z metres of an infinitely deep deposit.

Using the same principle for unloading as for loading, the heterogeneity may be represented by:

$$E_{u(e)} = E_{uo(e)} \left(1 + k \cdot \frac{z}{B}\right) \tag{20.13}$$

where $E_{uo(e)}$ is the undrained Young's modulus for unloading at the surface and k its rate of increase with depth.

The deformation at the centre can be obtained by symmetrically dividing the area into four rectangles forming a common point at the centre and adding the corner deformations for each subscribing rectangle, i.e.

$$\delta_{ec} = 4 \cdot \frac{q \cdot B \cdot I}{E_{uo(e)}} \qquad (20.14)$$

The values for the Influence Coefficient I for use in Butler's approach are given in Appendix E.

It is important to recognize that whereas the above formulae relate to surface loading, in the case of an excavation the negative load is applied at depth and so to get 'sensible' answers, depth correction factor such as that proposed by Fox (1948) for settlement calculations may be applied.

To allow for heterogeneity, the stratum should be divided into a number of equal discrete layers and the average E-value in each layer estimated with the use of Equation (20.13). Then using appropriate factors for the geometry of the soil below formation, total heave can be obtained by using the principle of superposition. The use of this procedure is illustrated in a solved example in Chapter 21.

Adaptation of modified Janbu et al. type solution

For linear, homogeneous, isotropic elastic half-space, by adopting the deformation parameter for unloading and making an allowance not only for the geometry of the problem but also for the depth and thickness of the 'elastic' material, it is suggested that the average deformation of a flexible foundation may be obtained by suitably modifying the expression proposed for settlement by Janbu *et al.* (1956) as follows:

$$\delta_e = \frac{q \cdot B}{E_{u(e)}} \cdot \mu_o \cdot \mu_1 \qquad (20.15)$$

where μ_o and μ_1 are dimensionless parameters, the former representing the effect of the embedment and the latter, that of the deforming layer itself.

For deformation of structures on elastic incompressible soils, Christian and Carrier (1978), proposed an improved version of the Janbu *et al.* chart using Giroud's (1968) results for the effects of depth (μ_1) and Burland's results for the effects of embedment (μ_o). These revised factors are obtained from Figure 20.5. In the context of heave calculations, the parameter H may be looked upon as the depth below formation over which the diffusion of stress relief reduces to zero (i.e. the seat of heave). The authors also observed that ignoring the effects of embedment completely, i.e. setting $\mu_o = 1$, gave reasonably satisfactory results in most cases.

It is possible to extrapolate the procedure to extend its application to a multilayered soil using the principle of superposition.

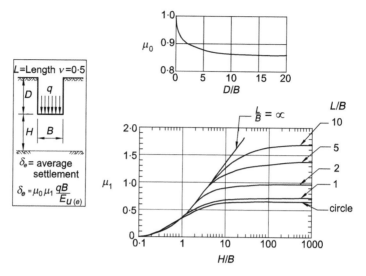

Figure 20.5 Parameters μ_o and μ_1 Parameters
(adapted from Christian and Carrier, 1978; Janbu, Bjerrum and Kjaernsli, 1956)

Strain modulus method

Based on his experience of highly compressible Mexican silty clays in which heave due to response of elastic soil elements takes place rapidly during excavation, Zeevaert (1973) proposes the following formula for elastic heave when porewater pressure equilibrium is restored:

$$\delta_e = \sum_1^n \left[M_{er} \left(\Delta\sigma_r' \right) \cdot d \right]_i \qquad (20.16)$$

As stated in Chapter 23,

$$M_{er} = v_c \cdot \rho_e \cdot M_{eo} \qquad (20.17)$$

$$v_c = \frac{(1+v)(1-2v)}{(1-v)} \qquad (20.18)$$

$$\text{and } \rho_e = \left(\frac{\sigma_r'}{\sigma_o'} \right)^{(c-1)} \qquad (20.19)$$

The parameter M_{er} is the equivalent secant strain modulus of elasticity for unloading corresponding to $\Delta\sigma_r'$, the average effective stress relief typically in the ith stratum of thickness d out of a total of n number of strata. Other parameters are as defined in Chapter 23.

For a stage of excavation where porewater pressure equilibrium has not been restored and, accordingly, a hydrodynamic condition might also exist, the expression

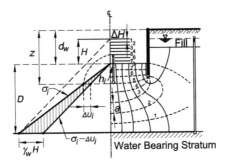

Figure 20.6 Flow Net to Estimate the Change in Effective Stress (after Zeevaert, 1973)

for elastic heave in Equation (20.16) can be modified (Zeevaert, 1973) as follows:

$$\delta_e = \sum_1^n \left[M_{er} \left(\Delta\sigma_r' + \Delta u \right) \cdot d \right]_i \tag{20.20}$$

$$\text{where} \quad \Delta u = \gamma_W \int_0^{(z-d_w)} i_z \cdot dz \tag{20.21}$$

The component Δu is the 'additional change in effective stress' at the centre of the ith stratum as shown in Figure 20.6 resulting from the hydrodynamic conditions in soil due to pumping from excavation; z is the depth below surface of the centre of the ith stratum; d_w is the depth below surface to which water table has been taken down because of pumping and i_z is the hydraulic gradient at depth z which can be obtained from a flow-net.

It should be recognized that, depending upon the type of confinement, the strain may be different for the same value of the linear strain modulus as given by Equation (20.17). In the case of total confinement, i.e. for $v = 0.5$, v_c and the strain will both be zero. That is to say, the material cannot heave when lateral displacements are totally restricted. For no restriction to lateral deformation, i.e. for $v = 0$, $v_c = 1$. So, in reality, the value of the parameter v_c could lie somewhere between 0 and 1, according as the Poisson's ratio v lies between 0.5 and 0.

Hashash–Whittle approach

Hashash and Whittle (1996) carried out a series of parametric studies using nonlinear finite-element analyses to investigate the effects of wall embedment depth, support conditions including different brace spacings, and stress history profile on ground movements caused by deep excavations where the diaphragm wall is embedded in a very deep layer of soft clay. Four soil stress history profiles are considered, i.e. normally consolidated ($OCR = 1$, $K_o = 0.53$), overconsolidated ($OCR = 2$ and 4; $K_o = 0.69$ and 1.00) and composite soil representing a succession of 27m of normally consolidated clay ($OCR = 1$), overlain by a 15m thick crust of overconsolidated clay ($OCR = 2$) and a further 12m of marine sand and cohesive fill ($OCR = 4$). The time frame considered is sufficiently short to ensure minimal drainage within the clay so that the resulting heave can be assumed, essentially, to be undrained.

Effective vertical stress at the surface is taken as $24 \cdot 5 kN/m^2$ and the rate of its increase per metre depth as $8 \cdot 19 kN/m^2$, such that $\sigma'_{vo} = 24 \cdot 5 + 8 \cdot 19 z k N/m^2$ and $\sigma'_{ho} = K_o \sigma'_{vo}$.

The study considers an idealized plane strain excavation geometry supported by 900mm-thick diaphragm walls with water table at $2 \cdot 5m$ below surface. Because of the symmetry, half the cut width (20m) is used in the analyses. The analyses assume a simplified construction sequence as shown in Figure 20.7 comprising the following steps:

- The soil is initially excavated to a depth h_u without lateral support.
- The wall is propped at the surface, and the excavation proceeds to a depth h_e in $2 \cdot 5m$ steps.
- Second level of support is installed at a spacing h ($\leq h_e$), and the excavation is advanced by a further interval h_e.
- The last step is repeated until either failure occurs or the excavation reaches the total depth H of 40m.

By varying the parameters h_u, h_e and h, a wide range of construction sequences are modelled. Figure 20.7 summarizes the ranges of the principal geometric parameters (D, h_u and h) considered in the study. The analyses consider the wall to be 'wished-into-place' so that its installation is assumed not to disturb the surrounding soil. Consequently, the movements likely to occur during the excavation for and construction of the wall are not considered.

The numerical study shows that the wall depth, particularly the penetration below formation, is an important parameter affecting the stability of braced excavations in a deep clay layer. It confirms that the conventional limit equilibrium methods for estimating the factor of safety against base heave do not consider the beneficial influence of wall embedment (for cases where $D \gg H_f$) and overestimate the safe depth (H_f) of excavation obtained in the numerical experiments (for the constant $OCR = 1$ profile). The study also reveals that, for overconsolidated clays with constant $OCRs$

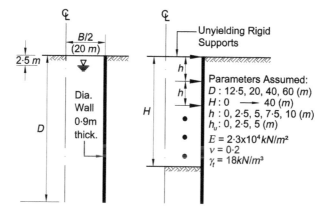

Figure 20.7 Model Geometry, Support Conditions (adapted from Hashash and Whittle, 1996)

Table 20.1 Coefficients for Maximum Heave

Stress history profile – OCR	Coefficients			
	a	$b\ (m^{-1})$	$c\ (m^{-1})$	$d\ (m^{-2})$
1·0	3·0449	−0·0244	0·3070	0·0304
2·0	4·4012	0·0180	0·0	0·0
4·0	2·6961	0·0161	0·0	0·0
Composite	0·0	0·0	0·3841	0·0200

(adapted from Hashash and Whittle, 1996)

of 2 and 4, the centreline heave is an approximately linear function of the excavation depth H, although the soil stiffness is nonlinear at small strains. For estimating the maximum undrained heave δ^h_{max} (mm), the authors have come up, based on curve-fitting, with an empirical relationship given in Equation (20.22), where a, b, c and d are constant dimensional coefficients as listed in Table 20.1.

$$\delta^h_{max} = H\left[ae^{bh} + (c + dh)H \right] \tag{20.22}$$

Other parameters, H (m) and h (m), are as defined previously. For overconsolidated profiles, where movements are linear functions of depth, $c = d = 0$.

The empirical relationship for the undrained elastic heave represented by Equation (20.22) above takes account, after a fashion, of the stress history of the soil, the depth of excavation, the depth of wall embedment and the vertical spacing of the stages of braces.

Progressive or 'consolidation' heave

Like the initial undrained heave, the time-dependent progressive heave may also be obtained with the use of Equations (20.12), (20.14) or (20.15) by replacing the pertinent undrained parameter $(E_{u(e)})$ with the corresponding drained one $(E'_{(e)})$. Alternatively, the following methods may be used:

Voids ratio method

Where the loaded area is large relative to the thickness of the expanding soil, the heave involves essentially one-dimensional strain. The stress distribution is also one-dimensional. Heave may be calculated simply from the e-log σ'_V plot (Lambe and Whitman, 1969 by the following formula:

$$\delta_c = \left[\frac{\sum H_o}{1 + e_o} \right] \cdot \Delta e \tag{20.23}$$

where, H_o is the thickness of the stratum, e_o is the initial voids ratio, and Δe is the change in voids ratio.

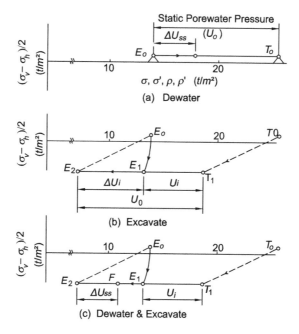

Figure 20.8 Stress Paths for Construction (adapted from Lambe, 1968)

Stress path method

Using the stress path method, it is possible to predict not only the initial, undrained heave and 'consolidation' heave upon porewater pressure dissipation, but also to identify contributions from the various stages, e.g. dewatering, excavation, etc. However, before the commencement of the test, it is essential to restore in the sample the undisturbed, *in situ* stresses existing prior to sampling and also to know the anticipated stress changes through which the soil around the cut is likely to pass during the various stages of excavation. Once again, the stress changes may be obtained from elastic theory.

Figure 20.8 shows stress paths in the p, p' *vs.* q stress space for dewatering, excavation, and combined dewatering and excavation (Lambe, 1968).

In figure (*a*), T_o and E_o represent the stress points for total and effective stresses, respectively, and the distance $T_o E_o$ the static porewater pressure u_o. Immediately upon dewatering, there is no change in the total stresses or porewater pressures. But as the dewatering approaches steady-state equilibrium, the time-dependent porewater pressure dissipation causes the effective stress point to move from E_o to F. Dewatering obviously increases p', thereby giving rise to settlement.

During the excavation phase, as shown in figure (*b*), the total stress point will move from T_o to T_1 whereas the effective stress will trace the path $E_o E_1$, generating, in the process, an excess negative porewater pressure Δu_i ($= E_1 E_2$). With time, porewater pressures will dissipate from E_1 to E_2, causing the effective stresses to decrease, resulting in heave. If dewatering and excavation are proceeded with simultaneously, the stress paths may look somewhat like those shown in figure (*c*). The *TSP* is represented by $T_o T_1$ and *ESP* by $E_o E_1 F$. The net movement of point E_1 can be towards the right

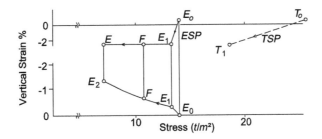

Figure 20.9 Stress Strain Data (adapted from Lambe, 1968)

(due to dewatering) or left (due to excavation), depending on whether the effect of the dewatering component outweighs that of the excavation or *vice versa*. Accordingly, the soil would experience either net settlement or net heave. It is interesting to observe that by judicious control of dewatering *vis-à-vis* excavation, i.e. making the equilibrium point F coincide with E_1, it is, in theory, possible to ensure that there is no net vertical displacement with time. A typical strain diagram is shown in Figure 20.9. The movement at any stage would be obtained by multiplying the strain at that stage by the thickness of the soil layer involved.

Figure 20.10 illustrates the versatility of representing the movement of stress points in terms of stress paths highlighting, in the process, the behavioural differences between

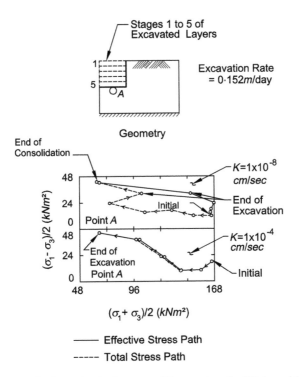

Figure 20.10 Stress Paths for Point in Bottom of Excavation for High and Low Permeability Soils (adapted from Osaimi and Clough, 1979)

the high- and the low-permeability cases. Total and effective stress paths have been traced for a point *A* below the excavation floor for both low- and high-permeability soils. The difference between the two cases is quite apparent.

For the high-permeability soil, as is to be expected, the *TSP* and the *ESP* are essentially identical throughout and after excavation, indicating complete dissipation of the excess porewater pressures apace with the excavation. It is also interesting to observe that, with the progress in excavation, as the drainage boundary becomes small towards the final stages of excavation, the excess negative porewater pressures dissipate quickly indicating consolidation. This is reflected by the sudden convergence of the *TSP* and the *ESP*.

Other mathematical tools

In order to model the various complex boundary conditions and the nonlinear, elasto-plastic nature of soil, mathematical techniques such as the finite element method, etc. which lend themselves to computer programming, are used more and more.

By the use of finite element techniques, it has been possible to show (Osaimi and Clough, 1979) that the conventional assumptions of undrained state at the end of excavation can be grossly wrong and only with the help of a refined technique such as the *FEM*, would it be possible to model with reasonable degree of confidence the indeterminate stress conditions around a cut and the moving drainage boundaries during the removal of soil through excavation. For details of some of these techniques, reference may also be made to Duncan and Chang (1970).

Illustrative examples amplifying the use of some of the methods for estimating heave outlined in this chapter can be found towards the end of Chapter 21.

21 Containment of Heave

21.1 Introduction

Cut-and-cover metro structures are generally located under busy thoroughfares in heavily built-up, high-density urban areas. In such environments, deep excavations, preparatory to the construction of such structures, can lead to heave of the formation which, if not adequately addressed, can cause:

- Total failure of the base if heave is excessive;
- Lateral movements of the surrounding ground and structures;
- In the case of fissured, overconsolidated clays, loss of ground due to softening of soil if it comes into contact with water;
- Surface settlement of the ground surrounding the excavation and the potential risk of damage to the structures in the vicinity.

Lowering of the water table in the ground surrounding the excavation, which is sometimes resorted to in order to reduce hydrostatic pressures that the ground support structure is called upon to withstand during excavation, can give rise to settlement of the surrounding ground. This can, in turn, also subscribe to additional undrained heave of base. Before proceeding with such a measure, all its implications should be carefully considered.

In view of the foregoing, it is imperative that, where large movements anticipated around an excavation are to be avoided, appropriate measures are taken to alleviate the situation by containing heave to within acceptable limits. In the case of deep cuts in soft soils, particularly, underestimating the potential of heave and its associated implications could be disastrous. In dealing with such soils, extreme care must be exercised to prevent failures and to ensure the integrity of the cuts.

21.2 Measures to contain heave

Realistically, heave due to stress relief cannot be completely eliminated; however, by controlling ground movements above and below the formation that are responsible for contributing to heave directly or indirectly, it may be possible to contain heave to within acceptable limits. Which particular measure, or a combination of measures, is likely to yield the desired result will depend on the nature of the site-specific problem. To assist the engineer in arriving at the appropriate decision under a given set of conditions and constraints, salient factors that need to be borne in mind have been identified and are briefly discussed in the following section.

Above the formation

In the case of deep cuts, the factors above the formation that can influence the extent of heave are as follows:

Size of excavation

It is commonly accepted that bigger the size $(L \times B)$ of a cut that is opened at any time, larger the expected heave. It would seem logical therefore that, in the case of metro stations where large areas are involved, consideration should be given, whenever possible, to carrying out excavation and construction piece-meal. The optimum size of excavation opened at any given stage can be defined by the extent of heave acceptable at that stage.

Staged excavation

The phenomenon of heave can be minimized if the change in the effective stress relief is made as small as possible. If it were possible to replace the foundation structure, part by part, by total substitution, i.e. replacing, step by step, the weight of the soil excavated by the weight of the structure, then the heave would be very small. In the case of deep metro excavations, this may not always be possible or practical. However, it may be possible to carry out the excavation in stages. The first stage depth to which the excavation may be made *en masse* can be fixed on the basis of the permissible heave and the second stage can then be proceeded with in step-by-step replacement. In this manner, the heave can be limited to that essentially due to the first stage.

Rate of excavation

The rate at which an excavation is carried out and the time over which it stays open can have a considerable influence on the amount of heave, particularly in soils of high drainage characteristics. Even occasional sand and silt lenses strategically disposed in an otherwise low-permeability soil can form an efficient drainage outlet for the dissipation of excess porewater pressures thereby changing the deformation characteristics of the soil. Under such circumstances, speed of operation can be of the essence. It is also important that the right type of plant is used and used efficiently so as to leave excavation open for as little time as is absolutely necessary. However, the extent of the excavation generally required in the construction of a metro structure can easily last from a year to two years or more and, depending upon its rate, significant progressive heave can also be expected.

Dryness of excavation

In the case of soils with high initial stresses, such as heavily overconsolidated clays, if external water is allowed to come into contact with the formation, softening and swelling can result. In such clays, dryness of the cut is therefore particularly important. It must be appreciated, however, that while blinding of the formation as soon as the excavation is opened will help keep unwanted water away and may be discourage softening and swelling, it is unlikely to prevent heaving of the floor resulting from the relief of the overburden pressure upon excavation.

Lowering of water table

Excavation of soil causes relief of effective stress resulting in heave, whereas lowering of the water table reduces the porewater pressures and increases the effective stress resulting in settlement. By using the expedient of the porewater pressure draw-down, i.e. by dewatering in advance of excavation, it is possible to control the amount of heave. Figure 21.1 illustrates the effect of dewatering upon excavation in two stages. If Stage 2 is carried out in step-by-step replacement, then the heave to contend with will be that due, essentially, to the first stage excavation carried out *en masse*.

The vertical effective stress at formation is given by the expression:

$$\sigma_b' = \gamma d_1 + \gamma_{sub} d_2 + \gamma_{sub} d_3 \qquad (21.1)$$

where, γ is the average unit weight of soil above water table, γ_{sub} is the submerged unit weight of soil $= (\gamma_S - \gamma_W)$, and γ_S is the unit weight of saturated soil below water table.

After first stage excavation, relief in the effective stress at level a–'a' is given by:

$$\Delta \sigma_a' = \gamma d_1 + \gamma_{sub} d_2 \qquad (21.2)$$

This relief in effective stress will tend to give rise to heave.

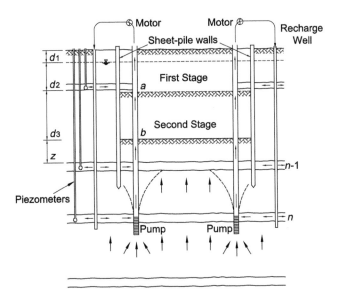

Figure 21.1 Dewatering of Excavation (adapted from Zeevaert, 1973)

If, at the end of Stage 1, the water level is lowered all the way down to a depth z below formation such that the resulting decrease in porewater pressure becomes equal to the relief in the total stress due to Stage 1 excavation, then the relief in the effective stress would have been compensated, i.e.

$$\gamma_W \left(d_3 + z\right) = \Delta\sigma'_a + \gamma_W d_2 \tag{21.3}$$

Upon solving the above three equations simultaneously, we obtain:

$$z = \frac{1}{\gamma_W} \left[\sigma'_b - \gamma_S d_3\right] + d_2 \tag{21.4}$$

So, for the effective stresses to remain roughly the same, the depth by which the water level must be lowered below the formation must be $\geq z$.

The lowering of the water level within the perimeter, besides minimizing heave, would also be needed to enable the excavation to be carried out in the dry. However, outside the perimeter, it may become necessary to recharge the ground by injecting water under pressure, as shown in the figure, to ensure that the change in the effective vertical stress is slowed down and maintained at a sensible minimum.

It is important to appreciate that, to eliminate the potential of deformation completely (i.e. balancing out settlement against heave), it is not enough to simply equalize the increase in the stress due to dewatering with the relief in it due to excavation; it is also necessary to achieve this equalization before either of these operations has had time to make their effect felt. The extent of time separating the two operations can, therefore, be crucial.

If the time lag is sufficient enough for the settlement to be realized under dewatering, then subsequent heave under a similar amount of pressure relief is unlikely to reverse the settlement in full. This is because of the difference in the values of the moduli for loading and unloading. However, since the modulus for unloading is generally much higher than that for loading, it can be safely assumed that at least the heave would be fully reversed even if some residual settlement might remain. It is therefore clear that, to eliminate deformation, the time lag between dewatering and excavation must be minimized. Otherwise, significantly greater pressure relief implying deeper excavation would be needed to fully neutralize the effect of settlement due to dewatering.

Below the formation

Factors outlined in the preceding section are those applicable above the formation level or the base of the cut. In the case of soft soils, particularly, it is imperative to ensure that the inward movement of the walls and the deformation of the ground below the formation are also effectively controlled. This can be achieved by adopting the following measures:

Piling

Piling has been found to work in two ways in controlling heave of the formation. Firstly, by acting in tension over the embedded depth, it can help control upward movement of the formation due to pressure relief upon excavation. Secondly, by increasing the stiffness of the soil below formation, it can also help inhibit laterally inward movement of the walls below the formation.

Lambe (1968) pointed out the influence of piles in counteracting heave in the case of Space Centre Building on the MIT campus in Boston, USA. It was observed that by piling, heave was reduced by about 33 per cent of what could have been anticipated had piles not been used.

Parkinson and Fenoux (1976) estimated a reduction of the order of 40 per cent in heave due to piling. Incidentally, they also indicated a total reduction of about 66 per cent in progressive heave due to the combined effects of piling and water table lowering.

Figure 21.2 shows friction piles placed in ground to a depth d. Sometime after the pile field has remained driven in the ground, if an excavation is made as shown, the soil will tend to heave upon stress relief, mobilizing shear strength in the upper parts of the piles (Zeevaert, 1973) producing a friction force to a depth of z_1 until:

$$q_{ex} \cdot \bar{a} = \int_0^{z_1} s_{z1} \cdot dz \qquad (21.5)$$

where, q_{ex} is the vertical stress relief, $\bar{a}$ is the area of excavation per pile (i.e. unit area of pile grid), and s_{z1} is the skin friction mobilized around per pile.

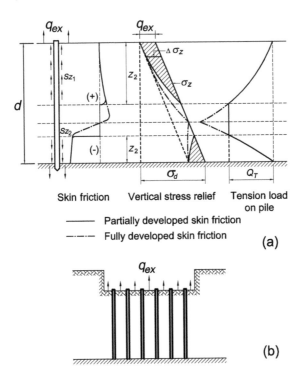

Skin friction   Vertical stress relief   Tension load on pile

—— Partially developed skin friction

—·—·— Fully developed skin friction

(a)

(b)

Figure 21.2 Stress Relief in Soil Mass, no Point Resistance, no Load on Pile
(after Zeevaert, 1973)

The tension load thus generated in the upper reaches, i.e. z_1, of the piles will be resisted by the friction mobilized in the lower portions of the piles to a distance z_2 above the pile tip. The subsoil over the extent z_2 is likely to experience higher shear strength and less compressibility in comparison to that over z_1. Accordingly, s_{z2} will be greater than s_{z1} and z_2 correspondingly smaller than z_1. If the pile penetration is such that:

$$\int_0^{z_1} s_{z1} \cdot dz = \int_0^{z_2} s_{z2} \cdot dz$$

then the piles will preserve the precompressed condition of the soil, thereby minimizing heave due to relief of load upon excavation.

In effect, the elastic heave without piles is:

$$\delta_e = \int_0^d M_{ez} \left(\Delta\sigma_{ex}\right)_z \cdot dz \qquad (21.6)$$

With piles, the elastic heave will be that due to shaded areas only, i.e.

$$\delta_{ep} = \int_0^{z_1} M_{ez} \left(\Delta\sigma_{ex}\right)_z \cdot dz + \int_0^{z_2} M_{ez} \left(\Delta\sigma_{ex}\right)_z \cdot dz \qquad (21.7)$$

where, M_{ez} is the Equivalent Strain Modulus.

To quantify the restraining effect of the piles, let us assume that M_{ez} is constant throughout the depth, then

$$\delta_e = M_e \cdot q_{ex} \cdot d \qquad (21.8)$$

$$\delta_{ep} \simeq \frac{1}{2} \cdot M_e \cdot q_{ex} \cdot z_1 + \frac{1}{2} \cdot M_e \cdot q_{ex} \cdot z_2$$

$$\simeq \frac{1}{2} \cdot M_e \cdot q_{ex} \cdot \left(z_1 + z_2\right) \qquad (21.9)$$

Dividing Equation (21.9) by Equation (21.8), we get:

$$\frac{\delta_{ep}}{\delta_e} = \frac{\left(z_1 + z_2\right)}{2d} \qquad (21.10)$$

If friction is fully mobilized, i.e. $z_1 + z_2 = d$, then, in theory, the ratio $\dfrac{\delta_{ep}}{\delta_e} = \dfrac{1}{2}$.

This implies a 50 per cent reduction in heave! The same principle can be adopted in assessing the containment of heave by driving timber stakes through the ground.

Pre-installing strut

The extent of the inward movement of the perimeter walls below the formation is typically a function, to varying degrees, of five main factors. These are:

- Flexibility of the perimeter walls;
- Extent of wall penetrations below formation;
- Flexibility of the wall support system above the formation;
- Stiffness of soil below formation providing lateral support to the walls;
- Extent of the soil strain necessary to mobilize the required support for the walls.

Bracing of the walls above the formation does not necessarily prevent wall movement below the formation; although, if the bracing system above the formation is less stiff, relatively greater movement below the formation can also be expected.

As also mentioned in Chapter 7, measurements taken on well designed and constructed sheeted excavations show that about 60–80 per cent of the total wall deflections take place below the excavation level (D'Appolonia, 1971). These movements depend upon the extent of the passive resistance that can be mobilized by the soil below the formation against the inward movement of the walls. In the case of excavation reaching a depth whereat bottom stability failure is approached, both the lateral movement and the ground settlement increase sharply.

The nature of the inward movement of the walls below the formation particularly in soft ground, even when the walls are adequately braced above the formation, is likely to be predominantly rotational – rotation taking place at or about the bottom-most brace depending upon its stiffness and whether or not it is pre-loaded. The extent of rotation, however, will depend upon the softness of the ground trapped between the embedments of the walls as well as their flexibility. In soft ground, the inward rotational movement of the embedded portion of the walls can cause upward movement or heave of the soil mass trapped between the walls below the formation. In so doing it is also likely to shed extra load on to the bottom-most brace which, if excessive, can cause its failure and, potentially, trigger progressive collapse of the excavation. The situation can be improved by increasing the stiffness and penetration of the walls. However, this measure may not always prove to be cost-effective. Besides, it may not be enough, on its own, to guarantee success.

It is important to recognize that deep-seated (i.e. below the final formation) movement of the walls is not dependent upon the excavation reaching the formation but may be initiated as early as the excavation is commenced. Thereafter, it will cumulate as the excavation advances downwards towards the formation. In view of this, the most positive way of limiting the wall movements below the formation to the very minimum can be achieved by installing the struts below the formation before commencing the excavation. The struts, in the form of discrete cross-walls, can be installed using the slurry-trench technique. However, in soft ground where, of necessity, the length of wall panels needs to be kept as short as the equipment and the method of excavation will safely allow, the ideal frequency of the cross-wall struts – one per panel – may not always prove to be the most effective.

Alternatively, a continuous strut extending from wall to wall could be preformed immediately below the formation by creating a jet grouted raft of the required thickness as shown in Figure 21.3. This would be intended to provide dual function – to act

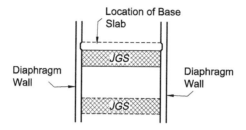

Figure 21.3 Typical Cross-Section (showing alternative locations of JGS)

as a buried strut maintaining the opposite walls the requisite distance apart and also as a base slab to resist uplift pressures – as the overburden is removed during excavation. In either case, whether the strutting action is provided by discrete cross-walls or a continuous raft, the strutting medium would need to be in place before the commencement of excavation. However, in the case of the jet grouted approach particularly, by the very nature of its construction and unreinforced composition, the raft would be subject to its inherent limitations of strength and stiffness. In view of this, the sensitivity of the raft to the action of combined axial compression and bending resulting from the lateral load, the rotational movement of the walls and the upward force of heave due to pressure relief, must not be overlooked or underestimated. Besides, the situation could also be exacerbated by the 'beam-column' (i.e. '$P - \Delta$') effect. As such, the success of such a measure would largely depend, *inter alia*, upon achieving the required:

- Quality and uniformity of grouting; and
- Strength and stiffness of the grouted raft.

However, it is important not to discount the fact that, in practice, achieving the aforementioned two requirements fully in a confined environment cannot be guaranteed. In the event of the grouted raft, when exposed, not quite measuring up to the requirements, it could experience excessive deformation. This could, in turn, cause extra load to be shed onto the bottommost brace above the formation. If this proves to be excessive, it could lead to buckling of the brace in the first instance and triggering progressive failure of the bracing system as a whole; the collapse of the entire cut thereafter could only be a matter of time. The consequences of such a scenario could be extremely serious. Design of deep cuts particularly in soft ground, even with the use of grouted rafts, must therefore be approached with utmost circumspection.

In view of the enormity of the consequences in the event of the grouted strut not functioning quite as anticipated, it would be prudent to forestall such a possibility by building-in generous safeguards both in the design of the grouted raft, say, by downgrading its strength, as well as in the design of the bottommost brace above the formation by incorporating an enhanced factor of safety. Furthermore, by profiling the grouted raft in the shape of a parabolically varying thickness (i.e. as a segment of a circle) slab such that it is relatively deeper mid-span than at the ends, it would be possible to invoke its inverted arch action against the upward soil pressure. In this manner it would be better able to counteract the effect of heave. Besides, the outward horizontal thrusts at the ends of the raft likely to result from the inverted arch action would also cancel some of the axial load. This would, in turn, control the inward movements of the walls.

In spite of employing the measures outlined in the foregoing section, if the doubt about the safety and stability continues to persist, it is perfectly feasible to consider installing more than one jet-grouted strut at different levels below formation. Alternatively, the possibility of using tension piles to anchor down the jet-grouted strut can also be explored.

Design of jet-grouted strut (JGS)

In the cross-section of Figure 21.3, as the excavation progresses down to the formation, preinstalled jet-grouted slab will tend to act as a deep seated strut between the walls. In the process, the typical forces that the *JGS* is likely to be subjected to are as shown in Figure 21.4. One of the likely levels at which to locate the strut is that directly below the intended base slab, although it can be argued that the toe of the wall which is likely to be subject to most lateral movement would be the location best served by the *JGS*. However, in extreme circumstances, employing more levels than one of such struts, if so required, should not be ruled out.

The uplift force on the *JGS* will be that due to undrained heave or the differential hydrostatic pressure, whichever may be the greater. The difference between this uplift pressure and the downward pressure due to the self-weight of the *JGS* and any other gravity load acting on it will represent the net, out-of-balance uplift pressure acting on the underside of the strut.

Consider *AB* as the idealized line diagram for the *JGS* with the end conditions assumed as hinged and the loads as shown in Figure 21.5.

Taking end *A* as the origin of the coordinate axes *x* and *y*, and ignoring the effects of shear deformation and the axial shortening, the equation defining the deflected axis of the member is given by:

$$EI\frac{d^2y}{dx^2} = M_x \qquad (21.11)$$

The parameter *y* represents the deflection of the axis of the member, *EI* its flexural stiffness (assumed constant) and M_x the bending moment at any cross-section which is given by the expression:

$$M_x = -P \cdot y - R_A \cdot x + w\frac{x^2}{2} = -P \cdot y - \frac{w \cdot L}{2} \cdot x + w\frac{x^2}{2}$$

where, '*w*' is the net uplift pressure on the strut.

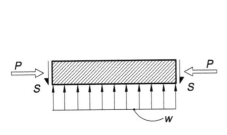

Figure 21.4 Typical Forces on JGS

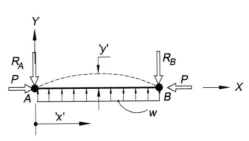

Figure 21.5 Idealized Line Diagram

Upon substitution and rearrangement, Equation (21.11) can be written as:

$$EI\frac{d^2y}{dx^2} + P \cdot y = \frac{w}{2}\left(x^2 - Lx\right)$$

Introducing the axial load factor $n = \sqrt{P/EI}$ (some authors represent it by k; however, n is preferred here to avoid confusing it with the permeability parameter), the above equation can be expressed in the form:

$$\frac{d^2y}{dx^2} + n^2 \cdot y = \frac{wn^2}{2P}\left(x^2 - Lx\right) \tag{21.12}$$

The general solution of this differential equation is:

$$y = A\cos nx + B\sin nx + \frac{w}{2P}\left(x^2 - Lx - \frac{2}{n^2}\right) \tag{21.13}$$

Parameters A and B are the constants of integration. By introducing the appropriate boundary conditions, i.e. $y = 0$ both at $x = 0$ and L, the two constants can be evaluated as:

$$A = \frac{w}{n^2P} \quad \text{and} \quad B = \frac{w}{n^2P}\tan\frac{nL}{2}$$

Upon making these substitutions, the general solution assumes the form:

$$y = \frac{w}{P}\left\{\frac{1}{n^2}\left[\cos nx + \tan\frac{nL}{2}\sin nx\right] + \frac{1}{2}\left[x^2 - Lx - \frac{2}{n^2}\right]\right\} \tag{21.14}$$

By substituting $x = L/2$ in the equation, the maximum upward mid-span displacement can be obtained as:

$$y_{\max} = \frac{w}{n^2P}\left[\sec\left(\frac{180°}{\pi}\cdot\frac{nL}{2}\right) - 1 - \frac{1}{8}n^2L^2\right] \tag{21.15}$$

Substituting this value in the moment equation and ignoring the sign, the magnitude of the maximum mid-span moment, numerically, works out to:

$$M_{\max} = EI \cdot \frac{w}{P}\left[\sec\left(\frac{180°}{\pi}\cdot\frac{nL}{2}\right) - 1\right] \tag{21.16}$$

Discussion of the assumptions

In the derivation of the above expressions, one of the principal assumptions has been that, typically, the end condition of the JGS at its junction with the wall is hinged. This implies freedom of rotation for the ends of the strut. In reality, high axial compression in the strut may simply not permit its separation from the walls in which case rotation is likely to be prevented and the fully hinged boundary conditions may not be realized.

However, any tendency on the part of the axial compression to inhibit rotation coupled with the resulting inability on the part of the strut to rotate freely would only subscribe to a reduction in the mid-span moment thereby improving the situation. So, treating the ends as hinged can only be a safer assumption for the design of the *JGS*.

The other assumption is that there is no differential movement vertically at the interfaces between the *JGS* and the walls. However, this may not always be the case if it is not practical, for instance, to toe down the walls into hard stratum or rock as a result of which they would remain embedded entirely in the soft ground itself. Under those circumstances, downward movement of the walls relative to the strut and the consequent increase in the mid-span moment in the strut cannot be ruled out. The enhancement in the moment will then be due not only to the differential slip at the interface but, potentially, the added '$P - \Delta$' effect also. Should this turn out to be excessive and beyond the capability of the *JGS* to resist safely, failure and even collapse of the excavation cannot be ruled out.

Estimate of the differential slip can be made by comparing the downward load through the wall against the counteracting frictional force likely to be mobilized at the soil–wall interface. If the former happens to be greater than the latter, as can be expected in a top-down sequence of construction, downward movement of the wall will ensue causing the enhancement in the moment as explained in the foregoing section. However, even if it is not so, downward movement of the wall cannot be avoided altogether since, for achieving equilibrium, friction to oppose the downward force can be mobilized only if there is some relative movement at the interface. So, depending upon the extent of such movement, some enhancement in the mid-span moment is only to be expected.

In order to obtain an idea of the extent of impact the differential slip can have on the moment in the *JGS*, assume a relative downward displacement of the end walls of Δ at the wall–*JGS* interfaces. Assuming also that fixed-end conditions prevail at the interfaces, the moments developed at the ends and the midspan of the *JGS* will each be $24EI\Delta/L^2$, where L is the span wall-to-wall of the *JGS*. If hinged conditions are assumed at the interfaces, the moments will be halved. However, even so, with the thickness of the *JGS* greatly in excess of $1m$, as is to be expected, the moments can turn out to be significant and must not be overlooked. This is illustrated in the solution to Example 21.7 at the end of the chapter.

To limit such enhancements to the very minimum, every avenue of minimizing the downward movement of the wall should be explored.

To sum up, measures generally available to contain the adverse effects of the potential deep-seated movements on *JGS* in soft ground may be listed as follows:

- Preinstallation of intermediate piles through *JGS*;
- Toeing the walls into a hard stratum or rock;
- Increasing the thickness of the *JGS*;
- Increasing stiffness of the walls;
- Treatment of the ground;
- Profiling the *JGS*.

Where circumstances so warrant and conditions are amenable, one or more of the measures outlined above may need to be considered.

Other measures

Excavation under water

In certain instances, it may be feasible to limit relief of stress upon excavation partially if open grabbing is carried out under a head of water. A thick plug of concrete can then be laid under water before resorting to dewatering. However, the operation has the disadvantage of necessitating the use of divers to inspect the formation besides the usual problems associated with underwater concreting.

21.3 Critique of the assumptions

In the available literature, a distinct paucity of properly documented, well-interpreted and extensively discussed case histories on heave has been noticed. Many more case histories need to be back analysed not only to improve knowledge about the relevant *in situ* deformation parameters but also to narrow the gap between prediction and observation. Meanwhile, however, it will be useful to critically examine the validity of the various assumptions employed in the assessment of heave as follows.

Undrained behaviour

During the period of excavation, it is generally assumed that the excess negative porewater pressures, out of balance with the ambient porewater pressure, are set up presenting an undrained case and eventually dissipating, with time, to their steady-state values. However, it is clear that the assumption that, during excavation, heave is exclusively undrained, can be extremely doubtful.

In practice, it is neither possible nor desirable to prevent drainage altogether and so the validity of the undrained assumption will depend, among other things, on:

- The rate of excavation
- Permeability of soil
- Drainage geometry.

If the rate of excavation is faster than the propensity of the soil to heave, unloading may be considered as undrained. However, Osaimi and Clough (1979) have pointed out that the classical undrained analysis provides, in most cases, only a design expedient. They were able to demonstrate that while with very low permeabilities, of the order of $1 \times 10^{-8} cms/sec$, the mode is essentially undrained regardless of the rate of excavation, with k-values encountered in practice, the assumption of the undrained behaviour is very rarely realized; they were able to conclude that the porewater pressure dissipation, and hence progressive heave, is very much a function of the rate of excavation. The authors observed the following for a wide excavation.

The presence of sand and silt lenses in an otherwise low-permeability soil can alter the porewater pressure movement characteristics considerably. We have also seen that, with the stages of excavation, the porewater pressure boundaries change. Rowe (1968, 1972) has established that the field permeability can easily exceed the value obtained in the laboratory by many times. Oteo (1979) has also indicated that the K-values measured in the laboratory could be underestimated by about six times due to the presence of thin sandy and silty layers inter-bedded with clay.

Table 21.1 Permeability versus Consolidation

Excavation rate (m/day)	k (cms/sec)	% Consolidation
0·2–1·5	1×10^{-6}	Little
- do -	1×10^{-4}	90

(adapted from Osaimi and Clough, 1979)

Although it may not be possible to define a criterion for an undrained analysis to be valid, it is important that very little porewater pressure dissipation takes place. Further field studies and research are necessary to isolate the effect and quantify the contributions of heave during excavation due to undrained behaviour and 'consolidation' as a result of porewater pressure dissipation.

Theory of elasticity

In heave calculations, the diffusion of stress relief is obtained using the theory of elasticity, whereas it is recognized that the real soils are elasto-plastic in behaviour. Only if the stress decrements due to excavation are small, will the theory of elasticity yield reasonable results, but when the soil yields or plastic zones develop, the method becomes suspect. It has been seen that, even at factors of safety as high as three, local yield and plastic zone formation can occur.

No method is available that allows for a redistribution of stresses consequent upon yield. Furthermore, by definition, the deformation of an elastic material is path independent. That is to say, between two stress points in stress space, strains are independent of the stress path followed between the two points. In a plastic material, on the other hand, the deformations between two stress states are path dependent (Drucker, 1964).

Threshold effects

Atkinson (1973) likened the apparent stiffening effect of a soil at small levels of stress changes to a threshold effect which he characterized as a soil property. Since the threshold effects due to depth (Stoll, 1971) or due to small levels of stress variations (Atkinson, 1973) cannot be properly modelled, the laboratory measured parameters may overestimate the actual strains for values of depth or stress relief lying within the threshold field.

Heave versus depth

Most heave originates in layers close to formation (Lambe, 1968; Butler, 1975; Skopek, 1976). But this cannot be attributed entirely to the increasing stiffness with depth. Lambe (1968) observed that 67 per cent of total foundation heave originated from the top heavily overconsolidated layer, whereas that in the underlying layers of overconsolidated and normally consolidated soil, it was approximately equal. This could well be due to the fact that, close to the formation, the drainage path is the shortest and consequently the deformation the greatest. Furthermore, a heavily overconsolidated formation is also particularly susceptible to softening and additional swelling if free water comes into contact with it (Henkel, 1970; Breth and Amann, 1974).

Stress path method

A good way of discovering how a soil element in the field will deform under stress is to observe how an identical soil element in the laboratory deforms when it is subjected to the stresses which will be imposed in the field. Although the basic idea is simple and sound, there are certain difficulties with the stress path method which must be appreciated. These are:

- It is extremely difficult to retrieve truly undisturbed samples from the ground. Disturbance of a sample usually reduces its stiffness somewhat.
- For want of a better method, distribution of stress relief is estimated by the use of elastic theory. However, this can only be approximate since soil is not perfectly elastic.
- Elastic analysis cannot allow satisfactorily for any change of total stresses that may occur during dissipation of excess porewater pressures.
- Actual stress path followed in the ground may be difficult to simulate in the laboratory.
- The stress history that is lost during sampling cannot be restored.

21.4 'Ideal' analytical model

If it were possible to evolve an ideal analytical model, it should be able to:

- Restore stress history;
- Recognize threshold levels;
- Represent soil–structure interaction;
- Redistribute stress relief upon soil yield;
- Simulate the elasto-plastic behaviour of soil;
- Model soil properties such as nonlinearity, anisotropy, heterogeneity;
- Permit variations in porewater pressure boundaries and drainage characteristics in the soil mass;
- Reflect the influence of the rate of excavation, swelling characteristics, permeability of the soil medium;
- Identify contributions due to:
  - Yield of soil
  - Undrained behaviour
  - Creep effects at constant effective stress
  - Consolidation due to dissipation of porewater pressures.

With, potentially, so many variables involved, it is not surprising that total confidence in the available methods of predicting heave is still lacking and considerable reliance is placed on the expensive field instrumentation and observations to provide the confirmation. Until the gap between the predictions and the observed values is significantly narrowed and is also consistently seen to be so, the uncertainty will continue to persist.

To evolve a semiempirical formula or an ideal analytical model, which incorporates all the aforementioned factors, is virtually impossible. However, well-planned and properly carried out instrumentations can furnish a wealth of information that can be extremely useful. In the case of a very deep excavation, such as that for a metro station structure, where excavation is carried out in stages, monitoring the excavation offers an ideal opportunity for adopting the 'observational method'. The data available from a series of such field studies must be used to refine, progressively, the methods of predicting heave by back analysing successive stages of excavation.

Example 21.1

A 300*m*-long metro station box structure is to be constructed in the conventional bottom-up sequence. Width of the excavation between the longitudinal sidewalls is 18*m* and the depth to the formation 15*m*. The embedment of the sidewalls below the formation is 10*m*. Examine the stability of the cut and ensure that a factor of safety of at least 1·5 is achieved assuming the following parameters:

- Average undrained shear strength around the base of the cut of $40kN/m^2$;
- Intensity of the surcharge load on the surface of $10kN/m^2$;
- Unit weight of soil of $18kN/m^3$;
- Adhesion factor of 0·8.

Solution

From the given data, we have: $D = 25\,m, B = 18\,m, H = 15m, h = 10m, s_u = 40kN/m^2$ and $q = 10\,kN/m^2$.

Case (a): Ignoring wall embedment

$H/B = 15/18 = 0.83$. The length of the cut is many times more than the width of the cut and so, for all practical purposes, B/L can be treated as zero. Correspondingly, from Figure 20.1(*b*), Chapter 20, $N_c = 6.4$. Factor of safety against base failure is given by the expression in Equation (20.2):

$$F = \frac{N_c \cdot s_u}{(\gamma H + q)} = \frac{6 \cdot 4 \times 40}{(18 \times 15 + 10)} = 0.91 - \text{Not adequate!}$$

Case (b): With wall embedment

$D/B = 25/18 = 1.4$. For $B/L = 0$, the corresponding value of $N_c = 6.85$. Recalling Equation (20.3) from Chapter 20, the factor of safety against base failure is:

$$F = \frac{s_u}{(\gamma H + q)}\left[N_c + (2h/B)\alpha\right] = \frac{40}{(18 \times 15 + 10)}\left[6 \cdot 85 + (2 \times 10/18)0 \cdot 8\right]$$

$$= 1 \cdot 1 - \text{Not enough!}$$

Case (c): With cross-walls and excavating under single cell

As the first trial, assume 1*m* thick (i.e. $T = 1$) cross-walls at 9*m* centres, so that $L = 8$. This makes $L < B$, so that D/L (instead of D/B) $= 25/8 = 3.125$. For $B/L = 8/18 = 0.44$, from the graphs of the Figure 20.1(b) the corresponding value of $N_c = 8$.

By substituting $n = 1$ in Equation (20.4), or simply recalling Equation (20.5), the factor of safety against base failure can be obtained from the expression:

$$F_{1c} = \frac{s_u}{(\gamma H + q)} \left[N_c + \frac{2(B+L)h}{B \cdot L} \cdot \alpha \right]$$

$$= \frac{40}{(18 \times 15 + 10)} \left[8 + \frac{2(18+8)10}{18 \times 8} \times 0 \cdot 8 \right] = 1 \cdot 56 - \text{ ok}$$

Case (d): With cross-walls and excavating under two cells

In this case, $L_{eff} = 2 \times 8 + 1 = 17$. This is $< B$, so that $D/L = 25/17 = 1 \cdot 47$. For $B/L = 17/18 = 0 \cdot 94$, from the graphs of Figure 20.1(b), the corresponding value of $N_c = 8$. Recalling Equation (20.4) from Chapter 20, the factor of safety against base failure for the n-cell case is:

$$F_{nc} = \frac{s_u}{(\gamma H + q)} \left\{ N_c + \frac{1}{[nL + (n-1)T]} \left[2nh \left(1 + \frac{L}{B} \right) \cdot \alpha + (n-1)TN \right] \right\}$$

By substituting $n = 2$ and $T = 1$, the factor of safety for the 2-cell case works out to:

$$F_{2c} = \frac{s_u}{(\gamma H + q)} \left\{ N_c + \frac{1}{(2L+1)} \left[4h \left(1 + \frac{L}{B} \right) \cdot \alpha + N \right] \right\}$$

$$= \frac{40}{(18 \times 15 + 10)} \left\{ 8 + \frac{1}{17} \left[40 \left(1 + \frac{8}{18} \right) (0 \cdot 8) + 7 \cdot 5 \right] \right\} = 1 \cdot 59 - \text{ ok.}$$

Case (e): With cross-walls and excavating under three cells

In this case, $L_{eff} = 26$. This is $> B$, so that $D/B = 25/18 = 1 \cdot 39$. For $B/L = 18/26 = 0 \cdot 69$, from the graphs of the Figure 20.1(b) the corresponding value of $N_c = 7 \cdot 62$. By substituting $n = 3$ and $T = 1$ in Equation (20.4) the factor of safety for the 3-cell case is:

$$F_{3c} = \frac{40}{(18 \times 15 + 10)} \left\{ 7 \cdot 62 + \frac{1}{26} \left[60 \left(1 + \frac{8}{18} \right) (0 \cdot 8) + 2 \times 1 \times 7 \cdot 5 \right] \right\}$$

$$= 1 \cdot 55 - \text{ ok.}$$

Proceeding on similar lines, the factor of safety for the 4-cell case works out to: $1 \cdot 44$.

Case (f): Alternatively, with cross-walls and excavating under multiple cells

For $D/B = 25/18 = 1 \cdot 4$ and $B/L = 0$, the corresponding value of $N_c = 6 \cdot 85$, as for Case (b). Recalling Equation (20.6) from Chapter 20, the factor of safety against base failure is:

$$F_{mc} = \frac{s_u}{(\gamma H + q)} \left\{ N_c + \frac{1}{(L+T)} \left[2h \left(1 + \frac{L}{B} \right) \cdot \alpha + TN \right] \right\}$$

$$= \frac{40}{(18 \times 15 + 10)} \left\{ 6 \cdot 85 + \frac{1}{(8+1)} \left[2 \times 10 \left(1 + \frac{8}{18} \right) (0 \cdot 8) + 1 \times 7 \cdot 5 \right] \right\}$$

$$= 1 \cdot 46$$

It is clear from the above solution and as is also to be expected that, with the progressive increase in the length of the excavation there is a commensurate reduction in the factor of safety. However, the marginal increase in the factor of safety in the 2-cell case can be attributed to the improvement in the geometry of the problem with the excavated area moving away from a plane strain condition towards an axi-symmetric (square) case. It is also evident that with the spacing of cross-walls as assumed (i.e. 9*m*) and excavating no more than three cells at a time, in this example, the required factor of safety can be achieved and so no further trials are necessary.

Example 21.2
An excavation for a metro station structure is 300*m* long, 20*m* wide and 15*m* deep to formation. Ambient water table is 2*m* below ground level. Assuming the soil to be homogeneous and isotropic (i.e. displaying a constant modulus) with unit weight of 20*kN*/*m*³ (idealized for simplicity) and an undrained, elastic, unloading modulus $E_{u(e)}$ of $5 \times 10^4 kN/m^2$, establish the undrained elastic heave profile against depth below the formation.

Solution
The length of the excavation is significantly (20 times) longer than its width. As such, it would be reasonable to regard the deformation of the formation as a two-dimensional plane strain problem. Heave profile below formation will be obtained using the Janbu *et al.*-type approach using the nondimensional parameters duly modified by Christian and Carrier.

Use of modified Janbu et al. type approach

Total stress relief at formation level,

$$q = 20 \times 15 = 300 \, kN/m^2$$

For the excavation, $D/B = 15/20 = 0.75$. For homogeneous, isotropic soil assuming constant unit weight throughout the depth, the depth below formation (*H*) at which the elastic heave reduces to zero will be 15*m*, so that $H/B = 15/20 = 0.75$. Then, from the graphs in Figure 20.5, Chapter 20, the dimensionless parameters are obtained as: $\mu_o = 0.93$, and $\mu_1 = 0.28$. Therefore, using Equation (20.15), Chapter 20, elastic heave at formation,

$$\delta_e = \frac{300 \times 20}{5 \times 10^4} \times 0.93 \times 0.28 \times 10^3 = 31 \cdot 2mm$$

In order to establish heave-depth profile, reduction of heave at 5*m* and 10*m* below formation corresponding to their respective reductions in the stress relief will also be estimated.

Reduction of heave at 5m below formation

Geometry of the excavation remaining the same as before, $\mu_o = 0.93$. However, $H/B = 5/20 = 0.25$. Therefore, from Figure 20.5, $\mu_1 = 0.08$. So, reduction in heave over 5*m* depth is given by:

$$\Delta\delta_{e(5)} = \frac{20 \times 5 \times 20}{5 \times 10^4} \times 0.93 \times 0.08 \times 10^3 = 3mm$$

Therefore, net heave $\delta_{e(5)} = 31.2 - 3 = 28.2mm$.

Reduction of heave at 10m below formation

As before, $\mu_o = 0.93$. Also, $H/B = 10/20 = 0.5$. From Figure 20.5, $\mu_1 = 0.17$. So, reduction in heave over 10m depth is given by:

$$\Delta\delta_{e(10)} = \frac{20 \times 10 \times 20}{5 \times 10^4} \times 0.93 \times 0.17 \times 10^3 = 12.7mm$$

Therefore, net heave $\delta_{e(10)} = 31.2 - 12.7 = 18.5mm$.

Example 21.3

Revisit Example 21.2, assuming, this time, the soil below formation to be layered. Taking unloading elastic modulus to be $6 \times 10^4 kN/m^2$ for the soil stratum up to 7.6m below the formation level and $3.27 \times 10^4 kN/m^2$ for the soil below this stratum, work out the revised heave profile.

Solution

Again, a modified Janbu *et al.*-type approach will be used. However, because of the layering with different elastic moduli, trial-and-error approach will be used to establish the seat of heave, i.e. the depth over which heave reduces to zero.

As a first trial, assume that the heave reduces to zero over a depth of 12m below the formation level and, as an approximation, ignore the layering and treat the soil as homogeneous, isotropic, with an elastic modulus equal to the weighted average of the moduli of the two layers. Then the ratio $H/B = 12/20 = 0.60$ and, from the graph in Figure 20.5, Chapter 20, the dimensionless parameter $\mu_1 = 0.2$; $\mu_o = 0.93$ as before. The weighted average for the unloading elastic modulus over the two layers is given by: $(7.6 \times 6 + 4.4 \times 3.27) 10^4 \div 12$, i.e. $5 \times 10^4 kN/m^2$. Using Equation (20.15), Chapter 20, elastic heave at formation,

$$\delta_{e(f)} = \frac{300 \times 20}{5 \times 10^4} \times 0.93 \times 0.2 \times 10^3 = 22.32mm$$

In order to establish heave-depth profile, reduction of heave at the depths of 4m, 7.6m and 12m below formation will be estimated. This is done in order to confirm the accuracy of the assumption.

Reduction of heave at 4m below formation

Parameter $\mu_o = 0.93$ as before, and for $H/B = 4/20 = 0.2$, from Figure 20.5, $\mu_1 = 0.05$. So, reduction in heave over $4m$ depth is given by:

$$\Delta\delta_{e(4)} = \frac{20 \times 4 \times 20}{6 \times 10^4} \times 0.93 \times 0.05 \times 10^3 = 1.24mm$$

Therefore, net heave at $4m$ below formation level is:

$$\delta_{e(4)} = 22.32 - 1.24 = 21.08mm$$

Reduction of heave at 7.6m below formation

Parameter $\mu_o = 0.93$ as before, and for $H/B = 7.6/20 = 0.38$, from Figure 20.5, $\mu_1 = 0.11$. So, reduction in heave over $7.6m$ depth is given by:

$$\Delta\delta_{e(7.6)} = \frac{20 \times 7.6 \times 20}{6 \times 10^4} \times 0.93 \times 0.11 \times 10^3 = 5.18mm$$

Therefore, net heave at $7.6m$ below formation level is:

$$\delta_{e(7.6)} = 22.32 - 5.18 = 17.14mm$$

Reduction of heave at 12m below formation

Over this depth, the effect of the layering of the soil in estimating the reduction of heave due to the lower layer can be worked out in two stages using the principle of superposition.

First, estimate the reduction in heave due to the combined weight of both the layers, i.e. over full $12m$ depth, based on the elastic modulus corresponding to the lower layer. Second, estimate the reduction due to the weight of the upper layer alone, again, using the elastic modulus corresponding to the lower layer. The difference between the two then represents the reduction in heave due to the self-weight of the lower layer on its own.

In the first case, as before, $\mu_o = 0.93$. Also, for the two layers combined, $H/B = 12/20 = 0.6$. Therefore, from Figure 20.5, $\mu_1 = 0.2$. So, apparent reduction in heave over $12m$ depth is given by:

$$\Delta\delta_{e(c)} = \frac{20 \times 12 \times 20}{3.27 \times 10^4} \times 0.93 \times 0.2 \times 10^3 = 27.3mm$$

In the second case, for the upper layer $\mu_o = 0.93$ and $\mu_1 = 0.11$, as before. Therefore, apparent reduction in heave due to the upper layer is given by:

$$\Delta\delta_{e(u)} = \frac{20 \times 7.6 \times 20}{3.27 \times 10^4} \times 0.93 \times 0.11 \times 10^3 = 9.5mm$$

So, reduction in elastic heave due to the self-weight of the lower layer alone is:

$$27 \cdot 3 - 9 \cdot 5 = 17 \cdot 8mm \ (cf. 17 \cdot 14mm)$$

Thus, with the effective depth as assumed in the beginning, for all practical purposes, the net heave at $12m$ depth reduces to zero; there is therefore no need to revise the computation assuming a different trial depth. The seat of heave can therefore be taken as $12m$.

Example 21.4

During a particular stage in the construction, partial excavation for a metro station structure is $100m$ long, $20m$ wide and $15m$ deep to current formation. Assume a unit weight of soil of $20kN/m^3$ and $E_{u(e)}$ varying linearly with depth from a value of $2 \times 10^4 kN/m^2$ at the surface to $6 \times 10^4 kN/m^2$ at $30m$ below surface and the stratum below $30m$ to be rigid. Compare the values of average heave as obtained using Butler's charts with that using the modified Janbu *et al.*-type approach.

Solution
Using Butler-type approach

From the relation $E_{u(e)} = E_{uo(e)} \left(1 + k \cdot \dfrac{z}{B} \right)$, upon rearrangement, the rate of change of E with depth can be obtained as:

$$k = \left(\frac{E_{u(e)} - E_{uo(e)}}{E_{uo(e)}} \right) \cdot \frac{B}{Z} = \left(\frac{6-4}{4} \right) \cdot \frac{10}{30} = 0 \cdot 33$$

Divide the plan area of the excavation into four subscribing subrectangles such that, for each sub-rectangle, $L = 50m$ and $B = 10m$. Then, for $L/B = 5$, $z/B = 15/10 = 1 \cdot 5$ and $k = 0 \cdot 33$, from Butler's graph, $I = 0 \cdot 12$. Elastic heave at the centre will be 4 times the heave at the corner of the typical subscribing sub-rectangle, i.e.

$$\delta_{ec} = 4 \cdot \frac{q \cdot B \cdot I}{E_{uo(e)}} = 4 \times \frac{(20 \times 15) \times 10 \times 0 \cdot 12}{4 \times 10^4} \times 10^3 = 36mm$$

$$\text{Average heave, } \delta_{eave} = 0 \cdot 85 \times 36 = 30 \cdot 6mm$$

Using modified Janbu et al. type approach

Divide the $15m$ deep stratum below the formation into three equal layers of $5m$ each. By interpolation, the average elastic moduli for the three layers, from top down, work out to: $4 \cdot 33 \times 10^4$, 5×10^4 and $5 \cdot 67 \times 10^4$, respectively. The ratio $D/B = 15/20 = 0 \cdot 75$ so that $\mu_o = 0 \cdot 93$. Also,

For the top layer, $H/B = 5/20 = 0 \cdot 25$, from graph $\mu_1 = 0 \cdot 07$
For top two layers, $H/B = 10/20 = 0 \cdot 50$, $\mu_1 = 0 \cdot 15$
For all three layers, $H/B = 15/20 = 0 \cdot 75$, $\mu_1 = 0 \cdot 25$

Then, by superposition, average heave,

$$\delta_{e(ave)} = \frac{300 \times 20}{10^4} \times 0.93 \left[\frac{0.07}{4.33} + \frac{(0.15 - 0.07)}{5} + \frac{(0.25 - 0.15)}{5.67} \right] \times 10^3$$

$$= 27.8mm \text{ (cf } 30.6mm)$$

The values of average heave obtained by the two methods, being within 10 per cent of each other, are not significantly different. Of the two, the modified Janbu *et al.*-type approach gives a marginally smaller value.

Example 21.5

In the case of Example 11.2 of Chapter 11, estimate the amount of elastic heave under the structure at the end of Stage 5 using the strain modulus method outlined in Chapter 20. Assume the following data:

- Soil strata under formation to be of high compressibility with unconfined compression test carried out on undisturbed samples for each one of the soil layers encountered.
- Typical values for the elastic response strain modulus for unloading, M_{eo}, as investigated in the laboratory from the hysteresis loops to be used as listed in Table 21.2.
- The value of Poisson's ratio v to be constant at 0.3 throughout the depth, except in layer 4 where it is to be taken as 0.25.
- An average value for the coefficient c of 1.5 for each soil layer.

Solution

Recalling Equation (20.16) of Chapter 20, elastic heave after porewater pressure equilibration can be obtained from:

$$\delta_e = \sum_1^n \left[M_{er} (\Delta\sigma_r') \cdot d \right]_i = \sum_1^n \left[v_c \cdot \rho_e \cdot M_{eo} (\Delta\sigma_r') \cdot d \right]_i$$

From Figure 11.6(*i*) of Chapter 11, it can be seen that the difference $(\sigma_o' - \sigma_5')$ between the initial effective stress σ_o' and the final effective stress σ_5' $(= \sigma_r')$ at the end of Stage 5 represents the uniform net effective stress relief $(\Delta\sigma_r')$ of $90kN/m^2$ throughout the depth from 10m downwards to 30m.

Table 21.2 Soil Strain Modulus Values (M_{eo}-unloading)

Layer	M_{eo} (m^2/kN)	Layer	M_{eo} (m^2/kN)
3	30×10^{-5}	7	10×10^{-5}
4	25×10^{-5}	8	6×10^{-5}
5	12×10^{-5}	9	4×10^{-5}
6	10×10^{-5}		

Table 21.3 Elastic Heave Calculation

Layer No.	d (mm)	M_{eo} (m^2/kN)	σ'_o	σ'_5	$\Delta\sigma'_r$	v_c	ρ_e	δ_e (mm)
				(kN/m^2)				
3	2000	30×10^{-5}	163	73	90	0·74	0·669	26·7
4	2000	25×10^{-5}	200	110	90	0·83	0·742	27·7
5	4000	12×10^{-5}	245	155	90	0·74	0·795	25·4
6	2000	10×10^{-5}	290	200	90	0·74	0·830	11·1
7	4000	10×10^{-5}	330	240	90	0·74	0·853	22·7
8	2000	6×10^{-5}	370	280	90	0·74	0.870	7·0
9	4000	4×10^{-5}	442	352	90	0·74	0·892	9·5

Total heave	For laterally confined conditions:	130·1
	Assuming unrestrained lateral deformation ($v_c = 1$):	171·6

Definition of elastic strain modulus is given in Chapter 23. Poisson's-ratio-related parameter can also be read directly from the graph of Figure 23.6 or obtained from Equation (23.7) of the same chapter as:

$$v_c = \frac{(1+v)(1-2v)}{(1-v)} = \frac{(1+0·3)(1-0·6)}{(1-0·3)} = 0·74$$

The corresponding value in layer 4 based on Poisson's ratio (v) of 0·25 may be read directly from the graph as 0·83.

Expansion parameter ρ_e can be evaluated with the help of the expression in Equation (20.19) of Chapter 20 using the given c-value of 1·5 as follows:

$$\rho_e = \left(\frac{\sigma'_r}{\sigma'_o}\right)^{(c-1)} = \left(\frac{\sigma'_5}{\sigma'_o}\right)^{(0·5)}$$

The elastic heave calculation for laterally restrained and unrestrained conditions is presented in Table 21.3.

Note: It is of no consequence what units are used for M_{eo} and $\Delta\sigma'_r$, as long as they are consistent with each other, which they need to be, as they cancel each other out owing to their reciprocity. In view of this, and since the parameters, v_c and ρ_e, are nondimensional, the unit of heave δ_e ends up being the same as that considered for the layer thickness d.

Example 21.6

For the safe construction of a cut-and-cover metro station structure in soft ground, it is required to ensure that the extent of the potential inward movement of the walls below formation is not such that it would trigger progressive collapse. It is therefore proposed to install, before the commencement of excavation, a jet grouted slab (*JGS*) directly below the level of the intended base slab to act as a deep-seated strut. Assuming an axial load of 600kN/m in the strut, design the thickness of the strut to suit the following parameters:

For the *JGS*:

- Unit weight, $\gamma = 14kN/m^3$
- Length of the strut, $L = 20m$
- Modulus of elasticity, $E = 6 \times 10^4 kN/m^2$
- Allowable bending stress, $f_b = 0.30N/mm^2$.

For the soil:

- Seat of heave, $D = 15m$
- Anticipated average heave, $\delta_{e(ave)} = 100mm$
- Undrained expansion modulus, $E_{u(e)} = 1 \times 10^4 kN/m^2$.

Also, estimate the maximum upward deflection of the strut under the loads.
   Assuming that the walls displace downwards by *5mm* relative to the *JGS*, examine what impact this could have on the integrity of the *JGS*.

Solution
Consider *1m* run of the structure. Recalling Equation (12.4), Chapter 12, the equivalent uplift pressure from ground heave is given by:

$$p_U = E_{u(e)} \cdot \delta_e(ave) \div D = 10^4 (0.1) \div 15 \simeq 67kN/m^2$$

Assume, as a first trial, *JGS* thickness of *4m*. Downward pressure from the self-weight of the strut $= 14 \times 4 = 56kN/m^2$. Therefore, net uplift pressure on the underside of the strut works out to:

$$w = 67 - 56 = 11kN/m^2$$

To allow for any potential uncertainty in the extent of the axial load in the strut from the total lateral ground pressures, assume a variation in the axial load of -100 to $+200kN/m$ yielding a range of $500–800kN/m$. It is proposed to ensure that the design is adequate for this range (i.e. -17% to $+33\%$).
   For the strut, moment of inertia, $I = 1 \times 4^3 \div 12 = 5.33m^4$ and $EI = 32 \times 10^4 kNm^2$.

Critical buckling load for pin-ended member under axial load is given by the following expression:

$$P_{cr} = \frac{\pi^2 EI}{L^2} = \frac{\pi^2 \times 32 \times 10^4}{20^2} = 800\pi^2 kN(\gg 800, \therefore \text{ ok})$$

(a) Axial load, $P = 500kN$
   For the strut, axial load parameter, $n = \sqrt{P/EI} = \sqrt{500/(32 \times 10^4)} = 0.0395$, and $nL/2 = 0.395$. Recalling Equation (21.16), the maximum mid-span moment

in the strut is given by:

$$M_{max} = EI \cdot \frac{w}{P} \left[\sec \left(\frac{180°}{\pi} \cdot \frac{nL}{2} \right) - 1 \right]$$

$$= 32 \times 10^4 \times \frac{11}{500} \left[\sec \left(\frac{180°}{\pi} \times 0.395 \right) - 1 \right]$$

$$= 587 kNm$$

Recalling Equation (21.15), the corresponding maximum upward mid-span deflection of the strut is given by:

$$y_{max} = \frac{w}{n^2 P} \left[\sec \left(\frac{180°}{\pi} \cdot \frac{nL}{2} \right) - 1 - \frac{1}{8} n^2 L^2 \right]$$

$$= \frac{11}{0.0395^2 \times 500} \left[\sec \left(\frac{180°}{\pi} \times 0.395 \right) - 1 - \frac{1}{8} \times 0.0395^2 \times 20^2 \right]$$

$$= 0.076 m = 76 mm$$

(b) Axial load, $P = 800 kN$ $(\ll P_{cr})$

With this axial load, the corresponding values for the various parameters, as before, are: $n = 0.0544, nL/2 = 0.544$, and $EI = 32 \times 10^4 kNm^2$

$$\therefore M_{max} = 32 \times 10^4 \times \frac{11}{800} \left[\sec \left(\frac{180°}{\pi} \times 0.544 \right) - 1 \right] = 742 kNm,$$

$$\text{and} \quad y_{max} = \frac{11}{0.0544^2 \times 800} \left[\sec \left(\frac{180°}{\pi} \times 0.544 \right) - 1 - \frac{1}{8} \times 0.0544^2 \times 20^2 \right]$$

$$= 0.096 m = 96 mm$$

Clearly, the extent of deflection indicated is exaggerated since the end conditions are unlikely to be fully hinged as assumed. It is important to ensure that, under the actual deflection, the *JGS* remains uncracked for the following calculation to be relevant.

Moment of resistance of the *JGS* is given by:

$$M_{R(max)} = f_b \cdot z = 0.30 \times 1000 \times 1 \times 4^2/6 = 800 kNm$$

So, *4m* depth of the *JGS* would appear to be adequate to cover the assumed range of axial loads; therefore no further trial is necessary. However, if it was felt necessary to enhance the factor of safety further, consideration could also be given to the possibility of profiling the *JGS* as discussed earlier in the chapter.

Assuming a downward displacement of *5mm* at the interfaces, additional moment per *metre* run in the *JGS* would be of the order of:

$$\frac{24 EI \Delta}{L^2} = \frac{24 \times 6 \times 10^4 \times 4^3 \times 5 \times 10^{-3}}{12 \times 24^2} = 66.7 kNm$$

Then total design moment: $742 + 66 \cdot 7 = 808 \cdot 7 kNm$ which is 1 per cent over the limit.

While this could be accepted as being reasonably close to the moment of resistance, such would not be the case if the relative displacements were to be significantly in excess of the *5mm* assumed.

Also note that, in the above computation, the 'beam-column' or the '$P - \Delta$' effect likely to arise as a result of the relative slip has not been taken into account. This should not be ignored. However, where the *JGS* is profiled to invoke its inverted arch action, the commensurate reduction in the axial load can be taken into account in computing the stress-resultants due to the '$P - \Delta$' effects. Furthermore, the moments in the *JGS* due to the interface slips would also need to be suitably modified to reflect the variability of its thickness along the span.

22 Flotation

22.1 Introduction

Evidence from the metro systems around the world suggests that the depth below ground level of the base of underground metro station structures ranges anywhere from around $12m$ to over $30m$ in some cases. The ambient groundwater table, likewise, has been known in some parts of the world to be at as shallow a depth below ground level as $300mm$, or even less. In view of this, it is only to be expected that the presence of groundwater and the rate of its movement are likely to play a significant role in the interaction between subsurface water and a deep cut-and-cover structure during the various stages of its construction and, potentially, its operational life time as well. It is therefore essential to take appropriate cognizance of the impact of the presence and movement of the subsurface water on design.

The most attractive feature of metro station structures constructed, typically, in the cut-and-cover sequence, as opposed to their bored tunnel counterparts, is their spaciousness and, inevitably, the flexibility it offers in their planning. However, this very aspect can sometimes give rise to problems and, potentially, lead to instability if the equivalent unit weight of the cellular box structure as a whole turns out to be comparable to the unit weight of water.

The various design parameters of water are accurately known or knowable; besides, its behaviour can be predicted with a reasonable degree of accuracy. By virtue of these properties it can, when existing exclusively on its own, present an ideal, most simple and predictable medium to deal with. However, in the case of typical subsurface environment water is never present as an entirely free medium. Since it is constrained to operate within the subsoil environment surrounding the underground structure and changing porewater pressure boundaries during the various stages of construction, the interaction of water and structure is not always simple or easily predictable.

In the subsoil environment, the free flow of water, and to a large extent its behaviour, are modified by the presence and the state of packing of the soil particles. Geological changes cause porewater migration and stratification of soil resulting in the unique current state of stress and groundwater regime in the soil mass. As a result, the different soil strata are able to display, potentially, different drainage characteristics both horizontally and vertically. This makes an accurate prediction of the global drainage characteristics of the soil mass purely on laboratory based testing and on, what can only be, unrepresentative samples, difficult and, as a result, not entirely reliable.

Whether the soil is granular thereby allowing free flow of water, or fine-grained causing viscous retardation to its free flow, and how long the dissipation of excess porewater pressures might take; whether the preferred orientation of soil particles has induced preferential directional permeabilities and, if so, what the global drainage characteristic of the soil mass is likely to be; whether the ambient water table is truly hydrostatic, artesian or perched – these are some of the factors which can have a significant influence on the response of subsoil water to construction activity and on the general stability of the structure as a whole.

22.2 Buoyancy

The magnitude of pressure exerted by water at a point in a mass of water is direction-invariant, i.e. it is equal in all directions. In conformity with this, a body, such as a cube, fully submerged in water would experience equal but opposite forces horizontally on its opposite sides maintaining lateral equilibrium. However, such is not the case vertically. Since the pressure increases with depth, the vertical pressures at different levels would not be the same; accordingly, the upward pressure at the bottom of the cube would be greater than the downward pressure on its top. The extent of this difference in the vertical pressures on the body represents the extent of buoyancy acting on it.

According to the Archimedes' principle, a body immersed in water experiences a loss of weight equal to the weight of water displaced by it. Whether the body sinks or floats depends upon whether the weight of water so displaced is less or more than the gross weight of the body. The weight of water displaced is a function of both the volume of the body and its unit weight, and is a measure of the force of buoyancy acting on it.

Consider a body of gross weight W and gross volume V immersed in water as shown in Figure 22.1. Let γ_W be the unit weight of water.

If $W > V \cdot \gamma_W$, the body will sink, and if $< V \cdot \gamma_W$, the body will float in a manner such that $W = V' \cdot \gamma_W$, where V' is the volume of that portion of the body which is below the water level and, as such, represents the volume of water displaced.

Within a uniform, homogeneous and free draining stratum of soil, it would be reasonable to assume the porewater pressure to be hydrostatic, i.e. varying linearly with depth. In reality, however, soils are rarely uniformly homogeneous and so, in a layered soil, the variation need not necessarily remain hydrostatic all the time. Absence of such information, therefore, calls for the use of certain simplifying but safe assumptions and approximations in the assessment of the loss of weight of the underground structure and the buoyant force likely to act on it.

Buoyancy of a structure immersed in water is solely a function of the weight of water displaced by it. It is a measure of the net upward pressure of water on the base of the structure and can be related to the 'wet depth' (D_W) of the structure. Wet depth may be defined as that depth of the structure over which it is in contact with

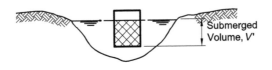

Figure 22.1 Archimedes Principle

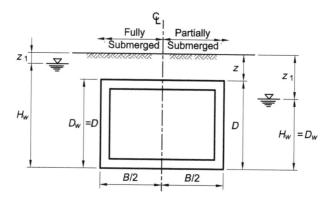

Figure 22.2 Geometry of Typical Submerged Structure

the groundwater. Clearly, for a fully submerged structure and irrespective of the extent of its submergence, wet depth is equal to, and can never be greater than, the overall depth of the structure.

In Figure 22.2, B (breadth) and D (depth) define the overall cross-section and z the depression of the structure below the ground level. Dimension H_W represents the height of the water table measured above the base of the structure. Left half of the cross-section represents a fully submerged structure whereas the right half, a partially submerged one. For the water table above the top $(H_W > D)$ of the structure, i.e. in the case of a fully submerged structure, the wet depth is independent of the groundwater level and carries a constant value of $D_W = D$. However, for water table below the top $(H_W < D)$, i.e. for a partially submerged structure, wet depth varies with the level of the water table, i.e. $D_W = H_W$. In other words, a structure that is fully submerged below groundwater table will be fully, and therefore more, buoyant than a partially submerged one.

22.3 Flotation potential

The force of buoyancy acting on an underground structure tends to displace the structure upwards; the gross weight of the structure, the effective weight of the backfill on top and the extent of the frictional forces which would be mobilized at the soil–structure interface around the perimeter of the structure, together, generally provide the counteracting force. Where the former exceeds the latter, there exists the potential for the structure to 'float'. In such a case, the use of additional holding-down measures to ensure stability may become necessary.

It should be clearly understood that, as long as the structure remains fully submerged below the water table, the force of buoyancy on it will remain constant and will not change with any variation in the extent of its submergence. However, against this, with the increase in the depression of the structure below ground, there is bound to be a commensurate increase in the depth of the effective backfill on top. This will add to the net gravity load counteracting flotation and thus lead to a correspondingly reduced potential for the structure to float.

22.4 Factor of safety

For an underground structure, factor of safety against flotation aims at providing, essentially, a measure of assurance and comfort against the incidence of its vertical instability under the action of buoyancy. In a given subsoil and groundwater environment, to ensure that the underground structure does not float upwards causing instability, the sum total of all the downward acting forces must exceed the upward acting force of buoyancy by a sensible margin. This margin must be sufficient enough to allow for the most adverse but realistically conceivable variations and uncertainties in the design parameters and boundary conditions. The desired margin of safety is usually measured as the ratio between all the downward acting forces counteracting flotation and the upward acting force of buoyancy. This ratio is commonly referred to as the factor of safety against flotation.

Figure 22.3 shows, typically, all the vertically downward and upward acting forces on a typical cut-and-cover metro station structure. Stated mathematically and using the notation of the figure, a general expression for the factor of safety (F) against flotation can be written as:

$$F = \frac{\sum W_G + \sum R_F + \sum P_A}{U_B} \qquad (22.1)$$

where, $\sum W_G$ = total gravity load = $W_A + W_B + W$, W_A = weight of backfill above the water table, W_B = effective weight of backfill below the water table, W = gross weight of the structure, $\sum R_F$ = total frictional resistance mobilized around the perimeter wall, $\sum P_A$ = sum total of other holding-down forces such as those from ground anchors, etc., and U_B = force of buoyancy or of net artesian pressure, whichever is greater.

All those other downward acting loads, such as the superimposed dead load, loads from ancillary walls, plant and pavement, live loads internal or external to the structure, surcharge loads, etc., which cannot be relied upon to be operative all the time throughout the design-life of the structure, are disregarded in computing the factor of safety against flotation.

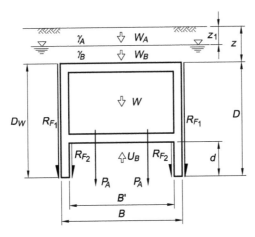

Figure 22.3 Vertical Loads on a Typical Cross-Section

Based on the foregoing approach, factors of safety against flotation ranging from 1·03 to 1·15 and even higher have been used in the designs of various metro structures in service in various parts around the world.

Being mindful of the gravity of the consequences should things go wrong, the need for safety cannot be overemphasized. In order to ensure, therefore, that the potential for instability is not underestimated, the designers often identify those cross-sections for investigation which tend to maximize the potential for flotation. Accordingly, a unit cross-section of the underground structure which is likely to offer the least dead weight to counteract flotation is generally selected to represent a typical unit for computation. Clearly, this approach disregards the potentially significant beneficial contribution, for instance, of the end cross-walls of the structure as a whole against global flotation. Examining the potential of flotation selectively in this manner which treats the lightest section identified discretely as being representative of the entire run of the structure is bound to exaggerate the effect of buoyancy and, as such, underestimate the available factor of safety against flotation. It would be more appropriate to check global flotation, i.e. the effect of buoyancy over the station structure as a whole. Because, if this is satisfactory then, clearly, there is unlikely to be any flotation problem even if computation based on the use of a discrete section locally might suggest otherwise.

In view of this, Equation (22.16) has been so formulated as to present a general expression for the calculation of the equivalent self-weight of the structure per unit run considered effective in counteracting flotation that incorporates the commensurate and beneficial contribution of the end cross-walls of the structure as well.

In the event that the factor of safety of a discrete section, ignoring the contribution from the end walls, turns out to be less than 1·0, it is reasonable to assume that the structure, acting as a linear box girder, would try to mobilize the necessary assistance and make up the shortfall from the adjacent areas to prevent the tendency of flotation locally. However, this is possible only as long as the global factor of safety is adequate and the structure is capable of mobilizing that assistance safely in the required manner. Given the size of cut-and-cover metro structures commonly encountered, enough inherent flexural and shear stiffnesses are generally available to feel sure that the load share-out in the required manner should present no problems generally. This is quite apart from any assistance that can be legitimately invoked from the soil–structure interface friction that is likely to exist around the perimeter of the structure in any case.

In the case of cut-and-cover metro structures, since the extent of the intimate soil–structure contact area is invariably substantial, a significant amount of interface friction has to be overcome before any concern in respect of flotation can be perceived as being real.

With the conservatism already built into the various parameters generating the gravity loads, such as in the weight of the structure, effective depth of overburden, weight of overburden, etc. and ignoring the beneficial contribution of the secondary elements within the box structure, a factor of safety of 1·03 should be perfectly acceptable. Besides, the use of $10kN/m^3$ (as against $9·81kN/m^3$) inherently carries nearly a 2 per cent overestimation in the unit weight of fresh water and therefore also in the buoyant force. In any case, numbers would confirm that 3 per cent of the buoyant force is, generally, likely to constitute a very significant downward force. A figure of 1·05 should therefore be more than adequate. In fact, it is now increasingly recognized that the measure of safety against flotation as a ratio can be unnecessarily too severe and wasteful and design against flotation is deemed to be satisfied as long as the gravity load counteracting flotation is in excess of the uplift force, i.e. measured as a positive difference between the restoring and the uplifting forces, by a sensible margin.

Notwithstanding the foregoing, the factor to be adopted for a particular structure that can be considered as appropriate in a given set of circumstances should be based on sound engineering judgement taking due cognizance of the following factors:

Accuracy of the design water level

Force of buoyancy, and so also the factor of safety against flotation, can be sensitive to variation in the level of the ambient water table especially in the case of a partially submerged structure. Appropriate investigation should be carried out to ascertain whether there is the likelihood of the structure experiencing a rise in the water table in the future. If so, and if it is likely to be a long-term feature, the maximum such rise in the water level over the operational lifetime of the structure should be identified so as to establish the highest credible water level for design. If, on the other hand, the rise is likely to be only seasonal and therefore short term, a case can be made out for accepting a short-term factor of safety against flotation of somewhat reduced value but only so long as the margin is not completely wiped out (i.e. F continues to remain greater than 1).

Furthermore, it can be reasoned with some justification that, during the seasonal rise in the water table, the structure is unlikely to be also stripped, simultaneously, of all those beneficial gravity loads (other than the live loads) which, though present, would have been routinely disregarded in the flotation calculation. As a result, the real factor of safety would be higher than the one computed conventionally. With the extra margin of safety already in hand in this way, it can be argued that there is no reason to aim for a higher factor. Adopting such an approach, therefore, is sensible and could save the client unnecessary extra costs.

Effect of construction tolerances

Construction tolerances of some magnitude can be expected, for example, in the setting out and the verticality of the perimeter walls of the structure during their installation. If it can be established that these can, due to the out-of-verticality of the perimeter wall, lead to an enlargement of the plan area of the base of the structure, it could potentially experience an increase in the upward force of buoyancy. This could, it may be argued, adversely affect the stability of the structure. Some designers have been known to take into account adverse effects of such tolerances with a view to maximizing the effect of buoyancy.

This approach is sensible if the increase in buoyancy is due to the adverse setting out tolerance leading to a larger structure while maintaining the verticality of the perimeter walls. However, if the increase in the buoyancy that results from the out-of-verticality of the perimeter walls leading to a larger base area is considered in the computations, then disregarding, at the same time, the compensating effect of the corresponding vertical component of the wedge of the soil against the out-of-vertical perimeter walls that would inevitably come into play if the structure were to displace upwards cannot be justified. Since the latter is likely, if anything, to more than compensate for the former, the effect of the construction tolerances in the verticality of the perimeter walls, if ignored, can lead to a safer assumption and should give no cause for concern.

Variability in the assessment of gravity loads

There is no doubt that reliance for providing most of the force counteracting flotation has to be placed on the permanent gravity loads, principally, own weight of the structure and weight of the permanent backfill on top. Therefore the possibility of variations in the unit weights of the concrete in the structure or that of the soil backfill on top, or both, which could influence the computations adversely should not be ruled out. For example, unit weight of *in situ* concrete may vary from 24 to more than $25kN/m^3$ and that of the soil backfill from 18 to $22kN/m^3$. To ensure that a potentially more onerous design case is not overlooked and that the factor of safety against flotation not overestimated, it may be prudent to realistically maximize the effect of buoyancy by a sensible selection of the lower-bound values of the unit weights to be adopted.

Allowance for accommodation of future services

Design Criteria in certain parts of the world have been known to require up to $1\cdot5m$ or more depth of backfill on top of the structure to be disregarded for the long-term stability calculations. This, it is argued, is to allow for relief in the gravity load that could be expected due to future excavations, for instance, to install any additional services. However, since the services are likely to be installed in discrete localized trenches, ignoring the effect of the backfill over the entire area of the structure permanently would appear to be an unnecessarily too onerous a condition and therefore unjustified and unwarranted. A safe approach that is realistic and can be sensibly justified should be adopted.

Frictional resistance mobilized

Excavation for the construction of an underground structure is bound to be accompanied, inevitably, by the relaxation of the surrounding ground. This will be reflected in terms of its movement inward towards the structure causing its global compression and attaining, in the process, an intimate contact with it. As a result, any differential movement vertically between the structure and the surrounding ground would bring into play friction at the soil–structure interface commensurate with, and as a function of, the magnitude of the lateral earth pressures generated. In the event that the gravity loads to counteract flotation are not sufficient enough, friction over the perimeter soil–structure interface would need to be overcome before the structure could be expected to displace upwards. The extent of the frictional resistance that can be mobilized around the perimeter of the structure depends upon, and can also vary with, the variation in the following factors:

- The geotechnical characteristics of the ground and the site-specific soil parameters, i.e. the effective stress shear strength parameters, c' and $\tan\phi'$;

- The effective *in situ* lateral pressures in the ground and their variation with depth;
- The method of perimeter wall construction.

Howsoever carefully may the soil sampling and testing be carried out, accurate quantification of the soil parameters cannot be guaranteed. If the parameters are underestimated, the lateral earth pressures on the perimeter walls would, accordingly, be overestimated. This would, in turn, lead to an artificial enhancement in the component of frictional resistance and therefore falsely exaggerate the factor of safety against flotation. Likewise, if the parameters are overestimated, the factor of safety is likely to be underestimated. It is therefore prudent to build in sensible safeguards against the potential variations in the soil parameters in order to obtain a sensibly realistic factor of safety. Where appropriate, a sensitivity analysis confirming the effect of parametric variation on the factor of safety may be useful.

Furthermore, if the perimeter wall is constructed, for example, under bentonite slurry, owing to the formation of a filter-cake at the soil–structure interface, with its inherent property of lubrication, the friction mobilized could be significantly reduced. Under such circumstances it would be advisable to sensibly downgrade the effect of the frictional resistance that could be mobilized.

Performance of ground anchors

Where holding-down anchors are used to enhance the factor of safety against flotation to the desired level, they will inevitably introduce new boundary conditions in design. In such a case, malfunction or failure of one or more of these could have a significant impact on design and would need to be appropriately investigated.

Where an underground structure is markedly asymmetric in a manner that could result in an equally markedly variable net hydrostatic uplift force profile, implications of potential rotational effects on the structure and the need for their containment should not be overlooked.

It is clear from the discussion of the various aspects thus far that a careful consideration should be given to the selection of the value for the factor of safety against flotation to be adopted under a given set of circumstances. It is also useful to remember that a more heavy structure is not necessarily going to be any more safe than a relatively less heavy structure; however, it will most certainly be more expensive. In other words, a costlier structure does not necessarily mean a safer structure, *per se*. Just as a low factor may not evince the desired degree of comfort or assurance about safety, an unnecessarily high factor, likewise, could lead to an overly conservative design and present an uneconomic structure. A sensible balance should be struck. As a guide, a factor of safety of the order of 1·05 for the long-term operational case is generally considered to be appropriate provided that the following conditions are complied with:

- Lower-bound unit weights for concrete and backfill to minimize the gravity loads counteracting flotation are used in the computations.
- Highest credible level of water table likely to obtain during the lifetime of the structure to maximize the force of buoyancy is used for design.
- Unit weight of water, other than that for sea water, of $10 kN/m^3$, is used in computations.

- Any benefit from the frictional resistance mobilized on the soil–structure interface along the perimeter of the structure is disregarded.

It should be recognized that even the residual (conservatively downgraded) frictional resistance along the soil–structure interface when the perimeter wall is constructed under bentonite can, owing to the enormous size of the contact area generally involved, provide a significant component of resistance against flotation. If this component is completely disregarded, the use of an even smaller factor of safety, say of the order of 1·05, can be justified.

If the frictional force along the perimeter wall, with due deference to its method of construction, is taken into account in the flotation calculations, a higher factor of safety of the order of 1·10 is generally stipulated by certain regulatory bodies. The difference between the higher and the lower bound factors of safety, i.e. 1·10 and 1·05, is intended, it is argued, to allow for the uncertainty in the ability to quantify accurately the magnitude of the frictional force that can be mobilized and can continue to be relied upon throughout the operational design life of the structure. However, if the extent of this uncertainty can be reduced with a degree of certainty, there is no reason why a factor less than 1·10, say of the order of 1·07 or even lower, should not be adopted. During the stages of construction, factor of safety against flotation of the order of 1·03 is generally considered acceptable.

The technical modalities as discussed above notwithstanding, some regulatory authorities have been known, at times, to display reluctance to shift away from their conservative stance and a tendency to stipulate higher factors of safety. This would appear to be so for no reason other than that, as a factor, a margin less than 10 per cent is perceived to be uncomfortably small and the one that, it is argued, could be lost by an adverse variation in the design parameters.

However, it is important not to lose sight of the fact that, by their very nature, metro structures are enormous in size carrying substantial self-weight and so even a seemingly small percentage can, in real terms, carry a significant cost element. Besides, it may often be the case that the antiflotation measures are not really adopted to counteract flotation so much, for instance when the gravity load on its own is able to neutralize the buoyancy, as to raise the margin of perceived comfort. It is in recognition of these reasons that some of the more progressive regulatory authorities around the world are sympathetically disposed towards accepting safety against flotation as proven as long as a positive difference (as distinct from a ratio) between the loads counteracting flotation and the force of buoyancy of a sensible magnitude can be shown to exist. This aspect is further elaborated in the solved example later on in the chapter.

22.5 Typical flotation calculation

Consider a typical cut-and-cover metro station structure shown in Figure 22.3 with the following geometry:

$$\text{Overall Length (Clear Length)} = L(L')$$
$$\text{Overall Width (Clear Width)} = B(B')$$
$$\text{Overall Depth (Penetration)} = D(d)$$

Let

$$\gamma_W = \text{Unit weight of water assumed} = 10 kN/m^3$$
$$\gamma_A = \text{Unit weight of soil backfill above water table}$$
$$\gamma_B = \text{Unit weight of soil backfill below water table}$$
$$\gamma_E = \text{Equivalent unit weight of soil above roof}$$
$$W_A = \text{Effective weight of soil backfill above water table} = \gamma_A \cdot B \cdot z_1$$
$$W_B = \text{Effective weight of soil backfill below water table} = (\gamma_B - \gamma_W) B (z - z_1)$$
$$W_S = \text{Effective weight of total soil backfill} = W_A + W_B = \gamma_E \cdot B \cdot z$$
$$W = \text{Gross permanent weight of structure}$$
$$U_B = \text{Force of buoyancy or uplift per unit run} = \gamma_W \left(B \cdot D - B' \cdot d \cdot \frac{L'}{L} \right)$$

Fully submerged case $(1 \ge n \ge 0)$

Effective weight of the backfill on top of the structure is, among other things, a function of the level of groundwater above the structure. A downward variation in its level leads to an increase in the effective weight of the backfill above the water level by an amount equal to the reduction in the weight of water. In the case of a fully submerged structure, as the force of buoyancy remains unaltered, the increase in the effective weight of backfill, correspondingly, subscribes to a commensurate increase in the factor of safety against flotation. Recalling Equation (22.1),

$$F = \frac{\sum W_G + \sum R_F + \sum P_A}{U_B}$$

If the frictional resistance is disregarded and no other means, such as anchors etc., either are used to hold down the structure, the expression reduces to:

$$F = \frac{\sum W_G}{U_B} = \frac{W_A + W_B + W}{U_B} = \frac{W_S + W}{U_B}$$

$$= \frac{\gamma_E \cdot B \cdot z + W}{\gamma_W \left(B \cdot D - B' \cdot d \cdot \dfrac{L'}{L} \right)} = \frac{0 \cdot 1 \gamma_E \cdot \dfrac{z}{D} + 0 \cdot 1 \dfrac{W}{B \cdot D}}{\left(1 - \dfrac{B'}{B} \cdot \dfrac{L'}{L} \cdot \dfrac{d}{D} \right)}$$

General expression for the factor of safety for a fully submerged structure shown above may be written as:

$$F = \frac{0 \cdot 1 (\gamma_E \cdot m + w_e)}{K} \tag{22.2}$$

where $\gamma_E = $ Equivalent unit weight of backfill above the structure

$$= \gamma_A \cdot n + (\gamma_B - \gamma_W)(1 - n) = \gamma_A \cdot n + \gamma_S (1 - n)$$
$$\gamma_S = \text{Submerged unit weight} = (\gamma_B - \gamma_W); \quad m = z/D;$$
$$K = 1 - \frac{B'}{B} \cdot \frac{L'}{L} \cdot \frac{d}{D}; \quad w_e = W/(B \cdot D); \quad n = z_1/z (\leqslant 1)$$

In the event that the water table does not occur at the ground level, there will be an enhancement in the effective weight of the backfill directly above the structure corresponding to the downward variation in the water level. Allowance for this is reflected in the parameter n, where n is the 'Groundwater–Structure' Depression Ratio. This parameter is subject to a maximum value of 1 when the level of water table is at or below the top of the structure.

Water table at ground level $(z_1 = 0, D_W = D)$

For water table at ground level, $n = 0$ and $\gamma_E = \gamma_B - \gamma_W = \gamma_S$. The expression for the factor of safety against flotation then becomes:

$$F = \frac{0 \cdot 1 \left[\gamma_S \cdot m + w_e \right]}{K} \tag{22.3}$$

As an approximation involving little loss of accuracy, if it is assumed that the unit weight of the submerged soil backfill $\gamma_B = 2\gamma_W = 20 kN/m^3$, then the above expression for the factor of safety simplifies to:

$$F = \frac{m + 0 \cdot 1 w_e}{K}$$

Water table flush with roof top $(z_1 = z, \gamma_A = \gamma = \gamma_E)$

When the level of water is at par with the top of the roof, $n = 1$ and $\gamma_E = \gamma_A = \gamma$. In such a case, the expression for the factor of safety becomes:

$$F = \frac{0 \cdot 1 \left(\gamma \cdot m + w_e \right)}{K} \tag{22.4}$$

Again, if it is assumed that $\gamma = 20 kN/m^3$, then the above expression further reduces to:

$$F = \frac{2m + 0 \cdot 1 w_e}{K}$$

Partially submerged case

For a downward variation of the water table below the roof-top of the structure, leading to a partially submerged case, the effective weight of the backfill directly on top of the roof does not undergo any further increase. However, being a function of the 'wet depth', the force of buoyancy is affected and this, in turn, influences the factor of safety also.

It should be recognized that with the downward variation of the water table below the roof-top of the structure, there will also be a commensurate increase in the lateral effective pressures on the perimeter wall. As a result, the potential for mobilizing a correspondingly higher frictional resistance along the perimeter of the soil–structure interface also exists. This identifies a further avenue of enhancing the factor of safety against flotation.

For partially submerged case, the expression for the factor of safety against flotation is similar to that of Equation (22.2) with the difference that the term D for the full depth of the structure must be replaced by D_W, its wet depth. In other words,

$$F = \frac{0 \cdot 1\left(\gamma \cdot m' + w'_e\right)}{K'} \qquad (22.5)$$

where, $K' = 1 - \dfrac{B'}{B} \cdot \dfrac{L'}{L} \cdot \dfrac{d}{D_W};$ $w'_e = W/\left(B \cdot D_W\right);$ $m' = z/D_W$

It can be seen that, for water table flush with the rooftop, i.e. when $D_W = D$, Equation (22.4) can be recovered.

Special case $(d = 0)$

For an underground station structure constructed conventionally within an open cut with stabilized side slopes or within a temporary sheet pile (or similar) cofferdam, integral perimeter walls are unlikely to extend below the track/base slab, i.e. $d = 0$. This is also likely to be the case with the construction of the relatively shallower entrance box structures. By making the substitution, $d = 0$, it can be seen that both the denominators K and K' become unity and the expressions on the right-hand side of all the aforementioned equations are reduced to their numerators only. So, for underground metro structures with zero penetrations of the perimeter walls below the track/base slab levels, the expressions for the respective factors of safety corresponding to the foregoing cases are given, approximately, by the following expressions:

$$F = 0 \cdot 1\left(\gamma_E \cdot m + w_e\right) \qquad (22.2a)$$

$$= 0 \cdot 1\left(\gamma_S \cdot m + w_e\right) \qquad (22.3a)$$

$$= 0 \cdot 1\left(\gamma \cdot m + w_e\right) \qquad (22.4a)$$

$$= 0 \cdot 1\left(\gamma \cdot m' + w'_e\right) \qquad (22.5a)$$

Typical use of the foregoing approach and expressions in providing a means of measuring the performance of the 'first guess' structure against flotation at the conceptual stage is illustrated in the solved examples at the end of the chapter.

22.6 Antiflotation measures

Where, for a given geometry and ground conditions, the available factor of safety against flotation is less than the minimum desired, appropriate measures capable of pushing the factor up to the required level need to be employed. Choice of such measures is influenced, among other aspects, by their feasibility within the existing ground conditions and the prevailing environment, practicalities of implementation, life-cycle maintenance, cost, etc. However, it should also be recognized that the treatment of the potential problem, if likely to present itself during the stages of construction, will be somewhat different from that of the permanent, in-service case. Accordingly, measures appropriate to each situation need to be identified.

During temporary construction stage

Construction of a typical cut-and-cover metro station structure is generally carried out within a preinstalled perimeter wall which forms the principal ground support system. The perimeter wall may either be treated as an outer cofferdam within which the main structure is constructed, or incorporated as an integral part of the permanent structure. In either case, dewatering within the perimeter is usually progressed a few metres ahead of the excavation so as to enable the construction works to proceed in a dry environment.

Following the base slab construction, for instance in the bottom-up sequence, if the dewatering is stopped, the pressure differential across the perimeter wall would begin to build-up progressively at a rate commensurate with the type of the ground and its drainage characteristics. As a result, the hydrostatic uplift pressure on the base of the structure would also increase progressively. If, at any stage during construction, sufficient downward weight is not available to counteract the uplift forces of buoyancy, 'flotation' may occur causing instability of the structure. Under such circumstances, it may be necessary to carry on pumping until the construction of the structure has advanced to a stage whereat pumping can be safely stopped.

However, continued pumping, if accompanied by loss of fines in the ground, can give rise to problems related to settlements of the surrounding ground and the structures in the vicinity. It is important to ensure that such settlements are avoided, if possible, or, at any rate, kept to within acceptable limits.

As an alternative, flotation can be prevented by providing pressure relief vents in the form of holes disposed at strategic locations in the base slab during its construction. However, the success of such a measure depends on the ease with which the water is able to flow through the holes thereby preventing any build up of excessive uplift pressures. This can be achieved by constructing the slab on a properly graded drainage blanket which needs to be so designed as to discourage the washing of fines up the relief holes. These holes can lend themselves to use for the purpose also of monitoring the porewater pressures below the base slab and can be suitably plugged when no longer necessary. However, in aggressive ground and groundwater environment, which may require the structure to be wrapped in a waterproof membrane, adequate sealing of pressure relief holes can present problems of detail. This should be appropriately addressed.

Other possible measures, which can be employed temporarily during construction, may involve the use of:

• Kentledge
• Holding-down anchors
• Flooding.

Kentledge in the form of steel ingots, precast concrete blocks, sand bags or similar can be used, if required, to enhance the margin of safety to the desired level. Soil anchors can also be used to hold the structure down temporarily. The anchors can be destressed when no longer required. Alternatively, maintaining partial flooding of the structure in a manner that does not interfere with the continuity of construction works can sometimes be a convenient way of reducing the differential head of water thereby alleviating the potential problem of flotation during construction stage. However, cost and logistics of such measures generally become the deciding factors for the selection of the most appropriate solution in a given set of circumstances.

It should be appreciated that it is not always possible to quantify the threshold levels and define boundary conditions, drainage characteristics and other allied parameters of the ground precisely. Of necessity therefore, flotation calculations are based on simplified, theoretical assumptions which inevitably lean on the conservative side. Bearing this in mind, every attempt should be made to avoid loading up further such construction costs through the application of antiflotation measures which, on closer and sensible scrutiny, may prove to be unnecessary.

During permanent in-service stage

From what has been discussed earlier, it can be inferred that, for a fully submerged structure, factor of safety against flotation can be increased to the desired level simply by increasing the depression of the structure below the ground level. This would not only increase the amount of the effective backfill on top of the structure and, potentially, the frictional resistance that could be mobilized around the perimeter – but both together while the buoyant force remains unchanged. However, it may not always be possible to carry out such a measure since the depth to the rails is likely to be fixed by other considerations including track alignment, etc. Height of the roof soffit level above the rails, likewise, is likely to be dictated by the planning, operational and structural requirements, and the depth of the rooftop below the ground by the extent of the accommodation necessary for the utilities, etc. So, given these constraints, the required increase in the margin of safety against flotation can be achieved through the following options:

(1) Permanent kentledge

In the context of underground metro structures, avenues for placement of permanent kentledge are generally limited. Under-platform space which could accommodate permanent kentledge is often used for housing ventilation and cable ducts, etc. However, where space, which would otherwise remain unused, can be made available, filling it up with mass concrete or sand may alleviate, partially or fully, a potential flotation problem. This should be the first option to be explored.

(2) Structure

In the event that the foregoing option does not yield the desired enhancement in the factor of safety or, simply, is not available, the most positive way of counteracting flotation, and therefore the next option to be explored, is to increase the gravity load by thickening one or more structural elements such as walls, floor slabs, etc., as necessary. This can be achieved by one or more of the following:

(a) Increasing the thickness of the perimeter wall: Constraints likely to be imposed by the proximity of obstructions to the perimeter of the structure will have a bearing on the extent of the possible outward thickening of the perimeter wall. In exploring this option, due consideration should be given to the nearness

of utilities, existing buildings – their foundations and overhangs, practicalities of construction, etc.

For a *fully submerged* case, with or without the perimeter wall penetration below the base slab, required increase in the perimeter wall thickness ΔT can be obtained from the positive root of the following quadratic equation:

$$u(2\Delta T)^2 + v(2\Delta T) + c = 0$$

$$\text{i.e.} \quad \Delta T = \frac{-v + \sqrt{v^2 - 4u \cdot c}}{4u} \tag{22.6}$$

where $u = \dfrac{1}{L \cdot B}\left[\gamma_W \cdot F_R - \gamma_E \cdot m - \gamma_C\right];$ $v = \left[u(L+B) - \dfrac{w_e}{L}\right];$

$K = 1 - \dfrac{B'}{B} \cdot \dfrac{L'}{L} \cdot \dfrac{d}{D} (= 1 \text{ for } d = 0);$ $c = K \cdot \gamma_W \cdot \Delta_F;$

$F_R =$ Factor of safety required; $F_A =$ Factor of safety available;

$\Delta F = F_R - F_A;$ $\gamma_C =$ Unit weight of concrete

All other terms are as defined previously.

For a *partially submerged* case, with or without the perimeter wall penetration below the base slab, required increase in the wall thickness can be obtained, once again, from Equation (22.6) but where:

$$u = \frac{1}{L \cdot B}\left[\gamma_W \cdot F_R - \gamma \cdot m' - \gamma_C \cdot K_D\right]; \quad K_D = \frac{D}{D_W}$$

$$v = \left[u(L+B) - \frac{w'_e}{L}\right]; \quad c = K' \cdot \gamma_W \cdot \Delta F$$

$$K' = 1 - \frac{B'}{B} \cdot \frac{L'}{L} \cdot \frac{d}{D_W} \quad (= 1, \text{ for } d = 0)$$

(b) *Increasing thickness of base slab downwards:* For a *fully submerged* case, with or without the perimeter wall penetration below the base slab, required increase in the thickness of the base slab (ΔT_B) can be obtained from the following expressions:

$$\Delta T_B = \left[\frac{B}{B'} \cdot \frac{L}{L'} - \frac{d}{D}\right] \frac{\gamma_W \cdot \Delta F \cdot D}{(\gamma_C - \gamma_W \cdot F_R)} \tag{22.7}$$

$$= \left[\frac{B}{B'} \cdot \frac{L}{L'} - \frac{d}{D}\right] \frac{\Delta F \cdot D}{(0 \cdot 1 \gamma_C - F_R)} \quad \text{for } \gamma_W = 10 kN/m^3$$

$$= \frac{\gamma_W \cdot \Delta F \cdot D}{(\gamma_C - \gamma_W \cdot F_R)} \quad \text{for } d = 0 \tag{22.8}$$

$$= \frac{\Delta F \cdot D}{(0 \cdot 1 \gamma_C - F_R)} \quad \text{for } \gamma_W = 10 kN/m^3$$

For a *partially submerged* case, with or without the perimeter wall penetration below the base slab, simply replace D with D_W and set the wall penetration d to zero as appropriate in the above expressions to obtain the required increase in the respective thicknesses of the base slab.

(c) *Increasing thickness of roof slab upwards*: If the depth of the backfill directly above the roof of the structure has some spare capacity beyond that required for accommodating the existing as well as any future utilities, increasing the thickness of the roof slab upwards to enhance the factor of safety against flotation is, in theory, possible. For a *fully submerged* case, the required increase in such thickness of the roof slab (ΔT_R) is given by the expressions:

$$\Delta T_R = \left[\frac{\gamma_W \cdot K \cdot \Delta F}{(\gamma_C - \gamma_E) - \gamma_W \cdot F_R}\right] D \tag{22.9}$$

$$= \left[\frac{K \cdot \Delta F}{0 \cdot 1(\gamma_C - \gamma_E) - F_R}\right] D \quad \text{for } \gamma_W = 10kN/m^3$$

$$= \left[\frac{\gamma_W \cdot \Delta F}{(\gamma_C - \gamma_E) - \gamma_W \cdot F_R}\right] D \quad \text{for } d = 0 \tag{22.10}$$

$$= \left[\frac{\Delta F}{0 \cdot 1(\gamma_C - \gamma_E) - F_R}\right] D \quad \text{for } \gamma_W = 10kN/m^3$$

For a *partially submerged* case, with or without the perimeter wall penetration below the base slab, simply replace D with D_W, K with K' and γ_E with γ in the above expressions to obtain the required increase in the thickness of the roof slab.

(d) *Increasing penetration of perimeter wall*: If the constraints specific to a given site are such that no modification which would increase the plan of the primary structure can be entertained, the need to enhance the factor of safety to the required level can also be met by increasing the penetration of the perimeter wall below the base slab. For a *fully submerged* case, required increase in the wall penetration (ΔP_W) can be obtained from the following expression:

$$\Delta P_W = \left[\frac{\gamma_W \cdot \Delta F}{\gamma_C - \gamma_W \cdot F_R}\right] \cdot \left[\frac{K}{K_A}\right] D \tag{22.11}$$

$$= \left[\frac{\Delta F}{0 \cdot 1\gamma_C - F_R}\right] \cdot \left[\frac{K}{K_A}\right] D \quad \text{for } \gamma_W = 10kN/m^3$$

$$\text{where, } K_A = 1 - \frac{B'}{B} \cdot \frac{L'}{L};$$

other terms are as defined previously.

For a *partially submerged* case, the corresponding expression can be obtained by substituting K' for K and D_W for D in the above expression.

(e) *Extending width of the roof slab beyond boundary walls:* Increase in the margin of safety to the required extent can also be achieved by extending the width of the roof slab transversely beyond the outside faces of the longitudinal boundary walls. In this way, the gravity load is increased commensurate with not only the increase due to the net weight of the concrete in the extensions themselves but also by mobilizing the additional effective weight of the backfill on top of the extensions.

For a *fully submerged* case, required increase in the width (ΔB) beyond the outside faces of the longitudinal boundary walls can be obtained from the following expression:

$$\Delta B = \left[\frac{0.5\, \gamma_W \cdot \Delta F \cdot K}{\gamma_E \cdot m + K_R \left(\gamma_C - \gamma_W \cdot F_R \right)} \right] B \qquad (22.12)$$

Corresponding value for the *partially submerged* case can be obtained from the following expression:

$$\Delta B = \left[\frac{0.5\, \gamma_W \cdot \Delta F \cdot K'}{\gamma_E \cdot m' + K'_R \left(\gamma_C - \gamma_W \cdot F_R \right)} \right] B \qquad (22.13)$$

$$\text{where } K_R = \frac{\text{Roof Slab Thickness} \left(= T_R \right)}{D}, \quad K'_R = \frac{T_R}{D_W}.$$

In the case of no penetration of the perimeter wall below the base, i.e. $d = 0$, substitute $K = 1 = K'$ in the above expressions to obtain the required increase in the width.

(f) *Extending roof slab all round the perimeter:* It may be feasible, in certain circumstances, to extend the roof slab by a uniform margin ΔS on all four sides. For a *fully submerged* case, with or without the perimeter wall penetration below the base slab, required extension on every side can be obtained from the positive root of the following quadratic equation:

$$(2\Delta S)^2 + b(2\Delta S) - c = 0$$

$$\text{i.e. } \Delta S = \frac{-b + \sqrt{b^2 + 4c}}{4} \qquad (22.14)$$

$$\text{where } b = L + B, \quad c = \frac{\gamma_W \cdot \Delta F \cdot K \cdot B \cdot L \cdot D}{T_R \left(\gamma_C - \gamma_W \cdot F_R \right) + \gamma_E \cdot z}$$

and all other terms are as defined previously.

For a *partially submerged* case, replace D with D_W and K with K' and, in the case of the perimeter walls with no penetration below the base slab, i.e. $d = 0$, set $K = 1 = K'$, to obtain the required values for the slab extension.

(g) *Extending 'width' of base slab to form toes*:   Extending the width of the base slab transversely beyond the longitudinal walls so as to form toes is a measure that can be considered only where the walls do not extend below the base slab. This is so where the structure is constructed either within a temporary outer cofferdam or conventionally in an open cut with stabilized side slopes. Under this option, the smaller component of the required additional factor of safety is provided by the net effective self-weight of the toes themselves and the larger component by the effective weights of the backfill on top of the toe extensions.

For a *fully submerged* case, the width of the toe (ΔB) required to achieve the desired enhancement in the factor of safety is given by the expression:

$$\Delta B = \left[\frac{0 \cdot 5 \times \gamma_W \cdot \Delta F}{\gamma_E (m+1) + (\gamma_C - \gamma_E) \cdot K_B} \right] B \qquad (22.15)$$

$$\text{where} \quad K_B = \frac{Base\ Slab\ Thickness\ (= T_B)}{D}$$

For a *partially submerged* case, replace m with m' and D with D_W in the above expression.

(3) Holding-down anchors (unstressed)

Steel or precast concrete piles may be driven or *in situ* concrete piles installed below, and adequately tied into, the base slab of the structure to assist in counteracting flotation. The rationale underlying this measure is that, in the event of the structure tending to float upwards, both the dead weight of the piles and the skin friction likely to be mobilized around their periphery will be called into play to assist in inhibiting the upward movement of the structure. The size, spacing and the depth of the piles will be governed by the type of soil and dictated by the requirements of design. In case they are closely spaced, it may not be possible to mobilize full frictional resistance around every individual pile. However, it may not be necessary anyway if the volume of soil trapped between the piles provides enough additional weight to achieve the desired enhancement in the factor of safety against flotation.

In the case of a highly aggressive soil environment, consideration should be given to the implications of the need to protect the piles. One way that such protection can be achieved is by coating the piles with an appropriate immune material. However, such treatment of the piles may reduce the skin friction that can be developed thereby limiting the advantage of such a measure. To remedy this, piles could be left unprotected but only if the extent of their potential deterioration over the design life were to be allowed for in design and construction by incorporating an extra sacrificial thickness in the size of the piles used.

(4) Holding-down anchors (stressed)

Where for reasons of practicality, such as the type of prevailing ground conditions which may not be conducive to the use of unstressed anchors, raising the factor of safety to the desired level can also be achieved by the use of stressed ground anchors. However, if stressed anchors are used in this way and incorporated as an integral part of the structure, they will introduce boundary conditions which are likely to influence the design of the structure in some way. It then becomes obligatory to investigate the implications on the performance of the structure of the potential failure of one or more of these anchors. In other words, if the vertical restraint is relieved, even if partially, due to failure or malfunction of one or more anchors, allowing the structure to displace upwards locally, the extent of modification in the resulting stress levels will need to be investigated. This may demand an increase in the amount of reinforcement in the structure or, in the event of it being excessive, may be impractical to be adequately met.

As the performance of the ground anchors is likely to influence the behaviour and performance of the structure, it is advisable not only to install a long-term monitoring system but also to have a pressure relief system as a fall-back measure should any anchor(s) become inoperative or unserviceable. Ideally, both the anchor loads as well as the vertical displacements should be monitored – the former to enable the measurement of any drop in the load and the latter to throw some light on the reason for the drop. For instance, a drop in the anchor load, if unaccompanied by any upward movement of the slab at anchor locations, would imply either consolidation or creep of soil, or nonrealization of the anticipated level of design uplift pressures, or a combination of the two, in which case restressing may not be necessary. A combination of upward movement and drop in load would, on the other hand, tend to suggest development of uplift pressures potentially close to their design values and malfunction or failure of the anchors or both, in which case, restressing or replacement of the anchors might be called for.

Installation of steel tubes with screw-on caps permanently concreted into the base slab in strategic locations can offer a simple system for venting excess hydrostatic uplift pressures thereby obviating the potential for the structure to float. The object of using such a fallback system is three-fold:

- To provide an interim measure to allow an immediate relief in the build-up of hydrostatic uplift pressures;
- Failing the above, to provide an easy access to enable tapping of the previously placed drainage layer under the track slab level;
- To act as a potential location to install a new anchor. Note that this may be possible only if the potential implications of such a measure have already been thought through and allowed for in the design, and not considered as an after-thought for remediation.

In view of the structural implications outlined thus far, the cost of installation, protection particularly in aggressive subsoil environment, ongoing monitoring and the potential necessity of restressing or replacing failed anchors, this option must be considered only as the very last resort. However, if sound rock is available at a reasonably shallow depth below the base slab, anchoring into it could make it possible to achieve deliberate reductions in the weight of the structure. This could lead to significant cost savings and therefore merit sympathetic consideration.

Example 22.1

Consider a typical cross-section of an underground metro station structure with the dimensions and geometry as shown in Figure 22.4. Assume overall length of the station to be 300*m* and that of the platform slab to be 180*m*. Investigate the stability of the

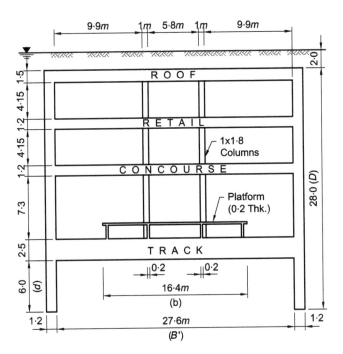

Figure 22.4 Typical Cross-Section of Station Structure

structure under the action of buoyancy ignoring, for this exercise, any voids in the floor slabs. Assume $\gamma_B = 2\gamma_W$.

Solution
Using same notation as before,

$$L = 300m, L' = 297\cdot6m, L_P = 180m, B = 30m, B' = 27\cdot6m, b = 16\cdot4m,$$

$$Z = 2m, D = 28m, d = 6m.$$

For a fully submerged structure and assuming the worst credible water level at ground level, factor of safety against flotation is given by the Equation (22.3):

$$F = \frac{0\cdot1\left[\gamma_S \cdot m + w_e\right]}{K}.$$

By substitution for $\gamma_S = 10kN/m^3$, $m = z/D = 0\cdot07$ and $K = 0\cdot805$ (K' if it were a partially submerged case) in the equation and for a desired factor of safety of, say, 1·05, we obtain the required minimum weight of the structure (W_R) as follows:

$$w_e = 8\cdot05\,F - 10m,$$
$$W_R = B.D\,(8\cdot05\,F - 10m)$$
$$= 30 \times 28\,(8\cdot05 \times 1\cdot05 - 10 \times 0\cdot07)$$
$$= 6,512\,kN \quad \text{per m run of the structure}$$

General expression for the equivalent, average area of cross-section of the structure per metre run, ignoring any voids in floors, assumed to counteract flotation, is given by:

$$A_{EQ} = B' \sum T_F \cdot \frac{L'}{L} + b \cdot \frac{L_P}{L} \sum T_P + 2 \cdot T_W \cdot D \left(1 + \frac{B'}{L}\right)$$

$$+ \frac{L_P}{L} \sum T_S \cdot H_P + A_C \cdot H_C \cdot \frac{n_r}{p} \tag{22.16}$$

where, B' = clear width between walls = 27·6m, L' = clear length of structure = 297·6m, L = overall length of structure = 300m, b = width of platform = 16·4m, L_P = length of platform = 180m, D = overall depth of perimeter wall = 28m, $\sum T_F$ = combined thickness of floor slabs = 6·4m, $\sum T_P$ = combined thickness of platform slabs = 0·2m, T_W = perimeter wall thickness = 1·2m, $\sum T_S$ = combined thickness of platform walls = 0·2 × 4 = 0·8m, H_P = height of platform walls = 1·0m, A_C = cross-sectional area of a typical column = 1·0 × 1·8 = 1·8m^2, H_C = net height of column between floor slabs = 4·15 + 4·15 + 7·3 − 0·2 = 15·4m, n_r = no. of longitudinal rows of columns = 2, p = longitudinal spacing of columns = 10·8m upon substitution, we get $A_{EQ} = 256·2m^2$ per m run, ignoring any floor voids.

If the voids are to be taken into account, the first term in Equation (22.16) needs to be adjusted accordingly by deducting the corresponding equivalent areas for the voids.

Assuming lower bound unit weight of reinforced concrete to be, say, 24kN/m^3, then weight per m run of the structure available to counteract flotation is given by:

$$W = 24 \times 256·2 \times 1 = 6,149 kN \left(< W_R\right)$$

With the parameters as assumed, it would appear that the gravity load alone is not enough to provide the desired factor of safety. The shortfall is of the order of (6512–6149), i.e. 363kN per meter run of the structure. If the shortfall in the gravity load were to be made up in terms of thickening of the structure alone, it would necessitate about 6 per cent increase in the weight of the structure. It is worth remembering that even this seemingly small percentage, given the enormous size and weight of such a structure, would reflect in a substantial increase in the cost. To avoid such unnecessary extra costs, contribution from the frictional resistance that can be safely mobilized along the soil–structure interface around the perimeter of the structure should be taken into account. However, in that case, conventionally, the required factor of safety to be adopted also needs to be increased to 1·10. The revised shortfall would then work out to be of the order of:

$$30 \times 28 (8·05 \times 1·1 - 10 \times 0·07) - 6149 = 701 kN/m$$

The average skin friction required to be mobilized to compensate for the shortfall is given by the expression:

$$f_S = \frac{Shortfall \times L}{2[(L+B)D + (L'+B')d]} \tag{22.17}$$

$$= \frac{701 \times 300}{2[(300+30)28 + (297·6+27·6)6]} = 9·4 kN/m^2$$

Even if the perimeter walls were to be constructed under bentonite, the required average value of the skin friction f_S as indicated above is small enough and can be easily mobilized.

Alternatively: Factor of safety available, ignoring frictional resistance around the perimeter, is given by the expression:

$$F = \frac{0 \cdot 1 \left[\gamma_S \cdot m + w_e \right]}{K} = \frac{0 \cdot 1 \left[10 \times 0 \cdot 07 + \dfrac{6149}{30 \times 28} \right]}{0 \cdot 805} = 0 \cdot 9963 \, (\simeq 1)$$

Shortfall in the factor of safety if frictional resistance is to be included is of the order of 0·1037. Component of factor of safety available from frictional resistance alone is given by the expression:

$$F_S = \frac{\sum R_F}{U_B} = \frac{0 \cdot 1 f_S \cdot K_P \left(1 + K_V \right)}{K} \qquad (22.18)$$

where f_S = unit skin friction.

$$K_P = \frac{Plan \; Perimeter}{Plan \; Area} = \frac{2 \, (300 + 30)}{300 \times 30} = 0 \cdot 073$$

$$K_V = \frac{(L' + B') d}{(L + B) D} = \frac{(297 \cdot 6 + 2 \cdot 76) 6}{(300 + 30) 28} = 0 \cdot 211$$

Whence, on substitution, the required unit skin friction to be mobilized is obtained as:

$$f_S = \frac{0 \cdot 1037 \times 0 \cdot 805}{0 \cdot 1 \times 0 \cdot 073 \, (1 + 0 \cdot 211)} = 9 \cdot 4 kN/m^2 \; \text{(as before)}$$

For a partially submerged case, Equation (22.18) assumes the form:

$$F_S = \frac{0 \cdot 1 f_S \cdot K_P \cdot K_D \left(1 + K_V \right)}{K'} \qquad \text{(for } D > D_W > d) \qquad (22.19)$$

where, $K_D = \dfrac{D}{D_W}$, as defined previously

Also note that, for a fully submerged structure, i.e. when $D_W = D$, K_D becomes unity and $K' = K$. On substituting these values in Equation (22.19), Equation (22.18) can be recovered.

Notes:

(1) If it is felt necessary to allow, in the weight of the backfill, relief that can be expected during the course of excavation of trench(es) to accommodate new services in the future, it can be achieved in one of two ways:

- By reducing the depth of the soil backfill, z, by the amount $k_b \cdot d_T$, and using the equivalent value in the computations, i.e. replacing z by $(z - k_b \cdot d_T)$;

- By downgrading the unit weight of the backfill by the factor $(1 - k_b \cdot k_d)$, i.e. using $\gamma(1 - k_b \cdot k_d)$ in place of γ but retaining the depth of the backfill as z.

In the foregoing, b_T and d_T represent the width and the depth of a typical longitudinal trench, k_b is the (combined) width of the trench(es) per unit width of the structure, i.e. $\sum b_T/B$, and k_d, likewise, represents the depth of the typical trench per unit depth of the backfill, i.e. d_T/z. The rest of the computations follow the same logic as before. If the depths of the trenches are different, then take the typical depth d_T as their average.

(2) In the foregoing example, available factor of safety on the considerations of gravity load alone, i.e. ignoring frictional resistance, is of the order of 1. Total buoyant force over the entire run of the structure is given by:

$$\sum U_B = \gamma_W \left(B \cdot D \cdot L - B' \cdot d \cdot L' \right)$$
$$= 10 \left(30 \times 28 \times 300 - 27 \cdot 6 \times 6 \times 297 \cdot 6 \right)$$
$$= 2{,}027{,}174 \cdot 4 kN = 202{,}717 t$$

Had the available gravity load been sufficient enough to provide a positive margin over the buoyant force of the order of even only 1 per cent, the margin in hand against flotation would have been:

$$0 \cdot 01 \times 202717 = 2{,}027t!$$

On the face of it, in terms of the 'factor' of safety that this figure would have represented, it would appear to be insignificant; nevertheless, there is no denying the fact that a positive weight of over 2000 *tonnes* in excess of the buoyant force would have represented a significant enough 'margin' of comfort in its own right and it would have been unreasonable to disregard it. So, even without invoking the benefit of frictional resistance the structure, even with such a seemingly insignificant 'factor' of safety, would have been unlikely to float. However, considering an average unit friction of even as low a value as $5kN/m^2$, a further reaction, at the very least, of 11,000 *tonnes* in terms of friction at the soil–structure interface would have been available to further enhance the margin of 'comfort' against flotation. Downgrading the frictional resistance in this manner by a factor of, say, 3 or more amounts to building-in a significant factor of safety in its own right. In view of this, there is no reason why a factor of even 1·01 on gravity load alone should not be accepted in the knowledge that, at the very least, 11,000 *tonnes* would also be available in terms of frictional resistance. This goes to show that, by following a sensible approach, it is possible to avoid waste and achieve cost savings without, in any way, compromising the safety.

Example 22.2
Revisit Example 22.1 and Figure 22.4 assuming, this time, zero penetration of the perimeter walls below the track slab.

Solution

As before, for the fully submerged structure, assuming the worst credible water level to be at ground level, the factor of safety against flotation is given by the equation:

$$F = 0{\cdot}1\left[\gamma_S \cdot m + w_e\right]$$

For the desired factor of safety of 1·05 and with $d = 0$, $D = 28 - 6 = 22$ and $m = 0{\cdot}09$, weight of the structure required to counteract flotation is given by:

$$W_R = 10B \cdot D\ (F - m)$$
$$= 10 \times 30 \times 22\ (1{\cdot}05 - 0{\cdot}09)$$
$$= 6{,}336\,kN \quad \text{per } m \text{ run}$$

Putting $D = 22$ in Equation (22.16), we obtain the gross cross-sectional area of the structure:

$$\sum A = 256{\cdot}2 - 2T_W \cdot d\left(1 + \frac{B'}{L}\right)$$
$$= 256{\cdot}2 - 2 \times 1{\cdot}2 \times 6\left(1 + \frac{27{\cdot}6}{300}\right) = 240m^2$$

So, available gross weight (W) of the structure $= 24 \times 240 = 5{,}760kN$ per m run, which is less than the required (W_R) and represents a shortfall of $6336 - 5760 = 576kN$ per m run.

However, if frictional resistance is relied upon to top up the factor of safety, the shortfall based on a factor of safety of 1·10 would be of the order of:

$$10 \times 30 \times 22\,(1{\cdot}10 - 0{\cdot}09) - 5760 = 906kN \quad \text{per } m \text{ run}$$

The value for the average skin friction required to be mobilized to compensate for this shortfall would be given by setting $d = 0$ and putting $D = 22$ in Equation (22.17), i.e.:

$$f_S = \frac{Shortfall \times L}{2\,(L + B)\,D} = \frac{906 \times 300}{2\,(300 + 30)\,22} = 19kN/m^2$$

Example 22.3

Repeat Example 22.1 to examine, for the fully submerged case, the implications of ignoring frictional resistance altogether and compensating for the shortfall in the factor of safety by the enhancement in the dead weight of the structure alone.

Solution

Recalling the parameters assumed and the various terms worked out previously, we have:

$$L = 300m,\ B = 30m,\ D = 28m,\ L' = 297{\cdot}6m,\ B' = 27{\cdot}6m,\ d = 6m,$$
$$F_A = 1{\cdot}00,\ F_R = 1{\cdot}05,\ \Delta F = 0{\cdot}05,\ \gamma_C = 24kN/m^3,\ K = 0{\cdot}805,\ m = 0{\cdot}07,$$
$$\gamma_B = 20kN/m^3,\ \gamma_W = 10kN/m^3,\ T_R = 1{\cdot}5m,\ \gamma_E = 10kN/m^3,\ W_R = 6{,}512kN.$$

(a) *Thickening perimeter walls*

$$\Delta T = \frac{-v + \sqrt{v^2 - 4u \cdot c}}{4u}$$

$$u = \frac{1}{L \cdot B} \left[\gamma_W \cdot F_R - \gamma_E \cdot m - \gamma_C \right]$$

$$= \frac{1}{300 \times 30} [10 \times 1 \cdot 05 - 10 \times 0 \cdot 07 - 24] = -0 \cdot 0016$$

$$v = \left[u(L + B) - \frac{w_e}{L} \right]$$

$$= \left[-0 \cdot 0016 (300 + 30) - \frac{6512}{300 \times 30 \times 28} \right] = -0 \cdot 495$$

$$c = K \cdot \gamma_W \cdot \Delta_F = 0 \cdot 805 \times 10 \times 0 \cdot 05 = 0 \cdot 4025$$

whence, $\Delta T = 0 \cdot 4m$

Volume of additional concrete:	$2(300 \cdot 4 + 30 \cdot 4) \times 0 \cdot 4 \times 28 = 7,410m^3$
Area of additional land-take:	$2(300 \cdot 4 + 30 \cdot 4) \times 0 \cdot 4 = 265m^2$
Volume of additional excavation:	$7410 \times 31 \cdot 5 \div 28 = 8,336m^3$
Volume of additional back-fill:	$8336 - 7410 = 926m^3$

(b) *Thickening base slab*

$$\Delta T_B = \left[\frac{B}{B'} \cdot \frac{L}{L'} - \frac{d}{D} \right] \frac{\Delta F \cdot D}{(0 \cdot 1 \gamma_C - F_R)}$$

$$= \left[\frac{30}{27 \cdot 6} \times \frac{300}{297 \cdot 6} - \frac{6}{28} \right] \frac{0 \cdot 05 \times 28}{(0 \cdot 1 \times 24 - 1 \cdot 05)} = 0 \cdot 914m$$

Volume of additional concrete:	$297 \cdot 6 \times 27 \cdot 6 \times 0 \cdot 914 = 7,507m^3$
Volume of additional excavation:	$7,507m^3$

(c) *Thickening roof slab*

$$\Delta T_R = \left[\frac{K \cdot \Delta F}{0 \cdot 1 \left(\gamma_C - \gamma_E \right) - F_R} \right] D$$

$$= \left[\frac{0 \cdot 805 \times 0 \cdot 05}{0 \cdot 1 (24 - 10) - 1 \cdot 05} \right] 28 = 3 \cdot 22m > 2m!$$

Not feasible at all!

(*d*) *Deepening perimeter wall*

$$K_A = 1 - \frac{B'}{B} \cdot \frac{L'}{L} = 1 - \frac{297\cdot6}{300} \times \frac{27\cdot6}{30} = 0\cdot087$$

$$\Delta P_W = \left[\frac{\Delta F}{0\cdot1\gamma_C - F_R} \right] \left[\frac{K}{K_A} \right] D$$

$$= \left[\frac{0\cdot05}{0\cdot1 \times 24 - 1\cdot05} \right] \left[\frac{0\cdot805}{0\cdot087} \right] 28 = 9\cdot59m$$

Increase in perimeter wall quantity: $9\cdot59 \times 100 \div 28 = 34\%$
[Volume of additional concrete: $2(300 + 30 - 2 \times 1\cdot2) \times 1\cdot2 \times 9\cdot59 = 7,540m^3$
Volume of additional excavation: $7,540m^3$]

(*e*) *Widening roof slab*

$$\Delta B = \left[\frac{0\cdot5\,\gamma_W \cdot \Delta F \cdot K}{\gamma_E \cdot m + K_R\,(\gamma_C - \gamma_W \cdot F_R)} \right] B$$

$$= \left[\frac{0\cdot5 \times 10 \times 0\cdot05 \times 0\cdot805}{10 \times 0\cdot07 + 0\cdot054\,(24 - 10 \times 1\cdot05)} \right] 30 = 4\cdot22m$$

Volume of additional concrete: $2 \times 300 \times 4\cdot22 \times 1\cdot5 = 3,798m^3$
Area of additional land-take: $2 \times 300 \times 4\cdot22 = 2,532m^2$
Volume of additional excavation: $3798 \times 3\cdot5 \div 1\cdot5 = 8,862m^3$
Volume of additional back-fill: $8862 - 3798 = 5,064m^3$

(*f*) *Increasing roof plan all round*

$$\Delta S = \frac{-b + \sqrt{b^2 + 4c}}{4}$$

$$b = L + B = 300 + 30 = 330m$$

$$T_R = \text{Thickness of roof slab } = 1\cdot5$$

$$c = \frac{\gamma_W \cdot \Delta F \cdot K \cdot B \cdot L \cdot D}{T_R\,(\gamma_C - \gamma_W \cdot F_R) + \gamma_E \cdot z}$$

$$= \frac{0\cdot05 \times 0\cdot805 \times 30 \times 300 \times 28}{1\cdot5\,(2\cdot4 - 1\cdot05) + 2} = 2520$$

whence $\Delta S = 3\cdot73m$

Volume of additional concrete: $2\,(303\cdot73 + 33\cdot73) \times 3\cdot73 \times 1\cdot5 = 3,776m^3$
Area of additional land-take: $2\,(303\cdot73 + 33\cdot73) \times 3\cdot73 = 2,517m^2$

Volume of additional excavation: $3776 \times 3.5 \div 1.5 = 8,811 m^3$
Volume of additional back-fill: $8811 - 3776 = 5,035 m^3$

Example 22.4

Repeat Example 22.2 to examine, for the fully submerged case, the implications of ignoring frictional resistance altogether and considering compensation for the shortfall in the factor of safety associated with the use of option (g), i.e. extending base slab to form toes along the two longitudinal walls.

Solution

From Example 22.2, $m = 0.09$ and $w_e = W \div B \times D = 5760 \div 30 \times 22 = 8.73$. Available factor of safety is given by:

$$F_A = 0.1 \left[\gamma_S \cdot m + w_e \right] = 0.1 \left[(20 - 10) \times 0.09 + 8.73 \right] = 0.96$$

Shortfall, $\Delta F = F_R - F_A = 1.05 - 0.96 = 0.09$

$$K_B = \frac{T_B}{D} = \frac{2.5}{22} = 0.114$$

Required toe, $\Delta B = \left[\dfrac{0.5 \times \gamma_W \cdot \Delta F}{\gamma_E (m+1) + (\gamma_C - \gamma_E) \cdot K_B} \right] B$

$$= \left[\frac{0.5 \times 0.09}{(0.09+1) + (2.4-1) \times 0.114} \right] \times 30 = 1.08 m$$

Volume of additional concrete: $2 \times 300 \times 1.08 \times 2.5 = 1,620 m^3$
Area of additional land-take: $2 \times 300 \times 1.08 = 648 m^2$
Volume of additional excavation: $2 \times 300 \times 1.08 \times 25.5 = 16,524 m^3$
Volume of additional back-fill: $16524 - 1620 = 14,904 m^3$

Conclusions

Comparing the results of the various measures worked out in the preceding section, following conclusions can be drawn:

- Increase in the thickness of roof slab upwards (case c) results in an increase in the net unit weight of the additional concrete of only $4kN/m^3$ which represents a very small return. This measure therefore results, as is to be expected, in large increases in the slab thickness for extremely modest gains in the factor of safety against flotation. The required increase in slab thickness in the above example, for instance, turns out to be well in excess of what is achievable. In view of this and the need to accommodate utilities within the available space above the roof of the structure, this measure is, in this particular case, impractical and, in general, the least attractive.

- Increase in the net unit weight of the additional concrete in the base slab works out to $14kN/m^3$. This is 3·5 times greater than that of the roof slab and is the best return that can be expected. This is so in cases (a), (d), (e) and (f) as well.

- In case (a), if the perimeter wall is constructed by a proprietary method, as would very often be the case, it is important to establish whether or not the required increase in the wall thickness corresponds to modular increment(s) achievable within the system. If yes, increase in thickness could, in theory, lead to savings in reinforcement. If not, a thickness bigger than that required would have to be adopted. In view of the likelihood that the wall penetration may be dictated by other factors in which case it may not be possible to reduce it to compensate for the unwanted component of increase in thickness, such a measure could prove somewhat wasteful.

- Measures (e) and (f) are essentially similar with only a marginal difference between the two. It can also be seen that both these measures require approximately only half the volume of additional concrete as compared to those of measures (a), (b) and (d). However, they involve significant additional land-take that may not always be on offer; even if the land were to be available, acquiring it could carry a significant cost penalty. Besides, intrusion into areas beyond the confines of the main box structure could fall foul of the constraints and easily lead to problems of interaction rendering the feasibility of these measures doubtful.

- Increasing the penetration of the perimeter wall, i.e. measure (d), compares favourably with measures (a) and (b). Out of all the aforementioned measures for increasing the factor of safety against flotation, these three are the most feasible which can merit consideration. It should be remembered, however, that measures (b) and (d) do not intrude beyond the plan of the box structure whereas measure (a) can be marginally intrusive. Eventual choice is likely to be influenced by logistics and considerations of cost.

- Whichever one of the three feasible options might be adopted, it will always be accompanied by an increase in cost. As it is possible to do so, it makes perfect sense to avoid such extra costs simply by seeking the required enhancement in the factor of safety in terms of the available frictional resistance along the soil–structure interface around the perimeter. Whether or not this is invoked, there is no denying the fact that it has to be overcome before the prospect of flotation could be considered as real, and it would be there on offer in any case.

- In view of the size and the self-weight of the metro structures generally encountered in practice, even a very small percentage in the enhancement of factor of safety can represent a significant load counteracting flotation.

- Enhancement of factor of safety by recourse to option (g) can be considered where the perimeter wall does not penetrate below the base slab and where it is constructed either within a temporary outer cofferdam or an open cut with stabilized side slopes. Even so, construction within an outer cofferdam would be feasible only if the toe width required is also sufficient enough to provide adequate working space; otherwise increasing the width beyond that dictated by design simply to provide enough working space would be wasteful. This option is therefore ideally suited only to construction within an open cut with stabilized side slopes.

- In order to come up with a structure that is most cost effective, it would be ideal to aim for a structural geometry that ensures that the gravity load (self weight plus backfill plus permanent kentledge, where applicable) yields a factor of safety of at least 1. In this way, the effect of buoyancy is at least neutralized. Beyond this, frictional resistance can be relied upon to make up for the desired margin of comfort. The rationale behind this reasoning is that, even ignoring the contribution of the frictional resistance which will be there to whatever extent in any case, those gravity loads which are routinely ignored in the flotation calculation but which are more than likely to be present nevertheless will ensure that the structure never floats. Based on this reasoning, it can be argued that a factor of safety, incorporating frictional resistance, of much less than 1·10 should be acceptable.

- It is perfectly feasible to aim for the factor of safety on the basic gravity load of even less than one as long as sufficient and safe (i.e. adequately factored) frictional resistance can be mobilized which can yield at least a minimum acceptable positive margin of comfort without the need to increase the basic gravity load. In any case, every situation should be treated on its merits.

In the event that the construction of a metro structure is carried out within an outer cofferdam, a clear distinction should be made as to whether it is to be treated as an integral part of the structure, in which case it could be relied upon to help counteract the potential of flotation, or only as a temporary measure for the construction of the main box structure. In the former case, consideration would need to be given to the durability of the cofferdam over its design life. In the latter case, however, it could be treated as sacrificial and its long-term durability would not be an issue as long as its deterioration was unlikely to subscribe in any way to the deterioration of the permanent structure.

23 Design Parameters

23.1 Introduction

Defining and assembling the various parameters and the related criteria for detailed design represents an important key stage in the design process. In the case of cut-and-cover structures, because of the extensive soil–structure interaction involved, this applies as much to the surrounding soil medium as it does to the structure itself. However, the structural parameters relate to materials which are man-made; these are routinely specified to suit design and manufactured under strict control. Since these parameters are generally invariant, easily predictable, well defined and can be readily obtained by reference to any standard textbook on structural design, they are not discussed here. Geotechnical parameters, on the other hand, are concerned with materials which occur naturally; since they are not man-made, they cannot be specified. They vary from site to site and are rarely constant. In fact, their variation from one location to the other even on the same site also cannot be ruled out. Appropriate selection and quantification of such parameters are therefore important and so, the type and the quality of these form the main focus of attention in this chapter.

It is not difficult to appreciate that, from an engineering perspective, no two sites, however close to each other, are ever likely to be identical in all respects. Even for identical structures, there may be dissimilarities, for instance, due to the changes in the size and stiffness of the surrounding structures, method and sequence of construction, the treatment required for dealing with the existing services and traffic, the structural requirements, the ambient stress fields, porewater pressure regimes, etc. Each site is therefore likely to present its own specific problems demanding its own specific design parameters and solutions.

Cut-and-cover metro structures tend to be exceptionally large underground structures. Even a single-level track station structure may, typically, comprise 300–500m-long, 20–26m-wide and 12–14m-deep box structure buried 2–4m below ground. Depending upon the sequence of construction adopted, a large volume of soil, of the order of 85,000 to over 230,000m^3 could easily need to be removed to make way for the construction of such a structure. In view of the sheer size of such a structure stretching over an enormous length, the possibility of variation in the soil parameters even over the extent of its length which could influence the design cannot be ruled out.

23.2 Identifying parameters

Cut-and-cover structures are in intimate contact with the surrounding soil over their entire operational lifetime. Accordingly, the nature and the extent of interaction between the structure and the surrounding ground are of primary consideration in the design and construction of the structure. The surrounding ground not only imposes loads on the structure but also provides all round support to it. Appreciation of the *in situ* ground environment, selection of the appropriate design parameters and their accurate assessment, closely reflecting the ground response at various stages of construction are of paramount importance. In view of this, the type and the quality of the parameters chosen will, to a large extent, reflect the quality of the design that can, potentially, be achieved.

Different geotechnical parameters are generally dictated and necessitated by the specific design requirements at the different stages of excavation and construction. In the process of identifying the associated parameters, therefore, it would be appropriate to recall the five principal stages in relation to the construction of a typical metro station structure. These are:

- Construction of the perimeter wall;
- Groundwater lowering and control;
- Excavation and support of ground;
- Construction of the main structure;
- Replacement of the backfill on top.

The parameters required at these stages are identified in the following section.

At first stage

During the installation of the perimeter wall, trench excavation can lead to varying degrees of movement in the adjacent ground and structures. However, excessive movements which could potentially lead to caving-in and catastrophic collapse of trench faces must be prevented by judicious design and use of adequate care during construction. Sensible design must reflect sound appreciation of the prevalent ground and groundwater conditions. It must also recognize the importance of trench stability in the given conditions and incorporate adequate margins of safety. For this purpose, the necessary design parameters are: undrained shear strength s_u, the effective stress shear strength parameters c' and $\tan\phi'$, coefficients of active (K_a) and at-rest (K_o) earth pressures and initial tangent modulus E_o.

Being a function of ϕ', K_a need not be treated as an independent design parameter. However, accurate assessment of K_o, empirically, purely on the basis of ϕ' may not always be possible; it may become necessary to obtain its value *in situ*. It has therefore been treated as an independent design parameter.

At second stage

Large volumes of soil need to be removed to make way for the construction of cut-and-cover structures. In order that the excavation and the construction activities can proceed in the dry, pumping out large volumes of groundwater from within the perimeter cofferdam also becomes necessary. This dewatering and its consequential effects around the site can potentially subscribe to a large proportion of settlements of the surrounding ground and structures. Ground movements, if not adequately controlled, can present major concerns to the engineer. Appropriate measures to control the movement of groundwater therefore need to be planned in advance and undertaken with due deference to safety and cost. For estimating the rate of leakage of groundwater into the excavation and the design of appropriate measures to control it, the knowledge of the *in situ* drainage characteristics of the soil medium is essential. The key parameter necessary for this exercise, assuming the soil to be homogenous and isotropic, is the coefficient of permeability k of the ground mass.

In reality, however, most soils are anisotropic to varying degrees and may often exhibit nonhomogeneity or layering resulting from preferred particle orientation. This may lead to higher coefficient of permeability (k_h) in the horizontal direction and lower (k_v) in the vertical direction. From these, the equivalent isotropic coefficient of permeability may need to be obtained for use in the design.

At third stage

Before the bulk excavation within the perimeter cofferdam is undertaken, it is necessary to establish the *in situ* stress history of the ground so that, against this threshold, the extent of the changes imposed by the different stages of excavation can be assessed and their implications on design quantified appropriately. The design parameter needed for obtaining this stress history is the coefficient of earth pressure at rest, K_o.

As the excavation is advanced downwards, movement of the ground support system and the ground, each to other in sympathy, inwards towards the excavation can be expected. Following the strain experienced by the surrounding soil in this manner, the lateral pressures would no longer be those at rest but would be expected to fall below this value; the size of the drop being a function of the extent of the strain undergone by the soil. However, in the limiting case, the pressures would drop to their active values. For estimating these minimum lateral pressures and their contribution to the design of the structure, the parameter needed would be the coefficient of active earth pressure K_a, or rather ϕ'.

If, in place of the perimeter cofferdam, the construction is intended to be carried out in an open cut within stabilized side slopes, it would be necessary to examine how stable the side slopes are likely to be during construction. For this, the knowledge of the inherent strength of the ground would be needed. Depending upon whether it is the short-term undrained or the long-term drained strength of the soil that is needed, the pertinent design parameter(s) required would be: the undrained shear strength s_u, or the effective stress shear strength parameters c' and $\tan\phi'$ (rather ϕ'), respectively.

Furthermore, porewater pressure, while not contributing to the strength of the soil directly, does, through the rate at which it moves through the subsurface, influence the way in which the ground responds to the various stages of excavation. For this, porewater pressure coefficient r_u would be required.

Removal of overburden during the excavation would cause pressure relief on the formation. Depending upon the type of the soil below formation, this could, in turn, give rise to short- and long-term heave of the floor of excavation. It would be necessary to estimate the extent of the anticipated heave and examine the effect of the associated equivalent uplift pressure on the base of the structure. The likely parameters to achieve this are: undrained shear strength s_u, coefficient of earth-at-rest pressure K_o, undrained expansion 'elastic' modulus (for unloading) $E_{u(e)}$, drained 'elastic' modulus for unloading $E'_{(e)}$, coefficient of volume change m_v and expansion secant strain modulus M_e.

At fourth and fifth stages

After the construction is completed and the backfill replaced, the lateral pressures would, in course of time, be expected to climb back towards their eventual at-rest values. It is therefore important to ensure that the structure is also capable of withstanding these pressures during their operational lifetime. For the estimation of the loads likely to be generated by the at-rest pressures and to assess their impact on the design of the structure, again, the coefficient of earth pressure at rest (K_o) would be needed.

In the modelling of the boundary conditions for the completed box structure during its behaviour in the in-service stage, the vertical soil reaction below the base of the structure and the horizontal soil reaction on the side wall against any out-of-balance lateral load acting on the opposite wall would need to be idealized in the form of closely spaced soil springs. To obtain the stiffness of the soil springs, the necessary parameters needed are the representative coefficients of the vertical and the horizontal subgrade reactions (k_{sv}, k_{sh}) for the soil. These may be obtained from appropriate plate load tests.

In areas susceptible to seismic activity, it is important to know the vertical and horizontal components of peak ground acceleration (k_V, k_H), velocities, displacements, design response spectra and, possibly, the time-history accelograms for the seismic design of the structure. Besides, it is also necessary to know the value of C_s, the apparent velocity of S-wave propagation in soil, and C_V, the seismic earth pressure coefficient.

To sum up, the geotechnical parameters for the design of cut-and-cover metro structures that may become necessary at the various stages of construction are listed as follows:

- Initial tangent modulus E_o
- Expansion secant strain modulus M_e
- Drained elastic modulus for unloading $E'_{(e)}$
- Undrained elastic modulus for unloading $E_{u(e)}$
- Effective stress shear strength parameters c', ϕ'
- Average porewater pressure coefficient r_u
- Undrained shear strength s_u
- Coefficient of volume change m_v
- Coefficients of permeability, k_h, k_v
- Coefficient of earth pressure at rest K_o
- Coefficient of vertical subgrade reaction k_{sv}
- Coefficient of horizontal subgrade reaction k_{sh}
- Components of earthquake acceleration (k_V, k_H)
- Apparent velocity of S-wave propagation, C_s
- Seismic earth pressure coefficient, C_V.

In addition to the assessment of the drainage characteristics as listed above, it is also important to establish the ambient porewater pressure regime in the ground accurately. This is particularly so where the water table is high and the porewater pressure is likely to figure as a dominant component of the lateral loading on the structure.

23.3 Factors influencing choice

The selection of the geotechnical parameters needed for the design of cut-and-cover metro structures from out of those listed previous section is mainly influenced by three factors – the first is related to the type of the soil medium, the second, to its drainage characteristics and the third, to the structure itself.

Construction of a cut-and-cover metro structure will, inevitably, disturb the *in situ* stress environment and the ambient porewater pressure regime in the surrounding soil mass. This disturbance will be dependent, *inter alia*, upon the type of the soil medium, whether free-draining, fine-grained or intermediate. With the relatively free-draining soils, such as sands and gravels, the effects of the various stages of construction are likely to be immediate. In the case of fine-grained soils, such as clays, due to the viscous retardation to free flow of water, the effects can be expected to be time-dependent. With the soils ranged in-between, such as silts, however, the response is likely to be mixed, i.e. both short-term (undrained) as well as long-term (drained).

In view of the foregoing, it is advisable to choose undrained parameters for design in fine-grained soil medium such as clay, drained and effective stress shear strength parameters in free draining medium such as sands and gravels, and both sets of parameters in a mixed medium such as silt.

The aforesaid notwithstanding, it is important to recognize that, in reality, fine-grained soils may not be absolutely undrained just as the granular soils may not, *per se*, be totally free draining. These idealized conditions are assumed for convenience and only to help simplify computations while erring on the side of safety. The behaviour of soil is essentially a function of the rate of flow of water through, or the drainage characteristic of, the soil mass. In establishing the response of the ground to a stage of excavation or construction, the extent of time over which that stage lasts and the associated movement of groundwater are, therefore, of great importance.

If, in a 'free-draining' soil, the works are likely to be of relatively short duration, as may be expected in the case of temporary works, there is the possibility that the conditions may be close to undrained. By the same token, conditions even for temporary works in fine-grained soils, if extending over a period over which perceptible movement of water could take place, may be close to being drained or at least partially so. Therefore, in selecting the appropriate parameters for design, it is important to consider all these factors carefully.

Note that in the text, generally, primed symbols or symbols carrying the subscript d are those representing drained and the unprimed symbols with the subscript u the undrained conditions. Furthermore, addition of subscript (e) to a parameter should be taken to denote 'expansion' due to unloading of soil caused by the removal of overburden upon excavation.

23.4 Assessment of design parameters

Important aspects related to the assessment of some of the key design parameters are discussed in the following section.

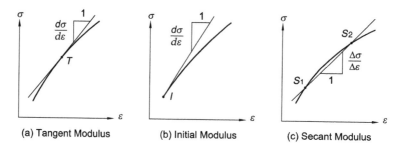

Figure 23.1 Modulus Related to Soil

Soil moduli and Poisson's ratio

In soils, the stress–strain relationship is nonlinear. However, over small ranges of stress, variations can be idealized as being linear. 'Linearizing' the actual nonlinear stress–strain curve thus makes it possible to employ the simple theory of elasticity in securing solutions to soil-related problems relatively easily. However, it also brings into focus the fact that this is an idealization since, for soils, modulus is not constant.

If, at any point such as T on an actual nonlinear stress–strain curve, as shown in Figure 23.1(*a*), a tangent is drawn, the slope of the tangent will represent what is appropriately known as the tangent modulus at that point; modulus being defined as the stress per unit strain, i.e. $d\sigma/d\varepsilon$. Obviously, the value of the tangent modulus will vary with the point selected on the curve. Initial tangent modulus, as shown in figure (*b*), is its value at the initial point such as I of the curve.

Slope of a line connecting two points such as S_1 and S_2 on the stress–strain curve as shown in figure (*c*) represents what is known as the secant modulus. Its value is a function of the location of the two points. As the two points approach each other towards a common point, clearly, the secant modulus will approach the tangent modulus at that point. When a value for modulus is quoted for soil, it is usually the secant modulus.

For a truly linear elastic material, there is no such diversity as outlined earlier for the values of the modulus parameter; there is only one value, and it is known as the elastic or the Young's modulus defined as the stress per unit strain.

In selecting the appropriate modulus, it is important to remember that different values of the modulus are likely to apply at different sets of stresses. Furthermore, in the case of granular soils, obtaining undisturbed samples is difficult. Since modulus depends upon void ratio, it is particularly difficult to measure the modulus of granular soils reliably. It would appear (Lambe and Whitman, 1969) that, during a laboratory test, second cycle of loading usually gives a better measure of *in situ* modulus. Apparently, the effects of sample disturbance are compensated by the effects of initial loading.

Poisson's ratio, v, is a measure of the extent by which the soil will undergo lateral strain corresponding to a unit vertical (axial) strain. It may therefore be defined as the ratio of the lateral strain to the axial strain. For soils, Poisson's ratio also, like the modulus, is not constant. During the early range of strains over which the concepts of theory of elasticity is applicable, Poisson's ratio is less than 0·5 and is found to vary with strain. For large strains implying failure, the Poisson's ratio for sand becomes constant with a value greater than 0·5. Because of such behaviour, it is difficult to quantify the value of v precisely. Fortunately, however, the effect of its value on engineering predictions is usually not significant.

For an idealized isotropic elastic soil, with E' and v' as the modulus and Poisson's ratio, respectively, appropriate to changes of effective stress, i.e. drained conditions, we can write:

$$K' = \frac{E'}{3(1-2v')} \quad \text{and} \quad G' = \frac{E'}{2(1+v')} \tag{23.1}$$

Symbols K' and G' are the drained bulk and the shear moduli. By removing the primes and adding the subscript u, the corresponding symbols and expressions for the undrained moduli can be obtained.

From the relationships of the stress and strain invariants with the bulk and shear moduli and noting that $v_u = 0·5$ for undrained conditions, it can be proved (Atkinson and Bransby, 1978) that $G' = G = E_u/3$. Hence,

$$E_u = 1·5\frac{E'}{(1+v')} \tag{23.2}$$

Strain modulus

Zeevaert (1973) uses the parameter 'strain modulus' in heave calculations. The determination of strain modulus, which is essentially a mechanical property of soil, is achieved by testing in the laboratory undisturbed samples representative of the subsoil materials. However, in spite of using the best sampling tools, undisturbed test specimens are often difficult to obtain by means of sampling. In the interpretation of test results, therefore, sampling disturbance must be taken into consideration. In the case of cohesive soils, better results are possible if undisturbed test specimens can be obtained from block samples. However, this is not always practical especially where the material to be investigated is deep seated and under high water table. In addition, in the case of granular soils, the process of obtaining an undisturbed sample is difficult and determining the value of modulus is complicated since it is necessary to examine the various states of compaction before the value for the modulus can be confirmed. The value that corresponds with the state of compaction and stress at which the soil is encountered in the field can then be taken to represent the strain modulus for the soil at that level.

The test generally used for obtaining the linear strain modulus in cohesive soils is the odometer test and that for the granular soils the triaxial test. However, the major drawbacks with the odometer test are that the sample is often too small to be considered representative of the soil mass, stresses get modified from anisotropic to isotropic during sampling and the *in situ* stresses are difficult to restore. Furthermore, porewater pressure levels during testing do not usually correlate with those obtaining in the field and there are no means available for monitoring the changes in the porewater pressures during testing.

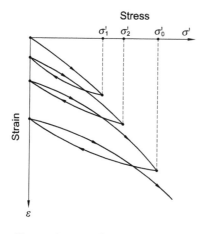

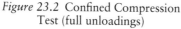

Figure 23.2 Confined Compression
Test (full unloadings)

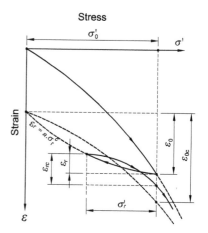

Figure 23.3 Confined Compression
Test (partial unloadings)

In the design of cut-and-cover metro structures, the linear strain modulus of interest would be that concerning the relief, rather than the increase, in the value of the effective stress. In order to correlate the state of stress in the field with that in the odometer or the triaxial test, one therefore needs to concern oneself with the relationship that exists between the stress relief in the field and the nature of the unloading curves obtained in the test.

Figures 23.2 and 23.3 show loading and unloading curves obtained in compression tests carried out at and below the effective overburden stress σ_o' for full and partial effective unloading, σ_r'. It can be seen that whenever full unloadings ($\sigma' \to 0$, i.e. $\sigma_r' = \sigma_o'$) are performed (Figures 23.2), the hysteresis loops are parallel to each other.

However, for partial unloadings (i.e. smaller hysteresis loops where $\sigma_r' < \sigma_o'$), that is not the case. Besides, the expansion strains are smaller although for the subsequent recompression phase the compression strains are relatively larger as shown in Figure 23.3. In other words, the expansion strain ε_r due to partial stress relief σ_r' is smaller than the corresponding compression strain ε_{rc} due to recompression under the same amount of stress increase σ_r'. The values ε_o and ε_{oc} would represent the corresponding expansion and recompression strains had complete unloading been performed.

The stress–strain relationship for the unloading curve (Figure 23.3) is given by: $\varepsilon_r = a \cdot \sigma_r'^c$ in which coefficients a and c are functions of the soil characteristics. Tests show (Zeevaert, 1973) that the relationship, when plotted on log-log-scale, represents a straight line as shown in Figure 23.4. Some typical values for parameters 'a' and 'c' are also indicated in the figure.

Note that, unlike those of the conventional tangent and secant moduli, the strain modulus is defined as the strain per unit stress. Therefore, the slope ε_r/σ_r' of the partial unloading curve in Figure 23.3 represents the secant strain modulus of expansion M_{er} corresponding to the stress relief σ_r' for totally unconfined conditions where lateral deformation is completely unrestrained, i.e. the Poisson's ratio $\nu = 0$ or $\nu_c = 1$ from

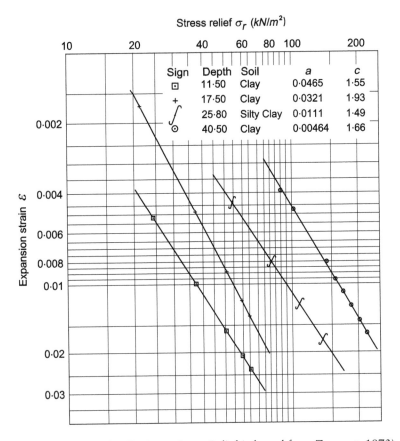

Figure 23.4 Expansion Strain vs. Stress Relief (adapted from Zeevaert, 1973)

Equation (23.7). Hence

$$M_{er} = \frac{\varepsilon_r}{\sigma'_r} \quad \text{or} \quad M_{er} = a \cdot \sigma_r^{\prime(c-1)} \tag{23.3}$$

Identifying M_{eo} as the expansion strain modulus obtained from the full hysteresis loops (i.e. for complete unloading where $\sigma'_r = \sigma'_o$) on undisturbed unconfined soil samples tested in the laboratory, the M-ratio of intermediate to full stress relief for totally unconfined conditions will be given by:

$$\frac{M_{er}}{M_{eo}} = \rho_e = \left(\frac{\sigma'_r}{\sigma'_o}\right)^{(c-1)} \tag{23.4}$$

$$\text{Or,} \quad M_{er} = \rho_e \cdot M_{eo} \tag{23.5}$$

The parameter ρ_e is the stress relief expansion factor defined as the ratio of expansion strain modulus obtained from hysteresis loops based on partial stress relief due to the

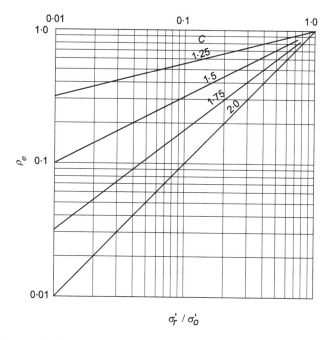

Figure 23.5 Stress Relief Expansion Factor ρ_e (after Zeevaert, 1973)

stage of excavation, and full stress relief (equal to the initial effective stress at that level). The ρ_e versus the stress ratio plots for different values of the soil parameter 'c' are given in Figure 23.5. For full unloading, i.e. when $\sigma'_r = \sigma'_o$, $\rho_e = 1$ and $M_{er} = M_{eo}$; otherwise, for partial unloading, with $\sigma'_r < \sigma'_o$, ρ_e will be less than 1 and, accordingly, M_{er} will be less than M_{eo}.

It should be recognized that the unit vertical deformation, as also the linear strain modulus, is a function of the Poisson's ratio which, in reality, is rarely zero. Therefore, introducing the effect of the Poisson's ratio, the general expression for the linear strain modulus can be expressed as:

$$M_{er} = v_c \cdot \rho_e \cdot M_{eo} \tag{23.6}$$

$$\text{where} \quad v_c = \frac{(1+v)(1-2v)}{(1-v)} \tag{23.7}$$

The relationship between v_c and v as represented by Equation (23.7) is presented graphically in Figure 23.6.

In the case of total confinement for $v = 0.5$, v_c and the strain will both be zero. That is to say, the material cannot heave when lateral displacements are totally restricted. For no restriction to lateral deformation, likewise, when $v = 0$, v_c will be 1. It follows that, in reality, the value of the parameter v_c could lie between 0 and 1, according as the Poisson's ratio v lies between 0.5 and 0. Therefore, depending upon the type of confinement, the strain can be different for the same value of the linear strain modulus

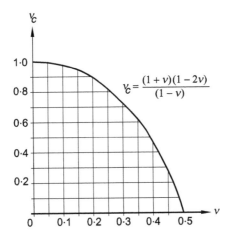

Figure 23.6 v_c vs. v (after Zeevaert, 1973)

as given by Equation (23.6). This is illustrated in the solution to Example 21.5 in Chapter 21.

Undrained shear strength

In order to appreciate the response of soil to various stages of excavation and construction, it is important to know its *in situ* shear strength as accurately as possible. However, particularly in soft clays with low strength and high compressibility, this presents somewhat of a problem. The strengths obtained by conventional means, such as that with the use of isotropic triaxial compression test where it is assumed that $\sigma_1 = \sigma_2 = \sigma_3$, or 'true' triaxial test where the stresses are not isotropic but can be varied, are not necessarily indicative of the *in situ* strength; they represent the two bounds. This is due to strength and stress–strain anisotropy and strain-rate effects (Ladd and Foott, 1974). However, by far the greatest contributor to the divergence in the value of the shear strength obtained from the conventional testing procedures from that of the actual *in situ* strength is due to the sampling disturbance generally associated with these tests.

Strength and stress–strain anisotropy

Undrained strength anisotropy may be either inherent resulting from the major differences in the soil structure resulting from its natural formation, such as that found in varved clays with their alternating layers of silt and clay; or, it may be induced due to rotation of principal stresses during shear at various locations along the shear plane and variations in the intermediate principal stress.

Strain rate effects

In triaxial compression tests, each log cycle decrease in strain rate is typically accompanied by a 10 ± 5 per cent decrease in s_u, the exact variation being a function of the plasticity and creep susceptibility of the soil (Ladd and Foott, 1974). The effect is due to undrained creep which accompanies shear in the sample. This results in the increase in porewater pressure, corresponding decrease in the effective stress and, of course, commensurate decrease in the strength.

Effect of sampling

In the course of sampling, the soil may experience disturbance which can be caused by the mechanics of sampling or stress relief or both. Mechanical disturbance generally accounts for up to 50 per cent of the sampling disturbance. The process of drilling holes for sampling changes the *in situ* stress regime locally. This brings about a change in the effective stress leading to deformation which, if not elastic, may well prove to be irreversible.

Likewise, insertion of sampling tube and extrusion in the laboratory can cause remoulding of the sample which can have profound effect on the results obtained, although the effect can be minimized by 'jacking-in' thin walled sampling tubes. However, the extent of the effect varies according to the soil type. The disturbance is the greatest in clays of low plasticity and low sensitivity. In the disturbed zone, there is the breakdown of the structure which causes an increase in the compressibility of the soil. Reconsolidation with a view to restoring the *in situ* stresses therefore leads to a large volume change and deformation.

In the highly sensitive quick clays, on the other hand, owing to their very small remoulded strength which virtually eliminates the friction between the clay and the sampling tube, the disturbance due to the insertion of the tube is the least.

Consider a hypothetical sample of soil at a certain depth under a vertical stress σ_V, a horizontal stress σ_H and porewater pressure u_o. If such a sample of soil is actually removed from this level in the ground, it will experience stress relief. The consequent disturbance from the stress relief will manifest in a change in the state of stresses from those of the anisotropic stresses *in situ* (i.e. in the ground) to a state of isotropic stresses in the laboratory as shown in Figure 23.7. Although the sampling is done in the undrained mode, i.e. maintaining volumetric strain $\Delta V/V = 0$, still some shear strain can be expected to occur (i.e. $\gamma_{xy} \neq 0$). This will cause shear deformation, a consequent change in the effective stress and, inevitably, a commensurate change in the strength.

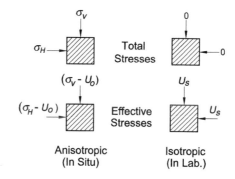

Figure 23.7 Stresses *in Situ* and in Laboratory

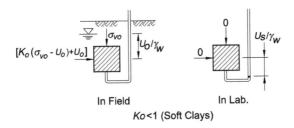

Figure 23.8 Stresses before and after Sampling

Effect of negative porewater pressure

The stress relief will induce negative porewater pressure in the sample and its influence can be significant. Let the porewater pressure in the sample, after stress relief, be u_s. Figure 23.8 compares the total stresses on a hypothetical sample *in situ* and on the actual sample in the laboratory.

Change in the ambient porewater pressure due to a change in the total stresses can be made up of two components: one that is B times the isotropic stress change $\Delta\sigma_3$, and the other that is $\overline{A}$ times the change in the principal stress difference $(\Delta\sigma_1 - \Delta\sigma_3)$. As explained in Chapter 20, this is presented by Skempton (1954) in the following form:

$$\Delta u = B[\Delta\sigma_3 + A(\Delta\sigma_1 - \Delta\sigma_3)] \tag{23.8}$$

In the above expression, $B \times A$ is generally represented by $\overline{A}$, where B and $\overline{A}$ are the experimentally determined porewater pressure parameters.

For a saturated soil $B = 1$ and so $\overline{A} = A$; the Equation (23.8) accordingly simplifies to:

$$\Delta u = \Delta\sigma_3 + A(\Delta\sigma_1 - \Delta\sigma_3) \tag{23.9}$$

By reference to figures (a) and (b) above, changes in the principal stresses and the porewater pressure can be obtained directly. These are:

$$\Delta\sigma_3 = -[K_o(\sigma_{V_o} - u_o) + u_o] \tag{23.10}$$

$$\Delta\sigma_1 = 0 - \sigma_{V_o} = -\sigma_{V_o} \tag{23.11}$$

$$\text{and} \quad \Delta u = u_s - u_o \tag{23.12}$$

Solution of the above equations simultaneously yields the value for the porewater pressure in the sample in terms of the following expression:

$$u_s = -[K_o + A(1 - K_o)](\sigma_{V_o} - u_o) \tag{23.13}$$

The second term $(\sigma_{V_o} - u_o)$ on the right hand side of the equation can be identified as the effective overburden pressure. The value of u_s in the above equation is thus presented as a proportion of this pressure.

For soft clays, generally, K_o is around 0·5 and the range for the porewater pressure parameter $\overline{A}(= A$ for $B = 1)$ is $-0·2$ to $+0·3$. By substituting these values, in turn, in the first term on the right hand side of the above equation, the theoretical range for the value of the porewater pressure in the sample can be obtained. This works out to 40–65 per cent of the effective overburden pressure. This may be compared with the range of 20–50 per cent suggested by Ladd and Foott (1974) which is relatively lower, nevertheless, still significant. In either case, it is clear that the major source of sample disturbance in soft clays is stress relief and that there is a significant potential for the imbibition of water.

It should be borne in mind that the theoretical u_s, as given by the expression in Equation (23.13), is greater than the real u_s. This is because there is a time-dependent migration of water due to the negative porewater pressure diffusion between the central, relatively undisturbed, core, and the more disturbed, and therefore more compressible, outer zone. Even if the outer zone is trimmed away, swelling of the mechanically undisturbed central core during the period of sampling-to-testing cannot be ruled out. This can give rise to two effects:

- First, there is the likelihood of a reduction in the effective stress such that the undrained, unconsolidated effective stress of the sample at failure is a fraction of that of the *in situ* condition at failure. The result is that the strength obtained from the test will be lower than that expected *in situ*.
- Second, the surplus water content in the clay can prevent the soil structure from offering any resistance against straining, which it would have done *in situ*. As a result, the deformability will be higher.

In view of the foregoing, it is important that the testing of the samples is so conducted as to ensure that the results obtained are truly representative of the ground conditions *in situ*. Towards this end, it is necessary to restore the ambient state of stress as well as the structure prior to shearing in the consolidated-undrained $(C - U)$ triaxial test. While it is possible to reconsolidate the test specimen anisotropically so as to recover its original state of stress as existing in the ground, it cannot be relied upon to restore the structure completely. This is because more water is likely to have been squeezed out than could possibly be expected to have been sucked in. Besides, the exercise requires prior knowledge of K_o, which may not always be available.

The effects on s_u of anisotropy, strain rate and sample disturbance may be partly self-compensating and yield reasonable average results. However, the method is highly empirical, and there is no certainty that the compensating factors can be controlled or are controllable. A rational method of accounting for these is therefore required.

To sum up, sampling alters the stress state from anisotropic to isotropic and is accompanied by lateral strain. Even though further strain is prevented, this strain remains locked-in during the K_o-consolidation test and so, the structure is not restored completely. The problem then is: how to achieve both – restoration of the *in situ* stresses as well as the structure?

Normalized behaviour

It has been observed (Ladd *et al.*, 1977) that clays with the same overconsolidation ratio (OCR) but different consolidation stresses $\left(\sigma'_{V_c}\right)$, and therefore different maximum past pressures $\left(\sigma'_{V_m}\right)$, exhibit similar (to within 10 per cent of the mean) strength and deformation characteristics in relation to σ'_{V_c}, provided that the clays do not have a high degree of structure as would generally be the case with quick clays, naturally cemented clays, varved clays, etc. K_o Consolidated-undrained shear tests with measurements of porewater pressures were carried out on a number of different soil samples including those in Boston Blue Clay. Figure 23.9(a) shows plots of the test results on samples with OCRs ranging from 1 to 8.

In each case, it can be seen in figure (b) that, upon normalization, different curves virtually coalesce into a single curve, thereby providing a unique way of obtaining 'normalized strength parameter' (*NSP*). For soils that exhibit normalized behaviour, it is possible to conduct laboratory tests at various OCR values and develop normalized plots for each of these. The *NSP* for each can then be obtained from these plots and applied to a wide range of *in situ* stress conditions. The most frequently used *NSP* is s_u/σ'_{V_o}, where σ'_{V_o} is the *in situ* vertical effective stress. Other *NSP* values used are: $s_{us}/\sigma'_{V_c}, E_u/s_u, K_o$ and porewater pressure parameters.

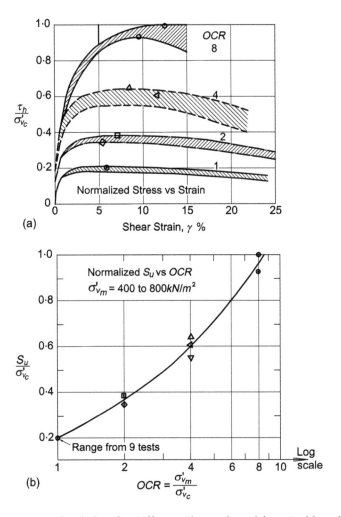

(a)

(b)

Figure 23.9 Normalized Plots for Different Clays (adapted from Ladd *et al.*, 1974)

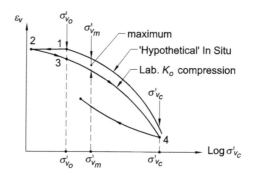

Figure 23.10 ε_V vs. $\log \sigma'_{V_c}$

Successful use of the *NSP* concept requires a prior knowledge of the *in situ* stresses and the σ'_{V_m} values. For this, high-quality odometer tests are essential. Good, undisturbed samples are, of course, also a major requirement. Fortunately, however, odometer tests do not seem to be highly sensitive to sample disturbance as are the undrained (U) and the unconfined-undrained (UU) tests.

Figure 23.10 compares 'hypothetical' *in situ* and 1-dimensional, K_o (odometer) compression, strain (ε_V) versus $\log \sigma'_{V_c}$ curves for a lightly overconsolidated soft clay. This plot is preferred over the conventional 'e versus $\log \sigma'_{V_c}$' plot as it is likely to give more reliable σ'_{V_m} values.

Point 1 represents the vertical *in situ* effective stress (i.e. overburden pressure) σ'_{V_o}. In the case of soft clays, the overburden pressure will not be far removed from the maximum past pressure σ'_{V_m}. An 'undisturbed' sample taken at point 1 will typically suffer a decrease in the effective stress even if the moisture content is maintained constant $(i.e. \varepsilon_V = 0)$. This is reflected by its movement to point 2. Reconsolidation back to the *in situ* stress σ'_{V_o} will take the sample to point 3 instead of retracing its path to point 1. The disparity between points 1 and 3 is indicative of the occurrence of the volumetric strain and confirms that the original structure has not been restored even if the *in situ* stress has.

Schmertmann (1953) suggests the maximum disparity between the *in situ* and the laboratory curves to be at the location of σ'_{V_m}. However, the following interesting observations can be made:

- When $\sigma'_{V_c} > (1\cdot5 \text{ to } 2\cdot0)\sigma'_{V_m}$, such as at point 4, the laboratory curve is close to the *in situ* virgin curve.
- Thus CK_oU (K_o-consolidation undrained) test specimens reconsolidated to $\sigma'_{V_c} \geq (1\cdot5 \text{ to } 2\cdot0)\sigma'_{V_m}$ should have structure similar to that of the normally consolidated *in situ* clays and hence give *NSP* $\left(i.e. \ s_{us}/\sigma'_{V_c}\right)$ values for the soil.
- If the *in situ* soil is overconsolidated, the sample would need to be reconsolidated to the virgin compression range to restore the *in situ* structure before unloading it to obtain the required *OCR* preparatory to the shear test.

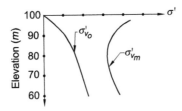

Figure 23.11 Variation of σ'_{V_o}, σ'_{V_m} with Depth (qualitative plot)

SHANSEP method

SHANSEP stands for stress history and normalized soil engineering properties (Ladd *et al.*, 1974). Based on the uniqueness of the normalized curves, it is possible to compute the values for the undrained shear strength s_u in accordance with the *SHANSEP* concept in three steps as follows:

- Using standard techniques commonly used in practice but a more thorough investigation than is customarily performed, establish the stress history of the deposit from high-quality constant rate of strain odometer tests with automatic logging. Variation through the deposit can be established from the σ'_{V_m} and σ'_{V_o} profiles on the lines shown in Figure 23.11. Then the *in situ* $OCR = \sigma'_{V_m}/\sigma'_{V_o}$.

- For the sample, establish normalized relationship i.e. $\left(s_{us}/\sigma'_{V_c}\right)$ versus OCR $\left(= \sigma'_{V_m}/\sigma'_{V_c}\right)$ based on the test results as shown in Figure 23.12. This is achieved by reconsolidating the CK_oU test samples to $\sigma'_{V_c} \geq (1\cdot5 \text{ to } 2\cdot0)\sigma'_{V_m}$ to restore the structure followed by K_o-swelling to the required OCR prior to shear. The parameter s_{us} is the shear strength as measured on the sample. Note that the OCR is plotted to log-scale.

- Compare the normalized parameter s_{us}/σ'_{V_c} for the sample with the *in situ* normalized parameter s_u/σ'_{V_o}. From the correspondence of the two NSPs, compute the *SHANSEP* s_u- values for different depths. For the purpose of illustration, the procedure typically for one depth has been schematically shown in Table 23.1.

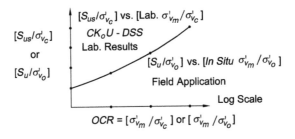

Figure 23.12 Normalized Strength vs. OCR

Table 23.1 Estimation of *in Situ* s_u Values

Depth	σ'_{V_o}	σ'_{V_m}	OCR	$s_{us}/\sigma'_{V_c} = s_u/\sigma'_{V_o}$	s_u *in situ*
1	2	3	$4 = 3 \div 2$	$5 = 6$	$7 = 2 \times 6$

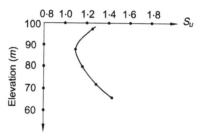

Figure 23.13 Strength vs. Depth Profile

From the columns 1 and 7 of the table, the *in situ* s_u profile against depth can then be easily drawn as shown, qualitatively, in Figure 23.13.

In conclusion, the practitioners of the *SHANSEP* concept claim that, for a soil exhibiting normalized behaviour, it will yield s_u versus depth profile very close to the *in situ* behaviour (i.e. as restored). The most contentious aspect of the approach, however, remains the possibility that by consolidating the sample beyond the *in situ* stresses, an important aspect of its structure could be destroyed. It is also worth remembering that if s_{us}/σ'_{V_c} varies consistently with stress, the *NSP* concept cannot apply to the clay.

Deformation parameters

From the point of view of their deformation characteristics, deep excavations for cut-and-cover structures in unconsolidated sediments can be of particular concern to the engineer. These sediments may be broadly classified as residual, eolian, alluvial and lacustrine and marine deposits. The deposits of volcanic and glacial origin may also be classified within this group (Zeevaert, 1973), especially with the erosion and transportation agents being the same, the difference being only in the respective pyroclastic or the clastic characteristics.

Residual soils are the product, *in situ*, of disintegration of the parent rock due to weathering extending several metres into it. The deformability may be high and, in some cases, very high with low shear strength. In humid regions, deep soil profiles are encountered with medium to high deformability and low shear strength.

Wind blown (eolian) sediments may result in the formation of dunes, loess and loess-ial-type deposits. They are typical of arid regions with water table encountered at great depth from the surface and may be found with low relative density, very low or no c' and medium to high deformability. Upon saturation due to changes in the water table or seepage conditions, they are prone to dramatic change in their mechanical properties suffering sudden breakdown in the structure resulting in compaction. They are, as such, also known as collapsible soils. After undergoing such a change, they display medium to low deformability and are identified as modified eolian deposits.

Water borne (alluvial) deposits are generally well-graded sediments occurring in medium to very compact state with grain sizes varying from large rock fragments to gravel, sand, silt and some clay. The coarser sediments display low to very low and the finer sediments medium deformability.

Lacustrine and marine deposits comprising fine and very fine sediments like silt and clay are deposited when running water comes to rest upon entry into lakes, marginal lagoons, estuaries and deltas. They may display medium to high and very high deformability and contain colloidal organic matter or comprise organic matter like peat totally. Also, because of their very low permeability, time-dependence of their deformation must not be overlooked.

Normalized behaviour

Just as it is used for the strength parameter, the normalized behaviour concept can also be used for obtaining the undrained and the drained normalized deformation parameters (*NDP*).

Poulos (1978) carried out tests on remoulded Sydney kaolin (liquid limit, 50 per cent; plastic limit, 34 per cent). Samples were tested in a special triaxial cell of the type described by Davis and Poulos (1963). In the majority of the tests, the samples were consolidated under K_o conditions to the required vertical effective preconsolidation pressure σ'_{V_m} and then, if necessary, allowed to swell to the required vertical effective stress σ'_{V_o}, again under K_o conditions. It was possible to determine the coefficient K_o by measuring the horizontal effective stress σ'_{V_h} at the end of this stage. After a period (generally, two days) of ageing under σ'_{V_o}, the samples were tested under triaxial conditions at a constant rate of strain until failure. From the tests on the undrained samples, E_u and the undrained shear strength s_u were determined; from the tests on the drained samples, E', v', and the effective stress shear strength parameters c' and ϕ' were obtained. In all cases, two-way end drainage was allowed where appropriate.

Definition of parameters

The deformation parameters were evaluated as follows:

$$E_u = \left(\Delta\sigma_V - \Delta\sigma_H \right) / \varepsilon_1$$

$$E' = \left(\Delta\sigma'_V - 2 \cdot v' \cdot \Delta\sigma'_H \right) / \varepsilon'_1$$

$$v' = \frac{\varepsilon'_1 \cdot \Delta\sigma'_H - \varepsilon'_3 \cdot \Delta\sigma'_V}{\varepsilon'_1 \left(\Delta\sigma'_H + \Delta\sigma'_V \right) - 2 \cdot \varepsilon_3 \cdot \Delta\sigma'_H}$$

where
$\Delta\sigma_V$ = increment of vertical stress,
$\Delta\sigma_H$ = increment of lateral stress,
ε_1 = undrained axial strain,
ε'_1 = drained axial strain,
ε'_3 = drained lateral strain = $0 \cdot 5 \left(\varepsilon_V - \varepsilon'_1 \right)$,
ε_V = volumetric strain.

Figure 23.14 illustrates the method of interpretation of the test results.

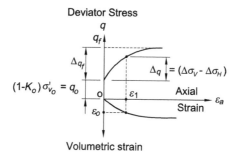

Figure 23.14 Definition Sketch (adapted from Poulos, 1978)

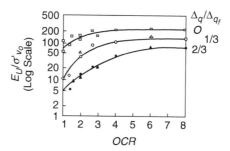

Figure 23.15 NDP vs. OCR (for Sydney Kaolin) (adapted from Poulos, 1978)

NDP vs. OCR

Figure 23.15 shows the plot of the undrained secant modulus normalized by σ'_{V_o} on the log-scale against the overconsolidation ratio (OCR). The relationships are shown for three incremental stress levels, $\Delta q / \Delta q_f = 0, 1/3$ and $2/3$ where q and q_f are as defined in the Definition Sketch, and

$$\frac{\Delta q}{\Delta q_f} = \frac{\Delta \sigma_V - \Delta \sigma_H}{\left(\Delta \sigma_V - \Delta \sigma_H \right)_f}$$

Subscript f denotes the state at failure. In the tests reported, $\Delta \sigma_H = 0$ and so the parameters for $\Delta q / \Delta q_f = 0$ represent the initial tangent values for the modulus. Somewhat similar results were obtained for the normalized drained parameters, i.e. E' / σ'_{V_o} vs. OCR.

Nonuniform variation in the $\Delta q / \Delta q_f$ ratio, as is to be expected, is indicative of the nonlinearity in the soil.

Effect of initial consolidation conditions

Tests have shown that the effect of the type of initial consolidation can be considerable. Figure 23.16 compares the normalized deformation parameters obtained

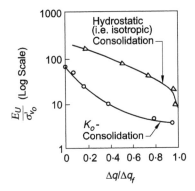

Figure 23.16 Effect of Initial Consolidation on *NDP* (normally consolidated Sydney Kaolin)
(adapted from Poulos, 1978)

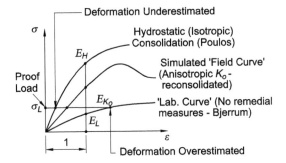

Figure 23.17 Comparison of Deformation Parameters

using hydrostatic (i.e. isotropic) consolidation with those of K_o initial consolidation; the difference is significant – the former being appreciably greater than the latter.

Comparison of the graphs suggests that the results obtained from isotropically consolidated samples for a given stress level may overestimate the modulus for normally consolidated clays significantly and thus emphasizes the desirability of following the correct field stress path in the laboratory.

Figure 23.17 shows a qualitative comparison of the moduli E_I obtained after isotropic consolidation of the sample (Poulos, 1978) and E_L from a laboratory curve without the sample undergoing any restoration measures (Bjerrum, 1973) against that of a simulated, anisotropically (i.e. K_o) consolidated 'field' curve represented by E_{K_o}.

Whereas the modulus obtained from laboratory test without any attempt at restoration of the *in situ* stresses is likely to be underestimated thereby overestimating the deformation, isotropic consolidation of the sample, once again confirms the potential overestimation of the modulus and, consequently, the underestimation of the deformation.

Coefficient of permeability: k

'Constant Head Permeameter' is used for permeabilities down to about $10^{-4} m/sec$. This covers the 'gravel-clean sand' range. 'Falling Head Permeameter' is used for permeabilities between 10^{-4} and $10^{-7} m/sec$, i.e. silts and fissured clays. For the range below $10^{-7} m/sec$, permeability is generally measured by indirect means, e.g. with the use of consolidation or dissipation tests.

Coefficient of earth pressure at rest: K_o

In the design of cut-and-cover metro structures, predictions of the *in situ* stresses and the evaluation of the coefficient of earth pressure at rest (K_o) are of major importance. Numerous investigators have examined these with varying degrees of success and various empirical expressions for the value of K_o have been presented.

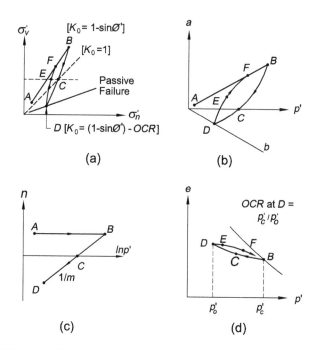

Figure 23.18 Typical Effective Stress Path *(ESP)* (soil consolidated one-dimensionally) (adapted from Wroth, 1975)

Wroth approach

Figure 23.18(*a–d*) gives a qualitative indication of typical effective stress plots for soil consolidated one-dimensionally. By plotting these in different stress spaces and paying due regard to the stress history, Wroth (1975) was able to make use of the 'elasticity' implied in their 'inherent', 'assumed' or 'acquired' linearity in relating K_o, empirically, to just the three parameters, viz.,

- Effective stress, shear strength parameter, ϕ'
- Overconsolidation Ratio, OCR
- Plasticity Index, I_P.

Definitions of the two principal types of soils considered, the normally and the overconsolidated soils, and the overconsolidation ratio are given in Appendix C.

Wroth divided the results into four categories of soils as follows:

(1) *Normally consolidated soil*: For the virgin consolidation path AB (figure a) which is linear and so over which K_o is constant, Jaky's theoretical relationship is used, i.e.

$$K_o^N = \left(1 + \frac{2}{3}\sin\phi'\right)\left(\frac{1 - \sin\phi'}{1 + \sin\phi'}\right) \tag{23.14}$$

For most engineering purposes and with little loss of accuracy, the following simplified empirical version (Jaky, 1944) is widely accepted for normally consolidated sand:

$$K_o^N = 1 - \sin\phi' \tag{23.15}$$

This may be compared with the recommendation by Brooker and Ireland (1965) for normally consolidated clay:

$$K_o^N = 0.95 - \sin\phi' \tag{23.16}$$

Based on their laboratory tests, Sherif *et al.* (1984) have shown that whereas Jaky's empirical expression for K_o, Equation (23.15), gives good results in the case of loose sand, in the case of dense sand, it grossly underestimates the lateral earth pressure at rest. This is attributed to the process of compaction of the backfill. For that reason, they have recommended the following relationship:

$$K_o^N = \left(1 - \sin\phi'\right) + 5.5\left[\frac{\gamma_d}{\gamma_{d(\min)}} - 1\right] \tag{23.17}$$

where γ_d: actual compacted dry unit weight of sand behind the wall, and $\gamma_{d(\min)}$: dry unit weight of sand in its loosest state.

For normally consolidated clays, Alpan (1967) recommends:

$$K_o^N = 0.19 + 0.233\log\left(I_P\right) \tag{23.18}$$

In this expression, I_P is the plasticity index in percentage. For the definition of plasticity index, see Appendix C.

For normally consolidated clays, the coefficient of earth pressure at rest may also be approximated as:

$$K_o^N = \left(K_a\right)^{0.72} \tag{23.19}$$

$$\text{where, } K_a = \tan^2\left(45° - \phi'/2\right)$$

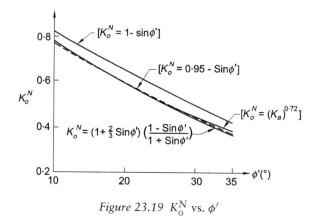

Figure 23.19 K_o^N vs. ϕ'

For comparison, Equations (23.14), (23.15), (23.16) and (23.19) have been plotted together in Figure 23.19. With the exception of Jaky's (1944) expression, the other three virtually coalesce.

(2) *Lightly overconsolidated soil*:   From the assumed linearity of the part of the unloading path BC (figure *a*), from $OCR = 1$ and $K_o = K_o^N$ at B, to $OCR \simeq 5$ and $K_o = 1$ at C, i.e. $K_o^N \leq K_o^L \leq 1$, or $1 \leq OCR \leq 5$, the following relationship is obtained:

$$K_o^L = OCR \cdot K_o^N - \frac{v'}{1 - v'} \cdot (OCR - 1) \tag{23.20}$$

The superscripts L and N refer to the lightly overconsolidated and the normally consolidated states of soil, respectively, and parameter v' represents the Poisson's ratio with respect to effective stresses. The Poisson's ratio is obtained from the graph in Figure 23.20.

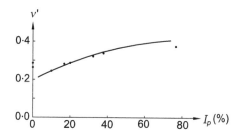

Figure 23.20 Values of Poisson's Ratio (lightly overconsolidated soils)
(adapted from Wroth, 1975)

(3) *Heavily overconsolidated soils:* Further unloading beyond the isotropic stress point C traces the curved path CD (Figure 23.18a). The curve can be straightened by plotting it in the η versus $\log p'$ space (figure c) where $\eta = q/p'$, and

$$p' = \frac{1}{3}\left(\sigma_1' + \sigma_2' + \sigma_3'\right) \quad \text{and} \quad q = \left(\sigma_1' - \sigma_3'\right)$$

In the expressions, p' is the mean effective stress and q the deviator stress. The value of K_o^N for $K_o^H > 1$ and $OCR > 5$ is presented in terms of the following relationship:

$$m\left[\frac{3\left(1 - K_o^N\right)}{1 + 2K_o^N} - \frac{3\left(1 - K_o^H\right)}{1 + 2K_o^H}\right] = \ln\left[\frac{OCR\left(1 + 2K_o^N\right)}{1 + 2K_o^H}\right] \tag{23.21}$$

In the expression, superscript H refers to heavily overconsolidated soil and m is the inverse gradient of BCD in figure (c). Plot of m against PI is shown in Figure 23.21.

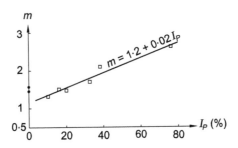

Figure 23.21 Gradient vs. Plasticity (Rebound line) (adapted from Wroth, 1975)

(4) *Reconsolidated soil:* The initial part DE (Figure 23.18a) during the reloading phase appears to be essentially linear just like the initial part BC of the unloading phase, and so an expression similar to that of Equation (23.20) can be obtained.

The aforementioned expressions for K_o do not reflect the effects of natural phenomena such as repeated cycles of loading whether caused by deposition and erosion or seasonal fluctuations in the water table. It is also essential to recognize that for a given current value of the effective overburden pressure σ_V' and the parameters ϕ', PI and OCR, K_o-value for soil is not unique but very much a function of the stress history. This can be clearly observed from Figure 23.18(a). Although points C and E are under the same current state of stress and the $OCR = \left(\sigma_V'\right)_B \div \sigma_V'$, yet they exhibit different values of K_o, because they relate to different phases of the stress history.

Wroth (1975) further points out the apparent anomaly that could present itself due to creep or delayed consolidation at constant effective stress. In Figure 23.22, points G and H in the p': e space represent voids ratios at the end of primary and secondary consolidation, respectively, whereas point I is the intersection at which the recompression curve HI rejoins the virgin consolidation curve GI. Although the

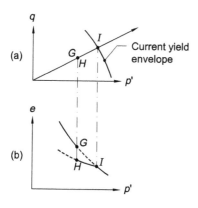

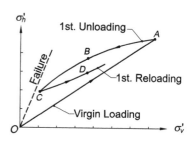

Figure 23.22 Change of State Caused
by Creep (adapted from Wroth, 1975)

Figure 23.23 Simplified Stress History
of Soil (under K_o conditions) (adapted
from Mayne and Kulhawy, 1982)

specimen at H could easily be taken as lightly overconsolidated, the stress at I being interpreted as the overconsolidation pressure, yet since effective stresses between G and H have not undergone any change, the K_o at H must have the same value as that at G.

Mayne and Kulhawy approach

Mayne and Kulhawy (1982) have considered the simplified stress history, plotted in the effective stress space, as depicted in Figure 23.23, for a homogeneous soil deposit under geostatic stresses. Stress path OA represents the virgin loading of the soil deposit indicative of the normally consolidated conditions. During this virgin consolidation phase, K_o remains constant at K_o^N value. Any relief in the effective overburden stress due to whatever reason (e.g. erosion, excavation, rise of groundwater table, or removal of surcharge loads, etc.), results in the overconsolidation of the soil. This will be represented by a stress path similar to that indicated by ABC in the figure. During the unloading phase, the overconsolidation ratio, $OCR = \sigma'_{v_{max}}/\sigma'_v$, has a marked effect on the value of K_o. Reloading after the simple rebound, would follow a path similar to CD. The stress paths for the subsequent unloading and reloading cycle are likely to remain within the loop $ABCDA$.

On the basis of their review of the laboratory data from over 170 different soils reported by various researchers, the authors have presented a common approach to clays, silts and sands. Furthermore, with the aid of a new stress history parameter OCR_{max}, they have proposed to represent K_o as a function of the stress history covering both the virgin consolidation and the first unloading phases by a single empirical expression as follows:

$$K_o = (1 - \sin\phi') \left[\left(\frac{OCR}{OCR_{max}^{(1-\sin\phi')}} \right) + \frac{3}{4} \left(1 - \frac{OCR}{OCR_{max}} \right) \right] \qquad (23.22)$$

The new parameter is, by reference to Figure 23.23, defined as:

$$OCR_{max} = \sigma'_{v_{max}}/\sigma'_{v_{min}}$$

For normally consolidated soils (stress path OA), $OCR_{max} = OCR = 1$. Upon substituting this into Equation (23.22), the expression for K_o^N as represented by Equation (23.15) can be recovered. During the stress relief or the unloading phase (stress path ABC), $OCR_{max} = OCR$. Upon making this substitution, Equation (23.22) yields a simple correlation between the coefficient of earth pressure at rest for overconsolidated soils and the overconsolidation ratio as given by the expression:

$$K_o^{o/c} = K_o^N \cdot (OCR)^{\sin \phi'} \tag{23.23}$$

However, it should be noted that this expression is not applicable if the deposit has been subsequently reloaded, such as that resulting, for instance, from the deposition of surface gravels over clay in many parts of London, as the effective stress path then tends towards the virgin consolidation path AB as indicated in the figure.

It is clear from the above that K_o can vary from location to location depending upon the respective stress history at each location. Due consideration should therefore be given to this fact in the selection of the K_o value appropriate to a given location.

Error in K_o

K_o is a function of three independent variables – σ_V, σ_H and u. Errors in the measurement of these parameters can give rise to errors in the magnitude of K_o. Using the theory of errors, Massarsch (1979) has expressed the error in the value of K_o as:

$$\Delta K_o = \frac{\partial K_o}{\partial \sigma_H} \cdot \Delta \sigma_H + \frac{\partial K_o}{\partial \sigma_V} \cdot \Delta \sigma_V + \frac{\partial K_o}{\partial u} \cdot \Delta u \tag{23.24}$$

where $K_o = (\sigma_H - u)/(\sigma_V - u)$ and $\Delta \sigma_H, \Delta \sigma_V$ and Δu are the potential errors in the three parameters.

Porewater pressures can vary considerably and to appreciate its influence, let us assume that $\Delta \sigma_H = 0 = \Delta \sigma_V$, then

$$\Delta K_o = \frac{\partial K_o}{\partial u} \cdot \Delta u = \frac{(\sigma_H - \sigma_V)}{(\sigma_V - u)^2} \cdot \Delta u = \frac{(K_o - 1)}{\sigma_V'} \cdot \Delta u$$

$$\text{Or,} \quad \frac{\Delta K_o}{K_o} = \frac{(K_o - 1)}{K_o} \cdot \frac{\Delta u}{\sigma_V'}$$

Massarsch (1979) has shown that, for normally consolidated clay, a rather small variation in porewater pressure results in a large range of K_o especially close to the surface. This is illustrated in Figure 23.24.

In view of the foregoing, the importance of obtaining an accurate ambient porewater pressure profile for use in design should not be underestimated. Where an accurate estimate is not available, it would be prudent to consider the effects on computations of a sensible range in the porewater pressure profile to allow for any potential variations.

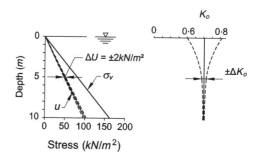

Figure 23.24 Effect of Variation in PWP on K_o (adapted from Massarsch, 1979)

Effective stress shear strength parameters: c', ϕ'

An indirect measure of shear strength in terms of the effective stress parameters c' and ϕ' can be obtained from consolidated-undrained or drained tests in a triaxial cell. This can be achieved by invoking the Mohr-Coulomb failure criterion in terms of the effective stress which requires the failure envelope to be tangential to the Mohr circles at failure. The definition of the parameters c' and ϕ' is given in Figure 23.25.

However, triaxial test does not present an ideal way of relating to values of stress in the field. Besides, the selection of those Mohr circles which can help establish the failure envelope and ignoring those which may represent deviations from the idealized 'fit' can often be confusing. One way of overcoming this problem is to enlist the help of an analytical method such as the stress path to plot the path for one point only. In this respect, it is useful to plot the topmost point of the Mohr circle.

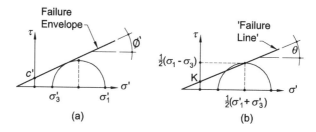

Figure 23.25 Definition of Parameters

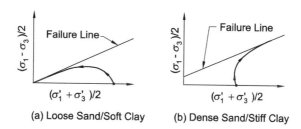

Figure 23.26 Typical Effective Stress Paths (U–D triaxial compression tests)
(adapted from Simons and Menzies, 1977)

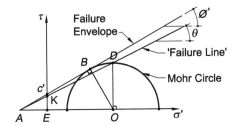

Figure 23.27 Failure Envelope and 'Failure Line' Geometry

To obtain the effective stress shear strength parameters, a 'failure line' tangential to the steady or the 'failure' portion of the stress path is drawn. Typical effective stress paths for soft and stiff soils in the consolidated-undrained triaxial test together with their respective 'failure lines' are shown (Simons and Menzies, 1977) in Figure 23.6.

By treating the slope of the 'failure line' as θ and its intercept on the maximum shear stress axis as K, and comparing its geometry with that of the failure envelope as shown in Figure 23.27, it can be seen that:

$$c' = K \sec \phi' \quad \text{and} \quad \phi' = \sin^{-1} (\tan \theta) \tag{23.25}$$

Plotting triaxial test results in this way gives a clear and unambiguous indication of failure.

24 Load Factors and Combinations

24.1 Introduction

There is no doubt that design and construction of cut-and-cover metro structures can often present complexities which may neither be readily appreciated nor the solution to which tractable. The complexities may result from uncertainties and variabilities in one or more of the following:

- Parameters for the design
- Loadings operative at different stages
- Structural modelling and the boundary conditions
- Progressive changes in ground and groundwater conditions
- Behaviour of the structure during different stages of its construction.

In the interests of safety and cost, it is essential to guard against the possibility of underestimating or overlooking some important aspects of design, consequences of which could be serious and rectification of which could prove to be prohibitively expensive. This, invariably and understandably, leads to caution and a tendency on the part of the designer to lean towards conservatism in order to compensate for any uncertainty, lack of information and understanding or ignorance.

Where there is genuine uncertainty, a safe approach is justified and to be welcomed. However, far too much conservatism which is unnecessary and avoidable will, inevitably, reflect in increased costs. It is worth remembering that higher costs do not necessarily imply a better structure, *per se*. Whereas a reasonable amount of caution may be justifiably necessary, piling up conservatism at every stage cumulatively resulting in an ultraconservative design may neither be justified nor warranted. However, all aspects need to be looked at sensibly, neither exaggerating concern nor underestimating caution. Towards this end, proper understanding of the principles of design and the appreciation of the soil–structure interaction are important in order to arrive at sensible load factors and load combinations for design.

24.2 Load factors

In order to avoid unnecessary expense and, at the same time, produce a safe and cost-effective design, it is important to choose load factors sensibly. In what follows, each load has been examined on its merits and a reasoned case made before assigning a factor to it.

Backfill on roof

Backfill over the metro structures, especially where it underlies a road surface, calls for its placement to be carried out in discrete layers which should be well compacted. In view of the fact that this can and has to be achieved under reasonably controlled conditions, a load factor greater than 1·2 on the weight of the backfill material is considered unnecessary. However, in exceptional circumstances where, for some reason, achieving such control is doubtful, a case may be made for the use of a load factor greater than 1·2.

Lateral earth pressures

With regard to lateral earth pressures, as pointed out in Chapter 13, the possibility that designing to active and at-rest values only could, potentially, overlook a genuine more onerous loading condition cannot be ruled out. This could be the case particularly where the gap between the active and the at-rest coefficients is large and the stiffness of the ground support system is unlikely to allow sufficient enough soil strain to realize active failure condition. Under these circumstances, it might be prudent to introduce an additional load case based on an earth pressure coefficient that represents, say, the average of active and at-rest values; or even opt for a value on the higher or the lower side of the average if need be. This would cover a broader spectrum of loading possibilities and significantly reduce the chances of a serious load case slipping through unaccounted for.

    With the approach outlined in the foregoing section, the extent of uncertainty usually associated with the identification of the most onerous load cases would be significantly reduced and a load factor of 1·4 would be justified.

Surcharge loads

In the case of vertical surcharge load from the road traffic overhead which is directly applied and felt by the structure, such as would be the case where the roof also forms the road surface or, to stretch a point, where it is within 600*mm* depth of the surface, a factor of 1·6 may be applied.

    However, for the roof located deeper than 600*mm* below the surface, it has been demonstrated in Chapter 12 that the blanket load of $10kN/m^2$, representing the effect of the traffic load surcharge, considering dispersion and even ignoring dissipation is, if anything, on the conservative side. In view of this and given the remoteness of the transient load, a factor of 1·6 on the live load surcharge would appear to be unnecessarily severe and a factor of 1·4 can be considered to be more than adequate.

    Lateral surcharge loads from a nearby structure can result from a combination of the permanent self-weight of the structure and the transient live load(s) acting on it. In the case of the surcharge load generated by the self-weight of the structure, being a permanent rather than a transient load, a load factor of 1·4 is appropriate. In the case of the lateral surcharge load resulting from the transient live load(s) acting on the structure also, in view of their dissipation and dispersal due to remoteness, there is no reason why a factor for the maximum surcharge load cannot be sensibly restricted to 1·4 instead of the usually stipulated value of 1·6.

Hydrostatic pressure

Unlike that for soil, the unit weight of water (including that of seawater) can be accurately known and so there is no need to apply a partial factor of safety on its unit weight as would be the case, for instance, with soil. The only uncertainty in regard to water would be that due, potentially, to its design level over the operational life time of the structure. However, as long as the worst credible water level is used, a maximum factor of 1·2 should be more than adequate. Besides, computation is invariably made using both the high and the low water tables to cover any uncertainty or fluctuation in the level. Notwithstanding this, if for some reason, it is not possible to obtain a realistic estimate of the worst credible water level for design, a case may be made for the use of a load factor greater than 1·2.

24.3 Load combinations

General

Underground structures are required to be designed for the envelopes of the stress-resultants resulting from the various combinations of load cases. For example in the case of a box section, maximum span moment in a perimeter wall element adjacent to the roof slab would occur under maximum lateral load on the wall taken coexistently with the minimum downward load on the adjacent span of the roof slab; whereas the maximum support moment at the junction between the roof slab and the adjacent wall would be obtained by imposing maximum loads both on the slab as well as the wall simultaneously.

Qualitatively, governing load combinations should be established on the basis of characteristic loads and not on artificially contrived differentially factored loads; otherwise, realistic deformation of the structure may neither be directly apparent nor their implication appropriately taken into account. For example, the location of true inflexion points cannot be established using the ultimate limit state analysis if different load factors are used with different loads in a particular combination. This could raise uncertainties with regard to the location and curtailment of bars in the various elements in the preparation of RC detail drawings. Whereas the ultimate limit state provides a sensible method for the design of the strength of the structure, it cannot be relied upon to reflect the true profile of its deformation. In recognition of this, it would be prudent, when considering curtailment, to be sensibly generous with rebar lengths or base its location on the serviceability limit state analysis.

Lateral earth pressures

Cut-and-cover metro structures are confined structures surrounded on all sides by the medium of soil. During the installation of the perimeter walls and the excavation for and the construction of such a structure, the surrounding soil mass must, inevitably, move inwards towards the excavation until equilibrium is restored. In so doing it will cause the elastic compression of the completed structure. During this process, a measure of friction at the various soil–structure interfaces would be mobilized as a matter of course. With the establishment of equilibrium and a preferential load path through it, the structure would then be held as if in a vice by the surrounding soil, unable to translate or rotate unless, of course, the entire surrounding soil mass were forced to deform for some reason. In the environment within which cut-and-cover metro structures are generally located and constructed and given the fact that major construction activity within their vicinity is generally restricted, the latter scenario would appear to be unlikely.

However, if an out-of-balance lateral load is set up due, say, to discrete surcharge loads acting on one side of the structure, every attempt will be made by the structure and its interaction with the surrounding soil to contain this and restore equilibrium once again as quickly as possible. The load creating the imbalance will tend to cause sway of the structure; however, for the sway to take place, a measure of the soil–structure interface friction over and above that already mobilized would have to be overcome not only over the outer vertical faces of the walls but also over the roof and below the base slabs of the structure. Given the enormous extents of the soil–structure interface areas, there is little doubt that minimal further strains would be needed to mobilize significant reactive forces. Therefore, in reality, very little sway of the structure is likely to occur.

Reaction to the force imbalance will be shared between the soil medium and the structure in proportion to their respective stiffnesses. Since the combined stiffness of the soil mass and that resulting from the friction mobilized at the soil–structure interfaces is likely to be many orders of magnitude higher than that of the structure itself, very little force imbalance is likely to be taken up in the flexural deformation of the structure. In other words, the inherent stiffness of the surrounding soil mass and the frictional resistance at the soil–structure interfaces together are likely to inhibit any significant rotation of the structure and contribute significantly to its stability.

Even with the use of waterproof membrane wrapped around the structure, or in the case of diaphragm wall constructed under bentonite, to expect the interface friction to be 100 per cent neutralized is unrealistic. One way of dealing with this realistically would be to stipulate limiting, in such a case, the use of interface friction to no more than, say, $5–15kN/m^2$ over the depth of the structure, or any other limiting values that may be deemed to be more appropriate for the given site-specific conditions. In any case, even if the lower-bound value of $5kN/m^2$ is used, when applied over the entire contact surface of the structure, it can represent a substantially large enough force which can make a significant difference. It would therefore seem unnecessarily wasteful to disregard this benefit which would be available legitimately.

Cut-and-cover metro structures are, generally, linear box structures with the length significantly greater than the width. In addition, the presence of the horizontal floor elements, including the generally much thicker roof and the base slabs, would contribute substantially to the inherent longitudinal stiffness of the box structure in the transverse direction. It would be reasonable to assume that any out-of-balance lateral load across the structure caused, for example, by traction or surcharge on one side of the box structure, would be operative over a discrete area locally. However, the inherent stiffness of the structure would bring into play significantly longer stretch of the structure to disperse the lateral load imbalance on to the opposite (reactive) side thereby reducing its intensity substantially. As a result, it would withstand the sway effects with minimal impact. Therefore, ignoring such beneficial aspects would artificially exaggerate the stress-resultants and could potentially load up construction costs unnecessarily. Besides, as explained previously, the soil–structure interface friction which is generally disregarded but which would have to be overcome in the event of the structure tending to sway would also inhibit the tendency to sway.

It is obvious from the foregoing discussion that both the stiffness of the surrounding soil mass as well as the inherent longitudinal (lateral) stiffness of the structure itself are generally capable of inhibiting and containing translation/rotation of the structure leaving, in reality, very little residual lateral load imbalance to be carried by the transverse flexural stiffness of the structure. Notwithstanding this, designers are known to use earth pressures on the opposite walls differentially for no reason other than to maximize moments. The justification usually advanced for such an approach is none other than one of opting for conservatism to enhance safety.

However, be that as it may, it is difficult to find any justification, on the basis of either specific geotechnical principles or general engineering logic, for using the concept of differential earth pressures on the opposite walls, i.e. 'earth-at-rest' on one wall together with anything other than the 'earth-at-rest' on the opposite wall, of the box structure. All that one can say about such an approach is that it is an artificially contrived situation that results in nothing other than grossly exaggerating the stress-resultants and for no reason other than to demonstrate that the approach is on the safe side, which for some elements it might well be, but unfortunately, to an unnecessarily expensive degree and also for the wrong reasons.

In the context of earth pressures, it is important to remember that their magnitude is a function of both the amount as well as the direction of soil strain. For instance, when there is no transverse movement of the wall, i.e. the lateral strain at the soil–wall interface is zero, such as would be the case if the structure were wished into place, the earth pressures existing are those at-rest. Upon excavation, when the wall deflects away from the soil towards the excavation and the soil moves in sympathy, the earth pressure will drop from its at-rest value towards, in the limiting case, its active value. However, when the wall is pushed away from the excavation and into the soil such as would be the case upon preloading the wall support system, the earth pressure would increase, in the limiting case, towards its passive value. It is also important to recall that, with the exception of heavily overconsolidated soils with significantly high K_o values where opposite might be the case, generally, relatively much smaller strain is needed to generate the active pressure than is necessary to mobilize the passive resistance.

Consider a simple, idealized box structure $ABCD$ as shown in Figure 24.1. Assume that it is subjected to 'at-rest' pressures on the left-hand wall and less than 'at-rest' but higher than 'active' pressures on the right-hand wall. 'Earth-at-rest' pressure implies zero soil strain along the vertical section, DAa, coincident with the plane of the left-hand wall. On the other hand, earth pressure somewhere in between the 'at-rest' and 'active' values on the opposite wall, CB, would suggest movement of the right-hand wall into the structure to an extent that would cause the pressure to drop below

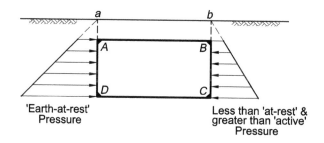

'Earth-at-rest' Pressure

Less than 'at-rest' & greater than 'active' Pressure

Figure 24.1 Box Structure Subject to Differential Lateral Loads

the 'at-rest' value but not low enough to realize the limiting active failure condition. However, the resulting disparity in the loads on the opposite walls cannot represent an equilibrium condition. Higher pressure on the left wall in comparison to that on the right will set up a force differential across the structure. This will cause the structure to sway towards the right, forcing the left-hand wall as also the soil in contact with it to move in sympathy until the equilibrium is restored. Clearly, with the movement of the left-hand wall and the associated strain in the soil, the pressures on the wall would drop away from, and would no longer be those, 'at-rest' thus contravening the assumption.

In view of this, it is not realistic to consider differential earth pressures on the opposite walls concomitantly. In any case, besides being wasteful, it does not make any engineering sense.

In the course of analysing metro box structures subject to a lateral force imbalance, it is not uncommon to use higher total applied pressure on one side and lower reactive pressure plus soil springs on the other. However, this is generally used purely as a mathematical expedient. One way of sensibly applying this principle would be as follows:

For any loading case with the possibility of out-of-balance lateral load due to, say, traffic surcharge on one side, an average of 'earth-at-rest' and 'active' pressures may be used symmetrically on both sides together with soil springs on the reactive side to counteract the extra over surcharge loads applied on the active side.

BD 31/87 approach

The old Highways Agency standard *BD 31/87* (now superseded by *BD 31/01*) gave advice in Clause 4.5 and subclause 4.5.1 on the use of minimum and maximum earth pressures in the design of 'buried' structures. It set the default values for the minimum and the maximum earth pressure coefficients at 0·2 and 0·6, and added:

> It is assumed that either the maximum or the minimum earth pressure can act on any side wall irrespective of the magnitude of the earth pressure on the other side wall.

This clause as it stood appeared, for its use in the design of cut-and-cover metro structures, much too onerous and inappropriate for the following reasons:

The coefficients of 0·2 and 0·6 appeared to be artificially contrived simply to exaggerate the sway effects on the buried structure which could, in certain cases, result in unrealistically severe stress-resultants. This would, *ipso facto*, reflect in increased member thicknesses or areas of reinforcement or both and, inevitably, in the increased costs as well. No reason or justification was given for the use of these stipulated coefficients.

The assumption that '... either the maximum or the minimum earth pressure can act on any side wall irrespective of the magnitude of the earth pressure on the other side wall' made it almost mandatory on the part of the designer to use the coefficients 0·2 and 0·6 in such a combination on the opposite sides as to maximize the effect of sway. In consonance with this, the designer would be obliged to use the coefficient 0·6 on earth pressure on the side that activates the disturbing force causing sway, and the coefficient 0·2 on earth pressure on the side that is supposed to mobilize passive resistance to oppose it! While this would not maximize the sway effects legitimately, it would certainly and very significantly exaggerate them artificially.

Clearly, this is not appropriate because it does not appear to make any distinction between the active force generated which would induce sway and potential rotation of the structure, and the reactive force mobilized which would counteract that tendency. In accordance with the engineering principles, the stipulated lower 'active' and 'reactive' coefficients of 0·2 and 0·6 should be used, if at all, only on the 'active' and the 'reactive' sides, respectively. However, even if these coefficients were to be so used, i.e. in their correct sense, the clause would continue to remain anomalous for the following reasons.

Under the influence of an out-of-balance lateral force, the structure would be expected to sway in the same direction as that of the 'disturbing force.' However, with the use of the 'lower active' coefficient of 0·2 and the 'relieving' coefficient of 0·6, and in spite of using differential γ_{f1} and γ_{f3} factors on the two sides, it is conceivable that the 'relieving' force, being three times as high as the 'disturbing' force, could exceed the combined effect of the 'disturbing' and the 'lower active' forces. This could, in that event, suggest movement of the structure in a direction opposite to that of the 'disturbing' force and would imply that the 'relieving' force instead of being a 'reactive' force mobilized behaves as an 'active' force generated. Clearly, that would be absurd. Such an anomaly would need to be corrected by progressively reducing the K-factor of 0·6 in an iterative way until the direction of sway became consistent with the reality. This would therefore imply the use of the reactive K-factor of a value potentially other than 0·6 which would be commensurate with that required to bring about equilibrium.

One way by which the anomaly identified in the foregoing section could be avoided would be by modelling the soil reaction necessary to be mobilized by a series of soil springs. With this approach, however, it would be necessary to ensure that the software used for analysis had the facility to model the springs as fully compressive with virtually zero tensile stiffness since, under tension, the structure would simply move away and lose contact with the soil. As such, it would not be possible to develop tension at the soil–structure interface. If this was not the case, then the results would be sensible only if, fortuitously, no spring went into tension; otherwise, 'tensile' springs would need to be eliminated iteratively until all the remaining active springs ended up being compressive – a cumbersome procedure, indeed!

Alternatively, 'active' pressures could be balanced out on the opposite walls of the structure and the soil springs modelled to mobilize only as much of the residual 'passive' reaction as necessary against the out-of-balance force. One would, of course, need to ensure that the sum total of the spring reaction and the applied 'active' pressure did not, at any level, exceed that extent of the passive resistance which could be safely mobilized without exceeding the acceptable levels of ground strain or structural movement.

The coefficient of 0·2 stipulated for the minimum horizontal earth pressure would imply an equivalent ϕ' value of about 42° for the compacted fill. In the event that this degree of compaction is either not achieved or not necessary, the 'disturbing force' is then likely to be underestimated. As an illustration, for a degree of compaction which would yield a ϕ' value of 35°, the 'disturbing force' would be underestimated by about 36 per cent and for a ϕ' value of 30°, the underestimation would be as much as 68 per cent! Clearly, this is significant and the choice of the value of the coefficient(s) to be adopted cannot, therefore, be taken lightly.

Since, for a given type of soil and the degree of compaction, the minimum earth pressure cannot, in theory, fall below that corresponding to its 'active' value, it would be more appropriate to use the conventional value for the coefficient K_a rather than 0·2 as stipulated for the minimum horizontal earth pressure in the *BD*. This approach would enable the designer to use the values of K_o and K_a which would be consistent with one identifiable value of ϕ'. It would also remove the anomaly in design between using K_o based on a certain value of ϕ' and that using the stipulated value of 0·2 which amounts to representing a totally different value of ϕ'.

BD 31/2001 approach

In recognition, among other things, of the anomaly in *BD 31/87* as discussed above, the document has been revised. The revised version (*BD 31/01*), which forms a mandatory requirement under Clause 3.1.3 (*a*) (*ii*) on the part of the Overseeing Organization, has dismissed the anomaly outlined above simply by stipulating that if, with the use of the coefficients 0·2 and 0·6, the structure is seen to sway in the direction opposite to that implied by the applied horizontal live load generating the imbalance, this load case need not be considered. In view of this, the clause as it stands, particularly in relation to the design of cut-and-cover metro structures, appears to be highly arbitrary and continues to be far from satisfactory.

With regard to the potential effects of negative arching (Figure 3.1 in the document) also, Clause 3.1.2 (*c*) (*ii*) in the document makes a stipulation of the minimum enhancement in the weight of the backfill ranging from 15 to 50 per cent depending upon the hardness of the founding stratum. For a box culvert with a width of up to $15m$ and a backfill depth of up to 11m, occurrence of such an enhancement is, under certain circumstances, conceivable. However, in the case of cut-and-cover metro structures, as explained in Chapter 12, negative arching from the backfill on the roof, if taken purely as a function of the depth of the backfill on top without any consideration to its aspect ratio or its method of placement, is likely to exaggerate the effect artificially. Once again, in its application to metro structures which are likely to be much larger than the box culverts for which the document seems to have been primarily intended, the clause would appear to be nonspecific.

In its older version, i.e. as *BD 31/87*, the document was extensively used by engineers in many parts of the world for the design of cut-and-cover metro structures also. However, in its revised version, i.e. *BD 31/01*, since it limits the scope by restricting the length of the buried structure between the walls to $15m$ only and by extending the backfill depth up to $11m$, it would appear that, in respect of such structures, this document was aimed, principally, at box culverts generally associated with highway structures. In view of this, it may be concluded that the document was drawn up specifically with conventional highway and related structures in mind. It can also be inferred that, in respect of the much larger and complex metro station structures which are generally $18m$ or more wide with relatively much less backfill on top and with aspect ratio unlikely to exceed 0·2., its application is nonspecific in intent. As a logical extension to this, it could be argued that such structures fall outside the scope of this document and are not expected to be specifically covered by it. As such, utmost caution must be exercised if use is made of this document in any way in the design of cut-and-cover metro structures.

25 Structural Modelling

25.1 Introduction

Cut-and-cover metro structures are generally located in heavily built-up urban areas and often in difficult ground conditions. Constraints imposed by such environments invariably dictate that the excavation for and the construction of such structures are carried out in carefully phased stages. In view of this and the intimate contact of the structure with the surrounding soil, the analysis and design of such structures demands taking into account the changes not only in the soil–structure interaction but also in the geometry of the structure, the boundary conditions and the loads that become operative on it during the various stages as the excavation for and construction of the structure progress.

In view of the all too often prevalent complexity inherent in the construction of such structures, careful attention needs to be given to the modelling of the geometry, boundary conditions and the loadings applicable at the different stages of construction. Furthermore, in order to ensure that the results obtained are valid, appropriateness of the principles of the chosen analytical methods also need to be examined and any inherent limitations identified. In view of this, modelling is discussed in this chapter with regard to four principal aspects. These are:

- Structural idealization
- Structural analysis
- Boundary conditions
- Applied loadings.

25.2 Structural idealization

By their very nature, cut-and-cover metro structures are of cellular box cross-section. In the analysis of such a structure, the cross-section is typically idealized by its line diagram representing the centre lines of the member elements. For assessing the overall structural behaviour, the section properties A, I and J related to the axial, flexural and torsional stiffnesses, respectively, are assigned to the line elements. Loads are then applied either individually to the various elements or in coexistent combinations to the structure as a whole. These give rise to commensurate reactions in a manner that ensures static equilibrium. Flow of loads through the structure right from their points of application to their mobilized reactions, together representing the external actions, defines the load path through the structure. The resulting internal actions of the structural elements, such as the moment, the axial force, the shear and the torsion, represent the stress-resultants in them. The external and the internal actions ('applied loads and mobilized reactions' and 'stress-resultants') in equilibrium together constitute (Fraser, 1981) the operational equilibrium force field for the structure.

In the structural idealization, the flexible lengths of the component elements are represented by their centre-to-centre distances between the adjacent joints which are treated as rigid and represented by nondimensional 'points' in the model. This, potentially, overestimates the flexibility of the box structure. As an alternative, the clear span lengths of the elements could be treated as the flexible lengths. However, this would, *ipso facto*, underestimate the box flexibility. In order to appreciate the implications of the centre-line idealization referred to above, consider the geometry of a simple cut-and-cover box structure as shown in Figure 25.1.

From the figure, the foregoing idealization can be seen to affect the analysis, in at least three possible ways:

- It ignores the effect of the weight of the backfill on the roof slab that falls outside the centre-lines of the outer walls. It also ignores the effect of the lateral earth pressures on the outer walls acting above and below the centre-lines of the roof and the base slabs, respectively. As a result, in both these cases, the axial loads on the respective elements of the structure are underestimated.
- It artificially increases the design lengths of all the structural elements increasing, correspondingly, the bending moments and shear forces in them.
- It does not make any allowance for the actual member sizes, i.e. widths and depths, of the structural elements thereby affecting the stress-resultants in them.

The resulting implications are discussed in the following section.

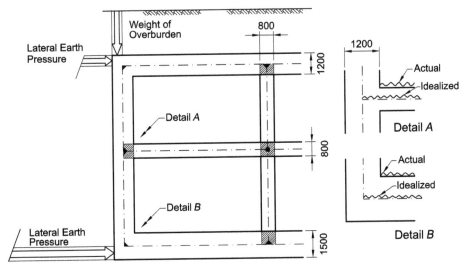

Figure 25.1 Typical Cut-and-Cover Structure (part section)

Axial compression

Underestimating the effects of the backfill and the lateral pressures as outlined earlier will, inevitably, also underestimate the axial forces in the perimeter wall, the roof slab and the base slab. This will, in turn, exaggerate the effects of flexure in these elements and push up the rebar requirements somewhat.

Consider Figure 25.2 representing a typical slab section under the action of a moment M and an axial load P. If the effect of the axial load were to be completely ignored, the area of the tensile reinforcement required to withstand the moment M would, by reference to the typical stress diagram in figure (*a*), be given by the expression:

$$A_s = \frac{M}{\left(0 \cdot 87\, f_y\right) z} \tag{25.1}$$

where f_y = characteristic strength of reinforcement (N/mm^2), z = lever arm = $a_1 \times d$, and d = effective depth.

For a given value of M, the corresponding value of z can be obtained from Table 25.1 in which f_{cu} = characteristic strength of concrete in N/mm^2.

To allow for the effect of an axial load P, it can be replaced by a combination of a similar load transferred to the centre-line of the tensile reinforcement and a coexistent moment $P \times e$. In other words, the axial compression can be seen to generate an additional component of moment, thereby pushing the total moment up to $(M + P \times e)$.

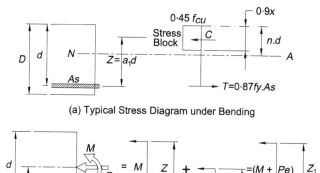

(a) Typical Stress Diagram under Bending

(b) Direct Force & Moment

Figure 25.2 Effect of Combined Axial Compression and Bending

Table 25.1 Lever Arm and Neutral Axis Depth Factors for Slabs

$K = M/bd^2f_{cu}$	·05	·06	·07	·08	·09	·10	·11	·12	·13	·14	·15
$a_1 = z/d$	·94	·93	·91	·90	·89	·87	·86	·84	·82	·81	·79
$n = x/d$	·13	·16	·19	·22	·25	·29	·32	·35	·39	·43	·47

Disregarding, for the time being, the effect of the direct load component, the tensile reinforcement necessary to resist the increased moment alone would be given by the expression:

$$A_{s1} = \frac{(M + P \times e)}{\left(0 \cdot 87 \, f_y\right) z_1}$$ (25.2)

where, z_1 is the new lever arm corresponding to the increased moment.

To include the effect of the direct load component as well, the amount of the tensile reinforcement, A_{s1}, as worked out above must now be adjusted to reflect this effect. Since the effect of the compressive force must be to neutralize some of the tension in the flexural reinforcement, this is achieved by reducing the A_{s1} by the amount $P/0 \cdot 87 f_y$ so that, finally, the area of tensile reinforcement required is given by the expression:

$$A_{st} = \frac{(M + P \times e)}{\left(0 \cdot 87 \, f_y\right) z_1} - \frac{P}{0 \cdot 87 \, f_y}$$ (25.3)

This expression can be rearranged in the form:

$$A_{st} = \frac{M}{\left(0 \cdot 87 \, f_y\right) z_1} + \frac{P}{\left(0 \cdot 87 \, f_y\right) z_1} \left(e - z_1\right)$$ (25.4)

As long as $z_1 > e$, the second part of the above expression on the right hand side of the equation will always end up being negative. Consequently, the decrease in the amount of reinforcement due to the axial compression component will always be greater than the corresponding increase due to the additional moment component. This implies that, the very marginal difference in the values of z and z_1 notwithstanding, a net reduction in the area of reinforcement as given by the expression A_s in Equation (25.1) will always be achieved if the effect of axial compression were to be taken into account. Therefore, ignoring the effect of axial compression altogether would, in theory, lead to a safer approach in the estimation of the area of reinforcement certainly in members where moment is the dominant principal action.

However, in reality, the extent of the residual compression likely to be effective in the base slab, after having to overcome the soil–structure interface friction and, where applicable, the resistance of wall penetration below the base slab, may prove to be difficult to quantify precisely. In view of this also, it would be sensible to ignore the beneficial contribution of the axial compression. Notwithstanding this, where its magnitude is significant, ignoring legitimate axial compression may not be advisable especially in the case of elements where compression predominates and the compressive stresses are likely to be critical.

Design lengths of elements

Artificially increasing the design lengths of the structural elements by using centre-line idealizations can affect the analysis in two ways:

- The effect of the increase in the span lengths of the elements, apart from the extreme corners where it is compensated, is to duplicate the component of the dead load of the structure at the junction between, and common to, the horizontal and the vertical members thereby increasing the axial load in the vertical support members. These areas are shown hatched in Figure 25.1. While in the case of the outer walls, this will, to some extent, compensate the loss of the weight of backfill on top of the roof outside the wall centre-lines, in the case of the intermediate columns, there will be no such compensation. However, the increase in the dead weight is likely to be only marginal and therefore the effect of the resulting approximation due to the duplication is unlikely to be significant and may be safely ignored.

- Idealized increase in the span lengths, particularly those of the intermediate and the base slabs, ignores the extent of thickness of the respective vertical support elements above these slabs over which live load cannot be physically applied. Application of live loads even in areas such as these will artificially add not only to the axial compression of the support elements but also to the total live load on the member considered as shown in Figure 25.1. However, since live loads are a small proportion of the total load and the thicknesses of the support elements, i.e. the areas over which the 'extra' live load is applied, are also very small, once again, the effects will be insignificant and can be safely ignored.

Actual member sizes

Analysis of a structure based on the centre-line diagram approach which idealizes a joint as a point can be adequate but only so long as the thicknesses of the members are small compared to their respective lengths; otherwise the peak moments at the joints would represent an overestimation of the required design moments. However, in the case of box structures comparable to those shown in Figure 25.1, for example, where the member thicknesses in relation to their respective lengths cannot be considered to be small, it would not be appropriate to consider the joint as a point. It would, in reality, be a block of material. In such a case, in order to model the structure reasonably accurately, it would be possible to undertake a finite element analysis to identify the flexible lengths of the member elements and define the boundaries of the joints, or come up with equivalent flexible lengths of uniform sections for the elements.

Modern computer methods of analyses do have the facility to deal with all these factors by including the finite widths of columns and other wall supports. However, in view of the approximations associated with the estimation of the loadings and the boundary conditions generally adopted, and the empiricism involved in the use of the various design parameters employed in the analysis of cut-and-cover metro structures generally, such a degree of sophistication and accuracy may not be warranted. Besides, centre-line diagram approach offers a simple analytical model, easy to deal with and, more importantly, the inherent assumptions on which it is based, are unlikely to push the results on to the unsafe side.

Nevertheless, to limit waste, a conscious attempt should be made to modify the results obtained from an idealized line diagram analysis to reasonably reflect, in some way, the effect of member thicknesses. In this context, to arrive at the maximum design hogging moments over supports the centre-line peak moments at joints may be duly modified to allow for the effects of the support widths as illustrated in Figure 25.3.

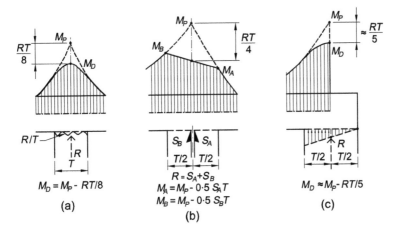

$$M_D = M_P - RT/8$$

(a)

$$R = S_A + S_B$$
$$M_A = M_P - 0.5\,S_A T$$
$$M_B = M_P - 0.5\,S_B T$$

(b)

$$M_D \approx M_P - RT/5$$

(c)

Figure 25.3 Reduction of Peak Moments over Supports

Figure 25.3(*a*) shows the proposed treatment of peak hogging moments in slabs directly over intermediate supports, such as columns, which are not too wide. To obtain the design moment M_D, the peak centre line moment, M_P, can be reduced by the amount $RT/8$, where 'R' is the reaction at, and 'T' the thickness of, the support considered. The rationale underlying this practice requires replacing the point load reaction, as assumed in and obtained from the centre-line frame analysis, by the corresponding equivalent upward *udl* over the full width of the support. The peak centre-line moment is then adjusted downwards by the moment due to the equivalent upward cantilever *udl* to one side of the centre line.

In the case of wider supports, such as thick walls, moments obtaining on either face of the support can be assumed as the respective design moments on the two sides. Figure 25.3(*b*) shows the proposed treatment for such a case. Moment M_F on the face of the support can be obtained by reducing the peak moment M_P, by the moment due to the respective shear force about the face under consideration, i.e. $S \times T/2$, as can be seen if a free body diagram between the face and the centre-line of the support were to be drawn. This approach is likely to yield different values for the moment on the two faces either side of the support. In such an event, practicality of detail would suggest that the amount of top reinforcement to be provided over the support be based on the greater of the two moments.

In the case of an end support such as that provided by a perimeter wall, due to the asymmetry of the geometry as well as the loading at this end, it would be reasonable to replace the idealized (point load) end reaction by a distributed upward load varying linearly, as an approximation, from zero at the outside face to twice the intensity at the inside face as shown in Figure 25.3(*c*). The peak moment may then be reduced by the moment due to the upward trapezoidal load on the inside half of the centre-line of the support to arrive at the design moment. This reduction works out approximately to $RT/5$, where R is the end wall reaction obtained from the centre-line diagram model. To allow for any adverse variation in the 'assumed' reaction profile, this reduction may be scaled down from $RT/5$ to $RT/6$.

25.3 Structural analysis

Access to all types of computer programmes is available these days, virtually at one's fingertips, and even to those without any knowledge of structural mechanics. The results of even the most complex of computations can therefore be available within a matter of seconds. This has no doubt made it possible to achieve enormous savings in the computational time and, in the hands of experienced engineers, use of computers has thus become an invaluable tool. However, obscurity of the software functionality i.e. the mechanics of the step-by-step progression towards the end results, being couched in complex mathematical processes and matrix operations may not often make it easy for a young, relatively inexperienced engineer to develop a 'feel' for the response of the structure particularly under the complex combination of varying loads and changing geometry and boundary conditions.

Lack of 'feel' can place the engineer at a disadvantage, for example, in not being able to interpret the results correctly or to readily recognize whether the results obtained make sense and unable therefore to decide whether they are right. Relying on the accuracy of input in terms of arithmetic alone cannot provide any guarantee against poor interpretation of results, inappropriate modelling or against the same errors being overlooked repeatedly and allowed to slip through undetected. Blind reliance upon the computer results, therefore, without any appreciation of their quality or relevance can be unsafe and dangerous and must be avoided.

It is for the foregoing reasons that in this and the next two chapters use has been made of hand computations without any recourse to computer analysis. However, this should not be construed, in any way, as a recommendation for replacing computer analyses by hand calculations. It is only hoped that the long-hand treatment followed herein goes some way in providing a young engineer with a better understanding and appreciation of the mechanics of the structure, in general, and its response to changing geometry, boundary conditions and loading at various stages of construction, in particular. It is also hoped that, for the engineer, such improved understanding of the structural behaviour will make the use of computer programmes in practice that much more meaningful.

Cut-and-cover metro structures can be safely categorized as major structures, generally founded in complex ground and groundwater conditions and located in high-density urban environments. For clearly understanding the influence of the changing geometry, boundary conditions, loading, etc. at every stage of construction, and how it differs from that of the in-service condition, it is important to appreciate the fundamental concepts of the various theories and principles commonly employed in the analysis and, where appropriate, the strengths and limitations inherent in their application. These are briefly discussed in the following section.

Equilibrium and compatibility

Static equilibrium constitutes an important fundamental concept in structural analysis. However, in the analysis of cut-and-cover metro box structures which are statically indeterminate, it is also essential to maintain structural continuity. This is achieved by ensuring compatibility as the structure deforms. Whatever be the method used, whether it is the force (or flexibility) method or the displacement (or stiffness) method, it is important that the demands of both equilibrium and compatibility are simultaneously satisfied. Fortunately, all the well-known force and displacement methods of structural analysis commonly employed do respect this fundamental principle.

Linear elastic theory

Most structures are analysed using linear elastic theory. While it greatly simplifies the analysis, it is based on the assumption that the deflections are small enough to maintain the stresses in the member elements within their elastic limits. In the case of very flexible structures, relatively large deflections can alter the load configuration to such an extent as to render the analysis nonlinear. However, in the case of a typical cut-and-cover metro structure, satisfying the serviceability requirements and the need to contain the potential of flotation do ensure that the member thicknesses are generally not insignificant. It is therefore not unreasonable to expect the deformations and deflections, particularly those due to shear, in the component elements of the structure to be relatively small. As a result, the geometry of the structure, irrespective of whether it is deformed or undeformed, can be assumed, without much concern, to be substantially the same. In other words, a particular state of equilibrium can be assumed to remain unchanged even though deformations and displacements are known to have taken place.

Metro structures that are confined underground can be subjected to significant amounts of compressive forces resulting from the surrounding ground pressures. In the course of time, these can give rise to a nonlinear material response, known as creep, in the reinforced concrete members. Fortunately, ultimate limit state methods commonly employed in the design for strength are independent of the stress history prior to failure and so nonlinear behaviour such as creep does not affect the predicted strength. However, over time, creep does affect deflection and should therefore be taken into consideration in assessing long-term deflections.

Superposition

In the analysis of underground metro structures, the principle of superposition is commonly employed to establish the effects of the combination of diverse loads on the stress-resultants of a typical member. However, this is appropriate in the case of only such loads where there exists a similarity of correspondence, between the applied loads and the respective structural responses evoked. Where such correspondence is absent, the principle of superposition cannot be applied. Whereas in the case of comparable structures above ground this principle can be safely employed, in the case of underground structures subject to significant axial compressive forces, this may not always be possible.

To examine such a case further, consider the equilibrium of a member AB under the action of an applied axial compressive force P and a clockwise couple M_a at the left end, and the support reactions R_A and R_B, as shown in Figure 25.4.

Axial force P will cause a bending moment in the beam but only if the beam experiences deflection under the gravity loads. However, bending moments in the beam, particularly due to force P, cannot be obtained until the deflections are

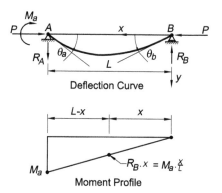

Figure 25.4 Member with Couple at Left End

determined. The member is therefore statically indeterminate and it is necessary to begin by solving the differential equation for the deflected curve of the member.

Taking end B as the origin of coordinates x and y, and ignoring the effects of shear deformations and the axial shortening, bending moment M_x anywhere in the member is given by the differential equation:

$$EI\frac{d^2y}{dx^2} = -M_x$$

$$= -P \cdot y - R_B \cdot x = -P \cdot y - \frac{M_a}{L} \cdot x \qquad (25.5)$$

The parameter y represents the deflection of the axis of the member and EI its flexural stiffness which is assumed constant over its entire length.

Equation (25.5) may be rearranged as:

$$EI\frac{d^2y}{dx^2} + P \cdot y = -M_a\left(\frac{x}{L}\right)$$

Introducing the axial load factor $n = \sqrt{P/EI}$ (also represented by the symbol k by some authors), the above equation can be expressed in the form:

$$\frac{d^2y}{dx^2} + n^2 \cdot y = -\frac{M_a}{EI}\left(\frac{x}{L}\right) \qquad (25.6)$$

The general solution of this differential equation is given by:

$$y = A\cos nx + B\sin nx - \frac{M_a}{P}\left(\frac{x}{L}\right) \qquad (25.7)$$

Parameters A and B are the constants of integration. By inserting the appropriate boundary condition, i.e. $y = 0$ both at $x = 0$ and L, these can be evaluated as:

$$A = 0 \quad \text{and} \quad B = \frac{M_a}{P} \cdot \frac{1}{\sin nL}$$

With the substitution of these values, the general solution assumes the form:

$$y = \frac{M_a}{P}\left(\frac{\sin nx}{\sin nL} - \frac{x}{L}\right) \tag{25.8}$$

This represents the equation for the deflection curve of the member. It can be inferred from this expression that the deflection at any section of the member is directly proportional to the couple M_a but not to the axial load P. Since there is no correspondence between M_a and P in relation to the deflection y, in such a case, the principle of superposition cannot ordinarily be used. However, it is possible to use this principle with respect to M_a but only so long as P remains constant or the variation in its value is deemed small enough to be insignificant.

Effect of axial force

Before dealing with the analysis of axial-load effects particularly in the floor elements of an underground metro box structure, it is useful to recognize that the presence of horizontal axial compression in a member results in two distinct actions. First, it causes elastic shortening in the member. The effect of differential elastic deformations in the vertical support elements such as the perimeter walls and columns is demonstrated in Chapter 26. Second, the effect of axial compression is to induce bending moment in the member which also causes a commensurate change in its rotational stiffness. Along with it, there will be a change in the carry-over factors as well. In general, the effect of 'change of length' (axial deformation) and the 'beam-column' ('$P - \Delta$') effect assume importance in the case of those slender members which are subjected to relatively high levels of axial forces. Furthermore, the '$P - \Delta$' effect is of consequence only when the axially loaded members are very slender, whereas the axial deformation tends to become important when the members are relatively stocky.

Elastic shortening

Axial deformation under a compressive load P causes elastic shortening in the length of an element in accordance with the expression $\Delta L = PL/AE$, where ΔL, L and A represent, the axial shortening, the length and the cross-sectional area, respectively, of the member and E the modulus of elasticity of its material. The expression can be rearranged in the form; $P/\Delta L = AE/L$. This represents the load necessary to cause a unit deformation and, as such, defines the axial spring stiffness for the element. Elastic shortening in members under axial compression causes displacement of joints in the structure. In the case of vertical members, the joint displacement will be in the vertical direction, whereas that for the horizontal members in the horizontal direction. The principle of treatment for the effect of horizontal joint displacement on the stress-resultants of members is similar to that for the vertical displacement as illustrated in Chapter 26; the only difference being in the magnitude of the spring stiffness and the direction of the spring supports.

Stiffness and carry-over factors

The effect of axial force in a member is to alter its rotational stiffness and carry-over factors. Tensile force has a stabilizing effect by increasing the rotational stiffness of the member whereas the compressive force has the opposite – that of reducing the stiffness. Since the elements of a typical metro box structure are unlikely to be under tension, only the effect of axial compression will be dealt with here.

As the compressive force in a member increases, the rotational stiffness will go on decreasing until the force reaches the critical buckling load when the stiffness will have become zero and the resistance to rotation at one end of the member will have been completely lost.

In arriving at the expressions for the rotational stiffness and carry-over factors, three stages will be followed. In Stage (1), a simply supported element subjected to an axial compressive force and a couple at one end only will be analysed to arrive at the rotations at the two ends. In Stage (2), by the use of analogy and the principle of superposition, end rotations due to the combined effect of a couple at each end will be established. Finally, from the expressions thus obtained, by substituting appropriate values for the rotations, required values for the stiffness and the carry-over factors for the corresponding boundary conditions can be obtained in Stage (3).

First Stage – Couple at One End Only, Far End Pinned: Consider the equilibrium of an element AB under the action of a couple at one end with the other end pinned as shown in Figure 25.4. Recalling Equation (25.8), the expression for the deflection curve of the member is given by:

$$y = \frac{M_a}{P}\left(\frac{\sin nx}{\sin nL} - \frac{x}{L}\right)$$

First derivative of the preceding equation will give the general expression for the slope of the deflection curve as:

$$\frac{dy}{dx} = \frac{M_a}{P}\left(\frac{n\cos nx}{\sin nL} - \frac{1}{L}\right) \tag{25.9}$$

From this expression, angles of rotation at the two ends of the member can be obtained by introducing the appropriate boundary conditions. Therefore, by setting $x = L$, and assuming the angle of rotation θ_a to be positive in the same direction as the clockwise end moment, rotation at end A is given by:

$$\theta_a = \left(\frac{dy}{dx}\right)_{x=L} = \frac{M_a L}{3EI}\cdot\frac{3}{nL}\left(\cot nL - \frac{1}{nL}\right)$$

$$= -\frac{M_a L}{3EI}\cdot\frac{3}{nL}\left(\frac{1}{nL} - \cot nL\right) \tag{25.10}$$

The expression for rotation has been arranged in two parts, deliberately in such a manner that the identification of the effect of the axial compression can be made easily. The first part of the expression on the right, i.e. $M_a L/3EI$, can be readily recognized as the value of rotation when there is no axial load present; the second part can therefore be interpreted as the amplification factor due to the presence of the axial load P.

It is also referred to as the stability function. Denoting this function as ψ, we have (Gere, 1963):

$$\theta_a = -\frac{M_a L}{3EI} \cdot \psi \tag{25.11}$$

$$\text{where} \quad \psi = \frac{3}{nL}\left[\frac{1}{nL} - \cot\left(\frac{180}{\pi} \cdot nL\right)^{\circ}\right] \tag{25.12}$$

As nL (measured in radians) approaches the value π, the stability function ψ approaches infinity. This suggests that the deflection increases without limit. In other words, at this value of nL and with both ends pinned, the axial load attains its critical buckling load, which is given by:

$$P = \frac{\pi^2 EI}{L^2} \text{ (both ends pinned)} \tag{25.13}$$

Using appropriate sign convention and setting $x = 0$, the angle of rotation at end B due to the couple at A can, similarly, be obtained from:

$$\theta_b = \frac{M_a L}{6EI} \cdot \phi \tag{25.14}$$

$$\text{where} \quad \phi = \frac{6}{nL}\left[\csc\left(\frac{180}{\pi} \cdot nL\right)^{\circ} - \frac{1}{nL}\right] \tag{25.15}$$

The graphs for the stability functions ψ and ϕ have been plotted in Figure 25.5. For the table of values for the stability functions, reference may be made to Timoshenko *et al.* (1961).

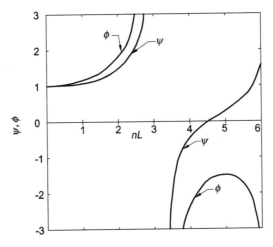

Figure 25.5 Graphs of Stability Functions for Beams with
Axial Compression (adapted from Gere, 1963)

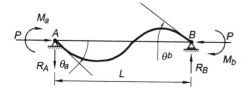

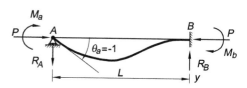

Figure 25.6 Member with Couples at   Figure 25.7 Member with Far End Fixed
Both Ends

Second Stage – Couples at Both Ends: Figure 25.6 shows couples at both ends of
the member.

Expressions in Equations (25.11) and (25.14) represent the values of rotations at
ends A and B, respectively, due to a couple M_a acting at end A only. Knowing these
values, the respective contributions to the end rotations corresponding to a couple M_b
at end B can, likewise, be obtained simply by analogy. Therefore, using the principle
of superposition, general expressions for end rotations due to couples at both ends can
be simply written down as:

$$\theta_a = -\frac{M_a L}{3EI} \cdot \psi + \frac{M_b L}{6EI} \cdot \phi \qquad (25.16)$$

$$\theta_b = \frac{M_a L}{6EI} \cdot \phi - \frac{M_b L}{3EI} \cdot \psi \qquad (25.17)$$

Third Stage – Rotational Stiffness and Carry-Over Factors: The ratio M/θ represents
the absolute rotational stiffness of a member. It can be defined as the moment required
to cause unit rotation. Given this definition, it is a simple matter to obtain the
expressions for the stiffnesses from Equations (25.16) and (25.17). For example, in
order to determine the absolute rotational stiffness of a member with its far end fixed
(Figure 25.7), all that is necessary is to substitute, for the rotation at the near end,
$\theta_a = -1$ and, at the far end, where rotation is prevented, $\theta_b = 0$, in the equations, and
then solving these simultaneously.

Solution of the equations enables the expression for the absolute rotational stiffness
for the member to be obtained (Gere, 1963) as:

$$K_{ab} = \frac{4EI}{L} \cdot \left(\frac{3\psi}{4\psi^2 - \phi^2} \right) \text{ (far end fixed)} \qquad (25.18)$$

Also, in the case of the member with its far end fixed, the ratio of the moment
induced at the far end and the applied or the inducing moment at the near end, i.e.
M_b/M_a, represents the moment carry-over factor from the near to the far end. Since
the rotation at the far end is prevented because of the fixity at that end, by setting

$\theta_b = 0$, the carry-over factor can easily be obtained from Equation (25.17) as:

$$CF_{ab} = \frac{1}{2}\left(\frac{\phi}{\psi}\right) \text{ (far end fixed)} \qquad (25.19)$$

The buckling load which, for the 'far end fixed' case, is critical at a value of $nL = 4\cdot493$, is given by:

$$P = \frac{20\cdot19EI}{L^2} \text{ (one end fixed)} \qquad (25.20)$$

In the case of the far end being 'pinned', $M_b = 0$. Upon making this substitution in Equation (25.16), the absolute rotational stiffness for the member is given as:

$$K_{ab} = \frac{3EI}{L} \cdot \left(\frac{1}{\psi}\right) \text{ (far end pinned)} \qquad (25.21)$$

Since Figure 25.4 can also be seen to represent the 'far end pinned' case, the above expression can also be directly obtained from Equation (25.11).

With regard to the carry-over factor, owing to its inability to prevent rotation, there can be no carry-over of moment to a pinned end. Graphs for the stiffness and carry-over factors have been plotted in Figures 25.8 and 25.9, respectively.

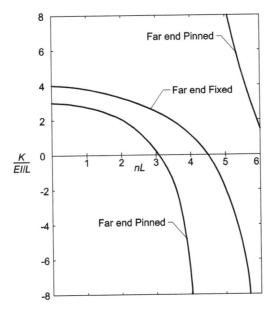

Figure 25.8 Stiffness Factors for Beams with Axial Compression (adapted from Gere, 1963)

The expressions within the parentheses on the right-hand side of Equations (25.18), (25.19) and (25.21) represent, in each case, the respective amplification factors for the effects of the axial compressive force P. Accordingly, in the absence of this force, i.e. when $P = 0$, the values of the stability functions ϕ, ψ and the amplification factors become unity and the stiffness and carry-over factors for the usual 'moment only' cases can be recovered.

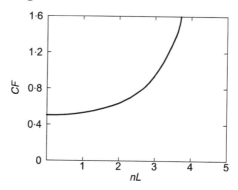

Figure 25.9 Carry-over Factors for Beams with Axial Compression
(adapted from Gere, 1963)

Fixed-end moments

Presence of axial forces in a member affects the fixed-end moments in it. A tensile force decreases the fixed-end moments whereas a compressive force causes an increase in their value. In the case of metro structures, owing to their confinement underground it is generally the compressive force that is encountered and therefore, potentially, of consideration to the engineer. In the computations that follow, it will be this force that will be considered.

In order to establish the extent of the effect of axial force in an element on its stress-resultants, consider a fixed-ended member subjected to vertical loads and an axial compressive force P as shown in Figure 25.10(*a*).

The fixed-end moments can be obtained by following a two-stage computation. In the first stage, moment restraints at the ends are removed as shown in figure (*b*) to allow the element to deflect under the action of the applied vertical and axial loads. The end rotations, θ_{a1} and θ_{b1}, are then evaluated by analysing the element *ab initio* following a similar procedure as before. In the second stage, the boundary conditions, i.e. fixity at the ends, are restored. This requires moments capable of reversing the rotations, i.e. M_{a1} and M_{b2}, to be applied at the two ends in turn; when moment is applied at one end, the other is held against rotation and *vice versa*, as shown in figures (*c*) and (*d*). Moments M_{b1} and M_{a2} represent the corresponding carry-over effects induced at the other ends in the two cases.

The moment required to be applied at end A is equal to the product of the stiffness and the angle through which that end needs to be rotated, i.e. $-\theta_{a1}$. In other words,

$$M_{a1} = K_{ab} \times (-\theta_{a1}) = -K_{ab} \times \theta_{a1} \tag{25.22}$$

The parameter K_{ab} represents the absolute rotational stiffness of the member with the other end fixed. The corresponding moment induced at the other end is obtained by multiplying the applied, or the inducing, moment by the carry-over factor, or

$$M_{b1} = M_{a1} \times \mathrm{COF}_{ab} = -K_{ab}\theta_{a1}\mathrm{COF}_{ab} \tag{25.23}$$

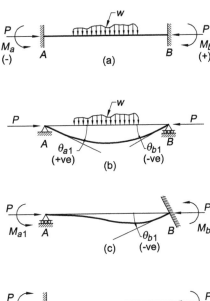

Figure 25.10 Fixed-ended Member with Axial Load

Likewise, the moment required to be applied to neutralize the rotation at end *B* is given by the expression:

$$M_{b2} = K_{ba} \times \theta_{b1} \qquad (25.24)$$

and the corresponding moment induced at end *A* by the expression:

$$M_{a2} = M_{b2} \times COF_{ba} = K_{ba}\theta_{b1}COF_{ba} \qquad (25.25)$$

For a member of uniform cross-section, $K_{ab} = K_{ba} = K$ and $COF_{ab} = COF_{ba} = COF$.

Finally, since the axial force has been maintained constant throughout the foregoing computation, the principle of superposition can be invoked to obtain the fixed-end moments. In other words,

$$M_a = M_{a1} + M_{a2} \quad \text{and} \quad M_b = M_{b1} + M_{b2}$$

So, with reference to the sign convention assumed in Figure 25.10, general expressions for the fixed-end moments may be written as:

$$M_a = -K\left(\theta_{a1} - COF\theta_{b1}\right) \qquad (25.26)$$

$$\text{and} \quad M_b = K\left(\theta_{b1} - COF\theta_{a1}\right) \qquad (25.27)$$

With the help of the appropriate values for the end rotations assuming pinned ends, together with the values of the stiffness and the carry-over factors, and by substituting these in the above general expressions, fixed-end moments for different load cases can be obtained. Fixed-end moments for four particular load cases generally encountered in the analysis of cut-and-cover metro structures are given in the following section:

Uniformly distributed load: By solving the basic differential equation, i.e. Equation (25.5), from first principles as before but with the member subjected to a uniformly distributed load, the following end rotations assuming pinned ends can be arrived at (Timoshenko *et al.*, 1961):

$$\theta_{a1} = \frac{wL^3}{24EI} \cdot \frac{24}{n^3L^3}\left(\tan\frac{nL}{2} - \frac{nL}{2}\right) = -\theta_{b1} \tag{25.28}$$

Substituting these and the values for the stiffness and the carry-over factors from Equations (25.18) and (25.19) into the general expressions above, the fixed-end moments obtained are:

$$M_a = -\frac{wL^2}{12}\gamma_1 = -M_b \tag{25.29}$$

$$\text{where} \quad \gamma_1 = \frac{12}{n^2L^2}\left[1 - \frac{nL/2}{\tan\left(\frac{180}{\pi} \cdot \frac{nL}{2}\right)^\circ}\right] \tag{25.30}$$

Graph in Figure 25.11(a) shows the plot of fixed-end moments against nL (measured in radians).

Point load anywhere: End rotations for a member subjected to a single point load W applied at a distance aL from the left and bL from the right end are given by the expressions (Timoshenko *et al.*, 1961):

$$\theta_{a1} = \frac{W}{P}\left(\frac{\sin bnL}{\sin nL} - b\right) \tag{25.31}$$

$$\theta_{b1} = -\frac{W}{P}\left(\frac{\sin anL}{\sin nL} - a\right) \tag{25.32}$$

Fixed-end moment at A is given by:

$$M_a = -WL\gamma_3 \tag{25.33}$$

$$\text{where} \quad \gamma_3 = \frac{bnL\cos nL - \sin nL + \sin anL + \sin bnL - nL\cos bnL + anL}{nL(2 - 2\cos nL - nL\sin nL)} \tag{25.34}$$

Note: In the expression, conversion factors from radians to degrees have been omitted for simplicity. However, in evaluating the parameter γ_3, the angles for every

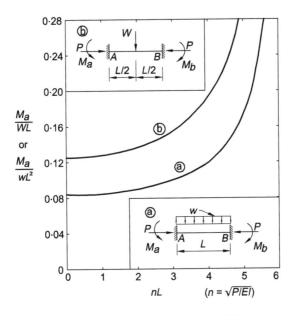

Figure 25.11 Fixed-end Moments with Axial and Transverse Loads
(adapted from Gere, 1963)

trigonometric function such as $\cos nL$, $\sin bnL$, etc. must be converted from radians to degrees with the use of the factor $(180/\pi)$ on similar lines as those shown for parameter γ_1 in Equation (25.30).

Figure 25.12 shows the plot of M_a against nL. By treating as abscissa b in place of a, the ordinate will correspondingly represent the values for M_b in place of M_a. **Cor.** With $a = 1/2 = b$,

$$\gamma_3 = \left[1 - \cos\left(\frac{180}{\pi} \cdot \frac{nL}{2}\right)^\circ \right] \div \left[2nL \sin\left(\frac{180}{\pi} \cdot \frac{nL}{2}\right)^\circ \right].$$

This expression represents the special case for a point load at mid-span. The plot of the fixed-end moments against nL for a centrally located point load has also been shown in Figure 25.11(b). However, this has been expressed as: $M_a = -M_b = -(WL/8)\gamma_2$, where $\gamma_2 = 8\gamma_3$.

Uniformly varying load: By following a rigorous mathematical procedure from first principles as for the uniformly distributed load case, fixed-end moments also for the uniformly varying load case incorporating the effect of the axial compression force can be obtained. However, without engaging in a complex mathematical treatment, it is still possible to get a rough idea of the effect of the axial force by employing a static equivalence of the uniformly varying load with a strategically located single point load.

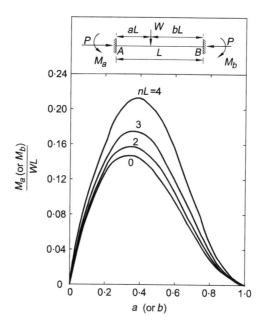

Figure 25.12 Fixed-end Moments with Axial Load and Point Load
(adapted from Gere, 1963)

In other words, a load varying from an intensity w at the near end to zero at the far end over an element AB of span L may be treated (Gere, 1963) as a single equivalent point load $W = wL/2.88$ located at $0.4L$ from the near end. Then, using $a = 0.4$ and $b = 0.6$, Equations (25.31) to (25.34) can be used to obtain some idea of the fixed-end moments.

Joint Translation: Consider a fixed-ended member AB of uniform cross-section and subjected to an axial load P. Assume that the right-hand end B is displaced downwards relative to end A by an amount Δ. Following the derivation from first principles as before, it can be shown that, in this case, fixed-end moments developed are no different from those obtained for joint translation and without the existence of axial compression. For the orientation of displacement as shown, the fixed-end moments are given by the expression:

$$M_a = -K(1+COF)\frac{\Delta}{L} = M_b \qquad (25.35)$$

However, the values for the stiffness and carry-over factors are those given by the expressions in Equations (25.18) and (25.19). Upon substituting these values, the above expression becomes:

$$M_a = -\frac{EI\Delta}{L^2} \cdot \gamma_4 = M_b \qquad (25.36)$$

where

$$\gamma_4 = \cfrac{(nL)^2 \sin\left(\dfrac{180}{\pi}nL\right)^\circ}{\left[2\sin\left(\dfrac{180}{\pi}nL\right)^\circ - nL\cos\left(\dfrac{180}{\pi}nL\right)^\circ - nL\right]} \tag{25.37}$$

Graph for M against nL (measured in radians) is plotted in Figure 25.13. It is clear from the orientation of the curve that, unlike that for the other two load cases above, the effect of compressive force on the joint-translation load case is to reduce the fixed-end moments. When axial load factor $n = 0$, that is to say, when the axial load is absent, the usual value for the fixed-end moment of $-6EI\Delta/L^2$ can be recovered. It can also be seen that at $nL = 2\pi$, the axial load attains its critical value of $P = 4\pi^2 EI/L^2$ at which the fixed-end moment due to joint translation is completely neutralized.

Effect of shear force

Internal actions for the elements of a metro box structure as indeed for most structures are, typically, bending moment, shear force and axial force. While both flexural and shear deformations contribute to the deflections of the elements, most of the deformations are principally caused by bending moments; shearing forces usually have only a secondary effect on the behaviour of the structures.

In arriving at the deflections, use can be made of the principle of virtual work and the well-known shear formula: $\tau = SAY/bI$, where S is the total shear force on the cross-section, A the cross-sectional area which is outside of the section where the shear stress is to be determined and Y its distance from the neutral axis such that the term AY represents the static moment of the area; I is the moment of inertia of the entire cross-sectional area about the neutral axis and b the width at the section

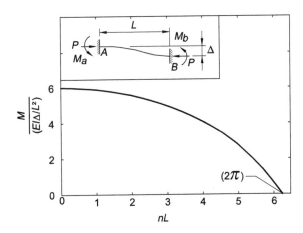

Figure 25.13 Fixed-end Moments with Axial Load and Joint Displacement
(adapted from Gere, 1963)

where the shear stress is to be determined. It should be recognized that the assumption of the plane sections remaining plane after bending employed in the derivation of the aforementioned formula is invalidated when shear stresses are present since they cause warping of the cross-sections. However, the results yielded by the use of the formula are deemed to be accurate enough for most cases and, in any case, to at least serve the important purpose of indicating how significant the contribution of the shear forces on the analysis might be.

Deflection

By reference to any standard text book on the theory of structures, it can be established that the expression for the deflection Δ_C at the mid-span of a uniformly loaded (w), simply-supported (span L) beam works out to:

$$\Delta_C = \frac{5wL^4}{384EI} + \frac{fwL^2}{8GA_C} \qquad (25.38)$$

where, f = form factor for the cross-sectional geometry (non-dimensional), = 6/5 (for rectangular cross-section), E = modulus of elasticity, G = shear modulus of elasticity = $E/2(1+\mu)$, A_C = total area of cross-section, μ = Poisson's ratio, = 0·20 for reinforced concrete, and $E/G = 2(1+\mu) = 2·40$ for reinforced concrete.

Shear modulus of elasticity is defined as the shear stress per unit shear strain and is also referred to as the modulus of rigidity.

The first term on the right hand side of Equation (25·38) is the component of deflection due to bending and the second term the corresponding contribution from shear. Relative importance of the contribution from shear deflection can be judged by the ratio β of the two contributions. Dividing the second term by the first, this ratio for a uniformly loaded member becomes:

$$\beta = \frac{48}{5} \cdot \frac{fEI}{GA_CL^2} \qquad (25.39)$$

For a rectangular cross-section, $f = 6/5$, $I = bd^3/12$ and $A_C = bd$. Taking $E/G = 2·4$, the ratio reduces to:

$$\beta = \frac{2·5}{(L/d)^2} \qquad (25.40)$$

For a span-depth (L/d) ratio within the range of 7–12 encountered in cut-and-cover metro structures generally, the corresponding β- values for a simply-supported case will range from 0·051 to 0·017. This suggests a contribution to deflection from shear of up to a maximum of mere 5 per cent. In the case of fully fixed case, the corresponding contribution would increase – fivefold, i.e. the maximum contribution reaching only up to 25 per cent. In reality, however, full fixity is rarely realized and so the contribution to deflection from shear is likely to be around 10–15 per cent, which is still negligible as compared to that of bending. Therefore, it would be reasonable to conclude that the behaviour of rigid-jointed box structures depends largely upon the magnitudes of the bending moments and axial forces in the members and the distribution of their flexural and axial stiffnesses. Besides, in view of the accuracy warranted also, the effects of shear deformations may be safely ignored.

Stiffness and carry-over factors

The effect of shear force upon the stiffness and the carry-over factors can also be evaluated from first principles. By following a similar approach as that adopted for the effect of the axial force previously, the expressions for the absolute rotational stiffnesses and the carry-over factor, noting that there is no carry-over to the pinned end, can be obtained (Gere, 1963) as:

$$K_{ab} = \frac{4EI}{L} \cdot \left(\frac{2+g}{2(1+2g)} \right) \text{ (far end fixed)} \tag{25.41}$$

$$CF_{ab} = \frac{1-g}{2+g} \text{ (far end fixed)} \tag{25.42}$$

$$K_{ab} = \frac{3EI}{L} \left(\frac{2}{2+g} \right) \text{ (far end pinned)} \tag{25.43}$$

where, g is the nondimensional 'shear factor' and is given by:

$$g = \frac{6f\,EI}{GA_C L^2} \tag{25.44}$$

which, for a rectangular cross-section reduces to:

$$g = \frac{1 \cdot 44}{(L/d)^2} \tag{25.45}$$

All other terms are as defined previously.

Again, for a span-depth ratio within the range of 7–12, g is very nearly equal to zero. As a result, the amplification factors for the rotational stiffness will approach unity and the carry-over factor its conventional 'moment only' value of 1/2.

Fixed-end moments

As far as the effect of shear on the fixed-end moments is concerned, there is no effect when the loading on the element is symmetrical. For an asymmetrically placed point load, the fixed-end moments are given by the expressions (Gere, 1963):

$$M_a = -\frac{Pab^2}{L^2} \left(\frac{1+(L/b)g}{1+2g} \right) \tag{25.46}$$

$$M_b = \frac{Pa^2b}{L^2} \left(\frac{1+(L/a)g}{1+2g} \right) \tag{25.47}$$

Also, for differential joint translation, the expressions for the fixed-end moments are:

$$M = -\frac{6EI\Delta}{L^2}\left(\frac{1}{1+2g}\right) \text{ (both ends fixed)} \tag{25.48}$$

$$M = -\frac{3EI\Delta}{L^2}\left(\frac{2}{2+g}\right) \text{ (one end pinned)} \tag{25.49}$$

In the context of metro structures with the shape factor g being generally close to zero and, consequently, the amplification factors for the fixed-end moments close to unity, once again, the above expressions revert back to their usual 'bending only' values.

In view of the foregoing, it can be concluded that, in the context of the elements of cut-and-cover metro structures where the span-depth ratios fall, typically, within the range of 7–12, the effect of shear on the stress-resultants is even less significant than that on the deflection and can therefore be safely disregarded.

In the case of deep beams, however, where the span-depth ratio is relatively small, i.e. 2 or less, contribution to deflection from shear deformation becomes more significant. In such cases, care should be exercised not to overlook the fact that as the deflection becomes excessive the analysis and the distribution of stress become nonlinear, and should therefore be addressed accordingly.

25.4 Boundary conditions

Underground structures are confined structures surrounded on all sides by the medium of soil. During the installation of the perimeter walls and the subsequent excavation for and the construction of such a structure, surrounding ground must, inevitably, move inwards causing, in the process, horizontal elastic compression of the completed structure until equilibrium is attained. During this process, a measure of friction at the various soil–structure interfaces would be mobilized as a matter of course. With the establishment of equilibrium and a preferential load path, the structure would then be held as if in a vice by the surrounding soil mass, impeding any tendency on its part to translate and rotate.

Subsequently, if an out-of-balance lateral load were to set up due, say, to surcharge loads on one side of the structure, every attempt would be made by the soil–structure interaction to contain this and restore equilibrium once again as quickly as possible. The load responsible for creating the imbalance would tend to cause transverse rotation/translation of the structure. However, for such rotation/translation to physically take place, a measure of the soil–structure interface friction over and above that already mobilized would have to be overcome not only over the outer vertical faces of the side walls but also over the roof and below the base slabs of the structure.

Minimal further strains are likely to be needed to mobilize these reactive forces. As a result, in reality, very little sway is likely to be caused and very little residual force imbalance is likely to be left to be taken up by the flexural deformation of the structure itself. In other words, the inherent stiffness of the surrounding soil mass together with the friction at the soil–structure interface are likely to inhibit the distortion of the structure and constitute the predominant contributors to the stability and equilibrium of the structure.

Limits of behaviour

In the case of a structure where the equations of statics of equilibrium alone are sufficient to establish the state of stress-resultants in it, the structure is statically determinate. In such a case, the computations are straightforward and relatively easy. Although cellular box structures generally associated with underground metros are, invariably, indeterminate and do not fall under this category, it is often possible to idealize them as such in order to obtain quickly some 'feel' for the behaviour and the capability of the structure at the preliminary design stage.

Knowledge and appreciation of the limits of behaviour or the bounds of a structure and its component elements are very useful tools in the repertoire of a competent analytical engineer. The ability to calculate with ease, using simplified analyses or standard formulae, answers which are greater than or less than the more exact solution enables the structural engineer to gauge the capability of the structure and its elements. With the ability to calculate the limits of behaviour of the structure and its component elements, i.e. the maximum and the minimum values of bending moments and deflections, the structural engineer should be able to develop a 'feel' for the behaviour of the structure. Also, with the appreciation that, in most cases, the true behaviour of the structure is likely to lie somewhere between these limits, the engineer can usually obtain reasonable estimates of the dominant stress-resultants or critical deflections without recourse to computer or other sophisticated analytical methods. This can be very useful particularly at the preliminary design stage.

As an illustration, consider the box structure *ABCD* of Example 27.4 in Chapter 27 under the action of a notional load of $10kN/m^2$ applied to the roof slab (*AB*) and an equal and opposite reaction developed at the base slab (*CD*). For the approximate solution, consider the boundary conditions at the ends of elements *AB* and *CD* as fixed and simply supported in turn and to obtain the moments, take the average of the values for the two conditions.

For the fixed-ended case,

End moments,

$$M_{AB} = -\frac{wl^2}{12} = -M_{BA} = -M_{DC} = M_{CD}$$

Span moments,

$$M_{S(AB)} = \frac{wl^2}{24} = -M_{S(CD)}$$

For the simply-supported case,

End moments,

$$M_{AB} = 0 = M_{BA} = M_{DC} = M_{CD}$$

Span moments,

$$M_{S(AB)} = \frac{wl^2}{8} = -M_{S(CD)}$$

Approximate moments

At ends,

$$M_{AB} = \frac{1}{2}\left(-\frac{wl^2}{12}+0\right) = -\frac{wl^2}{24}$$

$$= -\frac{10 \times 5 \cdot 05^2}{24} = -10 \cdot 63 kNm$$

$$= -M_{BA} = -M_{DC} = M_{CD}$$

In span,

$$M_{S(AB)} = \frac{1}{2}\left(\frac{wl^2}{24}+\frac{wl^2}{8}\right) = +\frac{wl^2}{12}$$

$$= +\frac{10 \times 5 \cdot 05^2}{12} = +21 \cdot 25 kNm = -M_{S(CD)}$$

The approximate moments as obtained above and those obtained from the relatively more rigorous solution of Example 27.4 are compared in Table 25.2:

From the results as tabulated, it can be seen that the moments from the approximate solution are within 20 per cent of those obtained by the more rigorous approach. This would suggest that, at the preliminary design stage, even without attempting to make up the shortfall, a sensible use of the approximate approach may yield satisfactory results where the members can be proportioned on the basis of the forces yielded by the simplified determinate system. In consonance with the 'Lower Bound Theorem', this assumes (Fraser, 1981) that the elements of a structure proportioned on the basis of a simple equilibrium force field within the actual, more complex force field, would be safe or, at worst, at their inherent limits of safety. While this may be true in the case of a simple structure such as the one in Example 27.4 where the lower bound theorem can be used with a reasonable degree of confidence, it cannot be generalized and care should be exercised where a significant redistribution of moments can be expected to occur due to differential deformation of the support elements such as that between the relatively more stiff perimeter walls and considerably less stiff discrete columns in a relatively complex metro station box structure.

Elastic supports

In the case of metro structures, generally, the outer support walls are, typically, an order of magnitude stiffer than the corresponding intermediate individual columns.

Table 25.2 Comparison of Moments

Solution	M_{AB}	$M_{S(AB)}$	M_{BA}	M_{DC}	$M_{S(DC)}$	M_{CD}
Rigorous	−13·29	+18·58	+13·29	+10·29	−21·58	−10·29
Approximate	−10·63	+21·25	+10·63	+10·63	−21·25	−10·63
Variation (%)	−20	+14.4	−20	+3·3	−1·5	+3·3

If the roof slab of such a structure were to be analysed on its own, the differentially stiff support system thus existing can be modelled by treating the supports as elastic springs with known spring stiffnesses. Axial spring stiffness, t, can be typically defined as the force required to cause unit axial deformation and can be determined from the load-displacement characteristics of the support element. However, the extents of the displacement, Δ, of the elastic supports are not known and cannot be determined in advance since the displacement is a function of the magnitude of the reactive force. The force, in turn, cannot be determined until the analysis of the structure is completed. Consequently, the determination of the compatible displacements and reactions becomes an inherent part of the analytical process itself.

The analysis of a continuous beam supported on an elastic spring can be carried out by using the method of successive approximations. To start off with, the approximate value of the initial displacement can be arbitrarily assumed and the corresponding reaction estimated. The new value for the approximate displacement can then be obtained from the formula $\Delta = R/t$, where R is the reaction. With this value of the displacement, a corresponding new value of the reaction is obtained. The successive cycles of approximations are continued until convergence is achieved yielding the required compatible values for the displacement and the reaction. Unless the initial guess happens, fortuitously, to be close to the real value, this process can be far too long. Computation can be relatively shortened by assuming the beam, initially, as rigidly supported and the corresponding reaction estimated. The approximate value of the initial displacement can then be obtained using this value of the reaction in the above formula and, subsequently, carrying out successive cycles until convergence is achieved. In any case, with either approach, the procedure of successive approximations can generally be too long.

Alternatively, a relatively quicker procedure involving the principle of superposition can be used. In accordance with this procedure, for a beam with one elastic support, two separate moment distribution analyses are carried out; the first one takes account of the effect of the applied loads on the structure assuming rigid, unyielding supports, and the second, the effect of arbitrary amount of joint displacement at the elastic support. In order to modify the arbitrary joint displacement to reflect the correct value, a correction factor is needed which is obtained from the equilibrium equation. This is then followed by an algebraic summation of the results of the two analyses obtained incorporating the appropriately corrected moments from the joint displacement case giving the final moments in the beam.

In general, in addition to the one distribution for the unyielding supports, there will be as many moment distribution analyses required to be carried out as the number of elastic supports and a corresponding number of the equilibrium equations which need to be solved simultaneously, although the conditions of symmetry may reduce the number of equations. The process is illustrated in the solved examples at the end of Chapter 26.

If the end span *AB* of a roof slab of a metro structure at its junction with the perimeter wall is deemed neither fully restrained nor fully free to rotate, the support can be considered to be partially restrained. The elastic restraint to reflect the boundary condition of the joint *A* can then be modelled by introducing the appropriate rotational spring stiffness, r, for the support element. This spring stiffness represents the moment required to produce unit rotation and can be obtained from the moment-rotation characteristics of the support element. At joint A, the distribution factor for the span AB of the slab can be obtained from the expression:

$$DF_{ab} = K_{ab}/(K_{ab} + r) \tag{25.50}$$

where, K_{ab} is the absolute rotational stiffness ($4EI/L$) for the span AB of the slab. Any moment imbalance set up at A due to the carry-over effect from B needs to be distributed into AB in proportion to the factor given in Equation (25.50). This will, in turn, cause a commensurate carry-over to support B triggering a new cycle of distributions and carry-overs. The process is continued until the completion of the moment distribution operation.

25.5 Applied loadings

Lateral earth pressures often constitute the dominant applied loads on cut-and-cover metro structures. Overestimation of such loads can lead to over-sized structural elements resulting, potentially, in an expensive structure. Underestimation, likewise, could easily lead to undersized members, resulting, at the very least, in serviceability-related problems or, potentially, an unsafe structure. Striking the right balance is important.

Besides, it is also important to remember that earth-pressure-related loads are not unique; they are dependent upon a number of factors, such as, the type of soil, its stress history, drainage characteristics, stiffness of and the strain undergone by the ground and wall support systems, etc. The loads also change with the changes in the geometry and boundary conditions of the structure at the different stages of construction. Therefore, in the interests of safety of the structure and to ensure that the serviceability-related problems which could be extremely expensive to put right, are avoided at all costs, it may be prudent, where so warranted, to carry out parametric studies to establish how sensitive the design might be to variations in the loadings resulting from lateral earth pressures.

As discussed in Chapter 13, the use of limit equilibrium theory or empirical design methods may significantly underestimate the lateral pressures in the upper reaches of the walls, particularly in rigidly braced cuts in deep layers of soft clay. It is suggested (Hashash and Whittle, 2002) that the lateral pressures can exceed the initial earth-at-rest pressures in the retained soil at elevations above formation level. Where containment of lateral wall movements embedded in soft clay is critical, necessitating the use of stiff (diaphragm) wall and rigid lateral bracing, the use of nonlinear finite element analysis for estimating lateral pressures should not be ruled out.

26 Structural Analysis: I

26.1 Introduction

Chapters 4 and 5 discuss the various methods and sequences that can be used for the construction of cut-and-cover metro structures and outline the considerations for design appropriate to each case. It would be beneficial for the reader, at this stage, to recall the contents of these chapters in order to relate to and better appreciate the salient aspects of the analysis of such structures as outlined in this chapter.

The single, most obvious feature that distinguishes a typical cut-and-cover structure from structures aboveground is, clearly, its confinement underground. In the case of an aboveground structure of comparable height, gravity loads represent the principal loads, and the stability of the structure depends, entirely, upon the stiffness and the strength of the structure itself. However, in the case of an underground structure, the surrounding ground, owing to its intimate contact with the structure, not only generates lateral soil pressures and mobilizes resistance constituting the dominant loadings on the structure, it also represents the medium providing the principal stability to it. How the changes in the boundary conditions and the loadings corresponding to the changing geometry of the structure come about during the various stages of excavation and construction and what influence these can have on the build-up of stress-resultants in the structural elements need to be fully appreciated. Experience has shown that, with the exception of those structures constructed conventionally in an open cut within stabilized side slopes, in the analysis and design of cut-and-cover metro structures, these factors invariably play a dominant role. It is therefore imperative that in the analysis of such structures such potential changes and their implications are fully appreciated and appropriately addressed.

It is equally important not to lose sight of the fact that, in view of the idealization of the geometry of the problem, empiricism of the loadings and the design parameters adopted, and the inherent approximations and assumptions, the degree of analytical sophistication to be aimed at needs to be tempered by the degree of accuracy warranted.

The key factors which can generally have significant impact on the analysis and design of cut-and-cover metro structures are:

- Response of the support system;
- Method of construction;
- Sequence of construction.

26.2 Response of support system

Apart from the particular nature of the applied loadings, the manner in which these are applied and the boundary conditions of the structure involved, perhaps the single most important feature that sets apart the behaviour of a deep-underground structure from that of an aboveground structure of a comparable size is the differential response of the support system and its effect on the stress-resultants of the structure. In an aboveground structure supported on columns, conventionally, the relative column stiffnesses reflect the share of the respective floor loads that they are called upon to carry. In other words, the intermediate columns carrying larger proportion of the dominant loads are proportionally stiffer than the perimeter columns carrying less. As a result, the tendency for the perimeter and the intermediate columns towards differential axial deformation is minimal and so the floor slabs behave as if they are 'rigidly' supported.

In the case of an underground metro structure, on the other hand, where, of necessity, the outer support structure has to be a relatively thick continuous perimeter wall, the behaviour is vastly different. The axial stiffness of the appropriate length of the perimeter support wall, also enhanced by the mobilization of the frictional resistance at the soil–structure interface, can easily be an order of magnitude higher than that of the corresponding individual, intermediate, support column. Notwithstanding this, the proportion of the floor loads required to be carried by the former is far less than that of the latter. Therefore, unlike the response of a structure on rigid supports, the supports in this case, as is to be expected, behave like springs with unequal stiffnesses subjected to a disproportionate distribution of loads. As a result, the columns are likely to undergo more axial deformation, elastically, than the perimeter walls. This would, inevitably, result in a vastly modified deflected profile leading to a significant redistribution of the stress-resultants in the structure. This is demonstrated in the solution to Example 26.1.

Furthermore, in a metro structure, if a beam-slab type of floor construction is adopted, the necessity to run the ventilation ducts across through the drop-beams imposes additional design constraints which can often lead to structural complexities. To keep such interaction problems to a minimum, floor slabs without drop beams, i.e. a flat-slab type of floor construction, is generally preferred. However, in that case, if the potential problems of shear in the floor slabs around the columns and the use of inordinate amount of shear reinforcement are to be avoided, the columns need to be sensibly sized and reasonably spaced.

It could be argued that, other things being equal, increasing the size of the column sections and reducing their spacing would, along with increasing the axial stiffness, also increase their flexural stiffness thereby attracting proportionally higher moments and shear forces. However, this need not be alarming when it is appreciated that the intermediate columns in an underground metro structure tend to be, principally, compression members subjected only to small levels of flexure. This is essentially because the rotational stiffnesses of the horizontal slabs are generally many orders of magnitude higher than those of the discrete column elements supporting them. As a result, the slab elements attract bulk of the unbalanced moments leaving very little to be carried by the columns themselves. The other factors, which also ensure that the column supports are subjected to no more than modest levels of flexure, can be listed as follows:

- Basic structural symmetry;
- Minimal out-of-balance lateral forces;

- Relatively higher values for the rotational stiffness of the perimeter walls;
- Rotational restraint offered by the friction mobilized at the soil interface around the perimeter of the structure.

In view of the foregoing, it can be concluded that, from a structural standpoint, the advantage of an increase in the axial stiffness of the columns far outweighs any adverse consequential effect of the corresponding increase in their rotational stiffness.

26.3 Method of construction

An underground metro structure may be constructed by the conventional 'bottom-up', the 'top-down', or a hybrid method of construction representing some combination of the two. Whichever be the method adopted, the structure is required to satisfactorily withstand all the loads to which it is likely to be exposed during its operational lifetime. Besides, it must also satisfy any additional design requirements that may arise during the various stages of construction. The analytical process adopted must also take into account changes in the geometry, boundary conditions and loadings to be expected at the various stages and the design of the structure must appropriately reflect this.

In the 'bottom-up' method of construction, the structural problems related directly to design are essentially conventional and relatively simple, and the additional requirements related to the method of construction are potentially minimal. This makes the design generally efficient. However, underground metro structures are usually located in high-density urban environments and often directly below busy thoroughfares. The need to limit the movements of the surrounding ground and the existing structures in the vicinity may therefore dictate the use of 'top-down' method of construction. This, almost invariably, has a predominant influence on design. It may, for instance, impose severe constraints on, and limit the choice of, the ground support and the temporary wall support systems to be adopted. It may also place additional demands on design in terms of having to build into the structure specific extra provisions. With the completion of the construction, however, the usefulness of the extra provisions, dictated by the specific method of construction and built into the structure, ceases and is rendered redundant for the remainder of its operational lifetime. This, therefore, has the potential of making the design less cost-effective and less efficient. However, this cannot always be helped.

In the 'top-down' method, temporary horizontal struts are generally positioned just enough clear of the respective floors to enable subsequent construction of the slabs to be carried out relatively unhindered. When the floor slabs are constructed and become operational also as permanent struts, the temporary struts can be removed. However, as the temporary struts are removed, the load hitherto borne by them is transferred on to the floor slabs as they assume the function of the permanent struts. The axial deformation of the floor slabs resulting from such transfer and assumption of loads can also subscribe to some inward movements of the perimeter walls which can modify the stress-resultants in them. The effects of such load transfers should therefore be duly assessed and allowed for in the design.

However, one way of avoiding altogether the effects of load transfer referred to above is to so locate and design the temporary struts that they need not be removed but may be built-in and accommodated as integral parts of the permanent structure. Where the thickness of the floor slab is $1m$ or more, for instance, this can be easily achieved by positioning the struts within the space between the top and the bottom mats of the slab reinforcement. Where this is not practical, and the struts need to be removed when no longer necessary, they should be so located as to avoid interference with the subsequent construction of the permanent floor slabs. In that case, however, the effect of the design peak moments and shear forces corresponding to the temporary support condition and the change in their magnitude and the shift in their location occasioned by the removal of the temporary struts and the installation of the permanent floors must be duly considered in building up the envelopes for the stress-resultants and allowed for in the design process.

It is also important to recognize the difference between the respective nature and location of the 'softer' boundary conditions when the formation level for a particular floor level is reached and the relatively 'harder' boundary condition offered by the installation of the temporary struts or after the floor slab is in place and operational as the excavation is advanced downwards. Allowances for their respective effects on design need to be made as appropriate.

26.4 Sequence of construction

The timing at which the temporary supports are removed, the manner of their removal and the sequence in which the various elements of the structure are constructed and subjected to the applied loads can modify the profiles of the stress-resultants to such an extent as to constitute a major design criterion. Whether the permanent vertical supports are constructed before or after the floor slabs are in place and, if after, the location of the temporary *vis-à-vis* those of the permanent supports, has an important influence on the design of the structure.

Consider a metro structure constructed within a nonintegral outer perimeter cofferdam by the 'top-down' method. Assume that, besides the supports provided by the cofferdam at the ends, there is also a single, longitudinal line of continuous temporary support along the centre-line of the structure. Assume also that the roof slab is constructed on ground in two halves in such a manner that it is discontinuous over the central temporary support and simply supported at the far ends. Subsequently, as the soil underneath is removed, the two halves of the slab are free to take up, under their own weight, the simply supported deflected profiles as they span between the central temporary support and the outer cofferdam walls on either side. In the process, dead load moments due to the self-weight of the slab become locked-in. Since the self-weight constitutes a substantial proportion of the total load, the level of the locked-in moments can be rather significant. This is also particularly so in this case because no relief in the free span moments is available owing to the absence of structural continuity over and across the central temporary support at this stage.

Subsequent construction of the permanent supports, i.e. perimeter walls and columns, structural integration of the two halves of the slab and the removal of the central temporary support alter the geometry and the boundary conditions of the structure. As the central temporary wall is removed and the permanent columns assume the support function in its place, the self-weight moment profile obtained hitherto gets modified in the light of the subsequent loading. In other words, any superimposed dead and live loads to which the structure may be subjected thereafter will induce additional stress-resultants based on the new geometry and boundary conditions, i.e. continuous slab structure. The implications of all these effects need to be addressed in design and are illustrated in the solved examples later on in the chapter.

Notwithstanding the foregoing, with careful consideration to detail, it is possible to neutralize some of the locked-in effects, for example, by permanently jacking the slabs up at the column locations so that the boundary conditions approximate, as closely as possible, those of the permanent in-service case. However, the sequence of jacking and its implications on design need to be carefully considered.

Moments are also enhanced in the vicinity of the temporary 'muck-out' holes where the adjacent trimming strips are called upon to carry extra share of the loads. The problem becomes further exacerbated where such holes adjoin the permanent discontinuities in the floor slabs such as those introduced by escalators and stairs.

Given the diversity in the sizes and stiffnesses of the various structural elements, the analysis and design of a metro structure can be sensitive to the method and sequence of construction adopted. Through a judicious selection and control of the sequence of construction and the location of the temporary line(s) of support, it may be possible to keep the deviations from the permanent geometry of the structure and the additional requirements arising out of the chosen sequence of construction to a minimum thereby avoiding a wasteful and less efficient design.

However, it may not always be possible to achieve 100 per cent efficiency in design; a certain amount of inefficiency may be inevitable. Nevertheless, at the planning and design stages, a conscious effort should be made to ensure that as much of the temporary works as possible can be incorporated into or accommodated as part of the permanent structure; more so, if it is doubtful that they can be sensibly salvaged for reuse. For example, as mentioned earlier, steel struts, if located at mid-depths of floor slabs clear of the top and bottom mats of reinforcement, could be concreted in. Likewise, temporary steel stanchion supports, if appropriately sized and located, could be concreted in as part of the permanent column supports. In this manner, the temporary supports with loads locked-in would remain undisturbed and stay in place permanently. Such an approach, therefore, can largely obviate the need to deal with the implications of load transfer.

In order to avoid serious and unexpected surprises during construction, the temporary support system should be so chosen and the sequence of construction so adopted as to possess, inherently, a degree of flexibility to adapt to changes should they become necessary. It is also prudent to give consideration to secondary measures to fall back on which can be put into commission without delay should the need arise. However, it is important that the implications of such measures are fully anticipated and thought through at the design stage. It is equally important that the necessary provisions are built-in in time as appropriate and not treated as an after-thought during construction. Otherwise, lack of adequate forethought could find the structure wanting and, at a late stage, unable to offer accommodation to the required changes. As a result, the required modifications could prove to be prohibitively expensive.

In order to appreciate, in principle, the significance of the various factors discussed in the foregoing section while keeping the computations relatively simple, in Examples 26.1 to 26.3 inclusive only the roof slab of a typical box structure is considered and the axial compression in the slab is ignored. Besides, for its great appeal to the engineers in enabling each step of the computation to be interpreted readily in terms of the behaviour of the structure being analysed, simple method of moment distribution has been employed. Since the basis of this method can be grasped easily from the physical point of view, it can help the analyst in developing a 'feel' for the behaviour of the structure right from the application of the load to the derivation of the stress-resultants induced in it.

Furthermore, for simplicity and achieving a significant economy in the hand computations, wherever applicable, symmetry of the structure has been used and the principle of symmetry and skew-symmetry of the applied asymmetric loading invoked to analyse only half the structure. The use of this approach is demonstrated through a solved example at the end of the following chapter.

The effects of axial compression on the stiffness, carry-over factors and the fixed end moments have already been dealt with in the previous chapter.

26.5 Distortion of box structure

Distortion is a phenomenon commonly associated with box or cellular-type structures. Any tendency on the part of the structure to distort has to be withstood by the structure of the box itself bringing into play the rotational or translational stiffness of the elements of the structure. Potentially, distortion can be caused by the tendency of the box structure to sway transversely. This can be brought about by one or more of the following factors:

- Asymmetry in the structure;
- Asymmetry in the applied loading;
- Change in the geostructural environment.

Typically, asymmetry in a cut-and-cover metro structure may be inherent in the structure which may be manifest even during the stages of construction. Alternatively, it may be imposed on the structure by the sequence of construction adopted at some stage while still under construction. In both the cases, it could very often be due to the asymmetry in the geometry of the structure. However, distortion in a box structure can result even with a symmetric geometry, for instance, if it is carried on an asymmetric foundation system. An example illustrating this case would be that of a box structure founded on piles end-bearing into bedrock with the profile of the rock horizon varying significantly from one end of the structure to the other leading to dissimilar pile lengths. If all the piles used were to be of the same diameter and disposed on a uniform grid, as the practicality of construction would demand, the variation in their lengths would result in a foundation system which would be differentially stiff from one end to the other. Such a differentially stiff foundation system could subscribe to the distortion of the structure even under symmetric loading unless, of course, the pile-grid were to be modified progressively so as to provide a uniformly stiff foundation system.

If the structure together with its support system is symmetric in every respect, the distortion of the box structure could still occur if the loading it is subjected to happens to be asymmetric. Asymmetry of the loading, like that of the structure itself, may also be inherent, as in the case of differentially applied permanent lateral surcharge load across the structure; or, it may be imposed during construction, as in the case of an uneven backfilling of soil against the opposite walls of the structure causing a temporary load differential.

Change in the geostructural environment due to new construction activity adjacent to and on one side of the structure setting up a differential effect across the structure can also cause distortion of the box structure. For example, if the new activity necessitates either excavation resulting in the removal of lateral soil support to, or pile driving causing lateral surcharge loads on, one side of the structure, a horizontal force imbalance would be set up transversely. This could cause sway of the box structure leading to its distortion.

It may neither be practical nor indeed necessary to eliminate distortion of the box structure altogether. Nevertheless, during the planning of the form and finalizing of the sequence of construction of the structure, every attempt should be made, as far as practicable, to minimize it. By controlling factors leading to distortion, it may be possible to achieve a measure of success in this direction.

Distortion of the box structure can also be caused due to ground shaking under a seismic activity; since such an activity cannot be controlled, it must be taken on board in design. Assessment of loads from a seismic activity are dealt with in Chapter 15 and Appendix F.

Example 26.1

Figure 26.1 shows a typical part floor plan of the roof slab of a metro structure. The slab is 1300*mm* thick and carries a 3·5*m* of backfill on top. The perimeter walls are 1000*mm* thick. Ignoring the presence of any intermediate level slab and assuming columns to be, typically, 1100*mm* x 900*mm* in cross-section and 10*m* high, draw the shear force and bending moment diagrams for the slab:

- Assuming idealized structure for analysis as shown in Figure 26.2;
- Ignoring surcharge effect from the traffic load on the surface;
- Assuming the supports to be unyielding;
- Allowing axial deformation of the supports.

Take unit weight of concrete as $24 kN/m^3$ and its E-value as $28 \times 10^6 kN/m^2$.

Solution

The shaded area represents the typical floor strip for analysis. For the purpose of this exercise, restraints against rotation are assumed at the junctions with the end walls. In reality, however, some rotation of the end joints, i.e. at the slab-wall junctions, can

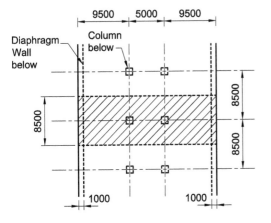

Figure 26.1 Part Plan of Typical Roof Slab

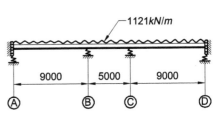

Figure 26.2 Idealized Structure for Analysis

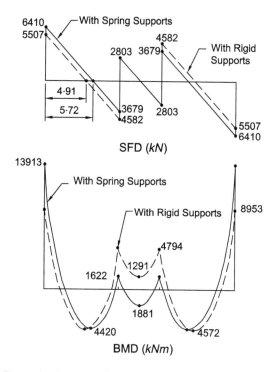

Figure 26.3 SFD and BMD for the Idealized Structure

be expected. Also, because of the symmetry of the structure and the loading, only half the structure will be analysed using appropriate member stiffnesses. For the sake of simplicity in this exercise and being a very small proportion of the total design load in any case, traffic load on the surface has been disregarded.

Using centre-line dimensions, consider the idealized structure as shown in Figure 26.2.

Spring stiffnesses at supports

For columns at B and C, area $A_C = 1 \cdot 1 \times 0 \cdot 9 = 0 \cdot 99$, say, $1m^2$

$$\text{Spring stiffness, } t_B(\text{or } t_C) = \frac{A_C \cdot E}{L} = \frac{1 \times 28 \times 10^6}{10} = 2 \cdot 8 \times 10^6 \, kN/m$$

For the perimeter walls, area $A_W = 1 \times 8 \cdot 5 = 8 \cdot 5m^2$

$$\text{Spring stiffness, } t_A(\text{or } t_D) = \frac{A_W \cdot E}{L} = \frac{8 \cdot 5 \times 28 \times 10^6}{10} = 8 \cdot 5 t_B kN/m$$

Allowing for the confined nature of the structure and soil friction at the walls, assume

$$t_A = 10 t_B = 28 \times 10^6 kN/m$$

Relative stiffness of slab

For span $AB = 1/9$

For span $BC = (1/2) \times$ Stiffness – for symmetry of structure

$= (1/2) \times 1/5 = 1/10$

Distribution factors

At Joint B, for Span $BA = \dfrac{1/9}{1/9 + 1/10} = 0.53$

Span $BC = 1 - 0.53 = 0.47$

Loading

Own weight slab $= 1.3 \times 8.5 \times 24 = 265.2 kN/m$
Weight of backfill $= 3.5 \times 8.5 \times 18 = 535.5 kN/m$
Total characteristic load $= 265.2 + 535.5 = 800.7 kN/m$
Ultimate Design Load $= 1.4 \times 800.7 = 1,121 kN/m$

Due to load alone

Fixed-end moments: (Supports assumed rigid)

Due to load, $m^f_{AB} = -1121 \times 9^2/12 = 7,566.75 kNm = -m^f_{BA}$

$m^f_{BC} = -1121 \times 5^2/12 = 2,335.42 kNm = -m^f_{CB}$

Stress-resultants

Joint	A	B	
member	AB	BA	BC
DF		0.53	0.47
m_o	−7567	+7567	−2335
Distribution	−1386	−2773	−2459
M_o (kNm)	−8953	+4794	−4794
Static shear	+5045	−5045	+2803
Elastic shear	+462	+462	0
$\sum$ Shear (kN)	+5507	−4582	+2803
Reaction, R_o (kN)	↑ 5507	7385 ↑	

Due to notional differential settlement of supports (A and D)

Fixed-end moments

Initially, a notional settlement of $0 \cdot 001m$ each is assumed at supports A and D.

$$I_{slab} = \frac{1}{12} \times 8 \cdot 5 \times 1 \cdot 3^3 = 1 \cdot 56m^4$$

Due to differential settlement, $m^f_{AB} = +6EI\Delta/L^2$

$$= +6 \times 28 \times 10^6 \times 1 \cdot 56 \times 0 \cdot 001/9^2 = +3,235 \cdot 56kNm = m^f_{BA}$$

Stress-resultants

Joint	A	B	
member	AB	BA	BC
DF		0·53	0·47
m_1	+3236	+3236	0
Distribution	−857	−1715	−1521
M_1 (kNm)	+2378	+1521	−1521
Elastic shear	−433	−433	0
Reaction, R_1 (kN)	↓ 433	433 ↑	

Due to notional differential settlement of supports (B and C)

Fixed-end moments

Again, initially, a notional settlement of $0 \cdot 001m$ each is assumed at supports B and C.

$m^f_{AB} =$ as above but with sign reversed $= -3,235 \cdot 56kNm = m^f_{BA}$

$m^f_{BC} = 0$

Stress-resultants

Magnitude of the stress-resultants will be the same as that for the previous differential settlement case but with the signs, or directions, reversed.

Equilibrium equations

Let α and β be the correction factors for the notional differential settlements at the supports A, D and B, C, respectively. Then the equilibrium equations will be given by:

$$R_A = R_{O_A} + \alpha R_{1_A} + \beta R_{2_A} = \alpha \Delta_A t_A \qquad (26.1)$$

$$R_B = R_{O_B} + \alpha R_{1_B} + \beta R_{2_B} = \beta \Delta_B t_B \qquad (26.2)$$

On substituting the values, we get:

$$5506 \cdot 6 + \alpha(-433 \cdot 21) + \beta(433 \cdot 21) = 0 \cdot 001 \times 28 \times 10^6 \alpha \qquad (26.3)$$

$$\text{and } 7384 \cdot 9 + \alpha(433 \cdot 21) + \beta(-433 \cdot 21) = 0 \cdot 001 \times 2 \cdot 8 \times 10^6 \beta \qquad (26.4)$$

Whence $\alpha = 0 \cdot 2289$ and $\beta = 2 \cdot 3147$

$$\therefore R_A = 0 \cdot 2289 \times 0 \cdot 001 \times 28 \times 10^6 = 6,409 \cdot 2 kN \text{ and}$$

$$R_B = 2 \cdot 3147 \times 0 \cdot 001 \times 2 \cdot 8 \times 10^6 = 6,481 \cdot 16 kN$$

$$\textbf{Check: } 2(R_A + R_B) = 2(6409 \cdot 2 + 6481 \cdot 16) = 25,781 kN$$

$$\sum \text{Load} = 1121(2 \times 9 + 5) = 25,783 kN - \text{ Close enough!}$$

Final stress-resultants: (With spring supports)

Ultimate shear forces:

$$S_{AB} = 5506 \cdot 6 + 433 \cdot 21(2 \cdot 3147 - 0 \cdot 2289) = 6,410 \cdot 19 kN$$

$$S_{BA} = -4582 \cdot 4 + 433 \cdot 21(2 \cdot 3147 - 0 \cdot 2289) = -3,678 \cdot 81 kN$$

$$S_{BC} = 2,802 \cdot 5 kN$$

Ultimate bending moments:

$$M_{AB} = -8953 \cdot 05 + 2378 \cdot 14(0 \cdot 2289 - 2 \cdot 3147) = -13,913 \cdot 37 kNm$$

$$M_{BA} = +4794 \cdot 15 + 1520 \cdot 71(0 \cdot 2289 - 2 \cdot 3147) = 1,622 \cdot 25 kNm$$

Ultimate span moments:

$$M^s_{AB} = 0 \cdot 5 \times 6410 \cdot 19 \times 5 \cdot 72 - 13913 \cdot 37 = 4,419 \cdot 77 kNm$$

$$M^s_{BC} = 1121 \times 5^2/8 - 1622 \cdot 25 = 1,880 \cdot 88 kNm$$

Shear force and bending moment diagrams for the idealized structure are shown in Figure 26.3.

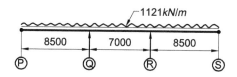

Figure 26.4 Idealized Structure

Example 26.2

Revisit the structure in Example 26.1 and assume that:

- The main structure is to be constructed within an outer cofferdam following a top-down sequence.
- The roof slab is initially supported temporarily on brackets (*P* and *S*) off the outer cofferdam walls and over 2 intermediate rows of steel H-piles (*Q* and *R*) as shown in the idealized Figure 26.4.
- Transversely, the brackets are 24*m* apart and the 2 rows of piles 7*m* apart centre to centre.

Compare the shear force and bending moment diagrams for the slab for both the rigid and the yielding support conditions and draw the design envelopes.

Solution

Ultimate design loading of 1,121*kN/m* run as obtained in Example 26.1 is considered.

(a) *Temporary case with rigid supports*

Using centre-line dimensions, consider the idealized structure as shown in Figure 26.4. Because of the symmetry in geometry and loading, only half the structure is analysed.

Relative stiffness of slab

$$\text{Span } PQ = \frac{3}{4} \cdot \frac{1}{8 \cdot 5} = \frac{3}{34}$$

$$\text{Span } QR = \frac{1}{2} \cdot \frac{1}{7} = \frac{1}{14} \text{ (for symmetry)}$$

Distribution factors

$$\text{At joint } Q: \text{ Span } QP = \frac{3/34}{3/34 + 1/14} = 0 \cdot 55$$

$$\text{Span } QR = 1 \cdot 00 - 0 \cdot 55 = 0 \cdot 45$$

Fixed-end moments: (Due to load alone)

$$m^f_{QP} = +1121 \times 8 \cdot 5^2/8 = +10,124 kNm$$

$$m^f_{QR} = -1121 \times 7^2/12 = -4,577 kNm$$

Stress-resultants: (*Due to load alone*)

Joint	P	Q	
member	PQ	QP	QR
DF		0·55	0·45
m_o	0	+10124	−4577
Distribution	0	−3051	−2496
M_o (*kNm*)	0	+7073	−7073
Static shear	+4764	−4764	+3924
Elastic shear	−832	−832	0
$\sum$ Shear (*kN*)	+3932	−5596	+3924
Reaction, R_o (*kN*)	↑ 3932	9520 ↑	

(b) Temporary case with yielding supports

Fixed-end moments: (*Settlement of supports A and D by 0·001m each*)

$$I_{slab} = 1 \cdot 56 m^4 \text{ as before}$$

Due to differential settlement, $m^f_{QP} = +3EI\Delta/L^2$

$$= +3 \times 28 \times 10^6 \times 1 \cdot 56 \times 0 \cdot 001 / 8 \cdot 5^2$$
$$= +1,814 kNm = m^f_{BA}$$

Stress-resultants: (*Due to notional differential settlement alone*)

Joint	P	Q	
member	PQ	QP	QR
DF		0·55	0·45
m_1	0	+1814	0
Distribution	0	−998	−816
M_1 (*kNm*)	0	+816	−816
Elastic shear	−96	−96	0
Reaction, R_1 (*kN*)	↓ 96	96 ↑	

In the case of supports Q and R, likewise, settling by the notional $0.001m$ each, same stress-resultants as above but with their directions, or signs, reversed will be obtained.

Equilibrium equations

$$R_P = R_{0_P} + \alpha R_{1_P} + \beta R_{2_P} = \alpha \Delta_P t_P \qquad (26.5)$$

$$R_Q = R_{0_Q} + \alpha R_{1_Q} + \beta R_{2_Q} = \beta \Delta_Q t_Q \qquad (26.6)$$

Assuming $t_Q = 2 \times 10^6 kN/m$, say, and $t_P = 20 \times 10^6 kN/m$, and on substituting the values in the above equations, we get:

$$3932 + \alpha(-96) + \beta(96) = 0.001 \times 20 \times 10^6 \alpha \qquad (26.7)$$

$$\text{and}\quad 9520 + \alpha(96) + \beta(-96) = 0.001 \times 2 \times 10^6 \beta \qquad (26.8)$$

$$\text{Whence }\alpha = 0.2174 \quad \text{and} \quad \beta = 4.5519$$

$$\therefore\ R_P = 0.2174 \times 0.001 \times 20 \times 10^6 = 4,348 kN$$

$$\text{and}\quad R_Q = 4.5519 \times 0.001 \times 2 \times 10^6 = 9,104 kN$$

$$\text{Check: } 2\left(R_P + R_Q\right) = 2(4348 + 9104) = 26,904 kN$$

$$\sum \text{Load} = 1121(2 \times 8.5 + 7) = 26,904 kN - \text{ Ok!}$$

Final stress-resultants: (With spring supports)

Ultimate shear forces:

$$S_{PQ} = 3932 + 96(4.5519 - 0.2174) = 4,348 kN$$

$$S_{QP} = -5596 + 96(4.5519 - 0.2174) = -5,180 kN$$

$$S_{QR} = +3,923.5 kN$$

Ultimate bending moments:

$$M_{PQ} = 0$$

$$M_{QP} = +7073 + 816(0.2174 - 4.5519) = +3,536 kNm$$

Ultimate span moments:

$$M_{PQ}^s = 0.5 \times 4348 \times 3.88 = 8,435 kNm \text{ (as computed from the SFD)}$$

$$M_{QR}^s = 1121 \times 7^2/8 - 3536 = 3,330 kNm$$

Shear force and bending moment diagrams for the idealized structure are shown in Figure 26.5.

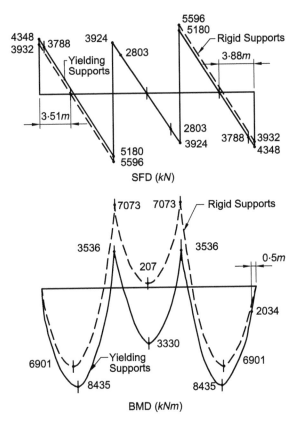

4348
3932
3788
3924
2803

Yielding
Supports

5596
5180
Rigid Supports
3·88m

3·51m

2803
3924
3788
3932
4348

5180
5596

SFD (*kN*)

7073   7073   Rigid Supports

3536   3536
0·5m

207

2034

3330

Yielding
Supports

6901   6901

8435   8435

BMD (*kNm*)

Figure 26.5 SFD and BMD for the Idealized Structure (slab with temporary supports in place)

(c) *In-service case*

Temporary supports are removed after the permanent supports are in place and operational. During this removal, loads will be transferred from the temporary on to the permanent supports. As a result, the stress-resultants previously obtained in the various elements of the structure will be modified. The removal of the temporary support, therefore, needs to be modelled carefully. This can be achieved by superimposing on the structure with the new permanent support system in place, downward load(s) equal in magnitude to the reaction(s) obtained from the previous temporary support condition. Shear forces and bending moments obtained from this case must be algebraically combined with those obtained from the temporary support condition, in order to arrive at the final stress-resultants.

Consider the idealized structure as shown in Figure 26.6.

Distribution factors: (*As those in Example 26.1*)

$$\text{At joint } B, \text{ for Span } BA = 0.53$$

$$\text{Span } BC = 0.47$$

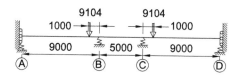

Figure 26.6 Idalized Structure and Loading

Fixed-end moments: (Supports assumed rigid)

Due to load, $m^f_{AB} = -9104 \times 8 \times 1^2/9^2 = -899 kNm$

$$m^f_{BA} = +9104 \times 8^2 \times 1/9^2 = +7,193 kNm$$

$$m^f_{BC} = 0$$

Stress-resultants: (Due to load alone)

Joint	A	B	
member	AB	BA	BC
DF		0·53	0·47
m_o	−899	+7193	0
Distribution	−1906	−3812	−3381
M_o (kNm)	−2805	+3381	−3381
Static shear	+1012	−8092	0
Elastic shear	−64	−64	0
$\sum$ Shear (kN)	+948	−8156	0
Reaction, R_o (kN)	↑948	8156 ↑	

Stress-resultants: (Due to notional differential settlement alone)

Stress-resultants due to the notional differential yielding of the various supports will be identical to those obtained in Example 26.1.

Equilibrium equations: (As in Example 26.1)

$$R_A = R_{O_A} + \alpha R_{1_A} + \beta R_{2_A} = \alpha \Delta_A t_A \qquad (26.9)$$

$$R_B = R_{O_B} + \alpha R_{1_B} + \beta R_{2_B} = \beta \Delta_B t_B \qquad (26.10)$$

On substituting the values, we get:

$$948 + \alpha(-433) + \beta(433) = 0.001 \times 28 \times 10^6 \alpha \qquad (26.11)$$

$$\text{and} \quad 8156 + \alpha(433) + \beta(-433) = 0.001 \times 2.8 \times 10^6 \beta \qquad (26.12)$$

Whence $\alpha = 0.0719$ and $\beta = 2.5324$

$$\therefore R_A = 0.0719 \times 0.001 \times 28 \times 10^6 = 2013 kN$$

$$\text{and} \quad R_B = 2.5324 \times 0.001 \times 2.8 \times 10^6 = 7091 kN$$

Check: $R_A + R_B = 2013 + 7091 = 9{,}104 kN = $ Load – Ok!

*Final stress-resultants: (**With spring supports**)*

Ultimate shear forces:

$$S_{AB} = 948 + 433(2.5324 - 0.0719) = 2{,}013 kN$$
$$S_{BA} = -8156 + 433(2.5324 - 0.0719) = -7{,}091 kN$$
$$S_{BC} = 0$$

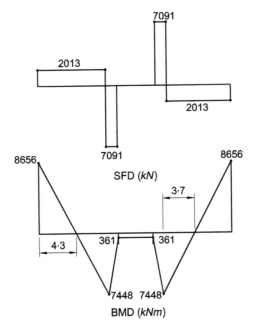

Figure 26.7 SFD and BMD for the Idealized Structure
(for Load Transfer Alone with Permanent Supports in Place)

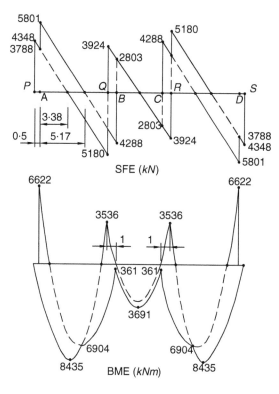

Figure 26.8 Final SF and BM Design Envelopes

Ultimate bending moments:

$$M_{AB} = -2805 + 2378\,(0 \cdot 0719 - 2 \cdot 5324) = -8,656kNm$$
$$M_{BA} = +3381 + 1521\,(0 \cdot 0719 - 2 \cdot 5324) = 361kNm$$

Ultimate span moment: (As computed from the shear force diagram)

$$M_{AB}^{s} = 2013 \times 8 - 8656 = 7,448kNm$$

Shear force and bending moment diagrams for the load transfer alone are shown in Figure 26.7, and the final design envelopes in Figure 26.8.

Example 26.3

For the part roof slab shown in Figure 26.1, outline a sequence of construction using 'top-down' method and adopting a suitable support system. List any assumptions made. Analyse the slab to obtain the stress-resultants and draw the shear force and bending moment diagrams.

Solution

Proposed sequence of construction

1. Install temporary perimeter cofferdam and longitudinal central temporary wall.
2. Construct roof slab either as two discontinuous halves either side of, or as one continuous slab over, the central temporary wall.
3. Excavate underneath the roof slab down to the underside of the floor level below. Construct the floor slab in two discrete halves supported on the outer cofferdam walls and off brackets either side of the central temporary wall. Advance the excavation down to the next level below and repeat the sequence until the base slab is constructed. Base slab will be permanently ground bearing and so no bracket support will be needed at this level.
4. Construct permanent columns and perimeter walls from base slab upwards in the bottom-up sequence.
5. Integrate the roof slab if constructed as discontinuous halves followed by the removal of one storey-height of the central temporary wall immediately below.
6. Repeat Step 5 at every slab level below except at the base slab.
7. At the base slab level, remove the central temporary wall still in place to sufficient depth below and backfill with sand (or some appropriate compressible material) up to the underside of the slab to remove the possibility of the formation of a hard spot underneath the slab when constructed and operational.
8. Integrate the base slab.
9. Backfill on roof, replace services and reinstate the road surface.

Assumptions

- The main structure is constructed within an outer perimeter cofferdam.
- During the 'top-down' sequence of construction, the slab is initially supported on three temporary supports – the outer longitudinal walls forming part of the temporary perimeter cofferdam, and the central wall.
- At the outer cofferdam walls, the slab is simply supported off brackets whereas over the central support, it is considered both 'discontinuous' and 'continuous', in turn, to illustrate the difference in their effect on the profile of the stress-resultants.
- Central temporary wall is axially twice as stiff as the outer temporary cofferdam walls.
- Spring stiffness of the permanent perimeter wall is 10 times that of a typical column.
- To illustrate the principle, only the roof slab is considered and analysed. For simplicity and keeping the hand calculations manageable, contribution from other floors is ignored.

Analysis

Analysis is carried out for three discrete stages:

- Stage I: Construction of the roof in two discontinuous halves (*a*) or as one continuous slab (*b*) and subsequent excavation of the ground below it such that the entire slab is suspended on the three temporary supports

- Stage II: Installation of the permanent supports (perimeter wall and columns), integration of the two halves of the slab as appropriate and removal of the central temporary support wall for the discontinuous (*a*) and the continuous (*b*) cases
- Stage III: Reinstatement of the backfill and completion.

Stage I(a): (Construction and excavation: 'discontinuous' slab case)

In Figure 26.9(*a*), T_1, T_2 and T_3 represent the temporary supports, whereas *A, B, C* and *D* indicate the locations of the eventual permanent supports.

Characteristic load from self-weight $= 265 \cdot 2 kN/m$ (from Example 26.1)
Ultimate design load $= 1 \cdot 4 \times 265 \cdot 2 = 375 kN/m$
Ultimate shear force $= 375 \times 6 = 2,250 kN$ (at central wall)
Ultimate reaction $= 2250 \times 2 = 4,500 kN$ (at central wall)
Ultimate SF at perimeter wall location $= \dfrac{5 \cdot 5}{6} \times 2250 = 2,062 \cdot 5 kN$
Ultimate SF at column location $= \dfrac{3 \cdot 5}{6} \times 2250 = 1,312 \cdot 5 kN$
Ultimate span moment $= 375 \times 12^2/8 = 6750 kNm$ (temporary mid-span)
Ultimate BM at perimeter wall location $= 0 \cdot 5 (2250 + 2062 \cdot 5) \times 0 \cdot 5 = 1,078 kNm$

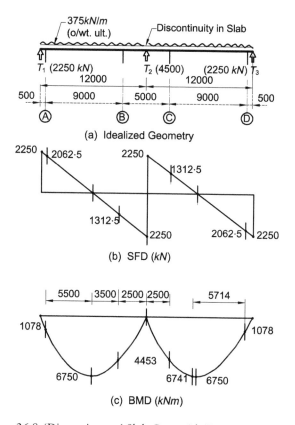

(a) Idealized Geometry

(b) SFD (*kN*)

(c) BMD (*kNm*)

Figure 26.9 'Discontinuous' Slab Case with Temporary Supports

Ultimate BM at column location$= 0.5(2250+1312.5) \times 2.5 = 4,453kNm$
For shear force and bending moment diagrams, see Figure 26.9 (*b* and *c*).

Stage I(b): (Construction and excavation: 'continuous' slab case)

Since the stiffnesses of the temporary supports are approximately proportional to the respective loads that they share, the supports can, without any significant loss in accuracy, be virtually treated as 'rigid'. In Figure 26.10(*a*), T_1', T_2' and T_3' represent the temporary supports, whereas *A*, *B*, *C* and *D*, as before, indicate the locations of the permanent supports.

Ultimate design load $= 375kN/m$ (as before)
Ultimate BM at central temporary wall location$= 375 \times 12^2/8 = 6,750kNm$
Ultimate SF at central temporary wall location $= 375 \times 6 + 6750/12 = 2,812.5kN$
Ultimate SF at temporary cofferdam location$= 375 \times 6 - 6750/12 = 1,687.5kN$
Ultimate reaction at central temporary wall location $= 2 \times 2812.5 = 5,625kN$
Point of zero shear $= 1687.5/375 = 4.5m$, as measured from the cofferdam
   support. This is also the point of maximum span moment.

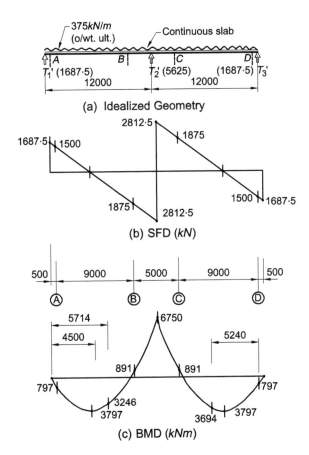

(a) Idealized Geometry

(b) SFD (*kN*)

(c) BMD (*kNm*)

Figure 26.10 'Continuous' Slab Case with Temporary Supports

Ultimate SF at perimeter wall location$= \dfrac{4}{4 \cdot 5} \times 1687 \cdot 5 = 1,500 kN$

Ultimate SF at column location $= \dfrac{5}{7 \cdot 5} \times 2812 \cdot 5 = 1,875 kN$

Ultimate BM at perimeter wall location$= 0 \cdot 5 \,(1687 \cdot 5 + 1500) \times 0 \cdot 5 = 797 kNm$
Maximum ultimate span moment $= 0 \cdot 5 \times 1687 \cdot 5 \times 4 \cdot 5 = 3,797 kNm$
Ultimate BM at column location $= 1687 \cdot 5 \times 9 \cdot 5 - 375 \times 9 \cdot 5^2 /2 = -891 kNm$

Shear force and bending moment diagrams are shown in Figure 26.10(*b* and *c*).

Stage II(a): (Support removal and load transfer: 'discontinuous' slab case)

When the permanent support system is in place and operational and the central temporary wall removed, load will be transferred from the temporary to the permanent supports. This removal of the central temporary wall can be mathematically modelled by superimposing in its place at the centre a point load equal in magnitude but opposite in direction to the central temporary reaction$(4500 kN)$, on the frame with the permanent supports in place as shown in Figure 26.11(*a*). Fixed-end moments

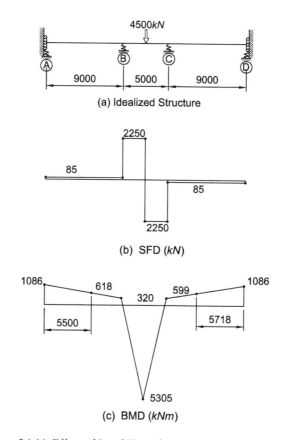

Figure 26.11 Effect of Load Transfer 'Discontinuous' Slab Case

under this load will be given by:

$$m^f_{BC} = -4500 \times 5/8 = -2{,}812{\cdot}5 kNm$$

Note that, whereas the central temporary wall has to be removed to accommodate the structure, the 'temporary' cofferdam does not have to be so removed. In any case, it would be impractical to achieve its removal and so, although not needed, it will continue to remain in place permanently. The supports provided by it at the ends will therefore also remain in place and untouched. As a result, the end reactions from the temporary support case, i.e. at T_1 and T_3, can be assumed to remain 'permanently' locked-in.

Owing to the symmetry of the structure and the applied loading, only half the structure is analysed.

- Stress-resultants: (Rigid supports)

Joint	A	B	
member	AB	BA	BC
DF		0·53	0·47
m_0	0·0	0·0	−2812·5
Distribution	+745·3	+1490·6	+1321·9
M_0 (kNm)	+745·3	+1490·6	−1490·6
Elastic shear, S_0	−248·4	−248·4	+2250·0
Reaction, R_0 (kN)	↓ 248·4	2498·4 ↑	

- Stress-resultants: (Elastic supports)

Stress-resultants due to the notional yielding of supports will be identical to those worked out under Example 26.1.

Equilibrium equations: (As in Example 26.1)

$$R_A = R_{O_A} + \alpha R_{1_A} + \beta R_{2_A} = \alpha \Delta_A t_A \qquad (26.13)$$

$$R_B = R_{O_B} + \alpha R_{1_B} + \beta R_{2_B} = \beta \Delta_B t_B \qquad (26.14)$$

On substituting the values, we get:

$$-248{\cdot}44 - 433{\cdot}21\alpha + 433{\cdot}21\beta = 28000\alpha \qquad (26.15)$$

and $\quad 2498{\cdot}44 + 433{\cdot}21\alpha - 433{\cdot}21\beta = 2{,}800\beta \qquad (26.16)$

Whence $\alpha = 0{\cdot}00304 \quad$ and $\beta = 0{\cdot}77315$

$$\therefore R_A = 0.00304 \times 0.001 \times 28 \times 10^6 = 85kN \text{ and}$$

$$R_B = 0.77315 \times 0.001 \times 2.8 \times 10^6 = 2,165kN$$

$$\textbf{Check: } 2\left(R_A + R_B\right) = 2\left(85 + 2165\right) = 4,500kN$$

$$\sum \text{Load, effective} = 4,500kN - \text{Ok!}$$

Final stress-resultants

Ultimate shear forces:

$$S_{AB} = -248.44 + 433\left(0.77315 - 0.00304\right) = 85kN = S_{BA}$$
$$S_{BC} = 2,250kN$$

Ultimate bending moments:

$$M_{AB} = +745.3 + 2378.14\left(0.00304 - 0.77315\right) = -1,086kNm$$
$$M_{BA} = +1490.63 + 1520.71\left(0.00304 - 0.77315\right) = +320kNm$$

Ultimate moment in span BC:

$$M_{BC}^s = 4500 \times 5/4 - 320 = 5,305kNm$$

Shear force and bending moment diagrams are shown in Figure 26.11(*b* and *c*).

Stage II(b): (*Support removal and load transfer: 'continuous' slab case*)

In respect of the effect of load transfer, the only difference, mathematically, between the 'continuous' and the 'discontinuous' slab cases is in the magnitude of the load transferred. Therefore, by replacing in the 'discontinuous' case above the loading to be transferred of 4,500kN with 5,625kN, as shown in Figure 26.12(*a*), and following the similar procedure as above, values for the stress-resultants etc. for the 'continuous' slab case can be obtained. However, the values can also be arrived at simply by scaling the corresponding values obtained in Stage II(*a*) by the factor: 5625/4500, i.e. 1·25. Accordingly,

$$\alpha = 1.25 \times 0.00304 = 0.0038, \beta = 1.25 \times 0.77315 = 0.9664$$
$$R_A = 1.25 \times 85 = 106kN, R_B = 1.25 \times 2165 = 2,706kN$$
$$S_{AB} = 1.25 \times 85 = 106kN = S_{BA}, S_{BC} = 1.25 \times 2250 = 2,813kN$$
$$M_{AB} = 1.25 \times (-1086) = -1,358kNm, M_{BA} = 1.25 \times (+320) = 400kNm$$
$$M_{BC}^s = 1.25 \times 5305 = 6,631kNm$$

Shear force and bending moment diagrams are shown in Figure 26.12(*b* and *c*).

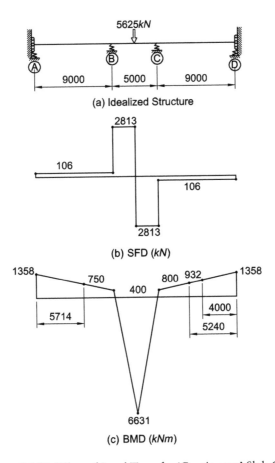

(a) Idealized Structure

(b) SFD (*kN*)

(c) BMD (*kNm*)

Figure 26.12 Effect of Load Transfer 'Continuous' Slab Case

Stage III: (Replacement of backfill on completed structure)

The stress-resultants for this stage can be obtained simply by appropriately scaling the values obtained in Example 26.1. By reference to the assessment of loading in Example 26.1, scale factor for the backfill load $= 535 \cdot 5/800 \cdot 7 = 0 \cdot 67$. Therefore, the stress-resultants for the Stage III case alone are:

$$S_{AB} = 0 \cdot 67 \times 6410 = 4{,}287 kN;$$

$$S_{BA} = -0 \cdot 67 \times 3679 = -2{,}465 kN;$$

$$S_{BC} = 0 \cdot 67 \times 2803 = 1{,}878 kN;$$

$$M_{AB} = -0 \cdot 67 \times 13913 = -9{,}322 kNm;$$

$$M_{BA} = 0 \cdot 67 \times 1622 = 1{,}087 kNm;$$

$$M_{AB}^{s} = 0 \cdot 67 \times 4420 = 2{,}961 kNm;$$

$$M_{BC}^{s} = 0 \cdot 67 \times 1881 = 1{,}260 kNm.$$

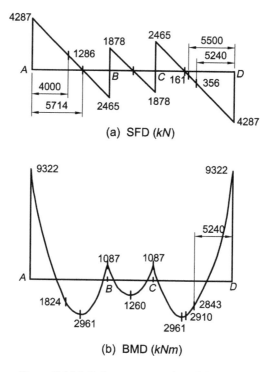

(a) SFD (*kN*)

(b) BMD (*kNm*)

Figure 26.13 Reinstatement of Backfill Only

For the corresponding shear force and bending moment diagrams, see Figure 26.13 (*a* and *b*).

For shear force and bending moment envelopes for the 'Discontinuous' and the 'Continuous' Slab cases, see Figures 26.14 and 26.15.

Conclusions

From the study of the foregoing three solved examples, the following conclusions can be drawn:

- Admittedly, the structure has not been analysed as a complete box cross-section; only the roof slab has been dealt with in isolation in every case. This simplification notwithstanding, the comparative study of the results does illustrate at least qualitatively and, to some extent, gives indication of the relative magnitude as well of the influence that can be expected on design from the chosen method and the sequence of construction.
- Stiffness normally to be expected of a continuous perimeter wall panel within the support system in a typical cut-and-cover metro structure can easily be an order of magnitude higher than that of the corresponding discrete intermediate columns as shown in Example 26.1. By comparing the corresponding values of the stress-resultants (Figures 26.3 and 26.5), particularly the bending moments, for the rigid and the yielding supports, significant influence of the differential elastic response of the support system is clearly in evidence. Accordingly, designing the floor

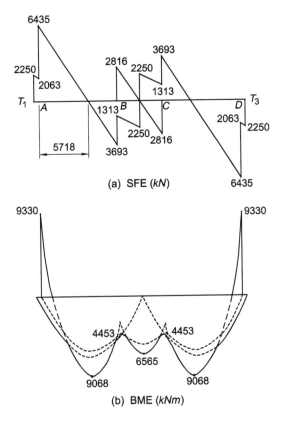

(a) SFE (*kN*)

(b) BME (*kNm*)

Figure 26.14 Final Design Envelopes 'Discontinuous' Slab Case

slabs assuming rigid supports could lead to serious serviceability-related problems particularly in the roof slab which is generally the heaviest loaded. In order to forestall such problems, it is essential to take due cognizance of the differential stiffness within the support system and its effect on the stress-resultants in the structure.

- In Example 26.1, no reference is made either to the method of construction adopted or to the manner in which the slab is temporarily supported during construction. Thus, the values of the stress-resultants (Figures 26.3 and 26.5) obtained are essentially those for a structure wished into place and would, to that extent, be no different from those that would be obtained if a conventional bottom-up sequence of construction were to be assumed.
- Overall length of the structure in Figure 26.4 is only marginally (i.e. 4·35 per cent) greater than that of Figure 26.2. Notwithstanding this, a fair idea of the effects of the boundary conditions and the disposition of the intermediate supports can be obtained by the comparison of the stress-resultants in Figures 26.3 and 26.5.
- Bending moment envelope in Figure 26.8 brings into focus the effect on the distribution of moments of the assumed end conditions during construction.
- On comparing the results of Figure 26.3 with those of the top-down sequence of construction as represented in Figure 26.8, the influence on the stress-resultants

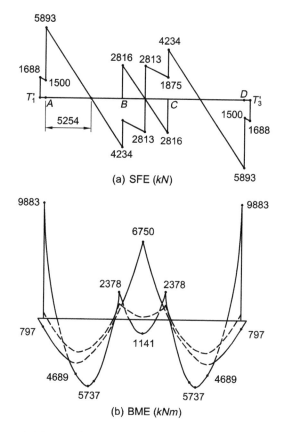

Figure 26.15 Final Design Envelopes 'Continuous' Slab Case

of the location of the temporary supports and the effect of load transfer due to their subsequent removal can be clearly seen.

*   Influence of the number and location of the temporary supports, their subsequent removal and installation of the permanent supports is also in evidence by comparing the results of Figure 26.8 with those of Figure 26.15. Differences in the magnitude and distribution of the moments are more pronounced than those of the shear forces.

*   Figures 26.9 and 26.10 compare the effects in the stress-resultants resulting from the methods of construction of the slab as a discontinuous or a continuous element over the temporary central support. As is to be expected, the reaction at the central temporary support in the case of the continuous slab case is significantly greater. This is also, in turn, reflected in the respective load transfer effects (Figures 26.11 and 26.12) when the permanent supports are in place and operational and the temporary supports are removed.

*   Comparison of Figures 26.14 and 26.15 reveals a significant difference, both qualitative and quantitative, in the stress-resultants for the 'discontinuous' and the 'continuous' slab cases. In the former case, with the configuration of the moments as indicated, the structure is clearly less efficient. However, in the latter,

the moments are more evenly distributed and so the structure is relatively more efficient. This underlines the impact that the sequence of construction can have on design and, by implication, potentially, on the associated costs as well.

In view of the foregoing, it is abundantly clear that, in the analysis and design of cut-and-cover metro structures, unlike those of conventional aboveground structures, the method and sequence of construction can play a predominant role and must be duly taken into account.

27 Structural Analysis: II

27.1 Introduction

In the case of structures, which are inherently simple and geometrically symmetrical, invoking the principle of symmetry of the structure and symmetry and skew-symmetry of the applied loading can help achieve significant economies in the computational efforts required in their analyses. Any asymmetric loading can also be split up into its symmetric and skew-symmetric components so that, for each component, only half the structure need be analysed. Stress-resultants so obtained for the component loadings can then be algebraically combined to arrive at the values for the required asymmetric loading. In relation to cut-and-cover metro structures, such an approach can be particularly useful in the hand analysis of box structures, such as those forming the access corridors (i.e. Entrances) from the surface above to the main station structure below and the running tunnels. However, for the application of this approach, understanding of the methods of translational and cantilever moment distributions is essential. These methods are briefly discussed in the following section.

27.2 Use of translational moment distribution

The mechanics of the translational moment distribution differs from that of its conventional counterpart in that the joint translation (or the side sway) is permitted to occur during the moment balancing process itself. In other words, it is not necessary to use artificial supports to prevent translation initially and to use two or more distributions and the principle of superposition to arrive at the final moments subsequently. The process of balancing the out-of-balance moments by releasing and relocking of the joints is followed in the same way as for the conventional method. The use of the translational moment distribution method makes it possible to determine the final moments in one distribution only thereby reducing the extent of computation. In the case of the geometry of the structure, which is symmetrical, the computational time can be reduced even further by solving only half the structure.

However, in the analysis of a structure by the method of translational moment distribution, since the joints are locked against rotation but are free to translate during the moment distribution process, the initial fixed-end moments, member rotational stiffnesses and the carry-over factors need to be modified to reflect this fact.

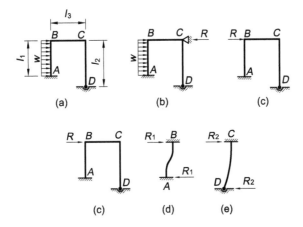

Figure 27.1 Determination of Fixed-end Moments

Translational fixed-end moments

In working out the fixed-end moments, the effects of joint translations need to be taken into account as appropriate. Fortunately, this can be usually accomplished without much difficulty. As an illustration, consider a frame loaded as shown in Figure 27.1.

Assuming, for all practical purposes, member BC to be infinitely stiff in its own plane, i.e. undergoing no elastic shortening, translations of joints B and C under lateral load will then be identical both in magnitude and direction. The fixed-end moments can be obtained easily in two stages. In the first stage, translation of the joints is prevented by introducing an artificial support at joint C as shown in figure (b). Fixed-end moments at A and B for this stage are then given by the conventional expression $\mp wl_1^2 \div 12$ and the support reaction R by the expression $wl_1 \div 2$. In the second stage, the restraint is released by removing the artificial support at joint C thereby allowing the joints B and C to translate. The effect of translation can be evaluated by applying a force equal in magnitude but opposite in direction to the reaction R at joint B as shown in figure (c). Force R will be shared by members AB and CD in proportion to their respective translational stiffnesses T_1 and T_2 as shown in figures (d) and (e). In other words, the forces in the two legs are given by the expressions:

$$R_1 = \left[\frac{T_1}{T_1 + T_2}\right] R \qquad (27.1)$$

$$\text{and} \quad R_2 = \left[\frac{T_2}{T_1 + T_2}\right] R \qquad (27.2)$$

The fixed-end moments for the second stage are then obtained from the expressions $-R_1 l_1 \div 2$ at A and B, and $-R_2 l_2$ at C.

For the frame in Figure 27.1, by combining the components for the two stages, the expressions for the translational fixed-end moments can thus be obtained from the following expressions:

$$(t)m^f_{AB} = -\left(\frac{wl_1^2}{12} + \frac{R_1 l_1}{2}\right) \tag{27.3}$$

$$(t)m^f_{BA} = \left(\frac{wl_1^2}{12} - \frac{R_1 l_1}{2}\right) \tag{27.4}$$

$$\text{and} \quad (t)m^f_{CD} = -R_2 l_2 \tag{27.5}$$

Translational stiffness

The translational stiffness of a beam is the measure of its resistance against translation while rotations are prevented. It may be defined as the force required to cause a unit translation of one end of the beam with respect to the other. For a member of length l, if one end translates by an amount Δ under the application of force F, the translational stiffness is given by the expression:

$$T = \frac{F}{\Delta} = \frac{12EI}{l^3} \text{ (both ends fixed)} \tag{27.6}$$

$$= \frac{3EI}{l^3} \text{ (one end pinned)} \tag{27.7}$$

The parameters I and l are member-specific; in the context of the frame in Figure 27.1, the translational stiffnesses for members BA and CD can be obtained by substituting the appropriate values for the respective I and l parameters in Equations (27.6) and (27.7) above.

Absolute rotational stiffness

In the frame of the Figure 27.1, since the joints B and C are free to translate, the horizontal member BC can be looked upon as providing an elastic support, acting as a spring, to the vertical member AB and, in a like manner, to DC. However, member BC itself is assumed to be infinitely stiff and so the spring stiffness attributed to it is that which derives from the translational stiffness of either member DC when providing support to AB or *vice versa*. The absolute rotational stiffnesses for the vertical members (Gere, 1963) are given by the expressions:

$$K_{ba} = \frac{4EI}{l_1} \cdot \frac{4T_2 + T_1}{4(T_2 + T_1)} \text{ (far end fixed)} \tag{27.8}$$

$$K_{cd} = \frac{3EI}{l_2} \cdot \frac{T_2}{(T_2 + T_1)} \text{ (far end pinned)} \tag{27.9}$$

The values for T_1 and T_2 can be obtained from the expressions given in Equations (27.6) and (27.7) above.

The absolute rotational stiffness for the horizontal member BC is the same as that in the conventional moment distribution, i.e.

$$K_{bc} = \frac{4EI}{l_3} \qquad (27.10)$$

Carry-over factors (COF)

In the translational moment distribution method, besides the carry-over factor from the near end of a member to its far 'fixed' end, such as that from B to A and B to C in Figure 27.1, there are also translational carry-over factors from BA to CD and *vice versa*. It is important to ensure that the translational carry-over of the moments, as appropriate, is not overlooked. The values of the carry-over factors and the translational carry-over factors are given (Gere, 1963) by the following expressions.

$$COF_{ba} = \frac{2T_2 - T_1}{4T_2 + T_1} \qquad (27.11)$$

$$COF_{bc} = \frac{1}{2} \text{ (far end fixed)} \qquad (27.12)$$

$$TCOF_{ba-cd} = -\frac{l_2}{l_1} \cdot \frac{3T_2}{4T_2 + T_1} \text{ (end D fixed)} \qquad (27.13)$$

$$TCOF_{ba-cd} = -\frac{l_2}{l_1} \cdot \frac{6T_2}{4T_2 + T_1} \text{ (end D pinned)} \qquad (27.14)$$

27.3 Use of cantilever moment distribution

Cantilever moment distribution is a special case of translational moment distribution. However, its application is limited to structures of symmetric geometry supporting skew-symmetric loads. In fact, it is for this very reason that the method can be particularly useful for cut-and-cover box structures, such as the one shown in Figure 27.3, where the computation is considerably simplified by invoking symmetry and skew-symmetry and solving half the structure only. For example, under skew-symmetric load, members parallel to the axis of symmetry, such as AD and BC, will deform identically with neither member required to offer any spring support to the other and thus behaving as vertical cantilevers with the upper ends virtually free. There is therefore no translational carry-over from DA to CB or *vice versa*; this can also be confirmed by setting $T_2 = 0$ in Equation (27.13). Also, by a similar substitution in Equation (27.11), the carry-over factor from D to A or A to D can be obtained which works out to '-1'.

A second simplification arising out of the symmetry of the geometry of the structure and the skew-symmetry of the applied loading is that the stiffness factors for members perpendicular to the axis of symmetry, such as AB and DC in Figure 27.3, can be taken as $6EI/l$, where l is the length of the member concerned. In addition, owing to symmetry, there is no carry-over along such a member and so moments need to be balanced for half the structure only.

It should also be recognized that, as long as the geometry of the structure is symmetrical, even asymmetric loading could be represented by a combination of appropriate symmetric and skew-symmetric loadings. In view of this, only half the structures need to be analysed for the symmetric loading case by the conventional moment distribution method and for the skew-symmetric case by the cantilever moment distribution method. Incidentally, for the symmetric loading case, in solving only half the structure, the usual absolute rotational stiffness ($4EI/l$) must be used for the member(s) parallel to the axis of symmetry but only half the value (i.e. $2EI/l$) for the member(s) perpendicular to the axis of symmetry, where I and l are member-specific parameters. Final moments can thereafter be obtained by combining, algebraically, the results of the two, i.e. symmetric and skew-symmetric, cases. The use of this procedure has been illustrated in the solution to Example 27.1.

The important aspects in the use of the method of cantilever moment distribution as it can be applied to a symmetric structure with skew-symmetric loads may now be summarized (Gere, 1963) as follows:

- The stiffness factor for a member parallel to the axis of symmetry is EI/l and the carry-over factor is $[-1]$ when the far end of the member is fixed. For a member with pinned ends, both these factors are zero.
- The corresponding factors for a member perpendicular to the axis of symmetry are $6EI/l$ and zero, respectively.
- The fixed-end moments are determined in the same manner as in the translational moment distribution outlined above.
- Moments are balanced for half the structure only.

The principles discussed in the foregoing section are illustrated in the solution to Example 27.1.

Example 27.1

Carry out a transverse analysis of the typical box cross-section of a metro station entrance structure shown in Figure 27.2(a) assuming the following:

- The foundations to the existing high-rise structures abutting either side of the site are carried on end-bearing piles penetrating well below the base slab of the entrance box structure.
- A surcharge load allowance of $50kN/m^2$ shall be considered for any future construction on either side considered effective over the full depth of the box structure.

Adopt the following parameters for design:

- Unit weight of concrete: $24kN/m^3$; Unit weight of water: $10kN/m^3$;
- Unit weight of soil above and below water table (for simplicity only): $20kN/m^3$;
- $\phi' = 25°$; $K_a = 0.4$; $K_o = 0.6$;
- Loading from traffic surcharge above: $10kN/m^2$;
- High water table: $1m$ below ground level;
- Low water table: $3m$ below ground level.

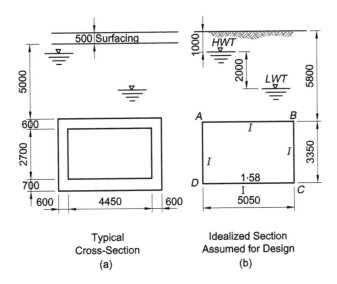

Typical
Cross-Section
(a)

Idealized Section
Assumed for Design
(b)

Figure 27.2 Entrance Box Structure

For the purpose of this exercise: (*a*) assume the structure to be stable against the effect of buoyancy and carrying out a check against flotation is not required and (*b*) heave calculation is not required.

Solution
Rationale: In the analysis of the structure, the principle of structural symmetry has been invoked to analyse only half the structure using the expedient of symmetry and skew-symmetry for the loadings. Distribution of moments has been carried out with the help of conventional and cantilever moment distributions and using notional unit loadings on members discretely. The stress-resultants have then been modified by the appropriate scale factors to represent contributions from the characteristic loadings. The results so obtained have been tabulated so as to enable the appropriate load factors and load combinations yielding the ultimate maximum and the minimum values of the stress-resultants to be identified easily.

In formulating load combinations, it has been possible to ensure, by inspection, that only such individual load cases are picked up for combinations as can be expected to coexist realistically.

It is assumed that the self-weight of, and the uniformly distributed live load on, the ground-bearing (base) floor slab are directly and uniformly transmitted to the subgrade below without inducing any stress-resultants in the slab or the structure. Accordingly, for strength calculations, these load cases have been disregarded. However, in working out the stress-resultants in the structure due to the hydrostatic uplift pressures acting on the underside of the base slab of the structure, appropriate reduction due to the self-weight of the base slab has been duly taken into account.

Under the influence of the out-of-balance lateral load acting on one sidewall of the box, the structure would tend to sway. The equilibrium would then be restored by invoking the stiffness of the structure together with the mobilization of the passive resistance against the opposite wall. Ideally, this would require modelling of the soil resistance in terms of a series of soil springs. However, in the analysis of this structure, to keep the hand computation simple, the mobilization of the passive resistance opposing the out-of-balance lateral load has been disregarded and the equilibrium assumed to be restored solely by the flexural resistance offered by the box structure itself. Clearly, since the soil–structure interaction has been disregarded, this approach could be too conservative especially where the out-of-balance lateral loads were to be significant. It must therefore be clearly understood that this approach is to be treated purely as indicative and not to be construed as a recommendation for its use. However, where the levels of ground movements and the corresponding moments induced by the out-of-balance lateral load are unlikely to be significant, such an approach may be used without the danger of involving too much conservatism.

The unique feature of the type of presentation of the results made in the solution to this example is that the magnitude and the nature of contributions from every discrete load case on the stress resultants of the entire structure are on display and can be readily identified. Presentation of the results in a tabulated form also makes it easy to watch out and ensure that no realistically onerous combinations are overlooked. Furthermore, the effect of change in any particular loading in isolation, if required, can be evaluated with ease by simply scaling-off the results as appropriate.

Box geometry

Idealized section for analysis has been based on the centre-line dimensions of the structural elements as shown in Figure 27.2(b). However, in computing the weight of the overburden on top of the structure and the hydrostatic uplift pressure on its base, actual geometry, and not the centre-line dimensions, has been considered.

Characteristic loadings

Dead load

> Self-weight roof slab: $0.6 \times 24 = 14.4 kN/m^2$
> Self-weight base slab: $0.7 \times 24 = 16.8 kN/m^2$
> Self-weight sidewall: $0.6 \times 2.7 \times 24 = 38.9 kN/m$ run per wall

Superimposed dead load

> Surfacing, etc.: $0.5 \times 24 = 12.0 kN/m^2$

Backfill (below surfacing) on top:

> Maximum Gross: $5 \times 20 = 100 kN/m^2$
> Maximum Effective: $100 - 10(5.5 - 3) = 75 kN/m^2$
> Minimum Effective: $100 - 10(5.5 - 1) = 55 kN/m^2$

Live load

Traffic surcharge on roof: $= 10kN/m^2$
On the invert of the box structure: (assume) $= 6kN/m^2$

Lateral soil pressures: 'active'

At 5·8*m* below GL, and high water table:

$$[12 + 20 \times 0\cdot5 + (20 - 10)4\cdot8] \times 0\cdot4 = 28\cdot0kN/m^2$$

At 9·15*m* below GL, and high water table:

$$[12 + 20 \times 0\cdot5 + (20 - 10)8\cdot15] \times 0\cdot4 = 41\cdot4kN/m^2$$

At 5·8*m* below GL, and low water table:

$$[12 + 20 \times 2\cdot5 + (20 - 10)2\cdot8] \times 0\cdot4 = 36\cdot0kN/m^2$$

At 9·15*m* below GL, and low water table:

$$[12 + 20 \times 2\cdot5 + (20 - 10)6\cdot15] \times 0\cdot4 = 49\cdot4kN/m^2$$

Lateral soil pressures: 'at-rest'

At 5·8*m* below GL and high water table:

$$[12 + 20 \times 0\cdot5 + (20 - 10)4\cdot8] \times 0\cdot6 = 42\cdot0kN/m^2$$

At 9·15*m* below GL and high water table:

$$[12 + 20 \times 0\cdot5 + (20 - 10)8\cdot15] \times 0\cdot6 = 62\cdot1kN/m^2$$

At 5·8*m* below GL and low water table:

$$[12 + 20 \times 2\cdot5 + (20 - 10)2\cdot8] \times 0\cdot6 = 54\cdot0kN/m^2$$

At 9·15*m* below GL and low water table:

$$[12 + 20 \times 2\cdot5 + (20 - 10)6\cdot15] \times 0\cdot6 = 74\cdot1kN/m^2$$

Lateral hydrostatic pressures: 'HWT'

At 5·8*m* below GL: $10 \times 4\cdot8 = 48kN/m^2$
At 9·15*m* below GL: $10 \times 8\cdot15 = 81\cdot5kN/m^2$

Lateral hydrostatic pressures: 'LWT'

At 5·8*m* below GL: $10 \times 2\cdot8 = 28kN/m^2$
At 9·15*m* below GL: $10 \times 6\cdot15 = 61\cdot5kN/m^2$

Hydrostatic uplift pressures on base slab (during construction)

To obtain the net design uplift pressure on the base slab, the gross uplift pressure must be appropriately reduced by the self-weight of the slab.

Gross maximum uplift pressure: $10(5\cdot0 - 0\cdot5 + 4\cdot0) = 85kN/m^2$
Gross minimum uplift pressure: $10(3\cdot0 - 0\cdot5 + 4\cdot0) = 65kN/m^2$
Net design uplift pressure (maximum): $85 - 16\cdot8 = 68\cdot2kN/m^2$
Net design uplift pressure (minimum): $65 - 16\cdot8 = 48\cdot2kN/m^2$

Hydrostatic uplift pressures on base slab (in-service)

Gross uplift pressure: $10 \times 4\cdot0 = 40kN/m^2$
Net design uplift pressure: $40 - 16\cdot8 = 23\cdot2kN/m^2$

Lateral surcharge loads

The buildings abutting the entrance box structure on either side are carried on end-bearing piles penetrating well below the base slab of the box structure and are unlikely to impose any effect on it. Surcharge load on entrance box structure from the adjacent buildings is, therefore, considered to be nil. However, lateral surcharge loading from

Future construction: $50 \times 0\cdot6 = 30kN/m^2$
Future traffic: $10 \times 0\cdot6 = 6kN/m^2$

Analysis

Theoretically, any asymmetry in the loading can cause sway of the structure. However, owing to the confined nature of the structure, this tendency will be resisted by the soil–structure interface friction around the perimeter of the structure and the mobilization of the requisite amount of passive pressure on the appropriate side necessary to restore equilibrium. The effect of the asymmetry of the loads on the variation of the pressure on the base of the structure is therefore unlikely to be significant and is, accordingly, ignored.

When analysing the box cross-section for any applied loads other than that of the self-weight of the structure itself, the structure is assumed to be weightless. In the modelling of the effects of the self-weight of the structure, own weight of the base slab of the structure is disregarded. This is because, being evenly distributed on the formation, it does not contribute to the stress-resultants of the structure as outlined in the rationale.

Base contact pressures

Total overburden pressure removed from the formation level when excavation is complete:

$$24 \times 0\cdot5 + 20(5\cdot5 - 0\cdot5 + 4) = 192kN/m^2$$

Pressure replaced on the formation by the self-weight of the box structure after the completion of its construction:

$$14\cdot4 + 16\cdot8 + 2 \times 38\cdot9 \div (4\cdot45 + 2 \times 0\cdot6) = 45kN/m^2$$

Pressure replaced by the reinstatement of the backfill and the road surfacing on top:

$$24 \times 0\cdot5 + 20(5\cdot5 - 0\cdot5) = 112kN/m^2$$

Total overburden pressure restored: $45 + 112 = 157kN/m^2$

Net residual pressure on formation: $157 - 192 = -35kN/m^2$

Even allowing for the transient traffic live load of $10kN/m^2$, the net pressure on the formation will continue to remain negative. The negative sign implies that the weight of the net volume of earth removed is in excess of that restored by the structure and the backfill, and so the formation will be under a net pressure relief. This implies that the foundation is 'floating'. This must not be confused with the flotation due to buoyancy. It simply means that the structure is unlikely to impose pressures which could be in excess of those that would have been experienced by the formation previously in its virgin state.

The pressure relief can also result in heave of the formation. In the normal course, it would have been necessary to establish the seat of heave, etc. in order to work out the equivalent upward load due to heave and to address its implications on the design of the box structure as appropriate. However, the exercise on heave related issues is not required to be carried out in this example. For details of how to deal with design related issues on heave, generally, reference may be made to Chapter 20 and the solved examples at the end of Chapter 21.

Stiffness and distribution factors

Because of the symmetry in the geometry of the structure, only half the box structure need be analysed while dealing with the symmetric and the skew-symmetric load cases. For the labelling of the appropriate member stiffnesses for the elements of half the structure to correspond with the symmetric and the skew-symmetric load cases, see Figure 27.3.

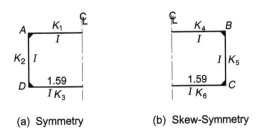

(a) Symmetry   (b) Skew-Symmetry

Figure 27.3 Member Stiffnesses

In the skew-symmetric load case, absolute rotational stiffnesses will be those commonly employed in the cantilever moment distribution method.

Case of symmetry

Rotational stiffnesses:

Symbol	Rotational stiffness	
	Absolute	Relative
K_1	$2I \div 5 \cdot 05$	1
K_2	$4I \div 3 \cdot 35$	3
K_3	$2 \times 1 \cdot 59I \div 5 \cdot 05$	$1 \cdot 59$

Distribution factors:

$$\text{At joint } A: \text{For member } AB = 1/4 = 0 \cdot 25$$
$$AD = 1 - 0 \cdot 25 = 0 \cdot 75$$
$$\text{At joint } D: \text{For member } DA = 3/4 \cdot 59 = 0 \cdot 65$$
$$DC = 1 - 0 \cdot 65 = 0 \cdot 35$$

Case of skew-symmetry

Rotational stiffnesses:

Symbol	Rotational stiffness	
	Absolute	Relative
K_4	$6I \div 5 \cdot 05$	4
K_5	$I \div 3 \cdot 35$	1
K_6	$6 \times 1 \cdot 59I \div 5 \cdot 05$	$6 \cdot 3$

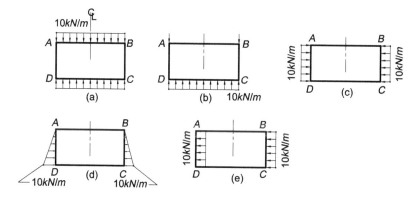

Figure 27.4 Notional Loadings

Distribution factors:

At joint *A*: For member $AB = 4/5 = 0.8$
$$AD = 0.2$$
At joint *D*: For member $DA = 1/7.3 = 0.14$
$$DC = 0.86$$

Distribution of notional moments around the box

The notional loadings assumed and as applied to the various elements of the box structure are shown in Figure 27.4(*a* to *e*).

Cases of symmetry

- Notional load of $10kN/m$ on the roof slab and the commensurate reaction from the ground on the base slab as shown in figure (*a*):

$$\text{Fixed-end moment, } m^f_{AB} = -10 \times \frac{5.05^2}{12} = -21.25kNm = -m^f_{DC}$$

- Distribution of notional moments

Joint	D		A	
Member	DC	DA	AD	AB
DF	0·33	0·67	0·75	0·25
Fixed-end moment	+21·25			−21·25
Distribution	−7·08	−14·17	+15·94	+5·31
		+7·97	−7·09	
	−2·66	−5·31	+5·32	+1·77
		+2·66	−2·66	
	−0·88	−1·78	+2·00	+0·66
		+1·00	−0·89	
	−0·34	−0·67	+0·67	+0·22
Moment (kNm)	+10·29	−10·29	+13·29	−13·29
Static shear (kN)	−25·25	0·00	0·00	+25·25
Elastic shear (kN)	0·00	−0·90	+0·90	0·00
Total shear (kN)	−25·25	−0·90	+0·90	+25·25

- Notional $10kN/m$ upward load on the base slab as shown in figure (*b*):

$$\text{Fixed-end moment, } m^f_{DC} = +10 \times \frac{5.05^2}{12} = +21.25kNm$$

• Distribution of notional moments

Joint	D		A	
Member	DC	DA	AD	AB
DF	0·33	0·67	0·75	0·25
Fixed-end moment	+21·25			
Distribution	−7·08	−14·17	−7·08	
		+2·66	+5·31	+1·77
	−0·88	−1·78	−0·89	
		+0·33	+0·67	+0·22
	−0·11	−0·22	−0·11	
			+0·08	+0·03
Moment (kNm)	+13·18	−13·18	−2·02	+2·02
Static shear (kN)	−25·25	0·00	0·00	0·00
Elastic shear (kN)	0·00	+4·54	−4·54	0·00
Total shear (kN)	−25·25	+4·54	−4·54	0·00

• Notional load of $10kN/m$ on the sides as shown in figure (*c*):

$$\text{Fixed-end moment, } m^f_{AD} = +10 \times \frac{3 \cdot 35^2}{12} = +9 \cdot 35 kNm$$

$$= -m^f_{DA}$$

• Distribution of notional moments

Joint	D		A	
Member	DC	DA	AD	AB
DF	0·33	0·67	0·75	0·25
Fixed-end moment		−9·35	+9·35	
Distribution	+3·12	+6·23	−7·01	−2·34
		−3·50	+3·12	
	+1·17	+2·33	−2·34	−0·78
		−1·17	+1·16	
	+0·39	+0·78	−0·87	−0·29

Joint	D		A	
Member	DC	DA	AD	AB
Distribution continued	+0·15	−0·44 +0·29	+0·39 −0·29	−0·10
Moment (kNm)	+4·83	−4·83	+3·51	−3·51
Static shear (kN)	0·00	+16·75	−16·75	0·00
Elastic shear (kN)	0·00	+0·40	+0·40	0·00
Total shear (kN)	0·00	+17·15	−16·35	0·00

- Notional triangular load (0 to 10kN/m) on the sides as shown in figure (*d*):

$$\text{Fixed-end moment, } m^f_{AD} = +WL/15 = +5 \times \frac{3.35^2}{15} = +3.74 kNm$$

$$m^f_{DA} = -WL/10 = -5 \times \frac{3.35^2}{10} = -5.61 kNm$$

- Distribution of notional moments

Joint	D		A	
Member	DC	DA	AD	AB
DF	0·33	0·67	0·75	0·25
Fixed-end moment		−5·61	+3·74	
Distribution	+1·87	+3·74 −1·41	−2·81 +1·87	−0·93
	+0·47	+0·94 −0·70	−1·40 +0·47	−0·47
	+0·23	+0·47 −0·18	−0·35 +0·24	−0·12
	+0·06	+0·12	−0·18	−0·06
Moment (kNm)	+2·63	−2·63	+1·58	−1·58
Static shear (kN)	0·00	+11·16	−5·58	0·00
Elastic shear (kN)	0·00	+0·31	+0·31	0·00
Total shear (kN)	0·00	+11·47	−5·27	0·00

Cases of skew-symmetry

- Notional $10kN/m$ load applied in the same direction on each side wall as shown in figure (*e*):

$$\text{Fixed-end moment, } (t)m^f_{DA} = +\left(\frac{wl_1^2}{12} + \frac{R_1l_1}{2}\right) = +\left(\frac{wl_1^2}{12} + \frac{wl_1^2}{4}\right) = +\frac{wl_1^2}{3}$$

$$= +\frac{10 \times 3\cdot35^2}{3} = +37\cdot4kNm$$

$$(t)m^f_{AD} = -\left(\frac{wl_1^2}{12} - \frac{R_1l_1}{2}\right) = +\frac{wl_1^2}{6}$$

$$= +\frac{10 \times 3\cdot35^2}{6} = +18\cdot7kNm$$

- Distribution of notional moments

Joint		D		A
Member	DC	DA	AD	AB
DF	0·86	0·14	0·20	0·80
COF		−1	−1	
Fixed-end moment		+37·40	+18·70	
Distribution	−32·16	−5·24	−3·74	−14·96
		+3·74	+5·24	
	−3·22	−0·52	−1·04	−4·20
		+1·04	+0·52	
	−0·90	−0·14	−0·10	−0·42
		+0·10	+0·14	
	−0·09	−0·01	−0·03	−0·11
Moment (kNm)	−36·37	+36·37	+19·69	−19·69
Static shear (kN)	0·00	−16·75	+16·75	0·00
Elastic shear (kN)	+14·40	−16·73	−16·73	+7·80
Total shear (kN)	+14·40	−33·48	+0·02	+7·80

Scaling-off stress-resultants

The foregoing distributions of moments are based on the notional loadings of $10kN/m^2$ or $10kN/m$, etc. as appropriate. The ratios of the relevant characteristic

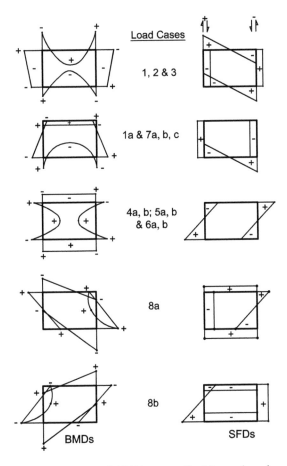

Figure 27.5 Qualitative BM and SF Diagrams (looking at box from inside-out)

loads and the corresponding notional loads represent the respective scale factors. The notional stress-resultants must be modified by these factors to arrive at the characteristic stress-resultants. Ultimate values for these are finally obtained by multiplying the characteristic values by the appropriate load factors as listed in Tables 27.1 and 27.2. Qualitative diagrammatic representation of the stress-resultants is shown in Figure 27.5(*a*) to (*e*).

(1) *Dead load only*

(*a*) *Roof slab*

Self-weight roof slab: $= 14 \cdot 4 kN/m^2$
Notional load assumed: $= 10 kN/m^2$
Scale Factor: $14 \cdot 4/10 = 1 \cdot 44$

Characteristic moments:

$$M_{AB} = 1.44(-13.29) = -19.14kNm$$
$$M_{DC} = 1.44(+10.29) = +14.82kNm$$
$$M_{S(AB)} = 14.4 \times 5.05^2/8 - 19.14 = +26.76kNm$$
$$M_{S(DC)} = -14.4 \times 5.05^2/8 + 14.82 = -31.08kNm$$

Characteristic shears:

$$S_{AB} = 1.44 \times (+25.25) = +36.36kN = -S_{BA}$$
$$= -S_{DC} = S_{CD}$$
$$S_{AD} = 1.44 \times (+0.90) = +1.30kN = -S_{BC}$$

Qualitative bending moment and shear force diagrams are presented in Figure 27.5 as those for load case 1.

(b) Walls

Self-weight walls: $= 38.9kN/m$ longitudinally per wall
Equivalent *udl* on base: $2 \times 38.9/5.65 = 13.8kN/m^2$ transversely
Scale factor: $13.8/10 = 1.38$

Characteristic moments:

$$M_{DC} = 1.38(+13.18) = +18.19kNm$$
$$M_{AB} = 1.38(+2.02) = +2.79kNm$$
$$M_{S(DC)} = -13.8 \times 5.05^2/8 + 18.19 = -25.80kNm$$
$$M_{S(AB)} = AsM_{AB} = +2.79kNm$$

Characteristic shears:

$$S_{DA} = 1.38 \times (+4.54) = +6.27kN = -S_{CB}$$
$$S_{DC} = 1.38 \times (-25.25) = -34.85kN = -S_{CD}$$

Qualitative bending moment and shear force diagrams are presented in Figure 27.5 as those for load case 1(a).

By combining the results from (a) and (b) above, the stress-resultants for the self-weight of the complete box structure can be obtained. However, it should be noted here that the contribution from the base slab has been disregarded. This is because the base slab is ground-bearing and its self-weight is uniformly borne by the subgrade without inducing any stress-resultants in the slab or the structure. This would be equally true also of the uniformly distributed live load on the base slab as outlined in the rationale.

(2) Superimposed dead load

Gross backfill on roof: $100 + 12 = 112kN/m^2$
Notional load assumed: $= 10kN/m^2$
Scale factor: $112/10 = 11.2$

Characteristic moments:

$$M_{AB} = 11.2(-13.29) = -148.85kNm$$

$$M_{DC} = 11{\cdot}2(+10{\cdot}29) = +115{\cdot}25kNm$$
$$M_{S(AB)} = 112 \times 5{\cdot}05^2/8 - 148{\cdot}85 = +208{\cdot}19kNm$$
$$M_{S(DC)} = -112 \times 5{\cdot}05^2/8 + 115{\cdot}25 = -241{\cdot}79kNm$$

Characteristic shears:

$$S_{AB} = 11{\cdot}2 \times (+25{\cdot}25) = +282{\cdot}8kN = -S_{BA}$$
$$= -S_{DC} = S_{CD}$$
$$S_{AD} = 11{\cdot}2 \times (+0{\cdot}90) = +10{\cdot}08kN = -S_{BC}$$

Qualitative bending moment and shear force diagrams are also as presented in Figure 27.5 as those for load case 2.

(3) Live load

Equivalent traffic surcharge on roof $= 10kN/m^2$ is of the same magnitude as the notional load and so, the Scale Factor $= 1$. By reference to the table for the distribution of moments corresponding to the notional load on the roof slab, the stress-resultants may be simply reproduced as follows:

Characteristic moments:

$$M_{AB} = -13{\cdot}29kNm$$
$$M_{DC} = +10{\cdot}29kNm$$
$$M_{S(AB)} = 10 \times 5{\cdot}05^2/8 - 13{\cdot}29 = +18{\cdot}59kNm$$
$$M_{S(DC)} = -10 \times 5{\cdot}05^2/8 + 10{\cdot}29 = -21{\cdot}59kNm$$

Characteristic shears:

$$S_{AB} = +25{\cdot}25kN = -S_{BA} = -S_{DC} = S_{CD}$$
$$S_{AD} = +0{\cdot}9kN = -S_{BC}$$

Once again, qualitative bending moment and shear force diagrams are as presented in Figure 27.5 as those for load case 3.

(4) Active earth pressure

(a) At high water table

Scale factor for uniform pressure: $28/10 = 2{\cdot}8$
Scale factor for triangular pressure: $13{\cdot}4/10 = 1{\cdot}34$
Characteristic moments:

$$M_{DA} = 2{\cdot}8(-4{\cdot}83) + 1{\cdot}34(-2{\cdot}63) = -17{\cdot}05kNm$$
$$M_{AD} = 2{\cdot}8(3{\cdot}51) + 1{\cdot}34(1{\cdot}58) = +11{\cdot}95kNm$$

General expression for maximum span moment, $M_{S(AD)}$, in the member AD:

Consider the statics of member AD under the action of the assumed loading and end moments as shown in Figure 27.6.

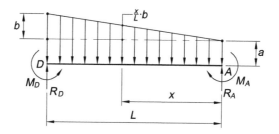

Figure 27.6 Position of Maximum Span Moment

Static shear at end A: $\dfrac{L}{2}\left(a+\dfrac{b}{3}\right)$

Elastic shear: $(M_D - M_A)/L$

Reaction at A, $R_A = \dfrac{L}{2}\left(a+\dfrac{b}{3}\right) - \dfrac{(M_D - M_A)}{L}$

At a section measured a distance 'x' from end A, component of the variable (triangular) load is given by: $\dfrac{x}{L}\cdot b$

General expression for the moment at this section is given by:

$$M_x = R_A \cdot x - \frac{a}{2}\cdot x^2 - \frac{b}{6L}\cdot x^3 - M_A$$

$$\frac{dM_x}{dx} = R_A - a\cdot x - \frac{b}{2L}\cdot x^2$$

For maxima, $\quad \dfrac{dM_x}{dx} = 0 = R_A - a\cdot x - \dfrac{b}{2L}\cdot x^2$

whence $\quad x = \dfrac{-aL \pm \sqrt{(a^2\cdot L^2 + 2R_A\cdot b\cdot L)}}{b}$

Using the positive root of the equation and making appropriate substitutions for R_A and M_A in the general expression for the moment, the value for the maximum moment $M_{S(AD)}$ can be obtained. This works out to $34\cdot18kNm$ at $x = 1\cdot69m$ below joint A.

Characteristic shears:

$$S_{DA} = 2\cdot8\,(17\cdot15) + 1\cdot34\,(11\cdot47) = +63\cdot39kN = -S_{CB}$$
$$S_{AD} = 2\cdot8\,(-16\cdot35) + 1\cdot34\,(-5\cdot27) = -52\cdot84kN = -S_{BC}$$

At low water table

Scale factor for uniform pressure: $36/10 = 3\cdot6$
Scale factor for triangular pressure: $13\cdot4/10 = 1\cdot34$

Characteristic moments:

$$M_{DA} = 3 \cdot 6(-4 \cdot 83) + 1 \cdot 34(-2 \cdot 63) = -20 \cdot 91 kNm$$
$$M_{AD} = 3 \cdot 6(3 \cdot 51) + 1 \cdot 34(1 \cdot 58) = +14 \cdot 75 kNm$$

From the geometry of the assumed loading and the support moment diagrams, following the same procedure as that used for the high water table case, maximum $M_{S(AD)}$ works out to $40 \cdot 96 kNm$ at $1 \cdot 68m$ below joint A.

Characteristic shears:

$$S_{DA} = 3 \cdot 6(17 \cdot 15) + 1 \cdot 34(11 \cdot 47) = +77 \cdot 11 kN = -S_{CB}$$
$$S_{AD} = 3 \cdot 6(-16 \cdot 35) + 1 \cdot 34(-5 \cdot 27) = -65 \cdot 92 kN = -S_{BC}$$

For both (*a*) and (*b*) above, qualitative bending moment and shear force diagrams are as presented in Figure 27.5 as those for load cases 4(*a*) and 4(*b*).

(5) *Earth-at-rest pressure*

(*a*) *At high water table*

Scale factor for uniform pressure: $42/10 = 4 \cdot 2$
Scale factor for triangular pressure: $20 \cdot 1/10 = 2 \cdot 01$
Characteristic moments:

$$M_{DA} = 4 \cdot 2(-4 \cdot 83) + 2 \cdot 01(-2 \cdot 63) = -25 \cdot 57 kNm$$
$$M_{AD} = 4 \cdot 2(3 \cdot 51) + 2 \cdot 01(1 \cdot 58) = +17 \cdot 92 kNm$$

From the geometry of the assumed loading and the support moment diagrams, maximum $M_{S(AD)}$ works out to $46 \cdot 59 kNm$ at $1 \cdot 67m$ below joint A.

Characteristic shears:

$$S_{DA} = 4 \cdot 2(17 \cdot 15) + 2 \cdot 01(11 \cdot 47) = +95 \cdot 08 kN = -S_{CB}$$
$$S_{AD} = 4 \cdot 2(-16 \cdot 35) + 2 \cdot 01(-5 \cdot 27) = -79 \cdot 26 kN = -S_{BC}$$

(*b*) *At low water table*

Scale factor for uniform pressure: $54/10 = 5 \cdot 4$
Scale factor for triangular pressure: $20 \cdot 1/10 = 2 \cdot 01$
Characteristic moments:

$$M_{DA} = 5 \cdot 4(-4 \cdot 83) + 2 \cdot 01(-2 \cdot 63) = -31 \cdot 37 kNm$$
$$M_{AD} = 5 \cdot 4(3 \cdot 51) + 2 \cdot 01(1 \cdot 58) = +22 \cdot 13 kNm$$

From the geometry of the assumed loading and the support moment diagrams, maximum $M_{S(AD)}$ works out to $58 \cdot 42 kNm$ at $1 \cdot 66m$ below joint A.

Characteristic shears:

$$S_{DA} = 5 \cdot 4(17 \cdot 15) + 2 \cdot 01(11 \cdot 47) = +115 \cdot 66 kN = -S_{CB}$$
$$S_{AD} = 5 \cdot 4(-16 \cdot 35) + 2 \cdot 01(-5 \cdot 27) = -98 \cdot 88 kN = -S_{BC}$$

For both (*a*) and (*b*) above, qualitative bending moment and shear force diagrams are as presented in Figure 27.5 as those for load cases 5(*a*) and 5(*b*).

(6) *Lateral hydrostatic pressure*

(*a*) *Maximum* (*HWT*)

Scale factor for uniform pressure: $48/10 = 4.8$
Scale factor for triangular pressure: $33.5/10 = 3.35$
Characteristic moments:

$$M_{DA} = 4.8(-4.83) + 3.35(-2.63) = -31.99kNm$$
$$M_{AD} = 4.8(3.51) + 3.35(1.58) = +22.14kNm$$

From the geometry of the assumed loading and the support moment diagrams, maximum $M_{S(AD)}$ works out to $63.51kNm$ at $1.7m$ below joint A.

Characteristic shears:

$$S_{DA} = 4.8(17.15) + 3.35(11.47) = +120.74kN = -S_{CB}$$
$$S_{AD} = 4.8(-16.35) + 3.35(-5.27) = -96.13kN = -S_{BC}$$

(*b*) *Minimum* (*LWT*)

Scale factor for uniform pressure: $28/10 = 2.8$
Scale factor for triangular pressure: $33.5/10 = 3.35$
Characteristic moments:

$$M_{DA} = 2.8(-4.83) + 3.35(-2.63) = -22.33kNm$$
$$M_{AD} = 2.8(3.51) + 3.35(1.58) = +15.12kNm$$

From the geometry of the assumed loading and the support moment diagrams, maximum $M_{S(AD)}$ works out to $44.1kNm$ at $1.73m$ below joint A.

Characteristic shears:

$$S_{DA} = 2.8(17.15) + 3.35(11.47) = +86.44kN = -S_{CB}$$
$$S_{AD} = 2.8(-16.35) + 3.35(-5.27) = -63.43kN = -S_{BC}$$

Again, for both the cases (*a*) and (*b*) above, qualitative bending moment and shear force diagrams are presented in Figure 27.5 as those for load cases 6(*a*) and 6(*b*).

(7) *Hydrostatic uplift pressure*

(*a*) *During construction* (*HWT*)

Design uplift pressure: $68.2kN/m^2$
Scale factor: $68.2/10 = 6.82$

Characteristic moments:

$$M_{DC} = 6{\cdot}82\,(+13{\cdot}18) = +89{\cdot}89kNm$$
$$M_{AB} = 6{\cdot}82\,(+2{\cdot}02) = +13{\cdot}78kNm$$
$$M_{S(DC)} = -68{\cdot}2 \times 5{\cdot}05^2/8 + 89{\cdot}89 = -127{\cdot}52kNm$$
$$M_{S(AB)} = \text{As } M_{AB} = +13{\cdot}78kNm$$

Characteristic shears:

$$S_{DA} = 6{\cdot}82 \times (+4{\cdot}54) = +30{\cdot}96kN = -S_{CB}$$
$$S_{DC} = 6{\cdot}82 \times (-25{\cdot}25) = -172{\cdot}20kN = -S_{CD}$$

(b) *During construction* (LWT)

Design uplift pressure: $48{\cdot}2kN/m^2$

Scale factor: $48{\cdot}2/10 = 4{\cdot}82$

Characteristic moments:

$$M_{DC} = 4{\cdot}82\,(+13{\cdot}18) = +63{\cdot}53kNm$$
$$M_{AB} = 4{\cdot}82\,(+2{\cdot}02) = +9{\cdot}74kNm$$
$$M_{S(DC)} = -48{\cdot}2 \times 5{\cdot}05^2/8 + 63{\cdot}53 = -90{\cdot}12kNm$$
$$M_{S(AB)} = \text{As } M_{AB} = +9{\cdot}74kNm$$

Characteristic shears:

$$S_{DA} = 4{\cdot}82 \times (+4{\cdot}54) = +21{\cdot}88kN = -S_{CB}$$
$$S_{DC} = 4{\cdot}82 \times (-25{\cdot}25) = -121{\cdot}70kN = -S_{CD}$$

(c) *In-service*

Design uplift pressure: $23{\cdot}2kN/m^2$

Scale factor: $23{\cdot}2/10 = 2{\cdot}32$

Characteristic moments:

$$M_{DC} = 2{\cdot}32\,(+13{\cdot}18) = +30{\cdot}58kNm$$
$$M_{AB} = 2{\cdot}32\,(+2{\cdot}02) = +4{\cdot}69kNm$$
$$M_{S(DC)} = -23{\cdot}2 \times 5{\cdot}05^2/8 + 30{\cdot}58 = -43{\cdot}38kNm$$
$$M_{S(AB)} = \text{As } M_{AB} = +4{\cdot}69kNm$$

Characteristic shears:

$$S_{DA} = 2{\cdot}32 \times (+4{\cdot}54) = +10{\cdot}53kN = -S_{CB}$$
$$S_{DC} = 2{\cdot}32 \times (-25{\cdot}25) = -58{\cdot}58kN = -S_{CD}$$

Qualitative bending moment and shear force diagrams for all the cases (*a*), (*b*) and (*c*) are presented in Figure 27.5 as those for load cases 7(*a*), 7(*b*) and 7(*c*).

(8) *Lateral surcharge load from future construction* (30kN/m²)

This is an asymmetric lateral load (applied to one side at a time but considered reversible) of $30kN/m^2$ intensity. This can be split into symmetric and skew-symmetric load cases of half the intensity (i.e. $15kN/m^2$) each. The scale factor for each case is therefore 15/10, i.e. 1·5, and note, not 3. The stress resultants for these two cases obtained by reference to the two relevant notional load cases, when combined, will represent the results for the asymmetric load case. The results are presented in Table 27.1 in the section that follows and their qualitative diagrammatic representation is given in Figures 27.5 as those for load cases 8(*a*) and 8(*b*).

It is assumed that the future construction could take place on either side and the resulting surcharge loads, likewise, could occur on either side; the results in Table 27.1 should therefore be treated as reversible.

Presentation of results

The results of ultimate moments and shear forces are presented in Tables 27.2 and 27.3.

Note: Maximum span moments in wall *DA* (and likewise in wall *CB*) for load cases 1, 2, 3, and 7 listed in the table should be those interpolated from the corresponding values at *D* and *A* (also *C* and *B* for wall *CB*) for locations at which, in combination with the moments from other coexistent load cases, the worst moments for different load combinations can be obtained. However, for laterally applied load cases, maximum span moments in the walls can, by reference to the relevant sections under 'Scaling-off stress-resultants', be seen to occur, in the case of this particular example, within the range 1·66–1·73m below joint *A*, i.e. practically close to the mid-height of walls. For simplicity, therefore, and without much loss of accuracy in this example, 'maximum' span moments in wall *DA* for load cases 1, 2, 3 and 7 have also been worked out at the mid-height of the wall where these are required to be combined with those of the coexistent lateral load cases in order to maximize the span moments in the walls.

Table 27.1 Stress-Resultants for Asymmetric Load Case

Member	Symmetric case		Skew-sym. case		Asymmetric case	
	M (kNm)	S (kN)	M (kNm)	S (kN)	M (kNm)	S (kN)
AB	−5·26	0·00	−29·54	+11·70	−34·80	+11·70
AD	+5·26	−24·53	+29·54	+0·03	+34·80	−24·50
BA	+5·26	0·00	−29·54	+11·70	−24·27	+11·70
BC	−5·26	+24·53	+29·54	+0·03	+24·27	+24·56
CB	+7·25	−25·73	+54·56	−50·22	+61·80	−75·95
CD	−7·25	0·00	−54·56	+21·60	−61·80	+21·60
DC	+7·25	0·00	−54·56	+21·60	−47·31	+21·60
DA	−7·25	+25·73	+54·56	−50·22	+47·31	−24·50

Table 27.2 Ultimate Bending Moments

No.	Load case	LF	Span	D		Span	A		Span
				DC	DA		AD	AB	
1a	Self-weight – Roof	1·4	−44	+21	−21	−24	+27	−27	+38
1b	Self-weight – Walls	1·4	−36	+26	−26	−11	−4	+4	+4
2	Backfill on Roof	1·4	−339	+161	−161	−185	+208	−208	+292
3	HA on Roof	1·4	−30	+15	−15	−17	+19	−19	+26
4a	Active Press.-HWT	1·4	+24	+24	−24	+48	+17	−17	−17
4b	Active Press.-LWT	1·4	+29	+29	−29	+57	+21	−21	−21
5a	At Rest Press.-HWT	1·4	+36	+36	−36	+65	+25	−25	−25
5b	At Rest Press.-LWT	1·4	+44	+44	−44	+82	+31	−31	−31
6a	Lat. Hyd. Press.-HWT	1·2	+38	+38	−38	+76	+27	−27	−27
6b	Lat. Hyd. Press.-LWT	1·2	+27	+27	−27	+53	+18	−18	−18
7a	Uplift-Const.-HWT	1·2	−153	+108	−108	−46	−17	+17	+17
7b	Uplift-Const.-LWT	1·2	−108	+76	−76	−32	−12	+12	+12
7c	Hyd. Uplift-In Serv.	1·2	−52	+37	−37	−16	−6	+6	+6
8a	Constn. Surcharge	1·4	−10	−66	+66	+9	+49	−49	−8
8b	-do- reversed	1·4	−10	+87	−87	+18	−34	+34	−8
	Max. Ultimate Moment (kNm):		−510	+421	−421	−142	+345	−345	+317
	Min. Ultimate Moment (kNm):		−132	+179	−179	+43	+50	−50	+15

Table 27.3 Ultimate Shear Forces

No.	Load case	LF	AB	BA	BC	CB	CD	DC
1a	Self-weight – Roof	1·4	+51	−51	−2	−2	+51	−51
1b	Self-weight – Walls	1·4	0·00	0·00	−9	−9	+49	−49
2	Backfill on Roof	1·4	+396	−396	−14	−14	+396	−396
3	HA on Roof	1·4	+35	−35	−1	−1	+35	−35
4a	Active Press.-HWT	1·4	0·00	0·00	+74	−89	0·00	0·00
4b	Active Press.-LWT	1·4	0·00	0·00	+92	−108	0·00	0·00
5a	At Rest Press.-HWT	1·4	0·00	0·00	+111	−133	0·00	0·00
5b	At Rest Press.-LWT	1·4	0·00	0·00	+138	−162	0·00	0·00
6a	Lat. Hyd. Press.-Max.	1·2	0·00	0·00	+115	−145	0·00	0·00
6b	Lat. Hyd. Press.-Min.	1·2	0·00	0·00	+76	−104	0·00	0.00
7a	Uplift-Const. HWT	1·2	0·00	0·00	−37	−37	+207	−207
7b	Uplift-Const. LWT	1·2	0·00	0·00	−26	−26	+146	−146
7c	Hyd. Uplift-In Serv.	1·2	0·00	0·00	−13	−13	+70	−70
8a	Constn. Surcharge	1·4	+16	+16	+34	−106	+30	+30
8b	-do- reversed	1·4	−16	−16	+34	+34	−30	−30
	Max. Ultimate Shear (kN):		+498	−498	+222	−423	+631	−631
	Min. Ultimate Shear (kN):		+51	−51	+117	−249	+246	−246

Conclusions

From the solved example in the preceding section, following broad conclusions may be drawn:

- In the construction of cut-and-cover structures, the net weight of the soil overburden removed is, invariably, greater than that of the structure replacing it, implying that the formation and the structure are more likely to be subject to the effects of heave rather than settlement. However, where commercial developments aboveground are integrated with the metro structure below-ground, this may not always be the case; the possibility of net increase in the effective stress at discrete locations cannot be ruled out.

- By reference to Tables 27.2 and 27.3, it can be seen that the most dominant loading case in this example is provided by the backfill on top. This represents the single most significant contribution to the stress-resultants accounting for up to 80 per cent. However, with the backfill being $5·5m$ deep, this is not surprising. The second most dominant load case is the lateral hydrostatic pressure, contributing up to 44 per cent to the stress-resultants in the various elements; lateral earth pressure, contributing up to 43 per cent of the stress-resultants at different locations, runs close third. The self-weight of the structure accounts for only up to 16 per cent contribution. This distribution is a function of the particular geometry of this example and must not be generalized for all such structures.

- Load case 7(*a*) represents the head differential, with the water level at the high water mark on the outside and drawn down to the formation on the inside of the box, causing a maximum uplift pressure on the base slab. This presents the worst uplift loading case. However, it should be considered as realistic only for the construction stage when the backfill has not been replaced and the ambient water table not restored. Accordingly, such a load case should be used in combinations only selectively and where appropriate.

- The stress-resultants as indicated in the tables, particularly in the vertical members of the box structure, arising out of the lateral surcharge load from future construction are also significant. This is because no soil restraint is assumed on the opposite side wall and the distortion caused by the sway effect due to this out-of-balance lateral load is assumed to be entirely resisted by the frame action of the box structure. Being only a potential load case for the future, it has been used only selectively in load combinations which maximize the stress-resultants.

Appendix A
Typical Loadings and Design Parameters

A.1 Typical loadings

Assessment of loads likely to be imposed on a cut-and-cover structure constitutes an important aspect in the design development process. Efficiency in design will, to a large extent, reflect how accurately the loads have been evaluated. In view of this, it should be particularly noted that the values for the various loadings given hereunder are purely indicative and may be used strictly for guidance only in preliminary design.

Self-weight of rails

Self-weight of a typical rail may be taken as $1 \cdot 5 kN/m$ run.

Ceilings

Self-weight of false ceilings may be taken as $2 \cdot 25 kN/m^2$.

Permanent way loads

Standard railway static loading (Figure A.1 *a* and *b*) to cover the rolling stock:

- Maximum normal axle load: $200kN$
- Minimum spacing of axles: $2m$
- Axle spacing per bogie at: $2m$, $13m$ and $2m$ centres
- Axle spacing between bogies: $4m$ centres
- Check bogie axle load: $240kN$
- Check bogie axle spacing: $1 \cdot 5m$.

The static loading shall be multiplied by an appropriate dynamic factor for spans along the length of the track as follows:

- For spans of $3m$ or less: $1 \cdot 55$
- For spans of $10m$ or more: $1 \cdot 20$
- For span L between 3 and $10m$: $1 \cdot 55 - 0 \cdot 35 \left(\dfrac{L-3}{7} \right)$.

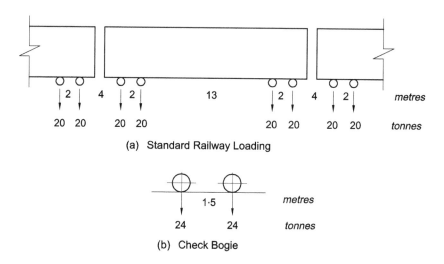

(a) Standard Railway Loading

(b) Check Bogie

Figure A.1 Train Loads for Design

For spans at right angles to the length of the track,

- Dynamic factor: 1·40.

Lurching

To allow for the temporary transfer of live load from one rail to the other due to lurching, 0·56 of the track load may be considered acting on one rail concurrently with the remainder, i.e. 0·44 of the track load, on the other.

Nosing

To allow for the effect of nosing, a single force of $100kN$ acting horizontally at right angles to and anywhere along the track (in either direction) at rail level may be considered.

Centrifugal force

On curved tracks (radius R in metres), an allowance for centrifugal force F_C per track acting radially at a height of $1·8m$ above rail level may be made in accordance with the following formula:

$$F_C = 3800/R \ (kN/m) \tag{A.1}$$

All tracks are considered occupied and, in principle, subjected to the action of centrifugal force.

Traction and braking

In considering the longitudinal effects of traction and braking loads, all the axles are considered driven or braked. So, all the axles which can possibly be accommodated within the metro structure are considered subjected to these loads as appropriate. Traction loads are taken at 30 per cent and braking loads at 25 per cent of the normal axle loads. In the case of two tracks operating in the same direction, both the tracks are considered occupied simultaneously and subjected to either traction or braking loads whichever gives the worst effect. In the case of the two tracks carrying trains in opposite directions, the force due to traction is applied to one track and that due to braking on the other, simultaneously, such that both are assumed to act in the same direction. In the event of more than two tracks being carried on the structure, the longitudinal forces are considered to be operative on two tracks only.

Escalator loadings

Approximate dead load reactions from escalators (carried on two or more supports as appropriate) as a function of the vertical rise H (measured in metres) are given in Table A.1.

Table A.1 Escalator Dead Load Reactions

Max. rise, H (m)	Escalator reaction $\times(2H+1)$ kN		
	Top	Inter.	Bottom
3·00	16·00	–	12·80
5·00	13·70	–	10·50
7·00	6·26	9·90	4·00
9·00	5·36	9·50	3·75
12·00	4·62	8·14	2·79
15·00	3·34	$u:4\cdot60$	2·02
		$l:4\cdot17$	

Notes: u: Upper; l: Lower; Inter.: Intermediate.

Live load reactions from escalators are generally based on crowd loading of $6kN/m^2$ acting on the plan area of the escalator treads. To enable handling of the escalator units for installation and, when required, replacement, provision of hoisting hooks with a carrying capacity of the order of $50kN$ may be needed in the soffites of concourse and the entrance roof slabs in the appropriate locations.

Floor loadings

Structural members may be designed for a live load combination, on concourse, platforms and all amenity areas, of a uniformly distributed load of $6kN/m^2$ and a concentrated load of $20kN$ acting on a square area of $300mm$ side in a location that presents the most onerous condition. However, in the ventilation plant rooms at concourse and platform levels, the uniformly distributed load component may be increased to $10kN/m^2$.

Loading from the plant and machinery shall be based on the actual dead weight of the assembled pieces of equipment duly enhanced by the appropriate dynamic factor to allow for the operation of the equipment in its design mode.

Impact loading

Impact loading from an errant train may be taken as follows:

*   For serviceability limit state, a pair of equivalent static line loads of *20kN/m* run, *1·4m* apart, applied vertically on to the suspended floor element carrying it anywhere within *2m* either side of and parallel to the track, or an individual concentrated load of *75kN* anywhere within the above defined limits, whichever yields the worse result.
*   For the ultimate limit state, four individual equivalent static concentrated loads, *120kN* each, arranged at the corners of a rectangle *2m* long (measured along the track) and *1·4m* wide (measured at right angles to the track) applied anywhere on the suspended floor element carrying it.
*   For the ultimate limit state, transverse component of the kinetic energy upon impact from an errant train taken as $16000\sin^2\theta$, where θ is the angle of impact.

Seismic loading

Evaluation of the seismic loading requires the knowledge of the vertical and horizontal components of earthquake acceleration. The intensity of these can vary from one geographical location to the other depending upon the seismic zone the location falls under and the corresponding seismicity. Accordingly, reference should be made to the particular code appropriate for the seismic zone considered. As a rough guide, the horizontal component of earthquake acceleration may fall within the 0·03–0·05g (may even reach much higher values) range, and the vertical component within the 0·01–0·03g range. However, in recent earthquakes such as those at Northridge and Kobe, measured vertical accelerations were equal to and sometimes even larger than the horizontal accelerations (Hashash *et al.*, 2001).

High vertical accelerations are responsible for generating large compressive loads on metro station columns. Together with lateral deformation, these compressive loads, given the right circumstances, can hasten buckling and, possibly, failure of the columns. Large vertical forces may well have been a contributory factor in the collapse of the Daikai Subway station in Japan. In view of this, the magnitude of the vertical component of ground motion has become an important issue in seismic design.

In the absence of site-specific data, Table A.2 may be used to relate the known peak ground acceleration to the estimate of peak ground velocity.

Seismologists have developed a standard magnitude scale for earthquakes that is completely independent of the type of the instrument used to measure it. It is known as the Moment Magnitude (M_w) and is commonly used for engineering applications. With the current state of knowledge, M_w is considered to be the most accurate magnitude for earthquakes measuring 6·0 or greater.

Moment Magnitude is determined from the amount of energy or strain released. It is a function of the Seismic Moment and is related to it by the expression:

$$M_w = \frac{2}{3}\left[\log_{10} M_0 - 16\right] \tag{A.2}$$

where M_0 = seismic moment (dyne-cm); (dyne = $10^{-8}kN$), = $\mu \times$ rupture (or fault) area $\times$ fault offset (or slip distance), μ = rock rigidity.

Table A.2 Ratio of peak ground velocity to peak ground acceleration at surface

	Moment magnitude (M_w)	(Peak gnd. velocity, cm/s) ÷ (Peak gnd. acceleration, g)		
		Source-to-site distance (km)		
		0–20	20–50	50–100
Rock*	6·5	67	80	90
	7·5	97	110	124
	8·5	120	130	145
Stiff Soil*	6·5	92	98	104
	7·5	130	135	140
	8·5	150	155	161
Soft Soil*	6·5	138	147	156
	7·5	195	203	210
	8·5	225	233	242

*In the table, the sediment types represent the following low-strain shear wave velocity (C_m) ranges *(m/s)*: rock: $\geq$ 750; stiff soil: 200–750; soft soil: < 200. The relationship between peak ground velocity and peak ground acceleration in soft soils is poorly constrained. (after Power *et al.*, 2006)

Magnitudes of earthquakes measured according to various scales have been compared in Figure A.2 (Idriss, 1985). As can be seen in the figure, it is possible to substitute magnitudes from the various scales directly for M_w within the following limits: $M_L < 6$; $m_B < 7\cdot5$; $M_S = 6$–8.

Furthermore, in the absence of more accurate (numerical) methods or data, Table A.3 may be used to determine the relationship of ground motion with depth.

From the table it can be inferred that, in the case of most cut-and-cover metro station structures with soil cover usually less than *6m*, the ratio of ground motions at tunnel depth and at surface is likely to be 1, and for running tunnels, it is unlikely to be less than 0·9.

A.2 Preliminary design soil parameters

Values of soil parameters for preliminary design listed in Table A.4 are based on German practice (*EAU 90*) and form part of their recommendations. The values are presented here purely for reference and comparative purpose only.

Clays

For clays ranging from heavily overconsolidated fissured to soft, typical index properties and effective stress, shear strength parameters collected from a limited number of case histories (Sevaldson, 1956; Bishop and Bjerrum, 1960; Kjaernsli and Simons, 1962; Skempton, 1964) are presented in Table A.5.

Hong Kong residual soils

Based on their experience, Morton *et al.* (1981) suggest the typical properties for the residual soils of Hong Kong as listed in Table A.6.

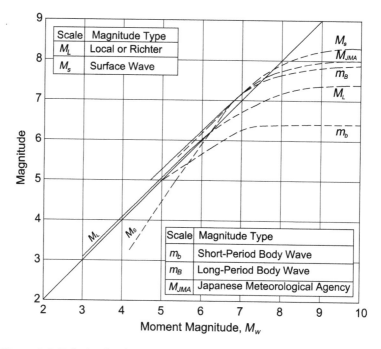

Figure A.2 Relationship between Moment M_w and other Magnitude Scales (adapted from Idriss, 1985, based on unpublished data, Heaton, 1982)

Table A.3 Ratio Between Ground Motion at Depth and at Surface

Tunnel depth (m)	Ratio of ground motions: at Tunnel depth to that at surface
≤6	1·0
6–15	0·9
15–30	0·8
>30	0·7

(after Power *et al.*, 2006)

Estimation of ϕ' values

The values of effective stress shear strength parameters c' and $\tan\phi'$ can be obtained from drained triaxial tests or undrained triaxial tests with porewater pressure measurements. Various aspects, such as the quality of sampling, stress levels, rate of strain for testing, etc. can influence the test results significantly.

Table A.4 Soil Parameters for Preliminary Design

Type of soil	Unit weight $\gamma\,(\gamma_{Sub})$ (kN/m^3)	U-D shear strength cal s_u (kN/m^2)	Effective stress shear strength parameters c'	ϕ'	Modulus of vol. change cal E_s (MN/m^2)
Granular:					
Sand, loose round	18 (10)	–	–	30	20–50
Sand, loose angular	18 (10)	–	–	32·5	40–80
Sand, med-dense round	19 (11)	–	–	32·5	50–100
Sand, med-den angular	19 (11)	–	–	35	80–150
Gravel, without sand	16 (10)	–	–	37·5	100–200
Coarse gravel sharp	18 (11)	–	–	40	150–300
Sand, dense angular	19 (11)	–	–	37·5	150–250
Cohesive: (Empirical values for undisturbed samples from the North German area)					
Clay, soft	17 (7)	10–25	10	17·5	1–2·5
Clay, stiff	18 (8)	25–50	20	20	2·5–5
Clay, semi-firm	19 (9)	50–100	25	25	5–10
Loam, soft	10 (9)	10–25	–	27·5	4–8
Loam semi-firm	21 (11)	50–100	10	27·5	5–20
Boulder clay, Intact	22 (12)	200–700	25	30	30–100
Silty / Organic:					
Soft, v/org, st-c, sea silt	14 (4)	10–20	15	15	0·5–3
Soft, org, sl-c, sea silt	17 (7)	10–25	10	20	2–5
Silt	18 (8)	10–50	–	27·5	3–10
Peat	11 (1)	–	5	15	0·4–1
Peat, under m-i loading	13 (3)	–	10	15	0·8–2

Cal s_u : strength calculated from undrained tests; *v/org*: very organic; *st-c*: strongly clayey; *sl-c*: slightly clayey; *m-i*: moderate initial (adapted from *EAU 90*, 1993)

Table A.5 Soil Parameters for Clays

Type of soil (clays)	Index properties (%)				$c'(kN/m^2)$	$\phi'(deg.)$
	w	LL	PL	PI		
O/C fissured	14–30	45–86	20–33	25–56	10–15	20–25
Brown London clay	30	80	28	52	15	20
O/C intact	30–43	35–60	18–35	18–27	8–12	20–32
Slightly O/C soft	30–35	35–36	18	17–18	1–3	27–32
Normally consolidated	55–65	55–60	25–29	26–35	0	23–32
N/C soft	25–31	30–32	18–19	11–12	0	32

Table A.6 Hong Kong Residual Soils – Typical Properties

Liquid limit (%)	Plastic limit (%)	Permeability (m/sec)	Compressibility (m^2/MN)	c'(kN/m^2)	ϕ' (deg.)
5–15	30–50	2×10^{-5}–10^{-7}	1×10^{-2}–6×10^{-2}	0–5	33–37

(after Morton, Leonard and Cater, 1981)

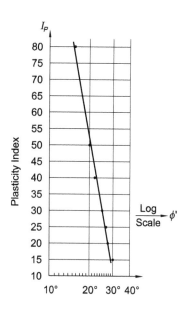

Figure A.3 Pasticity v/s ϕ' – Cohesive Soils

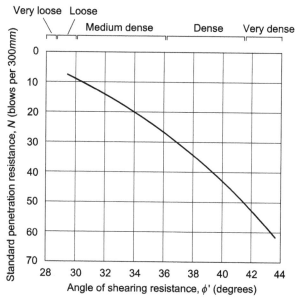

Figure A.4 Estimation of ϕ' for Granular Soils (adapted from Peck, Hanson and Thornburn, 1974)

At the preliminary design stage, however, in the absence of test results, indicative values of ϕ' for cohesive soils may be obtained conservatively from the ϕ'–I_p graph shown in Figure A.3. The points plotted on the graph represent those values for ϕ' as proposed by CIRIA, 1993.

Similarly, in the absence of *in situ* tests, the relationship between standard penetration resistance and ϕ' as represented by the graph in Figure A.4 may be used for granular soils.

For weak rocks, the indicative values of ϕ' as proposed by BS 8002 are given in Table A.7.

Coefficient of earth-at-rest pressure

Various methods of estimating K_o empirically have been outlined in Chapter 23. For normally consolidated clays, the K_o value lies in the range of 0·5–0·66 corresponding to ϕ' values in the range of 30–20°.

For overconsolidated clays, Brooker and Ireland (1965) have correlated K_o with overconsolidation ratio (OCR) for different values of plasticity index (I_p) as shown in Figure A.5. For some clays, such as London Clay, the maximum value of K_o may lie in the range of two to three closer to the surface, and reduce with depth, since *OCR* reduces with depth.

Table A.7 Estimates of ϕ' for Weak Rocks (BS 8002)	
Stratum	ϕ'°
Clayey Marl	28
Weak Mudstone	28
Sandy Marl	33
Chalk	35
Weak Siltstone	35
Weak Sandstone	42

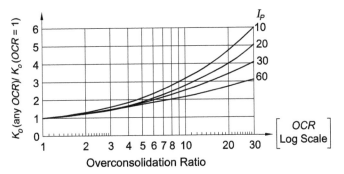

Figure A.5 K_0 v/s OCR
(adapted from Brooker and Ireland, 1965)

Coefficients of active and passive earth pressures

The most frequently used active and passive earth pressure coefficients are those due to Caquot and Kerisel (1948), based on logarithmic spiral failure surface.

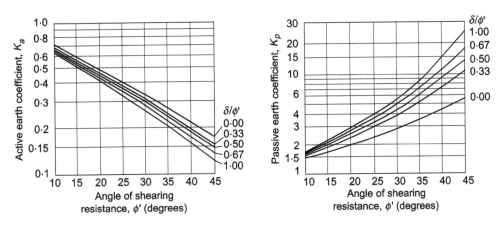

Figure A.6 K_a and K_p (Horizontal Component for Horizontal Retained Surface)
(adapted from Caquot and Kerisel, 1948)

Coefficients of active and passive earth pressures based on their horizontal components and assuming the surface of the retained soil to be horizontal can be obtained from graphs (*a*) and (*b*) in Figure A.6.

Modulus and Poisson's ratio

The values for the Secant Modulus for soils, corresponding to undrained conditions, typically range from 10^3 to $10^5 kN/m^2$ in clays, and 10^4 to $8 \times 10^4 kN/m^2$ in sands. Poisson's ratio for the undrained conditions, v, is 0·5, whereas that corresponding to the drained conditions, v', typically, ranges from 0·2 to 0·4 in clays and from 0·15 to 0·35 in sands.

Undrained shear strength

On the basis of field inspection, the relationship between consistency and strength may be generalized to give a rough guide for the undrained shear strength as shown in Table A.8.

Table A.8 Consistency-Strength Relationship

Consistency	Field indications	$s_u(kN/m^2)$
Very stiff	Brittle or very tough	> 150
Stiff	Cannot be moulded in the fingers	75–150
Firm	Can be moulded by strong pressure	40–75
Soft	Easily moulded in the fingers	20–40
Very soft	Exudes bet. fingers when squeezed	< 20

(from Field Inspection – BS8004)

For normally consolidated clays, Skempton (1970) proposed the following relationship between the undrained shear strength at a particular depth, effective overburden pressure at that depth and the plasticity index:

$$(s_u/p')_{N/C} = 0·11 + 0·0037I_P \tag{A.3}$$

Thus, with I_P in the range of 25–50 per cent for normally consolidated clays, undrained shear strength could be approximated to a proportion of the effective overburden pressure as:

$$(s_u)_{N/C} = (0·2 \ to \ 0·3)p' \tag{A.4}$$

Table A.9 Typical Ranges of Permeability

Soil type	k (m/sec)	Classification
Gravel	$>10^{-3}$	Free Draining
Sand and Gravel	10^{-4}–10^{-3}	Very High
Coarse Sand	10^{-5}–10^{-4}	High
Fine Sand	10^{-6}–10^{-5}	Moderately High
Very Fine Sand	10^{-7}–10^{-6}	Medium
Silt	10^{-8}–10^{-7}	Low
Sandy Clays, Peat	10^{-9}–10^{-7}	Moderately Low
Silty Clays	10^{-9}–10^{-8}	Very Low
Clays	$<10^{-9}$	'Impermeable'

Table A.10 Typical Values of Permeability

Granular and silt	k (m/sec)	Cohesive soils	k (m/sec)
Rockfill	5	N/C, Low Plasticity	$1 \cdot 5 \times 10^{-10}$
Gravel	5×10^{-3}	N/C, High Plasticity	1×10^{-10}
Sand	5×10^{-5}	O/C, Low Plasticity	1×10^{-10}
Silt	5×10^{-8}	O/C, High Plasticity	5×10^{-10}

For overconsolidated clays, Ladd *et al.* (1977) have demonstrated that the following relation holds approximately true:

$$\frac{(s_u/p')_{O/C}}{(s_u/p')_{N/C}} = (OCR)^{0 \cdot 8} \tag{A.5}$$

As an example, for an overconsolidated soil with plasticity index of 25, OCR of 4 and an effective overburden pressure of $80kN/m^2$ at a certain depth, the undrained shear strength for the soil at that depth would work out to be:

$$(s_u)_{O/C} = 0 \cdot 2 \times 80 \times (4)^{0 \cdot 8} = 48 \cdot 5kN/m^2$$

Permeability (k)

Typical ranges of permeability for natural soils are given in Table A.9.

'Constant Head Permeameter' is used for permeabilities down to about 10^{-4} *m/sec*; 'Falling Head Permeameter' is used for permeabilities between 10^{-4} and 10^{-7} *m/sec*; below 10^{-7} *m/sec*, permeability is generally measured by indirect means, e.g. consolidation or dissipation tests.

Typical values for permeability for various soil types may be assumed as given in Table A.10.

Appendix B
Design of Derailment Barrier

B.1 Introduction

The design of derailment barriers is influenced, essentially, by two main factors:

- The angle of impact of the errant train on the barrier
- The magnitude of the permissible deflection of the barrier.

The former governs the quantum of energy that may be necessary to be destroyed by the barrier and, as such, the amount of the impact load for which it needs to be designed; the latter defines the extent of the work the impact load should be capable of doing during the deformation of the barrier in order to achieve the desired result.

Angle of impact is a function of the unobstructed width across tracks. By restricting this width to a minimum through a judicious alignment of the barrier on the inside, or the introduction of a continuous horizontal fender beam along the outer station wall, or both, the angle of impact can be minimized. This can result in a significant reduction in the required thickness of the barrier.

The deformation of the barrier is a function of a number of factors, such as, its height and thickness, percentage of reinforcement provided and, of course, the magnitude of the design load. In the case of the side-platform stations, there is generally a central row of columns separating the up and down tracks which are protected by barriers on either side. If, additionally, these barriers are connected transversely by stiff cross-beams or strips of slab at regular intervals, stiffness/deformation of both the barriers can be brought into play in helping dissipate energy upon impact. Of the various factors mentioned, the height of the barrier is intrinsically related to the vertical alignment of tracks and is, therefore, unlikely to be open to modification.

Before dealing with the design proper of a typical derailment barrier, it is necessary to define the various parameters to be used and the general rationale on which the approach is to be based.

B.2 Definitions

Consider a typical idealized design stress–strain graph for reinforced concrete as shown in Figure B.1. Point '*a*' represents the strain, ε_{co}, at which the maximum stress, σ_{co},

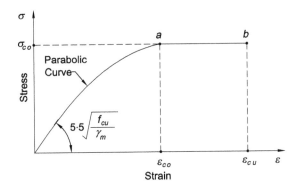

Figure B.1 Design Stress–Strain Curve

is first reached and point '*b*' the maximum limiting strain, ε_{cu}. Let the maximum compressive strain at the extreme fibre corresponding to a tensile strain of ε_s at any given time be represented by ε_{cc}.
Then, according to BS 8110,

$$\varepsilon_{co} = \frac{1}{4000}\sqrt{\frac{f_{cu}}{\gamma_m}}, \quad \varepsilon_{cu} = 0.0035 \quad \text{and} \quad \sigma_{co} = 0.67\left(\frac{f_{cu}}{\gamma_m}\right)$$

B.3 Rationale

To begin with, based on the considerations of practicality, values for both the size (i.e. the thickness) of the barrier and the amount of the main tension (and compression, if required) reinforcement in it are assumed. Then the work done by the residual impact load by its displacement through the deflection of the barrier is equated with the component of energy from the errant train which needs to be destroyed. For the cross-section of the barrier assumed, using a trial and error approach, the impact load and the deflections, both elastic and plastic, are estimated maintaining their compatibility. Adequacy of the assumed amount of reinforcement against the design moment based on the impact load is also ascertained. It is also important to ensure that, upon derailment and impact, the total deflection of the barrier remains comfortably within the permissible limits.

B.4 Design analysis

Consider a reinforced concrete cantilever element subjected at its free end to a transverse load *P*. A typical curvature profile adjacent to the fixed end is shown in Figure B.2.

The transverse component of the residual energy of an errant train upon impact can be destroyed in the work done by the residual impact load during its displacement through the deflection of the barrier. Depending upon the magnitude of the impact load, the resulting deflection can cause curvature of the barrier element, potentially, in three stages:

- Before cracking of concrete – Phase I
- After cracking but before the steel yields – Phase II
- After the steel has yielded – Phase III.

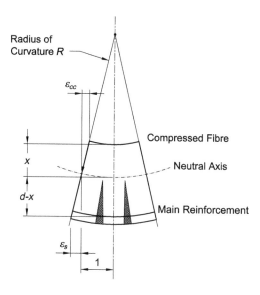

Figure B.2 Typical Curvature Profile (adapted from Regan and Yu, 1973)

The variations of strains and stresses in the concrete across the section of maximum moment for the three phases identified above are shown in Figure B.3.

In Phase I, stresses and strains in the concrete vary linearly across the section as shown in figure (*a*). Furthermore, since the concrete is uncracked, the contribution of the reinforcement does not come into play and its influence can therefore be disregarded in this phase. The end of this phase (Phase I), or the beginning of the next phase (Phase II), is characterized by the initiation of the first crack. The corresponding flexural cracking moment, M_{cr}, is given by:

$$M_{cr} = f_{tub} \cdot z \qquad (B.1)$$

where, f_{tub} is the tensile strength of concrete in bending $= 1 \cdot 5 f_{tu}$, z is the elastic section modulus $= bh^2/6$ for rectangular sections, f_{tu} is the uniaxial tensile strength of concrete taken (BS 8110) as $= 0 \cdot 22 (f_{cu})^{0 \cdot 67}$ or $0 \cdot 36 \sqrt{f_{cu}}$.

The limit state of deflection occurs in Phase II. In the case of derailment barriers particularly, the designer is concerned with the extent of acceptable deflection at the top of the barrier (without hitting a sensitive member) which is the effect of the cumulative curvature along the vertical cantilever, i.e. the integral of the curvature. Integration of curvature from the point of application of the load to the root of the cantilever incorporating both Phases I and II and using appropriate flexural rigidities

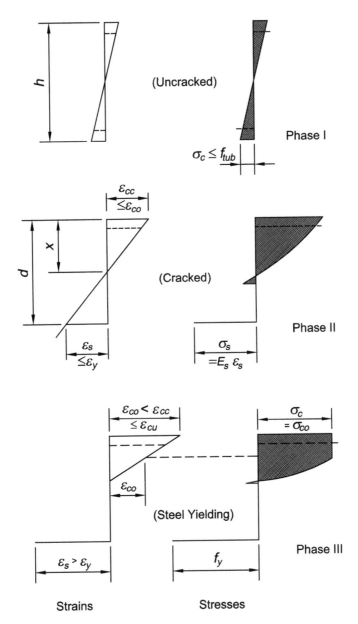

Figure B.3 Typical Stresses and Strains During Loading (adapted from Reqan and Yu, 1973)

(i.e. *EI*-values) corresponding to the cracked and the uncracked lengths would prove to be very complicated. It has been found in practice (Regan and Yu, 1973) that adequate estimates of deflection can be obtained by using a single equivalent value of the flexural rigidity for the entire length of the element as given by:

$$\frac{M_d}{EI_{equiv.}} = \left(\frac{M_{cr}}{E_c I_o}\right) + \left(\frac{M_d - M_{cr}}{0{\cdot}85 E_c I_c}\right) \tag{B.2}$$

The first expression on the right hand side is the contribution from the uncracked section and the second one that from the cracked section, where:

M_d is the design moment within Phase II, E_c is the initial tangent modulus of elasticity for concrete, to be taken $= 4500\sqrt{f_{cu}}(N/mm^2)$ in accordance with BS8110, and I_o is the second moment of area of the uncracked concrete section about its centroidal axis in the plane of bending.

Alternatively, Equation (B.2) can be rearranged to give the equivalent flexural rigidity as:

$$EI_{equiv.} = \frac{M_d}{\left(\dfrac{M_{cr}}{E_c I_o}\right) + \left(\dfrac{M_d - M_{cr}}{0 \cdot 85 E_c I_c}\right)} \tag{B.3}$$

The value of $E_c I_c$ to be used in the above expression is the one towards the end of Phase II and can be obtained from the expression:

$$E_c I_c = \frac{M(d - x)}{\varepsilon_s} \tag{B.4}$$

But with $\gamma_m = 1 \cdot 0$ for both steel and concrete, the steel is likely to yield well before ε_{cc} reaches ε_{co}. However, the rigidity of a member is insensitive to the stage for which it is calculated. As such, it is convenient to compute its value for a stage at which stress block parameters are predefined. In other words, all that is necessary to do is to ensure that the values chosen for M, x and ε_s for use in Equation (B.3) are those that are consistent with and correspond to one value of the steel stress, σ_s. The solved example that follows amplifies this process. Depth to the neutral axis for various percentages and dispositions of reinforcement can be read from the appropriate graphs of Figure B.4.

With the value of the equivalent rigidity thus obtained, estimation of elastic deflection to the end of Phase II becomes a simple procedure. It can be obtained from the expression:

$$\delta_e = \left(\frac{PL^3}{3EI_{equiv.}}\right) \tag{B.5}$$

This then leaves the component of deflection after the steel has yielded, i.e. the plastic deflection in Phase III, to be quantified.

Phase III of the beam's response to the applied load is initiated when the strain in the tensile reinforcement in the region of maximum bending moment reaches the value corresponding to the yielding of the steel. Once this happens, any increase in applied moment can no longer be balanced by an increase in the force in steel. Thus the very limited increase in load possible beyond this stage necessitates an increase in the lever arm between the longitudinal tensile and compressive forces. This, in turn, pushes the neutral axis outwards towards the extreme compression fibre thereby reducing the depth of the compression block. In this process, the area of the compression zone is soon reduced to such an extent that the limiting stress condition (through crushing of concrete) arises in it and this determines the maximum moment of resistance.

Typical moment-rotation curve to failure, along with its simple bilinear idealization, is shown in Figure B.5. It is not easy to define the rotation at its ultimate limit state

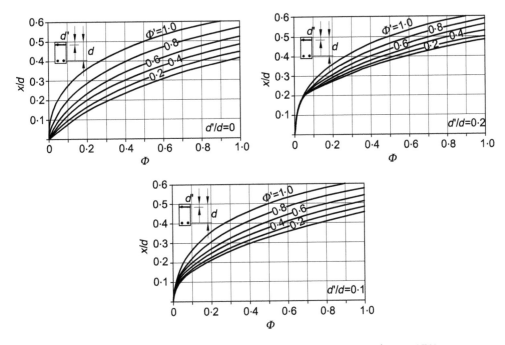

Figure B.4 Depth to Neutral Axis (adapted from Regan and Yu, 1973)

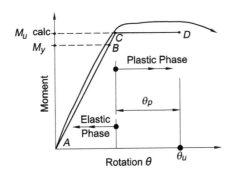

Figure B.5 Moment-Rotation Characteristic (adapted from Regan and Yu, 1973)

(*i.e.* θ_u) so as to quantify θ_p, since the fall of moment with increasing rotation is often very gradual. Although, in practice, if a drop in moment of five to ten per cent is accepted, the final moment corresponding to θ_u is close enough to the calculated M_u.

However, according to the Research committee Report of the Institution of Civil Engineers, London, (1964), rotation in the plastic phase (Phase III) may be obtained from the expression:

$$\theta_p = \left(\frac{\varepsilon_{cu} - \varepsilon_{co}}{x}\right) l_p \tag{B.6}$$

where
$$l_p = K_4 K_5 K_6 \left(\frac{z}{d}\right)^{0\cdot25} d$$

and, $\varepsilon_{cu} = 0\cdot0035$ $\varepsilon_{co} = 0\cdot002$, $K_4 = 0\cdot7$ for mild steel reinforcement, $= 0\cdot9$ for cold-worked reinforcement, $K_5 = 1\cdot0$ for members without axial loads, $K_6 = 0\cdot6$ when $f_{cu} = 40N/mm^2$ and $= 0\cdot9$ when $f_{cu} = 15N/mm^2$, and x is calculated for the condition with $\sigma_s = f_y$, $\varepsilon_{cc} = \varepsilon_{cu}$.

Here, z is the distance from the section of maximum moment to a section of contraflexure. However, for members free from loads along the length z, $M/Vd = z/d$.

With the value of θ_p thus obtained from Equation (B.6), plastic deflection in Phase III can be obtained simply from the expression:

$$\delta_p = L \cdot \theta_p \tag{B.7}$$

With both the elastic and the plastic deflections thus obtained and with the help of the idealized '$P - \Delta$' diagram, the value of the impact force can be obtained by trial and error and the design of the barrier confirmed.

Example B.1

For the residual (after losses) transverse component of energy of $13\cdot5mkN$ from an errant train as arrived at under Section $12\cdot5$ in Chapter 12, design a reinforced concrete derailment barrier of the vertical cantilever type listing, as appropriate, whatever assumptions are made.

Solution
Assumptions

- Thickness of barrier wall, h : $400mm$
- Effective height of barrier, L : $1500mm$
- Tension reinforcement, ρ_s : $1\cdot5\%$
- Compressive reinforcement, $\rho_{s'}$: $0\cdot25\%$
- Clear cover to reinforcement, d_c : $40mm$
- Characteristic cube strength of concrete, f_{cu} : $40N/mm^2$.

Concrete characteristics

For the limit state of deflection, with $\gamma_m = 1\cdot0$, the characteristics of concrete are as follows:

$$E_c = 4\cdot5\sqrt{f_{cu}} = 4\cdot5\sqrt{40} = 28\cdot46kN/mm^2 = 28\cdot46 \times 10^3 N/mm^2$$

$$\sigma_{co} = 0\cdot67f_{cu} = 0\cdot67 \times 40 = 26\cdot8N/mm^2$$

$$\varepsilon_{co} = \left(\sqrt{f_{cu}}\right)/4000 = \left(\sqrt{40}\right)/4000 = 0\cdot0016$$

$$f_{tub} = 1\cdot5f_{tu} = 1\cdot5 \times 0\cdot36\sqrt{f_{cu}} = 1\cdot5 \times 0\cdot36\sqrt{40} = 3\cdot42N/mm^2.$$

Design

The section will remain uncracked for moments up to M_{cr}, i.e. in Phase I.

$$M_{cr} = f_{tub} \cdot z = f_{tub} \frac{bh^2}{6} = 3\cdot42 \left(\frac{1000 \times 400^2}{6} \right) = 91\cdot2 \times 10^6 Nmm$$

$$E_c I_o = E_c \frac{bh^3}{12} = 28\cdot46 \times 10^3 \left(\frac{1000 \times 400^3}{12} \right) = 151\cdot8 \times 10^{12} Nmm^2.$$

In the cracked state, i.e. Phase II, with $\gamma_m = 1\cdot0$, for the limit state of deflection:

$$\Phi = \frac{1\cdot5\rho_s E_s \varepsilon_{co} \gamma_m}{\sigma_{co}} = \frac{1\cdot5 \left(1\cdot5 \times 10^{-2} \right) \left(200 \times 10^3 \right) \left(16 \times 10^{-4} \right) \times 1}{26\cdot8} = 0\cdot269$$

$$\Phi' = \frac{1\cdot5\rho_{s'} E_s \varepsilon_{co} \gamma_m}{\sigma_{co}} = \frac{1\cdot5 \left(0\cdot25 \times 10^{-2} \right) \left(200 \times 10^3 \right) \left(16 \times 10^{-4} \right) \times 1}{26\cdot8} = 0\cdot045$$

Assuming 40*mm* diameter bars, $d' = d_c + 20 = 40 + 20 = 60mm$

$$d'/d = d'/(h - d') = 60/(400 - 60) = 0\cdot18$$

From Figure B.4(c), $x/d = 0\cdot3$, $\therefore x = 0\cdot3 \times 340 = 102mm$.

With $\varepsilon_{cc} = \varepsilon_{co} = 0\cdot0016$, from the geometry of the strain diagram of Figure B.3, corresponding strain in steel,

$$\varepsilon_s = \left(\frac{d-x}{x} \right) \varepsilon_{cc} = \left(\frac{340 - 102}{102} \right) \times 0\cdot0016 = 0\cdot0037,$$

and the stress, $\sigma_s = \varepsilon_s \times E_s = 0\cdot0037 \times 200 \times 10^3 = 740N/mm^2$.

Corresponding to this hypothetical value of stress, co-existent:

$$\text{Tensile Force in steel, } T_s = \sigma_s A_s = 740 \times 0\cdot015bh$$

$$= 740 \times 0\cdot015 \times 1000 \times 400 = 4440 \times 10^3 N;$$

$$\text{Lever Arm, } a = d - 0\cdot375x = 340 - 0\cdot375 \times 102 = 302mm;$$

$$\text{Moment, } M = T_s \times a = 4440 \times 10^3 \times 302 = 1341 \times 10^6 Nmm.$$

Substituting the appropriate values in Equation (B.4), the rigidity per metre run of the barrier can be evaluated as follows:

$$E_c I_c = \frac{M(d-x)}{\varepsilon_s} = \frac{1341 \times 10^6 (340 - 102)}{0\cdot0037} = 86 \times 10^{12} Nmm^2,$$

and for the cracked section, $0\cdot85E_c I_c = 0\cdot85 \times 86 \times 10^{12} = 73 \times 10^{12} Nmm^2$.

In order to obtain the equivalent rigidity, the design moment for the deflection limit state is required. However, the impact load is not known at this stage and that is the problem. For the present and as a first guess, let us assume the equivalent rigidity to be approximately five per cent higher than the rigidity for the cracked section, i.e. $76·5 \times 10^{12} Nmm^2$. Based on this, the elastic deflection (in *mm*) of the cantilever barrier is given by:

$$\delta_e = \frac{1}{3}\left(\frac{PL^3}{EI_{equiv.}}\right) = \frac{1}{3}\left(\frac{P \times 10^3 \times 1500^3}{76·5 \times 10^{12}}\right) = 14·7 \times 10^{-3} \times P \text{ where } P \text{ is in } kN.$$

For the estimation of plastic deflection, let us recall Equation (B.6)

$$\theta_p = \left(\frac{\varepsilon_{cu} - \varepsilon_{co}}{x}\right) l_p$$

where
$$l_p = K_4 K_5 K_6 \left(\tfrac{z}{d}\right)^{0·25} d.$$

$$= 0·7 \times 1·0 \times 0·6 \times \left(\frac{1500}{340}\right)^{0·25} \times 340 = 207$$

Thus,
$$\theta_p = \left(\frac{0·0035 - 0·002}{102}\right) \times 207 = 0·003$$

Then, the plastic deflection is given by:

$$\delta_p = L \cdot \theta_p = 1500 \times 0·003 = 4·5 mm$$

From the idealized '$P - \Delta$' diagram:

$$P\left(0·5\delta_e + \delta_p\right) = \text{ energy to be destroyed}$$

Assuming that, on impact, a minimum of *2m* run of the barrier wall is brought into play to neutralize the energy, then for *1m* run of the wall:

$$P\left(0·5 \times 14·7 \times 10^{-3}P + 4·5\right) = 0·50 \times 13·5 \times 10^3 mmkN, \text{ where } P \text{ is in } kN.$$

Solution of the quadratic equation yields: $P = 700kN$ per *m* run.

[Note: By cross-connecting the barriers on either side of the central row of columns, it would be possible, in the event of impact, to bring into play both the barriers. Such a connection, assuming the cross-connecting members to be 'infinitely' stiff, would yield a lesser *P*-value per wall. On this basis, in the present example, a *P*-value of the order of 437kN per barrier wall would be suggested. This will cause a commensurate reduction in the thickness and, possibly also, in the amount of reinforcement required in each of the walls. However, it is important to recognize that, in order to cater to the potential of derailment to exist both in the up as well as the down tracks, each face of both the barrier walls would need to be identically reinforced.]

  With the value of *P* so obtained and the help of back-analysis, it is obligatory to check whether the assumption of the value of the equivalent rigidity was correct or whether further trials are necessary.

Check value of equivalent rigidity

The design moment for the deflection limit state for which $\gamma_f = 1.0$ is, in the case of the barrier, the cantilever moment per m run

$$M_d = P \cdot L = \left(700 \times 10^3\right) \times 1500 = 1050 \times 10^6 \, Nmm$$

Also recall that: $M_{cr} = 91 \cdot 2 \times 10^6 \, Nmm, 0 \cdot 85 E_c I_c = 73 \times 10^{12} \, Nmm^2$ and $E_c I_o = 151 \cdot 8 \times 10^{12} \, Nmm^2$.

Upon substituting these values in Equation (B.3), we obtain the equivalent rigidity as follows:

$$EI_{equiv.} = \frac{M_d}{\left(\dfrac{M_{cr}}{E_c I_o}\right) + \left(\dfrac{M_d - M_{cr}}{0 \cdot 85 E_c I_c}\right)}$$

$$= \frac{1050 \times 10^6}{\left(\dfrac{91 \cdot 2 \times 10^6}{151 \cdot 8 \times 10^{12}}\right) + \left(\dfrac{1050 \times 10^6 - 91 \cdot 2 \times 10^6}{73 \times 10^{12}}\right)}$$

$$= 76 \cdot 5 \times 10^{12}$$

This turns out to be exactly as assumed! So, no further trial is necessary.

Check adequacy of reinforcement

$$K = M_d / bd^2 f_{cu} = 1050 \times 10^6 / \left(10^3 \times 340^2 \times 40\right) = 0 \cdot 23$$

Elastic deflection, $\delta_e = 15 \times 10^{-3}. P = 15 \times 10^{-3} \times 700 = 10 \cdot 3 mm$

Plastic deflection, $\delta_p = 4 \cdot 5 mm$

Total deflection, $\sum \delta = \delta_e + \delta_p = 10 \cdot 3 + 4 \cdot 5 = 14 \cdot 8$, i.e. of the order of $15 mm$

This is by no means excessive.

By repeating the entire exercise with the barrier wall thickness of 300 instead of the 400*mm* and similar reinforcement percentages as before, the elastic and plastic deflections turn out to be 42 and 5*mm*, respectively, and the total deflection equal to 47*mm*, i.e. of the order of, say, 50*mm*. In this case also, the equivalent rigidity turns out to be, similarly, very close to the rigidity for the cracked section.

Appendix C
Identification of Soil Type

C.1 Normally and overconsolidated soils: definition

Based on its stress history, clay may be classified as normally consolidated or overconsolidated. The clay which has not suffered any erosion and whose current effective overburden pressure, therefore, represents the maximum pressure that it has ever been subjected to is said to be normally consolidated. Great majority of clays are very old which, during their geological history, would have been subjected to a very considerable pressure corresponding to the deposition of several hundred metres of sediments. However, if subsequent removal of part of the overburden through erosion has left the clay in the state whereby the current effective overburden pressure is less than that of the pressure experienced by it in the past, then the clay is overconsolidated. Overconsolidation can also result from groundwater level fluctuations or chemical and weathering effects. The maximum past effective overburden pressure is also known as the preconsolidation pressure (p'_c), and its ratio with the current effective overburden pressure (p'_o) represents the overconsolidation ratio (OCR). Stated mathematically,

$$OCR = \frac{p'_c}{p'_o} \tag{C.1}$$

It is important to recognize that preconsolidation can also result from delayed compression of young, normally consolidated soils. Clay which has been deposited recently and has attained equilibrium under its own weight but has not undergone any significant secondary consolidation may be classified as young normally consolidated clay. Young clay is characterized by the fact that it is capable of carrying just the overburden weight of the soil, but any additional load will result in relatively large settlements. Such a clay soil, if left under constant effective stresses for hundreds or thousands of years, would go on settling.

C.2 Index properties

Interpretation of the index properties or the stress–strain relationship can help establish whether a soil is normally consolidated or overconsolidated. Any index property in isolation may not conclusively define the type of soil although it may offer certain indications. Collectively, however, they can shed significant light.

Table C.1 Sensitivity of Clays

Sensitivity $[S_t]$	Type of clay
Circa 1·0	*Insensitive*
1–2	*Low sensitivity*
2–4	*Medium sensitivity*
4–8	*Sensitive*
8–16	*Extra-sensitive*
>16	*Quick*

(adapted from Skempton, 1953)

Index properties of soils are generally identified by the following:

Moisture content (w), also referred to as the water content, is defined as the weight of water as a proportion of the weight of the 'dry' soil. It is often expressed as a percentage.

Plastic limit (PL) is the moisture content expressed as a percentage at which the soil changes from its semisolid to the plastic state.

Liquid limit (LL) is the moisture content expressed as a percentage at which the soil begins to flow as a liquid, i.e. changes from its plastic to the liquid state.

Atterberg limits is the collective name for the above two limits, i.e. the liquid and the plastic limits.

Plasticity index (PI or I_P) is the difference between the liquid limit and the plastic limit $(PI$ or $I_P = LL - PL)$.

Liquidity index (LI) is expressed as the ratio:

$$LI = \frac{w\% - PL}{I_P} = \frac{w\% - PL}{LL - PL} \tag{C.2}$$

Sensitivity (S_t) is the ratio of the undisturbed to remoulded undrained shear strength of soil, i.e.

$$S_t = \frac{undisturbed \ s_u}{remoulded \ s_u} \tag{C.3}$$

From the point of view of their sensitivity to remoulding, clays may be classified (Skempton, 1953) as listed in Table C.1.

C.3 Preconsolidation pressure

Since, for a given increment or decrement in effective applied load, $\Delta p'$, the deformation (settlement or heave) is much greater for a normally consolidated than an overconsolidated clay, it is important to determine whether or not the clay is overconsolidated. For example:

- If $\Delta p' > (p'_c - p'_o)$, then deformation is likely to be large and so, the clay is likely to be in the normally consolidated range.

- If $\Delta p' < \left(p'_c - p'_o\right)$, then deformation is likely to be small and so, the clay is likely to be in the overconsolidated range.

Whether the clay is normally or overconsolidated can be established with the help of the following:

Casagrande's construction

Use Casagrande's construction to determine preconsolidation pressure p'_c. Also find out the current effective overburden pressure p'_o. Then, if:

- $p'_c = p'_o$, clay is normally consolidated
- $p'_c > p'_o$, clay is overconsolidated
- $\left(p'_c - p'_o\right)$ is constant with depth, there is clear indication of overconsolidation
- p'_c/p'_o is constant with depth, then the 'apparent overconsolidation' is in fact delayed consolidation (Bjerrum, 1967).

Overconsolidation ratio (OCR)

For normally consolidated soils, obviously, $OCR = 1$;
For slightly overconsolidated soils, $OCR > 1{\cdot}0$ and up to $2{\cdot}0$;
For heavily overconsolidated soils, $OCR > 2{\cdot}0$.

Moisture content

w (%)	Soil type (Moisture content grade)
>50	Normally consolidated (High)
30–45	Overconsolidated – In tact (Medium)
20–30	Overconsolidated – fissured (Low)
<20	Heavily overconsolidated – fissured (Very low)

(adapted from Bishop and Bjerrum, 1960)

Moisture content vs. plastic limit

w v/s PL	Soil type
$w \gg PL$, say, 60–30	Normally consolidated
$w > PL$, say, 40–30	Overconsolidated, intact
$w \le PL$, say, 12–13	Heavily overconsolidated, fissured

(adapted from Bishop and Bjerrum, 1960)

Liquidity index

- When w is greater than PL but less than LL, i.e. when $LI < 1$, the soil is likely to be lightly overconsolidated.

- If $w = LL$, then $LI \to 1$, clay is normally consolidated
- If $w = PL$, then $LI \to 0$, clay is overconsolidated
- For 'quick' clays, $LI > 1$
- For normally consolidated clays, $LI = 0.6$–1.0
- For overconsolidated clays, $LI = 0$–0.6
- For heavily overconsolidated fissured clays, $LI < 0$.

Porewater pressure parameter

If the porewater pressure parameter at failure:

- $A_f \to 1.0$, clay is normally consolidated
- < 0.8, clay is overconsolidated.

Other methods

- Compare measured undrained shear strength with that to be expected of normally consolidated clay having a similar Plasticity Index. If measured strength is greater than that anticipated for the normally consolidated clay, the clay is probably overconsolidated.
- Compare the Compression Index C_c corresponding to p'_o with that predicted for a normally consolidated clay by the expression: $C_c = 0.009(LL - 10)$. If C_c at p'_o (i.e. the former) is less than that expected for a normally consolidated clay (i.e. the latter), the clay is probably overconsolidated.

Appendix D
Stability of Simple Slopes

D.1 Spencer's solution for stability of simple slopes with steady-state seepage

Bishop (1955), in his investigation into the stability of slopes using the method of slices, presented a method of analysis in terms of effective stress which satisfied the conditions of force and moment equilibrium and took full account of the inter-slice forces. However, in its application, such a rigorous approach proves to be extremely lengthy. Simplified version of the method, on the other hand, presents an expression for the factor of safety which assumes the inter-slice forces to be horizontal. As a result, the expression, while satisfying the equilibrium with respect to the moments, does not do so with respect to the forces.

Spencer (1967) came up with an alternative, useful and relatively quicker method of analysis, also in terms of the effective stress, for determining the factor of safety which takes into account all the inter-slice forces and satisfies the force as well as the moment equilibrium conditions. For the derivation of the expression used in this method, the reader is advised to look up the appropriate reference. However, graphical representation of the final results of Spencer's work and the step-by-step procedure for following his analysis are presented in the following section.

Consider the geometry of a typical slope as shown in Figure D.1 under the conditions of steady-state seepage. Let the average soil parameters be as follows: unit weight of soil, γ; effective stress shear strength parameters, c' and $\tan\phi'$.

Mobilized angle of shearing resistance is symbolized by ϕ'_m and is given by the expression: $\phi'_m = \tan^{-1}\left(\dfrac{\tan\phi'}{F}\right)$. The construction porewater pressure at any point is expressed as a function of the total weight of the column of soil above that point, i.e. $u = r_u \gamma h$. The nondimensional parameter r_u at a point, known as the porewater pressure coefficient, is defined as the ratio of the construction porewater pressure and the total weight of the column of height h of soil above that point to the surface, i.e. $u/\gamma h$. Variations in the value of Stability Number, $S_n = c'/\gamma HF$, for different values of slope angle β, ϕ'_m and r_u, for use in the Spencer's analysis are shown in the graphs of Figure D.2(a to c).

By comparison of the graphs in the three figures a, b and c, the factor of safety can be clearly seen to be sensitive to variations in r_u. Reasonably accurate results can be obtained if r_u is constant. For minor variations in r_u, however, an average value may be used. In the case of steady-state seepage, a weighted average of r_u must be adopted. While this approach may not lead to the most accurate result, it can at least prove useful in quickly guiding the designer to the most likely critical failure surface.

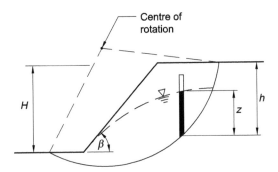

Figure D.1 Geometry of a Typical Slope With Steady-State Seepage

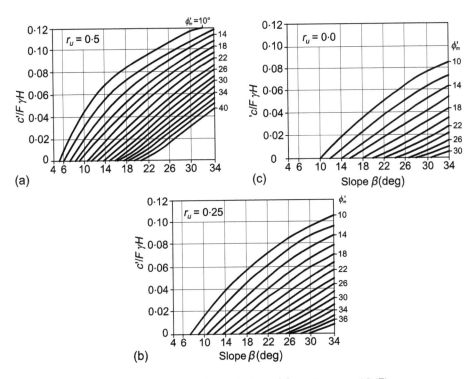

Figure D.2 Stability Charts (adapted from Spencer, 1967)

Given the geometry of the slope, the soil parameters and porewater pressure coefficient as identified earlier, the following step-by-step procedure for estimating the available factor of safety can be adopted:

(1) 'Assume' a value for F_a.

(2) Calculate the stability number, $S_n = \dfrac{c'}{\gamma HF_a}$.

(3) With the value of the stability number so calculated and the slope angle β, enter the appropriate graph in Figure D.2 to obtain the value of ϕ'_m. Note that since this

value is picked up from the appropriate r_u graph it reflects, *ipso facto*, also the effect of the porewater pressure coefficient. This is the value to be used in step (4) below. Note also that the graphs cover only three values, $0.00, 0.25$ and 0.50, of r_u; for other intermediate values, interpolation may be used.

(4) 'Calculate' $F_c = \dfrac{\tan\phi'}{\tan\phi'_m}$.

(5) If the 'assumed' and the 'calculated' values of the factor of safety, i.e. F_a and F_c, are not identical, repeat the steps 2 to 4 for different assumed values of F_a until convergence representing the 'available' factor of safety F $(i.e.\ F_a = F_c = F)$ is achieved.

Where convergence is not achieved in three or four trials, it is useful to plot a graph between the calculated and the assumed F values. This can readily help identify the trend in their variation. The point of intersection at which a $45°$ line drawn from the origin of the axes hits the graph defines the point of convergence representing the 'available' factor of safety against failure for the given slope.

The ease with which the Spencer's charts and the trial-and-error procedure outlined in the foregoing section can be used is demonstrated by the solutions to Examples D.1 and D.2.

Unlike the procedure for estimating the available factor of safety for a given slope which entails the use of trial-and-error process, the procedure for finding a safe slope corresponding to a stipulated factor of safety is much simpler. It can be achieved rapidly in three quick steps with the help of the stability charts as demonstrated in the solution to Example D.3.

Example D.1

For the slope in Figure D.1, find the factor of safety against failure assuming steady-state seepage and the following parameters:

$$\text{Height, } H = 15m;\ \beta = 33.69°;\ \gamma = 18.5 kN/m^3;\ c' = 20 kN/m^2;$$

$$\phi' = 25°;\ r_u = 0.25.$$

Solution

(1) For first trial, assume a factor of safety, $F_a = 1.1$

(2) Stability number, $S_n = \dfrac{c'}{\gamma HF_a} = \dfrac{20}{18.5 \times 15 \times 1.1} = 0.066$

(3) Enter graph (*b*) of Stability Charts with $S_n = 0.066$ and $\beta = 33.69°$, to obtain the mobilized angle $\phi'_m = 18.7°$

(4) Then the calculated factor of safety,

$$F_c = \frac{\tan\phi'}{\tan\phi'_m} = \frac{\tan 25°}{\tan 18.7°} = 1.38 \neq F_a$$

(5) Subsequent trials: Repeat the steps (2) to (4) for further assumed F_a-values of 1.2, 1.3 and 1.4. For $\beta = 33.69°$, $r_u = 0.25$ and the different assumed F_a-values, the results are tabulated as follows:

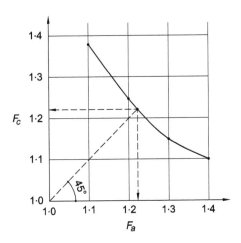

Figure D.3 Plot of F_c vs. F_a with $r_u = 0.25$

Table D.1 Trials for F

F_a	S_n	ϕ'_m	F_c
1·1	0·066	18·7	1·38
1·2	0·060	20·5	1·25
1·3	0·055	22·0	1·15
1·4	0·051	23·0	1·10

Plot of F_c against F_a is shown in Figure D.3.

From the intersection of the $45°$ line drawn from the origin of the axes with the foregoing graph, the available factor of safety $F = 1·22$ is indicated suggesting a safe slope.

Example D.2
Repeat Example D.1 assuming, this time, a porewater pressure coefficient of 0·50.

Solution
Following the similar procedure as above, for $\beta = 33·69°$ and $r_u = 0·50$, the results are tabulated as follows:

Table D.2 Trials for F

F_a	S_n	ϕ'_m	F_c
0·8	0·090	22·5	1·13
0·9	0·080	26·0	0·96
1·0	0·072	29·5	0·82

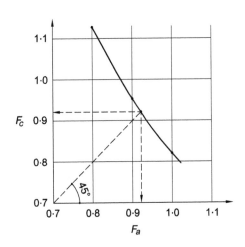

Figure D.4 Plot of F_c vs. F_a with $r_u = 0·50$

Plot of F_c against F_a is shown in Figure D.4.

In comparison with the finding of the previous example, the intersection of the $45°$ line with the graph, in the present case, indicates an available factor of safety $F = 0.92$ (i.e. <1), suggesting an unsafe slope!

Comparison of the results of the two examples confirms the significant influence of the porewater pressure coefficient on the stability of slopes and the need, therefore, of obtaining the porewater pressure profile in the ground as accurately as possible.

Example D.3

For a cutting depth (H) of $15m$ and steady-state seepage conditions, find the safe slope with a factor of safety of 1.5 against failure assuming the following soil parameters:

$$\gamma = 18.5 kN/m^3; c' = 20 kN/m^2; \phi' = 25°; r_u = 0.25.$$

Solution

(1) Stability number, $S_n = \dfrac{c'}{\gamma HF} = \dfrac{20}{18.5 \times 15 \times 1.5} = 0.048$

(2) Mobilized angle, $\phi'_m = \tan^{-1}\left(\dfrac{\tan\phi'}{F}\right) = \tan^{-1}\left(\dfrac{\tan 25°}{1.5}\right) = 17.22°$

(3) By reference to graph (b) of the Stability Charts (i.e. for $r_u = 0.25$), with $S_n = 0.048$ and $\phi'_m = 17.22°$, the required slope to give a factor of safety of 1.5 against failure is approximately $1:2$ (i.e. $\beta = 26°$).

For a value of $r_u = 0.50$, the required slope to give the same factor of safety of 1.5 against failure works out to, approximately, $1:3$ (i.e. $\beta = 18°$). Once again, the comparison of the results underlines the sensitivity of the stability of slopes to the variation in the porewater pressure coefficient.

Appendix E
Diffusion of Stress with Depth

E.1 Butler's method

Values of Influence Coefficients for undrained, elastic $(v = 0.5)$ deformation of the corner of a flexible, uniformly loaded rectangle on the surface of a saturated clay with elastic modulus increasing linearly with depth have been presented in Figure E.1 for L/B ratios of 1, 2 and 5 and various k-values. For intermediate ratios, appropriate values can be obtained by interpolation.

Value of Poisson's ratio (v)

In a three-dimensional case, linear strains can be expressed as follows:

$$\varepsilon_1 = \frac{1}{E}\left[\Delta\sigma_1 - v\left(\Delta\sigma_2 + \Delta\sigma_3\right)\right] \tag{E.1}$$

$$\varepsilon_2 = \frac{1}{E}\left[\Delta\sigma_2 - v\left(\Delta\sigma_3 + \Delta\sigma_1\right)\right] \tag{E.2}$$

$$\varepsilon_3 = \frac{1}{E}\left[\Delta\sigma_3 - v\left(\Delta\sigma_1 + \Delta\sigma_2\right)\right] \tag{E.3}$$

Volumetric strain, $\Delta V/V_o = \varepsilon_1 + \varepsilon_2 + \varepsilon_3$ approximately

For undrained mode, $\Delta V/V_o = 0$. So, for conditions of zero volume change, i.e. saturated clay loaded 'rapidly', $\left(\Delta\sigma_1 + \Delta\sigma_2 + \Delta\sigma_3\right) - 2v\left(\Delta\sigma_1 + \Delta\sigma_2 + \Delta\sigma_3\right) = 0$

whence, $v = 0.5$.

E.2 Fadum's method

Fadum's design curves (Figure E.2) give the stress effect under the corner of a uniformly loaded, rectangular, flexible area. The effect at any other point is obtained by dividing the area into a number of rectangles so as to form a common corner at the point under consideration and superposing contributions from the subscribing rectangles. The data are obtained by integration using the Boussinesq equation for a point load at the surface. Similarly, the technique can be extrapolated for a point outside the area, care being taken to ensure that the effect of the fictitious extension areas is appropriately neutralized.

The use of the chart is illustrated in Figure E.3.

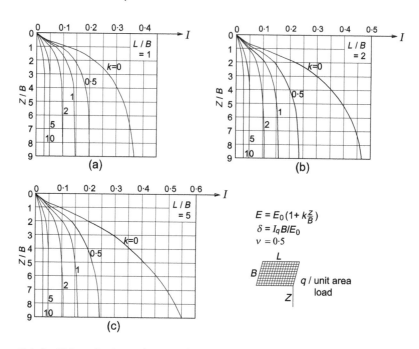

Figure E.1 I – Values for Immediate Settlement of Corner of a Flexible Uniformly Loaded Rectangle on the Surface of Clay; E – values increasing linearly with depth (adapted from Butler, 1975)

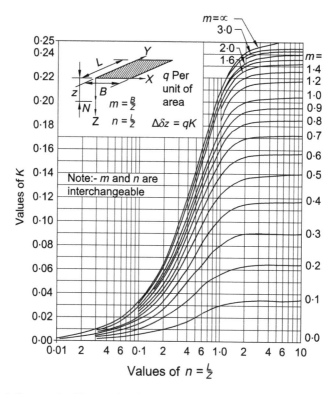

Figure E.2 Influence Coefficients for the Increase in Vertical Stress Under the Corner of Uniformly Loaded Flexible Rectangular Footing (adapted from Fadum, 1948)

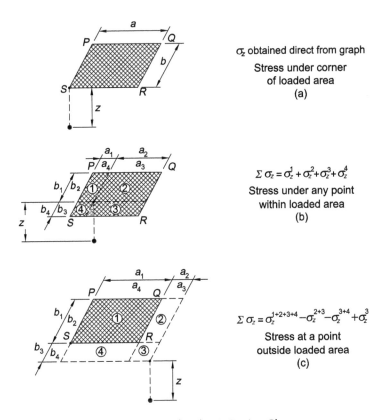

σ_z obtained direct from graph

Stress under corner
of loaded area
(a)

$\Sigma\,\sigma_z = \sigma_z^1 + \sigma_z^2 + \sigma_z^3 + \sigma_z^4$

Stress under any point
within loaded area
(b)

$\Sigma\,\sigma_z = \sigma_z^{1+2+3+4} - \sigma_z^{2+3} - \sigma_z^{3+4} + \sigma_z^3$

Stress at a point
outside loaded area
(c)

Figure E.3 Use of Fadum's Design Charts

E.3 Janbu, Bjerrum and Kjaernsli method

Janbu *et al* (1956) produced useful charts for estimating variation in stress under the centre of uniformly loaded, flexible footings, rectangular or circular in shape. These are reproduced in Figure E.4.

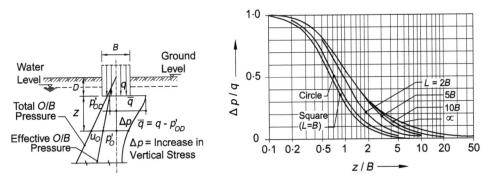

Figure E.4 Determination of Increase in Vertical Stress Under the Centre of Uniformly Loaded Flexible Footing (adapted from Janbu, Bjerrum and Kjaernsli, 1956)

E.4 Jarquio and Jarquio method

Using Boussinesq's basic derivation and further integration, Jarquio and Jarquio (1984) have come up with the variation in stress at the corners of a rectangular area *ABCD* of length L and breadth B under the application of linearly varying load (0 to q/unit area) as shown in Figure E.5.

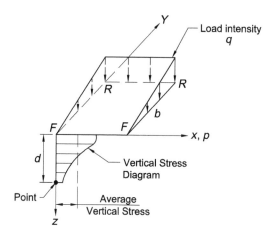

Figure E.5 Vertical Stress Under Linearly Varying Load
(adapted from Jarquio and Jarquio, 1984)

The intensity at depth z is given by:

- Under Front Corners F and F

$$\Delta p_{(F)} = \frac{qz^3}{2\pi L^2 B} \left\{ \frac{L\left(z^2 + A'^2\right)}{z^2 A'} + \ln\left[\frac{(C'+L)(A'-L)}{(C'-L)(A'+L)}\right] - \frac{L\left(B'^2 + C'^2\right)}{B'^2 C'} \right\} \qquad (E.4)$$

- Under Rear Corners R and R

$$\Delta p_{(R)} = \frac{q}{2\pi}\left[\frac{LBz\left(A'^2 + B'^2\right)}{A'^2 B'^2 C'} + \frac{\pi}{2} - \tan^{-1}\left(\frac{zC'}{LB}\right)\right] - \Delta p_{(F)} \qquad (E.5)$$

where

$$A' = \sqrt{\left(Z^2 + L^2\right)}, \quad B' = \sqrt{\left(Z^2 + B^2\right)} \quad \text{and} \quad C' = \sqrt{\left(Z^2 + L^2 + B^2\right)}.$$

For $L/z > 12$, the following approximations may by used without much loss of accuracy:

$$\Delta p_{(F)} = \frac{q}{2\pi}\left(\frac{Bz}{z^2 + B^2}\right) \text{ and } \Delta p_{(R)} = \frac{q}{2\pi}\left[\frac{\pi}{2} - \tan^{-1}\left(\frac{z}{B}\right)\right] \tag{E.6}$$

E.5 Steinbrenner's method

Steinbrenner (1934) expressed the stresses under a foundation in terms of Influence Factor, I_P, given by the expression:

$$I_P = A \cdot F_1 + B \cdot F_2 \tag{E.7}$$

where

$$A = \left(1 - v^2\right) \tag{E.8}$$

$$B = \left(1 - v - 2v^2\right) \tag{E.9}$$

Factors F_1 and F_2 are functions of geometry and the depth of influence and can be obtained from the graphs of Figure E.6, and v is the Poisson's ratio. For a Poisson's ratio of 0·5, $A = 0·75$ and $B = 0$.

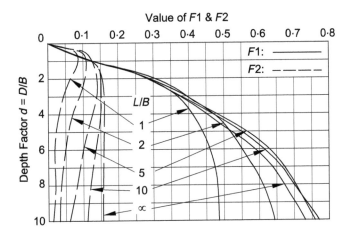

Figure E.6 Factors F1 and F2 (adapted from Steinbrenner, 1934)

E.6 Newmark's method

For uniformly loaded irregular shapes, Newmark (1942) presented charts similar to that shown in Figure E.7.

The loaded area is drawn on transparent paper to a scale such that the depth at which the change in stress is required equals the scale datum on the chart. The transparency

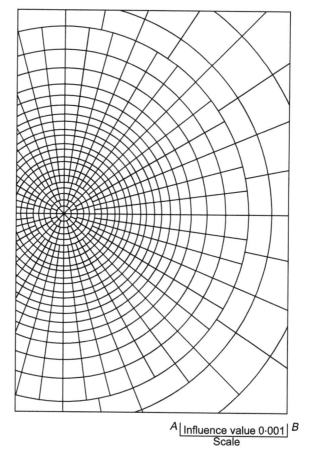

$$A \lfloor \text{Influence value 0·001} \rfloor B$$
Scale

Figure E.7 Influence Chart for Increase in Vertical Stress below a Uniformly Loaded Flexible Area (adapted from Newmark, 1942)

is then superposed on the chart with the point in question, be it inside or outside the area, coincident with the centre of the chart. The number of segments, including the fragments, enclosed within the plan area is estimated. Then the required stress is given by the expression:

$$\sigma_z = I \cdot N \cdot q \tag{E.10}$$

where, I = scale factor for the particular chart used, N = number of segments, and q = load intensity.

Appendix F
Seismic Design Considerations

This section should be read in conjunction with Chapter 15. In dealing with the assessment of seismic loads for design in that chapter, two possible approaches – one based on the inertial force method and the other on seismic deformation method using the racking deformation – have been discussed. Of the two, the former may be applied in the case of open U-shaped running tunnels over the transitional ramped sections where they emerge out of their confinement below ground, whereas the latter is more appropriate to cut-and-cover box structures completely confined below ground.

Considerations for seismic design of cut-and-cover box structures taking into account axial deformation, longitudinal bending and distortional effects are outlined in the following section.

Axial deformation and longitudinal bending

Free-field axial and curvature deformations caused by a harmonic wave propagating at a given angle of incidence in a homogeneous, isotropic, elastic medium as shown in Figure F.1 may be estimated by a simplified method as proposed by Newmark (1968) and Keusel (1969). However, to guard against the adverse uncertainties of seismic predictions, typically, the most critical angle of incidence yielding the maximum deformation is used. This approach, while requiring only a minimal input, provides an order of magnitude estimate of wave-induced strains thereby making it useful both as an initial tool as well as a method of design verification (Wang, 1993).

St. John and Zahrah (1987) used Newmark's approach to develop solutions for free-field axial and curvature strains due to compression, shear and Rayleigh waves. It is often difficult to determine which type of wave will dominate a design. However, strains produced by R-waves tend to govern only in shallow structures and at sites far from the seismic source (Wang, 1993).

Combined axial and flexural deformations can be obtained by treating the cut-and-cover linear structure as an elastic beam. Using elastic beam theory, total free-field axial strains, (ε^{ab}), are found by combining the longitudinal strains generated by axial and bending deformations (Power et al., 2006) as follows:

$$\text{For } P\text{-waves:} \quad \varepsilon^{ab} = \left[\frac{V_P}{C_P} \cdot \cos^2\phi + r \cdot \frac{a_P}{C_P^2} \cdot \sin\phi \cdot \cos^2\phi \right] \qquad \text{(F.1)}$$

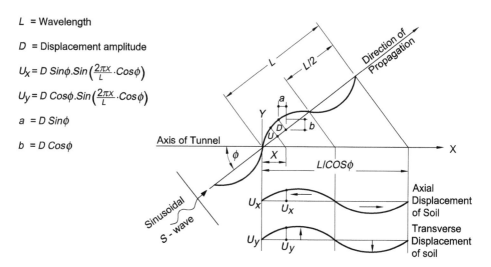

Figure F.1 Sinusoidal Shear Wave Oblique to Tunnel Axis (adapted from Wang, 1993)

$$\text{For } S\text{-waves:}\quad \varepsilon^{ab} = \left[\frac{V_S}{C_S}\cdot\sin\phi\cdot\cos\phi + r\cdot\frac{a_S}{C_S^2}\cdot\cos^3\phi\right]\qquad\text{(F.2)}$$

For *R*-waves (compressional component):

$$\varepsilon^{ab} = \left[\frac{V_R}{C_R}\cdot\cos^2\phi + r\cdot\frac{a_R}{C_R^2}\cdot\sin\phi\cdot\cos^2\phi\right]\qquad\text{(F.3)}$$

The axial stress within the elastic range for *P*-, *S*- and *R*-waves, due to the combined axial and curvature deformations is: $\sigma^{ab} = E\varepsilon^{ab}$, where:

ϕ: angle of incidence with respect to tunnel axis, r: distance from neutral axis to extreme fibre of box section, a_P: peak particle acceleration associated with P-wave, a_S: peak particle acceleration associated with S-wave, a_R: peak particle acceleration associated with R-wave, V_P: peak particle velocity associated with P-wave, V_S: peak particle velocity associated with S-wave, V_R: peak particle velocity associated with R-wave, C_P: apparent velocity of P-wave propagation, C_S: apparent velocity of S-wave propagation, C_R: apparent velocity of R-wave propagation, and E: Young's Modulus of the structure.

The apparent *S*-wave velocities fall in the range of 2–4*km/s* whereas the apparent *P*-wave velocities in the range of 4–8*km/s* (Power *et al.*, 2006).

In Equations (F.1), (F.2) and (F.3), in each case, the first expression within the parenthesis represents the contribution to the longitudinal strain from the axial strain and the second that from the flexural strain. Clearly, as the parameter *r* increases, the contribution of flexural deformation to longitudinal strain also increases. However, computations using free-field equations indicate that the bending component of strain is, in general, relatively small as compared to the axial strains for tunnels under seismic loading. Structures built in rock or stiff soils can be designed using the free-field deformations.

The presence of a cut-and-cover structure and its stiffness modify the free-field ground deformations. Again, the beam-on-elastic foundation approach is used but, in this case, to model quasi-static soil–structure interaction effects. Under seismic loading, the cross-section of a tunnel will experience axial bending and shear strains due to free-field axial, flexural and shear deformations. The maximum structural strains for use in case of structures built in soft ground (St. John and Zahrah, 1987) are as follows:

The maximum axial strain caused by a $45°$ incident shear wave is:

$$\varepsilon^a_{max} = \frac{\left(\frac{2\pi}{L}\right) \cdot A}{2 + \frac{E_t A_c}{K_a} \cdot \left(\frac{2\pi}{L}\right)^2} \leq \frac{fL}{4E_t A_c} \tag{F.4}$$

where, L: wavelength of an ideal sinusoidal shear wave [see Equation (F.9)], A: free-field displacement response amplitude of an ideal sinusoidal shear wave [see Equations (F.11) and (F.12)], E_t: elastic modulus of tunnel material, A_c: cross-sectional area of tunnel structure, K_a: longitudinal spring coefficient of medium [in force per unit deformation per unit length of tunnel; see Equation (F.8)], and f: ultimate soil–structure interface friction force per unit length.

The maximum frictional forces that can be developed at the soil–structure interface limit the axial strain in the tunnel structure. The maximum frictional force, Q_{max}, can be estimated (Sakurai and Takahashi, 1969) as the ultimate frictional force over one quarter of the wave length as shown in Equation (F.4).

The maximum bending strain caused by a $0°$ incident shear wave is:

$$\varepsilon^b_{max} = \frac{\left(\frac{2\pi}{L}\right)^2 \cdot A}{1 + \frac{E_t I_t}{K_t} \cdot \left(\frac{2\pi}{L}\right)^4} \cdot r \tag{F.5}$$

where, I_t: moment of inertia of tunnel cross-section, K_t: transverse spring coefficient of medium [in force per unit deformation per unit length of tunnel; see Equation (F.8)], and r: as defined previously.

The maximum shear force acting on the tunnel structure can be written as a function of this maximum bending strain as:

$$V_{max} = \frac{\left(\frac{2\pi}{L}\right)^3 \cdot E_t \cdot I_t \cdot A}{1 + \frac{E_t I_t}{K_t} \cdot \left(\frac{2\pi}{L}\right)^4} = \left(\frac{2\pi}{L}\right) M_{max} = \left(\frac{2\pi}{L}\right) \left(\frac{E_t I_t \varepsilon^b_{max}}{r}\right) \tag{F.6}$$

For structures built in soft ground, a conservative estimate of the total axial strain (and stress) is obtained by combining the strains from the axial and bending forces (modified from Power *et al.*, 2006 as:

$$\varepsilon^{ab} = \varepsilon^a_{max} + \varepsilon^b_{max} \tag{F.7}$$

Obtaining maximum sectional forces for design involves maximization of forces and displacements with respect to wavelengths. *JSCE* (1975) suggests substituting the values of wavelength that will maximize the stress, back into each respective equation to yield maximum sectional forces. St. John and Zahrah (1987) use similar approach with the exception that the spring coefficients K_a and K_t are considered functions of the incident wavelength:

$$K_a = K_t = \frac{16\pi\, G_m\left(1 - v_m\right)}{\left(3 - 4v_m\right)} \cdot \frac{d}{L} \tag{F.8}$$

where, G_m and v_m are the shear modulus and the Poisson's ratio of the soil medium, and d is the height of the rectangular box structure.

The spring coefficients represent the ratio of the pressure between the structure and the surrounding medium and the reduced displacement of the medium due to the presence of the structure. The coefficients must not only be representative of the dynamic modulus of the ground, but their derivation must also consider (Wang, 1993) the fact that the seismic loading is alternately positive and negative due to the assumed sinusoidal wave.

According to Matsubara *et al.* (1995), the incident wavelength may be estimated as:

$$L = T \cdot C_s \tag{F.9}$$

where, T is the predominant period of the shear wave in the soil deposit on site, or the period at which maximum displacements occur (Dobry *et al.*, 1976; Power *et al.*, 1996).

For a soil deposit of thickness h, if ground motion can be attributed primarily to shear waves and the medium is assumed to consist of a uniform soft soil layer overlying a stiff layer (St. John and Zahrah, 1987), Idriss and Seed (1968) recommend that:

$$T = \frac{4h}{C_S} \tag{F.10}$$

The ground displacement response amplitude, A, represents the spatial variation of ground motions along a horizontal alignment and should be derived considering the site-specific subsurface conditions; it increases with increasing wavelength (*SFBART*, 1960). The displacement amplitude, A, can be computed from the following equations:

For free-field axial strains: $\quad \dfrac{2\pi A}{L} = \dfrac{V_s}{C_s} \cdot \sin\phi \cos\phi \tag{F.11}$

For free-field bending strains: $\quad \dfrac{4\pi^2 A}{L^2} = \dfrac{a_s}{C_s} \cdot \cos^3\phi \tag{F.12}$

Distortional effects

For cut-and-cover box structures subjected to seismic activity, good practice demands particular consideration to the requirements of design and attention to detail. If the analyses indicate that the member elements are unable to withstand the effects of racking distortion or seismic earth pressures safely, appropriate structural modifications need to be considered. Some possible strategies include increase and redistribution of reinforcement and building-in flexibility of the structure. A number of investigators (Kuesel, 1969; Owen and Scholl, 1981; Desai *et al.*, 1997; Hashash *et al.*, 2001) have identified, either directly or by implication, a number of considerations for the seismic design of underground structures and their elements. These are:

1. For the preliminary design of the structure and the initial sizes of the members, use may be made of static design and associated design requirements. The structure so obtained can then be checked for seismic considerations.

2. In a cut-and-cover structure, ground shaking under a seismic activity induces distortion which can be dealt with in two ways. In the pseudo-static approach, the ground displacements are imposed as static loads and the soil–structure interaction disregards dynamic or wave propagation effects. In the dynamic analysis approach, a dynamic soil–structure interaction is carried out using numerical analysis tools such as finite element or finite difference methods. Such an analysis is warranted only when highly variable ground conditions exist on the site and the pseudo-static approach would yield results that are too conservative. The method adopted in the book confines itself to the former approach whereby the seismically-induced racking displacement is imposed on the structure using a simple frame analysis. The extent of the distortion is a function of the member stiffnesses (*EI*s); stiffer the member, greater the effect of distortion. This, unfortunately, cannot be controlled significantly by strengthening the structure. To minimize the effect of distortion, ideally, the structure needs to be inherently flexible.

3. The structure must also be able to carry the static loads safely, and this calls for adequate stiffness and strength. Therefore, under a seismic activity, cut-and-cover structures need to satisfy, seemingly, contradictory requirements, i.e. flexibility and ductility on the one hand, and stiffness and strength on the other. Under the circumstances, the key to achieving efficiency in design lies in ensuring that the structure possesses sufficient ductility to absorb the induced distortion without losing the capacity to carry the static loads safely.

4. The extent by which a box structure can deform is a function of its inherent flexibility which is, in turn, dependent upon the structural details. With the increase in the flexibility, the forces induced in the structure as a result of the seismic activity, correspondingly, decrease. Consequently, from the seismic design standpoint, it is preferable to have a structure that is flexible rather than stiff. On the other hand, increasing the thickness of the member elements increases the stiffness and makes the structure less flexible.

5. The structure should be redesigned if the strength and the ductility requirements are not met and/or, depending upon the performance goals of the structure, the resulting inelastic deformations exceed allowable limits.

6. Flexibility of the structure can be increased by either or both of the following:

   • Introducing ductility in the reinforcement details of the elements
   • Incorporating pinned connections at the slab/wall junctions

7. It is important to recognize that strengthening an overstressed member by increasing its size, i.e. cross-sectional dimensions, may prove to be counterproductive;

although it will increase its stiffness and strength capacity, it may not result in any reduction in the induced forces. More often than not, a flexible configuration with adequate reinforcement to provide sufficient ductility may be a more desirable measure to aim for.

8. If the elastic distortional capacity of the most rigid corner joint exceeds the imposed shearing distortion then, clearly, no further provisions are necessary. Where this may not be the case, pinned joints at slab-wall junctions are often adopted to provide sufficient flexibility. However, in so doing, care must be exercised to avoid the formation of a mechanism which could trigger failure and lead to total collapse of the structure.

9. Where the imposed shearing distortion is in excess of the elastic distortional capacity of the most rigid corner joint of the structure, plastic distortion will be imposed on the less rigid member at that joint. The elastic rotation of the other member may be deducted from the imposed soil distortion to determine the maximum end rotation of the plastically deformed member. If the imposed rotation exceeds this value for a single member, the joint may be designed to distribute plastic yielding to both members of the joint, by equalizing their elastic stiffnesses (this will only be necessary in most unusual circumstances). Care should be exercised in ensuring that shear failures are prevented in members experiencing plastic yielding.

10. Adjacent to the ends of a typical cut-and-cover metro station structure, the end cross-walls will act as rigid diaphragms against the relatively flexible structural frame thereby inhibiting the transverse distortions of the box structure at these locations. Moving away from the cross-walls, however, the deformations will progressively increase attaining their maximum values some distance away. Over the transitional length over which the distortion progresses to its maximum value, special construction details incorporating a modicum of flexibility at the slab-wall junctions to absorb the movements need to be in place. Alternatively, in lieu of the flexibility, the amount of reinforcement could be appropriately increased to take care of the effects likely to be induced by displacements.

11. The junctions between the roof slab and the end cross-walls must be able to withstand the transverse differential motion equivalent to the imposed shearing displacement between the roof and the base slabs. The intermediate junctions must accommodate similar but proportionately smaller displacements. The junction between the longitudinal sidewalls and the end cross-walls must be able to withstand the anticipated racking distortion of the structure. The deformation joints between the walls and floors should preferably be located in the longitudinal floor slabs. The joints should be accessible to permit the necessary repairs of the overstressed members after a seismic event.

12. The prime consideration in the location of deformation joints is that no collapse be imminent because of the plastic deformation of the structural frame forming a mechanism. At all joints where plastic deformation is anticipated or special joints are used especially in the structural elements which are likely to be in contact with soil and water, adequate provision to prevent water leakage must be made.

13. Where the imposed ground shearing distortion does not strain the main structural frame beyond its elastic capacity, all the minor appurtenant structures may be treated as rigidly attached and may be designed as integral parts of the main structure. However, where plastic deformation of the main framework is

anticipated, major appendages (such as the running tunnels, entrance boxes, vent shafts, etc.) should be designed, preferably, as loosely attached (the joint must be designed to be easily repairable or to accommodate differential movement).

14. In the static design, the vertical reinforcement in the inside faces of the outer longitudinal walls is required, principally, over the mid-height regions. However, the worst effects of distortion are likely to be felt both at the top and the bottom and in both the faces (because of the reversibility of racking displacement) of the walls and so, the reinforcement in both the faces of the walls needs to be properly anchored into the top and bottom slabs. Similar principle applies to the requirements of the reinforcement details in the top and the bottom faces of the roof and the base slabs at their junctions with the outer longitudinal walls except that the top and the bottom mats of reinforcement, in this case, need to be anchored into the walls.

15. In the case of vertical members, such as columns, in cut-and-cover subway station structures, the predominant stress-resultant is axial compression. In view of this, the influence of the high vertical accelerations during ground shaking also causing, additionally, significant enhancement of the compressive loads in combination with the lateral racking displacements can be critical to design and safety. Large vertical forces generated by unusually high vertical accelerations may have been a key factor in the failure of the columns and the collapse of the Dakai Subway station in Japan. Accordingly, the extent of the vertical ground acceleration, which is often the subject of much controversy and debate, and the effects of the racking displacements need to be carefully considered. The worst locations are the junctions of the (vertical) columns with the slabs acting as continuous (horizontal) diaphragms. The detailing of these areas must be in consonance with the design assumptions.

16. In general, interior columns, walls, beams and slabs should also be designed to withstand dynamic forces normal to their respective longitudinal axes in addition to the imposed conventional static loads.

17. When structural elements that have no direct contact with the surrounding soil are continuous with the stiff outer shell elements that are strained beyond their elastic rotational capacity, these internal members may also suffer plastic rotation. In such cases, ductile sections or hinges should be designed and incorporated in the connections between these elements.

18. Given the typical geometry of the metro station structures and the extent of the longitudinal reinforcement generally provided, the effects of the axial and the global longitudinal bending deformations are unlikely to be significant. Nevertheless, this should be duly confirmed. However, at the cross-sectional changes like those at the junctions of station structures with running tunnels (i.e. 'step-plate' locations), concentration of stresses can be expected. To forestall any problems, such junctions need to be designed to account for the effects of the longitudinal axial deformations adequately.

19. At interfaces between subway station structures and entrance structures, running tunnels and vent-shafts, owing to the differential stiffnesses between the structures on account of abrupt changes, very large stress-resultants will be generated if a continuous design is used. At such and similar junctions, provision of flexible joint detail is preferable to allow differential movement. Preparatory to the design of the flexible joint, however, it is necessary to establish the allowable longitudinal and

transverse movements and the relative rotations. The joint must also be designed to withstand the static and dynamic earth and water pressures and must remain watertight during an earthquake.

20. Where the soil medium surrounding a cut-and-cover subway structure is susceptible to liquefaction under seismic activity, in theory, one of the problems that the structure can experience is flotation. This tendency is more pronounced in the relatively smaller bored tunnel sections. In the relatively larger cut-and-cover subway structures with the penetration of the perimeter wall well below the main box structure, however, it is more difficult to cause uplift of the structure. In any case, even if the flotation potential of the subway structure itself is mitigated, the use of flexible joints with the bored running tunnels may still be required.

21. In the location of a fault, it is preferable not to straddle the fault but to locate the main subway structure to one or the other side of it. The location of the joint between the subway station and the running tunnel can be made to correspond with the natural location of the fault. Alternatively, a specifically designed flexible joint detail in the running tunnel can be made to correspond with the location of the fault. In the case of tunnels crossing active faults potentially subject to fault displacement, the concept of enlargement of the tunnel to accommodate the displacement is commonly used for retrofit strategies. The length of the tunnel over which it is widened is a function of the extent of the fault displacement as well as the permissible operational curvature of the tracks. For further information on these, reference may be made to the available literature (Rosenbleuth, 1977; Hradilek, 1977; Brown *et al.*, 1981; Owen and Scholl, 1981; Desai *et al.*, 1989; Power *et al.*, 2006).

It is to be expected and should be recognized as such that, in the case of fully confined structures, the profile of the stress resultants based on the boundary displacement due to the racking of the structure which invokes the flexural stiffness of the wall elements can be vastly different from that resulting from the inertial force method using an enhanced lateral earthquake load which disregards the soil–structure interaction. For such structures, therefore, seismic deformation method which takes cognizance of the soil–structure interaction should be used.

References and bibliography

Aas, G. (1976). 'Stability of Slurry Trench Excavation in Soft Clay', *Proceedings, 6th European Conference on SMFE*, Wien, Vol. 1.1, 103–110.

Alpan, I. (1967). 'The Empirical Evaluation of the Coefficients K_o and K_{or}', *Soils and Foundations*, Vol. 7, No. 1, 31.

Atkinson, J. H. (1973). 'The Deformation of Undisturbed London Clay', *Ph. D. Thesis*, University of London, UK.

Atkinson, J. H. and Bransby, P. L. (1978). *The Mechanics of Soils – An Introduction to Critical State Soil Mechanics, University Series in Civil Engineering*, McGraw-Hill Book Company (UK) Limited.

Banayi, M. (1984). 'Stabilization of Earth Walls by Soil Nailing', *Proceedings, 6th Conference on SMFE*, Budapest, Hungary.

BD 31/87. 'The Design of Buried Structures', *Highways Agency*, London, UK.

BD 31/2001. 'The Design of Buried Concrete Box and Portal Frame Structures', *Highways Agency*, London, UK.

Benjamin, A. C., Endicott, L. J. and Blake, R. J. (1978). 'The Design and Construction of some Underground Stations for the Hong Kong Mass Transit Railway System', *The Structural Engineer*, 56A, No. 1, 11–20.

Beresford, J. J. *et al.* (1989). 'Merits of Polymeric Fluids as Support Slurries', *Proceedings, Conference on Piling and Deep Foundations*, London.

Bishop, A. W. *et al.* (1945). 'The Theory of Indentation and Hardness Test', *Proceedings Physical Society*, Vol. 57, p. 147.

Bishop, A. W. (1952). 'The Stability of Earth Dams', *Ph.D. Thesis, University of London*.

—— (1955). 'The Use of Slip Circle in the Stability Analysis of Earth Slopes', *Geotechnique*, Vol. 5, No. 1, 7–17.

—— (1959). 'The Principle of Effective Stress', (Text of lecture to the Norwegian Geotechnical Society, in 1955), *Teknisk Ukeblad*, 106, No. 30, 859–863.

Bishop, A. W. and Bjerrum, L. (1960). 'The Relevance of the Triaxial Test to the Solution of Stability Problems', *Research Conference on Shear Strength of Cohesive Soils, Boulder, Colorado, ASCE*, New York, 437–501.

Bjerrum, L. (1967). 'Engineering Geology of Norwegian Normally Consolidated Marine Clays as related to Settlement of Buildings', *7th Rankine Lecture, Geotechnique*, Vol. 17, 81–118.

—— (1973). 'Problems of Soil Mechanics and Construction on Soft Clays', *8th International Conference on Soil Mechanics and Foundation Engineering*, Moscow, Vol. 3, 111–159.

Bjerrum, L. and Eide, O. (1956). 'Stability of Strutted Excavation in Clay', *Geotechnique*, Vol. 6, No. 1, 32–47.

Bolt, B. A. (1978). *Earthquakes: A Primer*, W. H. Freeman and Company.

Bolton, M. D. (1986). 'The Strength and Dilatancy of Sands', *Geotechnique*, Vol. 36, No. 1, 65–78.

—— (1991). 'Geotechnical Stress Analysis for Bridge Abutment Design', *Transport and Road Research Laboratory*, Contractor Report 270. London: HMSO.

Boussinesq, J. (1883). *Application des Potentials a L'Etude de L'Equilibre et du Mouvement des Solides Elastiques*, Gauthier-Villars, Paris.

—— (1885). '*Application des Potentials a L'étude de L'équilibre, et du Movement des Solides Élastiques avec des Notes Etendues sur Divers Points de Physique, Mathématique et D'analyse*'. Paris, Gauthier-Villais.

Breth, H. and Rhomberg, W. (1972). 'Messungen an Einer Verankerten Wand', *Vortrage der Baugrundtagung 1972 in Stuttgart*. Deutsche Gesellschaft fur Erd-und Grundbau, Essen, 807–823.

Breth, H. and Wanoschek, H. R. (1972). 'The Influence of Foundation Weights upon Earth Pressure Acting on Flexible Strutted Walls', *5th European Conference on SMFE*, Vol. 1, Madrid.

Breth, H. and Amann, P. (1975). 'Time-Settlement and Settlement Distribution with Depth in Frankfurt Clay', *COSOS, British Geotechnical Society*, Cambridge, UK, 141–154.

Brink, F. (1990). 'Large Reinforced Concrete Caissons', *Civil Engineering for Underground Rail Transport*, Edited by J. T. Edwards, Butterworths, 233–251.

Brooker, E. W. and Ireland, H. O. (1965). 'Earth Pressure at Rest Related to Stress History', *Canadian Geotechnical Journal*, Vol. 2, No. 1, 1–15.

Brown, I., Brekke, T. and Korbin, G. (1981). 'Behavior of the Bay Area Rapid Transit Tunnels through the Hayward Fault', *UMTA Report No. CA-06-0120-81-1, USDOT*.

Bruce, D. A. and Jewell, R. A. (1986). 'Soil Nailing: Application and Practice – Part 1', *Ground Engineering*, Nov., Vol. 19, No. 8, 10–15.

—— (1987). 'Soil Nailing: Application and Practice – Part 2', *Ground Engineering*, Jan., Vol. 20, No. 1, 21–33.

Burland, J. B. and Hancock, R. J. R. (1977). 'Underground Car Park at the House of Commons, London: Geotechnical Aspects', *The Structural Engineer*, February, No. 2, Vol. 55, 87–100.

Burland, J. B., Simpson, B. and St. John, H. D. (1979). 'Movements around excavations in London Clay', *Proceedings, 7th European Conference on SMFE*, Brighton, 13–29.

Burland, J. B. *et al.* (2004). *Design and Construction of Deep Basements Including Cut-and-Cover Structures*. The Institution of Structural Engineers, London. March.

BS 5400 (1987). Steel, Concrete and Composite Bridges. Part 2: Specification for Loads, British Standards Institution, London.

BS 8002 (1994). *Code of Practice for Earth Retaining Structures*, British Standards Institution, London.

BS 8004 (1986). *Code of Practice for Foundations*, British Standards Institution, London.

BS 8110 (1985). *Structural Use of Concrete*, Part 1: *Code of Practice for Design and Construction*, Part 2: *Code of Practice for Special Circumstances*, British Standards Institution, London.

Bush, D. I. (1994). *Highways Agency, Private Communication*, May, London, UK.

—— (2000). *Highways Agency, Private Communication*, August, London, UK.

Butler, F. G. (1975). General Report and State-of-the-Art review, Session 3: 'Heavily Overconsolidated Clays', *Proceedings, COSOS, British Geotechnical Society*, Cambridge, UK, 531–578.

Byrne, J. R. (1992). 'Soil Nailing: A Kinematic Analysis', *Proceedings, Conference on Grouting, Soil Improvement and Geosynthetics, ASCE*, Geotechnical Special Publication, New Orleans, Vol. 2, 751–764.

Caquot, A. and Kerisel, J. (1948). *Tables for the Calculation of Passive Pressure, Active Pressure and Bearing Capacity of Foundations*, Gauthier-Villars, Paris, France.

—— (1956). '*Traite de Mecanique des Sols*', 3rd edition, Gauthier-Villars, Paris.

Cartier, G. and Gigan, J. P. (1983). 'Experiments and Observations on Soil Nailing Structures', *Proceedings, 8th European Conference on SMFE*, Helsinki, 473–476.

Cedergren, H. R. (1967). *Seepage, Drainage and Flow Nets*, John Wiley and Sons, New York.

Chang, C. Y. and Duncan, J. M. (1970). 'Analysis of Soil Movement around a Deep Excavation', *Journal SMF Division, Proceedings, ASCE*, Vol. 96, SM5, Sept., 1655–1681.

Christian, J. T. and Carrier, W. D. (1978). 'Janbu, Bjerrum and Kjaernsli's Chart Reinterpreted', *Canadian Geotechnical Journal*, Vol. 15, No. 1, 124–128.

Clough, G. W. and Duncan, J. M. (1971). 'Finite Element Analysis of Retaining Wall Behaviour', *Journal of the Soil Mechanics and Foundation Division, Proceedings, ASCE*, December, SM12, 1657–1673.

Clough, G. W. and O'Rourke, T. D. (1990). 'Construction Induced Movements of *In Situ* Walls', *Proceedings of a Specialty Conference on the Design and Performance of earth Retaining Structure. Cornell, ASCE*, Geotechnical Special Publication No. 25, 439–470.

Clough, G. W. and Tsui, Y. (1974a). 'Finite Element Analysis of Cut-and-Cover Tunnel Constructed with Slurry Trench Walls', *Duke University, Durham, N. C. Soil Mechanics*, Ser. No. 29.

—— (1974b). 'Performance of Tied-back Walls in Clay', *ASCE Journal Geotechnical Division*, Vol. 100, Dec.

Coulomb, C. A. (1773). 'Essai Sur Une Application des Regles de Maximums et Minimis a quelques Problemes de Statique, relatifs a l'Architecture'. *Memoires de Mathematique et du Physique, Presentes, a l'Academie Royale des Sciences*, Paris, Vol. 3, 38.

Creed, *et al.* (1980). 'Back Analysis of the Behaviour of a Diaphragm Wall Supported Excavation in London Clay', *Ground Movements*, Editor Geddes, *Proceedings, 2nd International Conference*, University of Cardiff, Wales.

D'Appolonia, D. J. (1971). 'Effects of Foundation Construction on Nearby Structures', *4th Pan-American Conference on SMFE*, San Juan, Puerto Rico, Vol. 1, 189–236.

—— (1973). Cut-and-Cover Tunnelling. U.S. Dept. Transp., *Federal Highway Administration, San Francisco, Project Review Meeting*, Sept.

D'Appolonia, D. J. and Lambe, T. W. (1970). 'Method for Predicting Initial Settlement', *Journal of Soil Mechanics Foundation Division, Proceedings, ASCE*, Vol. 96, SM2, March, 523–544.

—— (1971a). 'Performance of Four Foundations on End Bearing Piles', *ASCE, Journal of the Soil Mechanics and Foundation Division*, SM1, Jan.

—— (1971b). 'Floating Foundations for Control of Settlement', *Journal SMF Division, Proceedings, ASCE*, Vol. 97, SM6, March, 899–915.

Darcy, H. (1856) *Les Fontaines Publiques del la Ville de Dijon*, Dalmont, Paris.

Das, B. M. (1982). *Fundamentals of Soil Dynamics*, Elsevier.

Das, B. M. (1990). *Principles of Geotechnical Engineering*. Second Edition, PWS-Kent Publishing Company, Boston.

Davis, E. H. and Poulos, H. G. (1963). 'Triaxial Testing and Three-dimensional Settlement Analysis'. *Proceedings of the 4th Australia–New Zealand Conference on Soil Mechanics and Foundation Engineering*, Vol. 1, pp. 233–243.

—— (1968). 'The Use of Elastic Theory for Settlement Prediction under Three-Dimensional Conditions', *Geotechnique*, Vol. 18, 67–91.

Davies, R. and Henkel, D. J. (1980). 'Geotechnical Problems Associated with Construction of Chater Station', *Proceedings, Conference on Mass Transport in Asia*, Hong Kong, paper J3, 1–31.

Desai, D. B., Chu, C-T. and Redd, R. E. (1997). 'Impervious Tunnel Linings in a Land of Shakes', *Proceedings of the 1997 Rapid Excavation Tunneling Conference*, 45–55.

Desai, D. B., Merritt, J. L. and Chang, B. (1989). 'Shake and Slip to Survive – Tunnel Design', *Proceedings of the 1989 Rapid Excavation Tunneling Conference*, June 11–14, Los Angeles, CA, USA.

Dibiagio, E. L. (1966). 'Stresses and Displacements around an Unbraced Rectangular Excavation in an Elastic Medium', *Ph.D Thesis*, University of Illinois.

Dibiagio, E. and Myrvoll, F. (1972). 'Full Scale Field Tests of a Slurry Trench Excavation in Soft Clay', *5th European Conference on SMFE*, Madrid, Vol. IV-3, 461–471.

Dibiagio, E. and Roti, J. A. (1972). 'Earth Pressure Measurements on a Braced Slurry Trench Wall in Soft Clay', *Proceedings, 5th International Conference on SMFE*, Vol. 1, Madrid.

Dobry, R., Oweis, I. and Urzua, A. (1976). 'Simplified Procedures for Estimating the Fundamental Period of a Soil Profile', *Bulletin for the Seismological Society of America*, 66, 4, 1293–1321.

Dowding, C. H. and Rozen, A. (1978). 'Damage to Rock Tunnels from Earthquake Shaking', *Journal Geotechnical Engineering Division, ASCE* 104, (GT2), 175–191.

Drake, D. A., Jackson, M. I. and Doubell, C. I. (1999). 'Design and Construction of Canary Wharf station on the Jubilee Line Extension'. *Proceedings of the Institution of Civil Engineers, Civil Engineering, Jubilee Line Extension*, 132, 47–55.

Drucker, D. C. (1964). 'Stress–Strain-Time Relations and Irreversible Thermodynamics', in *Second-Order Effects in Elasticity Plasticity and Fluid Dynamics*, M. Reiner and D. Abir (Eds), pp. 331–351, Pergamon Press, Oxford.

Duncan, J. M. and Chang, C. Y. (1970). 'Non-linear Analysis of Stress and Strain in Soils', *Journal of Soil Mechanics Foundation Division, Proceedings, ASCE*, Vol. 96, SM5, Sept., 1629–1653.

EAU 90. (1993). 'Recommendations of the Committee for Waterfront Structures, Harbours and Waterways', *Ernst und Sohn*, Berlin, *6th* English Edition.

Egger, P. (1972). 'Influence of Wall Stiffneess and Anchor Prestressing on Earth Pressure Distribution', *Proceedings, 5th International Conference on SMFE*, Vol. 1, Madrid.

Eide, O., Aas, G. and Josang, T. (1972). 'Special Application for Cast-in-Place Walls for Tunnels in Soft Clay in Oslo', *Proc. 5th European Conference on Soil Mechanics and Foundation Engineering*, Madrid, 485–498.

Fadum, R. E. (1948). 'Influence Values for Estimating Stresses in Elastic Foundations', *2nd International Conference on Soil Mechanics and Foundation Engineering*, Rotterdam, Vol. 3, 77–84.

Feld, J. (1968). *Construction Failure*, John Wiley and Sons, New York.

Fernie, R. (1990). 'Concrete Piling Walls', *Civil Engineering for Underground Rail Transport*, Edited by J. T. Edwards, Butterworths, 185–208.

Fraser, D. J. (1981). *Conceptual Design and Preliminary Analysis of Structures*, Pitman Publishing Inc. Massachusetts.

Fraser, R. A. and Jenkins, E. V. (1990). 'Hand-dug Caissons or Wells', *Civil Engineering for Underground Rail Transport*, Edited by J. T. Edwards, Butterworths, 209–231.

Fox, E. N. (1948). The Mean Elastic Settlement of a Uniformly Loaded Area at Depth below the Ground Surface. *Proceedings of the 2nd ICSMFE*, Rotterdam, Vol. 2, pp. 236–246.

Gerber, E. (1929). *Untersuchungen uber die Druckverteilung im Orlich belasteten Sand*, Technische Hochschule, Zurich.

Gere, J. M. (1963). *Moment Distribution*. D. Van Nostrand Company, Inc., Princeton, New Jersey.

Giroud, J. P. (1968). 'Settlement of a Linearly Loaded Rectangular Area', *Journal of the Soil Mechanics and Foundation Division*, ASCE, 94, SM 4, *Proceedings* Paper 6021, Juillet, pp. 813–831.

Gould, J. P. (1970). 'Lateral Stresses on Rigid Permanent Structures', *Proceedings, ASCE Specialty Conference*, Cornell University, NY.

Goldberg, D. T., Jaworski, W. E. and Gordon, M. D. (1976). *Lateral Support Systems and Underpinning, Vol. 1, Design and construction, Vol. 2, Design Fundamentals, Vol. 3, Construction Methods*. Federal Highway Administration, Washington, D.C., FAWA-RD-75, 128.

Golder, H. Q., Gould, J. P., Lambe, T. W., Tschebotarioff G. P. and Wilson, S. D. (1970). 'Predicted Performance of Braced Excavation', *Journal of Soil Mechanics and Foundations Division, Proceedings, ASCE*, May, SM3, 801–815.

Hagerty, D. J. (1969). 'Some Heave Phenomena Associated with Pile Driving', *Ph.D. Thesis*, University of Illinois, Urbana, Illinois.

Hansen, J. Brinch (1961). 'The Ultimate Resistance of Rigid Piles against Transversal Forces', *Danish Geotechnical Institute*, Bulletin No. 12, p. 5.

Harr, H. E. (1962). *Groundwater and Seepage*, McGraw-Hill, New York.

—— (1966). *Foundations of Theoretical Soil Mechanics*, McGraw-Hill, New York.

Hashash, Y. M. A. (2001). 'Seismic Behaviour of Underground Structures and Site Response', *China-US Millennium Symposium on Earthquake Engineering*, Nov. 2000, Beijing, A. A. Balkema.

—— (2006). Private Communication.

Hashash, Y. M. A., Hook, J. J., Schmidt, B. and Yao, J. I-C. (2001). 'Seismic Design and Analysis of Underground Structures', *Tunnelling and Underground Space Technology*, 16, 247–293.

Hashash, Y. M. A., Tseng, W. S. and Krimotat, A. (1998). 'Seismic Soil–Structure Interaction Analysis for Immersed Tube Tunnels Retrofit', *Geotech. Earthquake Eng. Soil Mech.* III 2, *ASCE Geotechnical Special Publication*, No. 75, 1380–1391.

Hashash, Y. M. A. and Whittle, A. J. (1992). Analysis of Braced Diaphragm Walls in Deep Deposits of Clay. Research Report R92-19, MIT Dept of Civil Engineering, 192 p.

—— (1996). 'Ground Movement Prediction for Deep Excavations in Soft Clay', *Journal of Geotechnical Engineering, ASCE*, Vol. 122, No. 6, June.

—— (2002). 'Mechanism of Load Transfer and Arching for Braced Excavations in Clay', *Journal of Geotechnical and Geoenvironmental Engineering*, Vol. 128, No. 3, March 1, 187–197.

Hencky, H. (1923). 'Uber einige statisch bestimmte Falle des Gleichgewichts in plastischen Korpern', *Z. ang. Math.*, Vol. 3, p. 241.

Henkel, D. J. (1960). The Shear Strength of Saturated Remoulded Clay. *Proceedings of the ASCE Research Conference on Shear Strength of Cohesive Soil*, Boulder, Colorado, USA, 533–544.

—— (1970). 'Geotechnical Considerations of Lateral Stresses', *Specialty Conference, ASCE, Cornel University*, Ithaca, NY, 1–50.

—— (1971). 'The Calculation of Earth Pressure in Open Cuts in Soft Clays', *The Arup Journal*, Vol. 6, No. 4, 13–15.

Hokugo, H. (1967). 'Observation of Soil Movement Due to Pile Driving', *Building Engineering News*, Nippon Telegraph and Telephone Public Corporation (in Japanese).

Hradilek, P. J. (1977). 'Behavior of Underground Box Conduit in the San Fernando Earthquake', *The Current State of Knowledge of Lifeline Earthquake Engineering*. ASCE, New York, 308–319.

Huder, J. (1969). 'Deep Braced Excavations with High Groundwater Level', *Proceedings, 7th International Conference on SMFE*, Mexico City.

—— (1972). 'Stability of Bentonite Slurry Trenches with some Experiences in Swiss Practice', *Proceedings, 5th European Conference on SMFE*, Madrid, Vol. IV-9, 517–522.

Hulme, T. W., Potter, L. A. C. and Shirlaw, J. N. (1989). 'Singapore M. R. T. System: Construction', *Proceedings, Institution of Civil Engineers, Part 1*, 86, Aug., 709–770.

Idriss, I. M. (1985). 'Evaluating Seismic Risk in Engineering Practice', *Proceedings, 11th International Conference on SMFE*, San Francisco, Vol. 1, 255–320.

Idriss, I. M. and Seed, H. B. (1968). 'Seismic Response of Horizontal Soil Layers', *Journal Soil Mechanics Foundation Division, ASCE*, 94, (SM4), 1003–1031.

Iida, H., Hiroto, T., Yoshida, N. and Iwafuji, M. (1996). 'Damage to Daikai Subway Station', *Soils and Foundations, Special Issue on Geotechnical Aspects of the January 17, 1995 Hyogoken-Nambu Earthquake. Japanese Geotechnical Society*, 283–300.

Ingold, T. S. (1979). 'The Effects of Compaction on Retaining Walls', *Geotechnique*, 29, No. 3, 265–283.

Jacobson, B. (1958). 'On Pressure in Silos', *Proceedings, 58th Conference on Earth Pressure Problems*, Brussels, Vol. 1, 123–136.

Jaky, J. (1944). 'The Coefficient of Earth Pressure at Rest', *Journal of the Society of Hungarian Architects and Engineers*, Vol. 7, 355–358.

James, E. L. and Phillips, S. H. E. (1971) 'Movement of a Tied Diaphragm Retaining Wall during Excavation', *Ground Engineering*, 4, July, 14–16.

Janbu, N., Bjerrum, L. and Kjaernsli, B. (1956). Veiledning ved Losning av Fundamentering-soppgaver. *NGI Publication*, No. 16, p. 93.

Jarquio, R. (1981). 'Total Lateral Surcharge Pressure Due to a Strip Load', *Journal of the Geotechnical Engineering Division, ASCE*, Vol. 107, No. GT10, Oct., 1424–1428.

Jarquio, R. and Jarquio, V. (1984). 'Vertical Stress Formulas for Triangular Loading', *Journal of the Geotechnical Engineering Division, ASCE*, Vol. 110, No. GT1, Jan., 73–78.

Japanese Society of Civil Engineers (JSCE) (1975). 'Specifications for Earthquake Resistant Design of Submerged Tunnels'.

JSCE (1988). 'Earthquake Resistant Design for Civil Engineering Structures in Japan', *Japanese Society of Civil Engineers*, Tokyo.

Karlsrud, K. (1983). 'Performance and Design of Slurry Walls in Soft Clay', *Norwegian Geotechnical Institute*, Publication NR. 149, Oslo.

Karlsrud, K. and Myrvoll, F. (1976). 'Performance of a Strutted Excavation in Quick Clay in Deep Foundatiopns and Deep Excavations', *Proceedings of the 6th European Conference on Soil Mechanics and Foundation Engineering*, Vienna. ISSMFE Austrian National committee, 1, 157–164.

Karlsrud, K. *et al.* (1980). 'Experience with Slurry Walls in Soft Clay', *Symposium on Design and Construction of Slurry Walls as part of Permanent Structures*, Cambridge, Mass. 1979. *Proceedings US Department of Transportation, Federal Highway Administration*, Washington, Report TS-80-221, 383–393.

Katagiri, M., Saitoh, K., Masuda, T., Aizawa, F. and Ugai, K. (1997). 'Shape Effect on Deformation Behaviour and Stability of Slurry Trench Walls Constructed in Sandy Ground', In: *Proc. IS-Nagoya '97*, Tokyo. Pergamon, 665–670.

——(1998). 'Measurement of the Confining Pressure Around Slurry Trenches in Sandy Ground', In: *Proceedings, Centrifuge '98*, Tokyo, Balkena, 655–660.

Kaul, P. K. (1980). 'State-of-the-Art on Heave in Excavations', *M.Sc. Dissertation*, University of Surrey, UK.

Keusel, T. R. (1969). 'Earthquake Design Criteria for Subways', *Journal Structural Division, ASCE*, ST6, 1213–1231.

——(1969). Discussion Presented at Specialty Session. *7th International Conference on SMFE*, Mexico.

Kitagushi, H. (1976). 'Observed Performance of Diaphragm Walls at Joto Site', Osaka, Aoki Construction Co., Osaka, Apr.

Kjaernsli, B. and Simons, N. (1962). 'Stability Investigations of the North Bank of the Drammen River'. *Geotechnique*, Vol. 12, 147–167.

Kramer, S. (1996). *Geotechnical Earthquake Engineering*. Prentice-Hall, Upper Saddle River, NJ, USA.

Ladd, C. C. and Foott, R. (1974). 'New Design Procedure for Stability of Soft Clays', *Journal of the Geotechnical Engineering Division, ASCE*, Vol. 100, No. GT7, 763–786.

Ladd, C. C., Foott, R., Ishihara, K., Schlosser, F. and Poulos, H. G. (1977). 'Stress Deformation and strength Characteristics', *Proceedings, 9th International Conference on Soil Mechanics and Foundation Engineering*, Tokyo, Vol. 2, 421–494.

Lambe, T. W. (1967). 'Stress Path Method', *ASCE, Journal of the Soil Mechanics and Foundation Division*, SM6, Nov.

——(1968). 'The Behaviour of Foundations during Construction', *Journal SMF Division, Proceedings, ASCE*, Vol. SM1, Jan., 93–130.

——(1970). 'Braced Excavations', *Specialty Conference, ASCE*, Cornel Univ., Ithaca, NY, 149–218.

Lambe, T. W., Wolfskill, L. A. and Wong, I. H. (1970). 'Measured Performance of Braced Excavation', *ASCE, Journal of the Soil mechanics and Foundation Division*, SM3, May.

Lambe, T. W. and Whitman, R. V. (1969). *Soil Mechanics*, John Wiley & Sons Inc., New York.

Lambe, T. W. *et al.* (1972). 'The Performance of a Subway Excavation', *Specialty Conference on the Performance of Earth and Earth-Supported Structures.*

Littlejohn, G. S. (1970). 'Soil Anchors', *Proceedings, Conference on Ground Engineering*, Institution of Civil Engineers, 33–44 (Discussion, 115–120).

Lumb, P. (1964). 'Report on the Settlement of Buildings in the Mongkok District of Kowloon', Hong Kong, *H. K. Government Publications*, Hong Kong.

Lynch, T. J. (1960). 'Pile Driving Experience at Port Everglades', *ASCE, Journal of the Soil Mechanics and Foundation Division*, SM2, Apr.

Mansur, C. I. and Alizadeh, M. (1970). 'Tie Backs in Clay to Support Sheeted Excavation', *ASCE, Journal of the Soil Mechanics and Foundation Division*, SM2, Mar.

Marston, A. (1930). 'The Theory of External Loads on Closed Conduits in the Light of the Latest Experiments', *Bulletin No. 96, Iowa Engineering Experiment Station*, Ames, Iowa.

Massarsch, K. R. (1979). 'Lateral Earth Pressure in Normally Consolidated Clay', *7th European Conference on Soil Mechanics and Foundation Engineering*, Vol. 2, London.

Matheson, D. S. and Koppula, S. D. (1971). 'Discussion of "Analysis of Soil Movement around a Deep Excavation"'. *Journal of the Soil Mechanics and Foundations Division*, Vol. 97, No. 7, 1034–1036.

Matsubara, K., Hirasawa, K. and Urano, K. (1995). 'On the Wavelength for Seismic Design of Underground Pipeline Structures', *Proceedings of the First International Conference on Earthquake Geotechnical Engineering*, 587–590.

Matsuo, H. and O'Hara, S. (1960). 'Lateral Earth Pressures and Stability of Quay Walls During Earthquakes', Proceedings, *2nd World Conference on Earthquake Engineering*, Japan, Vol. 1.

Mayne, P. W. and Kulhawy, F. (1982). 'K_0-OCR Relationships in Soil', *Journal of the Geotechnical Engineering Division*, ASCE, 108, No. GT6, June, 851–872.

McIntosh, D. F., Imamura, M., Walker, A. J. R., Doherty, H. and Eastwood, D. J. (1980). 'Hong Kong M. T. R. Modified Initial System: Design and Construction of Underground Stations and Cut-and-Cover Tunnels', *Proceedings, Institution of Civil Engineers, Part 1*, 68, Nov., 599–626.

Menard, L. (1965). Rules for Calculation of Bearing Capacity and Foundation Settlement Based on Pressuremeter Tests, *Proceedings, 6th International Conference on Soil Mechanics and Foundation Engineering*, Montreal, Canada, Vol. II, 295–299.

Meyerhof, G. G. (1951). 'The Ultimate Bearing Capacity of Foundations', *Geotechnique*, Vol. 2, p. 301.

—— (1972). 'Stability of Slurry Trench Cuts in Saturated Clay', *ASCE, Proceedings of The Specialty Conference on the Performance of Earth and Earth-Supported Structures*, Purdue University.

—— (1985). Private Communication.

Menzies, B. K. (1973). 'The Measurement of the Drained Shear Strength of Soils', *Short Course in Slope Stability*, University of Surrey, 3, 1–11.

Mohr, O. (1900). 'Welche Umstande Bedingen Die Elastizitatsgrenze und den Bruch eines Materiales?', *Zeitschrift des Vereines Deutscher Ingenieure*, Vol. 44, 1524–1530, 1572–1577.

Mononobe, N. (1929). 'On the Determination of Earth Pressures during Earthquakes', *Proceedings, World Engineering Conference*, Vol. 9, 274–280.

Morgenstern, N. and Amir-Tahmasseb, I. (1965). 'The Stability of Slurry Trenches in Cohesionless Soils', *Geotechnique*, 15(4), 387–395.

Morton, K., Leonard, M. S. M. and Cater, R. W. (1980a). 'Observed Settlements of Buildings Adjacent to Stations Constructed for the Modified Initial System of the MTR, Hong Kong', *6th South East Asian Conference on Soil Engineering*, Taipei, 415–429.

—— (1980b). 'Building Settlements and Ground Movements associated with Construction of Two Stations of the Modified Initial System of the Mass Transit Railway, Hong Kong',

Ground Movements, Ed. Geddes, *Proceedings, 2nd International Conference, University of Cardiff*, Wales.

—— (1981). 'Building Settlements and Ground Movements Associated with Construction of Two Stations of the Modified Initial System of the Mass Transit Railway, Hong Kong', *Proceedings, International Conference on Ground Movements and Structures*, Cardiff, 788–801.

Nash, J. K. T. L. (1974a). 'Stability of Trenches Filled with Fluids', *Proceedings, ASCE*, Vol. 100, No. CO4, 533–542.

—— (1974b). 'Diaphragm Wall Construction Techniques', *Proceedings, ASCE*, Vol. 100, No. CO4, 605–620.

Nash, J. K. T. L. and Jones, G. K. (1963). 'The Support of Trenches using Fluid Mud', Proceedings, *The Symposium on Grouts and Drilling Muds in Engineering Practice.* Butterworths, London, 177–180.

Nendza, H. and Klein, K. (1973). 'Bodenverformung Beim Aushub Tiefer Baugruben', *Vortragsveroffentlichungen, Haus der Technik*, Essen, 314, 4–18.

Newmark, N. M. (1942). 'Influence Charts for Computation of Stresses in Elastic Soils', *University of Illinois Engineering Experiment Station, Bulletin No. 338.*

Newmark, N. M. (1968). 'Problems in Wave Propagation in Soil and Rock', *Proceedings of the International Symposium on Wave Propagation and Dynamic Properties of Earth Materials.*

Oblozinsky, P., Ugai, K., Katagiri, M., Saitoh, K., Ishii, T., Masuda, T. and Kuwabara, K. (2001). 'A Design Method for Slurry Trench Wall Stability in Sandy Ground based on the Elasto-Plastic FEM', *Computers and Geotechnics*, Vol. 28, No. 2, Elsevier.

Okabe, S. (1926). 'General Theory of Earth Pressure', *Journal of Japanese Society of Civil Engineers*, Tokyo, Vol. 12, No. 1.

O'Rourke, T. D. (1981). 'Ground Movements Caused by Braced Excavation', *Journal of the Geotechnical Engineering Division*, ASCE, 107, (9), 1159–1178.

Osaimi, A. E. and Clough, G. W. (1979). 'Pore Pressure Dissipation during Excavation', *Journal of Geotechnical Engineering Division, Proceedings, ASCE*, Vol. 105, No. GT4, Apr., 481–498.

Ostadan, F. and Penzien, J. (2001). 'Seismic Design of Cut-and-Cover Sections of the Bay Area Rapid Transit Extension to San Francisco Airport', *2nd US-Japan SSI Workshop*, March, Tsukuba, Japan.

Ostermayer, H. (1974). 'Construction, Carrying Behaviour and Creep Characteristics of Ground Anchors', *Proceedings, Diaphragm Walls Anchorages*, Institution of Civil Engineers, London.

Oteo, C. S., Sagaseta, C., Sainz, J. A. and Ballester, F. (1979). 'Determination of Consolidation Parameters of Soft Clays from a Large Scale Load Test', *Proceedings, 7th European Conference on SMFE*, Brighton, 121–127.

Owen, G. N. and Scholl, R. E. (1981). 'Earthquake Engineering of Large Underground Structures', Report No. FHWA/RD-80/195, *Federal Highway Administration and National Science Foundation.*

Padfield, C. J. and Mair, R. J. (1984). *Design of Retaining Walls Embedded in Stiff Clay.* CIRIA, Report 104, London.

Parkinson, J. J. and Fenoux, G. Y. (1976). 'Bottom Heave of Excavation in Urban Site', *Proceedings, 6th European Conference on SMFE*, Vienna, Vol. 1, 641–646.

Pavlovsky, N. N. (1956). *Collected Works.* Akad. Nauk USSR, Leningrad.

Peck, R. B., Hanson, W. E. and Thornburn, T. H. (1974). *Foundation Engineering*, Second Ed. Wiley, New York.

Peck, R. B. (1969). 'Deep Excavation and Tunnelling in Soft Ground', *State-of-the-Art Report, 7th International Conference on SMFE*, Mexico City, 225–290.

Penzien, J. (2000). 'Seismically Induced Racking of Tunnel Lining', *Earthquake Engineering and Structural Dynamics*, Vol. 29, 683–691.

Piaskowski, A. and Kowalewski, Z. (1965). 'Application of Thixotropic Clay Suspension for Stability of Vertical Sides of Deep Trenches Without Strutting', *Proceedings, 6th International Conference on SMFE*, Montreal, Vol. II, 526–529.

Potts, D. M. (1992). 'The Analysis of Earth Retaining Structures', *Proceedings, Conference on Retaining Structures*, Institution of Civil Engineers, London, 167–186.

Poulos, H. G. (1978). 'Normalized Deformation Parameters for Kaolin', *Geotechnical Testing Journal, GTJODJ*, Vol. 1, No. 2, June, 102–106.

Poulos, H. G. and Davis, E. H. (1974). *Elastic Solutions for Soil and Rock Mechanics*. John Wiley & Sons: New York.

Power, M. (Lead Author), Fishman, K., Makdisi, F., Musser, S., Richards, R. and Youd, T. L. (2006). Seismic Retrofitting Manual for Highway Structures: Part 2 – Retaining Structures, Slopes, Tunnels, Culverts and Roadways.,*Technical Report* MCEER-06-SP11, Buffalo, New York.

Prandtl, L. (1923). 'Compression of Plastic Blocks between Rough Rigid Plates', *Z. Angew. Math. Mech.*, 3, p. 401.

Prentis, E. A. and White, L. (1950). *Underpinning*, Columbia University Press, Second Edition.

Puller, M. (1996). *Deep Excavations – A Practical Manual*, Thomas Telford.

Regan, P. E. and Yu, C. W. (1973). *Limit State Design of Structural Concrete*, Chatto & Windus, London.

Rendulic, L. (1937). 'A Fundamental Principle of Soil Mechanics and its Experimental Verification', (in German). *Bauingenieur*, Vol. 18, p. 459.

Research Committee Report. (1964). 'Ultimate Load Design of Concrete Structures', *Institution of Civil Engineers*, 104.

Roscoe, K. H. and Burland, J. B. (1968). 'On the Generalised Stress–Strain Behaviour of "Wet" Clay,' *Engineering Plasticity*. J. Heyman and F. A. Leckie, eds., Cambridge University Press, Cambridge, England, 535–609.

Rosenbleuth, E. (1977). 'Soil and Rock Mechanics in Earthquake Engineering', *Proceedings of the International Conference on Dynamic methods in Soil and rock mechanics.*

Rowe, P. W. (1968). 'The Influence of Geological Features of Clay Deposits on the Design and Performance of Sand Drains', *I.C.E. London Proceedings*, Supplement Paper 70585, p. 72.

—— (1972). 'The Relevance of Soil Fabric to Site Investigation Practice', *12th Rankine Lecture, Geotechnique*, Vol. 22, No. 2, 195–300.

Rowe, R. (1992). 'Tunnelling in Seismic Zones'. *Tunnels and Tunnelling*, 24, 41–44.

Sakurai, A. and Takahashi, T. (1969). 'Dynamic Stress of Underground Pipeline During Earthquakes', *Proceedings of the Fourth World Conference on Earthquake Engineering.*

Sandhu, B. S. (1974). Earth Pressure on Walls due to Surcharge. *Civil Engineering*, ASCE, Vol. 44, No. 12, 68–70.

Schlosser, F., Unterreiner, P. and Plumelle, C. (1992). 'French Research Program CLOUTERRE on Soil Nailing', *Proceedings, Conference on Grouting, Soil Improvement and Geosynthetics*, ASCE, Geotechnical Special Publication, New Orleans, Vol. 2, 739–750.

Schmertmann, J. H. (1953). 'Estimating True Consolidation Behaviour of Clay from Laboratory Test Results', *Proceedings ASCE*, Vol. 79, Separate No. 311.

Schnabel, P. B., Lysmer, J. and Seed, B. H. (1972). 'SHAKE – A Computer Program for Earthquake Response Analysis of Horizontally Layered Sites'. *Report No. EERC 72-12*, University of California at Berkeley, California.

Schneebeli, G. (1964). 'Le Stabiliti des tranchees profondes forees en presence de boue', *Houille Blanche*, 19(7), 815–820.

Seed, H. B. and Whitman, R. V. (1970). 'Design of Earth Retaining Structures for Dynamic Loads', *Specialty Conference on Lateral Stresses in the Ground and Design of Earth Retaining Structures, Proceedings*, ASCE, 103–147.

Serota, S. and Jennings, R. A. J. (1959). 'The Elastic Heave of the Bottom of Excavations', *Geotechnique*, Vol. 9, No. 1, 62–70.

Sevaldson, R. A. (1956). 'The Slide in Lodalen, October 6th, 1954'. *Geotechnique*, Vol. 6, No. 4, 167–182.

SFBART. (1960). *Technical Supplement to the Engineering Report for Trans-Bay Tube*, July.

Sherwood, D. E. *et al.* (1989). 'Recent Developments in Secant Bored Pile Wall Construction', *Proceedings, Conference on Piling and Deep Foundations*, London, 211–219.

Sherif, M. A., Fang, Y. S. and Sherif, R. I. (1984). 'K_a and K_o Behind Rotating and Non-Yielding Walls', *Journal of the Geotechnical Engineering Division*, ASCE, Vol. 110, No. GT1, 41–56.

Simons, N. E. and Menzies, B. K. (1974). 'A Note on the Principle of Effective Stress', *Geotechnique*, Vol. 24, 259–261.

—— (1977). '*A Short Course in Foundation Engineering*', Newnes-Butterworths, London, UK.

Skempton, A. W. (1948) 'The $\phi = 0$ Analysis of Stability and its Theoretical Basis', *2nd International Conference on SMFE, Rotterdam*, Vol. 1, 72–77.

Skempton, A. (1953). Discussion, *3rd International Conference on Soil Mechanics and Foundation Engineering, Zurich*, Vol. 3, p. 172.

—— (1954). 'The Pore Pressure Coefficients A and B', *Geotechnique*, Vol. 4, 143–147.

—— (1970). 'The Consolidation of Clays by Gravitational Compaction', *Quarterly Journal of the Geological Society*, London, UK.

—— (1977). 'Slope Stability of Cuttings in Brown London Clay', *Proceedings, 9th International Conference on Soil Mechanics and Foundation Engineering*, Tokyo, 261–270.

Skempton, A. W. (1964). 'Long-term stability of clay slopes', Fourth Rankine Lecture. *Geotechnique*, Vol. 14, No. 2, 77–101.

Skopek, J. (1976). 'Elastische Hebung der sohle von Tiefen Baugruben', *Proceedings, 6th European Conference on SMFE*, Vienna, 657–662.

Spangler, M. G. (1933). 'The Supporting Strength of Rigid Pipe Culverts', *Bulletin No. 112, Iowa Engineering Experiment Station*, Ames, Iowa.

—— (1938). 'Horizontal Pressures on Retaining Walls Due to Concentrated Surface Loads', *Iowa State University Engineering Experiment Station*, Bulletin No. 140.

—— (1947). 'Underground Conduits – An Appraisal of Modern Research', *Proceedings, ASCE*, June.

Spangler, M. G. and Handy, R. L. (1960). *Soil Engineering*. 3rd Edition. International Textbook Company, Scranton, Pa.

Spencer, E. (1967). 'A Method of Analysis of the Stability of Embankments Assuming Parallel Inter-Slice Forces', *Geotechnique*, Vol. 17, No. 1, 11–26.

Steinbrenner, W. (1934). *Tafeln zur Setzungberechnung*. Die Strasse, Vol. 1, p. 121.

St. John, C. M. and Zahra, T. F. (1987). 'Aseismic Design of Underground Structures', *Tunnelling Underground Space Technology*, 2(2), 165–197.

Stoll, U. W. (1969). Discussion on 'The Behaviour of Foundation during Construction', *Journal SMF Division, Proceedings, ASCE*, Vol. 95, SM1, Jan., 420–424.

—— (1971). Discussion on 'Floating Foundations for Control of Settlement', *Journal SMF Division, ASCE*, Vol. 97, SM 10, October, 1508–1511.

Swiger, W. F. (1948). 'Effect of Vibration on Piles in Loose Sand', *Proceedings, 2nd International Conference on SMFE*, Vol. II, Rotterdam.

Tamano, T., Fukui, S., Suzuki, H. and Ueshita, K. (1996). 'Stability of Slurry Trench Excavation in Soft Clay', *Soils and Foundations*, Vol. 36, No. 2, June, *Japanese Geotechnical Society*, 101–110.

Tavenas, F. A. (1975). 'In Situ Measurement of Initial Stresses and Deformation Characteristics', *Proceedings, Specialty Conference on 'In Situ Measurement of Soil Properties'*, ASCE, 263–270.

Terzaghi, K. (1922). 'Der Grundbruch an Stauwerken und seine Verhutung', *Die Wasserkraft*, Vol. 17, 445–449.

—— (1943). *Theoretical Soil Mechanics*, Wiley, New York.

—— (1948). Closing Discussion on: 'Foundation Pressure and Settlement of Buildings on Footings and Rafts', *2nd International Conference on Soil Mechanics and Foundation Engineering*, Rotterdam, Vol. 6, p. 118.

Terzaghi, K. and Peck, R. B. (1967). *Soil Mechanics in Engineering Practice*. 2nd Edition, John Wiley & Sons, New York.

Timoshenko, S. P. and Gere, J. M. (1961). *Theory of Elastic Stability*. 2nd Edition, McGraw-Hill Book Co., Inc., Princeton, New Jersey.

Tsai, J. S. (1997). 'Stability of Weak Sub-layers in a Slurry Supported Trench', *Canadian Geotechnical Journal*, Vol. 34.

—— (2002). *Private Communication*.

Tsai, J. S. and Chang, J. C. (1996). 'Three-Dimensional Stability Analysis for Slurry-Filled Tench Wall in Cohesionless Soil', *Canadian Geotechnical Journal*, Vol. 33.

Tschebotarioff, G. P. (1967). *Proceedings, 3rd Pan-American Conference on SMFE*, Caracas. *General Report Division 4*, 301–322.

UK Department of Transport, (1994). 'Design Methods for the Reinforcement of Highway Slopes by Reinforced Soil and Soil Nailing Techniques', *Design Manual 4, Section 1, HA 68/94*.

Verruijt, A. (1970). *Theory of Groundwater Flow*, Macmillan and Co. Ltd.

Wang, J. N. (1993). *Seismic Design of Tunnels: A State-of-the-Art Approach*, Monograph 7, Parsons Brinckerhoff Quade & Douglas, Inc., New York.

Westergaard, H. M. (1933). 'Water Pressure on Dams During Earthquakes', *Transactions of ASCE*, Vol. 98, 418–472.

Whittle, A. J. and Hashash, Y. M. A. (1994). 'Soil Modeling and Prediction of Deep Excavation Behavior', *Proc., Int. Symposium on Pre-Failure Deformation Characteristics of Geo-Materials (IS-Hokkaido '94)*, Vol. 1, Balkema, Rotterdam, The Netherlands, 589–595.

Wong, I. H. (1981). 'Ground Settlement Related to Dewatering for a Subway Excavation', *Proceedings, International Conference on Ground Movements and Structures*, Cardiff, 778–787.

Wong, K. S. (1987). 'A Method to Estimate Wall Deflection of Braced Excavations in Clay', *Fifth International Geotechnical Seminar, case histories in soft clay*. Singapore, Nanyang Technological Institute.

Wood, J. H. (1973). 'Earthquake-Induced Soil Pressures on Structures', *Report No. EERL 73–05*, California Institute of Technology.

Wroth, C. P. (1975). 'In Situ Measurements of Initial Stresses and Deformation Characteristics', *Proceedings, ASCE Specialty Conference on In Situ Measurements of Soil Properties*, Raleigh, 181–230.

Xanthakos, P. P. (1974). 'Underground Construction in Fluid Trenches', *Colleges of Engineering*, University of Illinois, Chicago.

—— (1979). *Slurry Walls*. McGraw-Hill, New York.

Yong, P. M. F. (1985). 'Dynamic Earth Pressures Against A Rigid Earth Retaining Wall', *Central Laboratories Report 5-85/5*, Ministry of Works and Development, New Zealand.

Zeevaert, L. (1950). Discussion of 'Effects of Driving Piles into Soft Clay', *Transactions, ASCE*, Vol. 115.

—— (1973). *Foundation Engineering for Difficult Subsoil Conditions*. Van Nostrand Reinhold Company, New York.

Index

accidental load 229, 239–40; derailment
impact 239–41, 243, 245
active earth pressure 88, 91, 94, 100; at-rest
earth pressure 88, 91, 94, 100
adhesion factor 422–3, 426, 455
admixture 69
advances in trench stability 342, 344;
3-dimensional approach 342, 347, 370;
elasto-plastic *FEM* approach 344; weak
sub-layers 349, 352
airlock 64
alignment 7, 9, 13, 16, 17, 206; horizontal
16, 17; utilities *see* services; vertical 16, 17
angle of impact 622
anisotropy/ic 377, 391–4, 396, 400, 402, 503
antiflotation measures 472, 475; permanent
in-service stage 477–82; temporary
construction stage 476–7
apparent cohesion 214
apparent velocity 303, 305, 312
approach to stability: Aas 338–9; Huder 329,
333–4, 338, 361–3, 365–9; limit
equilibrium 317–8, 362, 369–70;
Meyerhof 325, 329, 335–7, 361–3;
Morgenstern, Amir-Tahmasseb 365, 369
Piaskowski and Kowalewski 329, 331–2,
360, 362, 365–9; Schneebeli 329–31,
357–60, 362, 365, 367–9; Tsi and Chang
365, 369–70; Waltz-Prager 365, 369
aquifer 127–9, 132
arch action 26–7; 99
Archimedes' principle 465
arching 142, 232; 318, 328–30, 332, 334,
338, 342, 350, 352, 357–63, 365, 370,
529; positive 233; negative 233–4
aspect ratio 232–4
assessment of coefficient of earth pressure at
rest: empirical formulae 253–4; *in situ*
testing 253–4; laboratory testing 253;
influence of errors 519; Mayne and
Kulhawy approach 518–9; Wroth
approach 514
assessment of surcharge load: line 273, 277;
orthogonal line 273; pile 273, 282–3,
289–90; point 273, 275; strip 273,

279–80; uniform 273–5, 283; uniform
rectangular 276
asymmetry 562, 594
Atterberg limits 633; liquid and plastic limits
see Atterberg limits

backfill: compaction of 33, 38–9, 70
balanced stress 331, 356, 360–2, 365–7, 369
basement 157–8, 170, 188, 190–1, 194–5
beam-column effect 446, 463, 539, 628, 630;
'$P - \Delta$' diagram *see* beam-column effect
beam-column frame 187, 189, 191
bearing capacity factor 336, 364
bending moment 447, 461, 463; *see also*
moment; *see also* stress-resultants
Benoto 109, 113
bentonite 66–8
Bernoulli's theorem 373
bilinear idealization 626
bleeder wells 325
blow-up 90, 103, 110; 128–9; 136, 387–9;
piping *see* blow-up; quicksand *see* blow-up
bored filter wells 127
bottom stability 141, 152–3
boundary conditions 30, 33, 43, 53, 59,
60–1, 64, 68, 372, 375, 378–80, 380, 390,
557–61, 583; impervious 378–9, 392, 399;
potential 379; virtual 379
box-outs 97, 105
brace(ing) 50, 52–3, 56, 59, 60–1, 64, 75;
145, 149–50, 155–9, 161–5, 167, 171,
174–6, 181–4
brittle behaviour 214
bulkhead 66–7
burster slab 15

capping beam 49, 50, 54
cavity 103
choice of analysis 260; effective stress 260–3;
finite element 257; total stress 252, 260–2
choice of earth pressure coefficient for
design 254
choice of ground support system 82–8, 90,
99, 100, 104, 108, 112–15, 119, 122;
Berlin wall *see* soldier pile wall; contiguous